**Reviews in Modern Astronomy**

*Edited by*
*Siegfried Röser*

## Related Titles

Dvorak, R. (ed.)
### Extrasolar Planets
Formation, Detection and Dynamics

2008
ISBN: 978-3-527-40671-5

Shore, S. N.
### Astrophysical Hydrodynamics
An Introduction

2007
ISBN: 978-3-527-40669-2

Horneck, G., Rettberg, P. (eds.)
### Complete Course in Astrobiology

2007
ISBN: 978-3-527-40660-9

Christensen, L. L., Fosbury, B.
### Hubble
15 Jahre auf Entdeckungsreise

2006
ISBN: 978-3-527-40682-1

Rüdiger, G., Hollerbach, R.
### The Magnetic Universe
Geophysical and Astrophysical Dynamo Theory

2004
ISBN: 978-3-527-40409-4

Foukal, P. V.
### Solar Astrophysics

2004
ISBN: 978-3-527-40374-5

Röser, S. (ed.)
### Reviews in Modern Astronomy
Vol. 19: The Many Facets of the Universe – Revelations by New Instruments

2006
ISBN: 978-3-527-40662-3

Röser, S. (ed.)
### Reviews in Modern Astronomy
Vol. 18: From Cosmological Structures to the Milky Way

2005
ISBN: 978-3-527-40608-1

Schielicke, R. E. (ed.)
### Reviews in Modern Astronomy
Vol. 17: The Sun and Planetary Systems – Paradigms for the Universe

2004
ISBN: 978-3-527-40476-6

Schielicke, R. E. (ed.)
### Reviews in Modern Astronomy
Vol. 16: The Cosmic Circuit of Matter

2003
ISBN: 978-3-527-40451-3

# Reviews in Modern Astronomy

Vol. 20: Cosmic Matter

*Edited by*
*Siegfried Röser*

WILEY-VCH Verlag GmbH & Co. KGaA

**The Editor**

*Dr. Siegfried Röser*
Astronomisches Rechen-Institut
Universität Heidelberg
ag@ari.uni-heidelberg.de

**Cover picture**

The distribution of the dark matter in simulations constrained by the observed distribution and motion of galaxies in the real Universe. An analog of our Galaxy is a tight small halo in the center of the plot. Positions of known clusters of galaxies are marked on the plot. The distribution of dark matter is a hierarchy of filaments, voids, and halos.

Graphics by Anatoly Klypin, New Mexico State University

All books published by Wiley-VCH are carefully produced. Nevertheless, authors, editors, and publisher do not warrant the information contained in these books, including this book, to be free of errors. Readers are advised to keep in mind that statements, data, illustrations, procedural details or other items may inadvertently be inaccurate.

**Library of Congress Card No.:**
applied for

**British Library Cataloguing-in-Publication Data**
A catalogue record for this book is available from the British Library.

**Bibliographic information published by the Deutsche Nationalbibliothek**
The Deutsche Nationalbibliothek lists this publication in the Deutsche Nationalbibliografie; detailed bibliographic data are available on the Internet at <http://dnb.d-nb.de>.

© 2008 WILEY-VCH Verlag GmbH & Co. KGaA, Weinheim

All rights reserved (including those of translation into other languages). No part of this book may be reproduced in any form – by photoprinting, microfilm, or any other means – nor transmitted or translated into a machine language without written permission from the publishers. Registered names, trademarks, etc. used in this book, even when not specifically marked as such, are not to be considered unprotected by law.

**Composition** Uwe Krieg, Berlin

**Printing** Strauss GmbH, Mörlenbach

**Bookbinding** Litges & Dopf GmbH, Heppenheim

Printed in the Federal Republic of Germany
Printed on acid-free paper

**ISBN:** 978-3-527-40820-7

# Preface

The annual series *Reviews in Modern Astronomy* of the ASTRONOMISCHE GESELLSCHAFT was established in 1988 in order to bring the scientific events of the meetings of the Society to the attention of the worldwide astronomical community. *Reviews in Modern Astronomy* is devoted exclusively to the Karl Schwarzschild Lectures, the Ludwig Biermann Award Lectures, the invited reviews, and to the Highlight Contributions from leading scientists reporting on recent progress and scientific achievements at their respective research institutes.

The Karl Schwarzschild Lectures constitute a special series of invited reviews delivered by outstanding scientists who have been awarded the Karl Schwarzschild Medal of the Astronomische Gesellschaft, whereas excellent young astronomers are honoured by the Ludwig Biermann Prize.

Volume 20 continues the series with sixteen invited reviews and Highlight Contributions which were presented during the International Scientific Conference of the Society on "Cosmic Matter", held at Würzburg, Germany, September 24 to 29, 2007. The Karl Schwarzschild medal 2007 was awarded to Professor Rudolf Kippenhahn, Göttingen, Germany. His lecture with the title "Als die Computer die Astronomie eroberten" opened the meeting.

The Ludwig Biermann Prize was awarded twice in 2007, because no assembly has taken place in 2006. The two winners are Henrik Beuther, Heidelberg and Ansgar Reiners, Göttingen. The title of Henrik Beuther's talk was: "Massive Star Formation: The Power of Interferometry", and Ansgar Reiners gave a lecture on: "At the Bottom of the Main Sequence. Activity and magnetic fields beyond the threshold to complete convection".

Other contributions to the meeting published in this volume discuss, among other subjects, cosmology, high-energy astrophysics, astroparticle physics, gravitational waves, extragalactic and stellar astronomy. The roadmap for astroparticle physics in Europe, as presented at the conference, is also published in this volume.

The editor would like to thank the lecturers for their stimulating presentations. Thanks also to the local organizing committee from the Institut für Theoretische Physik und Astrophysik of the Julius-Maximilians Universität Würzburg, Germany, chaired by Karl Mannheim.

Heidelberg, April 2008 *Siegfried Röser*

The ASTRONOMISCHE GESELLSCHAFT awards the **Karl Schwarzschild Medal**. Awarding of the medal is accompanied by the Karl Schwarzschild lecture held at the scientific annual meeting and the publication. Recipients of the Karl Schwarzschild Medal are

- 1959 Martin Schwarzschild:
  Die Theorien des inneren Aufbaus der Sterne.
  Mitteilungen der AG 12, 15

- 1963 Charles Fehrenbach:
  Die Bestimmung der Radialgeschwindigkeiten
  mit dem Objektivprisma.
  Mitteilungen der AG 17, 59

- 1968 Maarten Schmidt:
  Quasi-stellar sources.
  Mitteilungen der AG 25, 13

- 1969 Bengt Strömgren:
  Quantitative Spektralklassifikation und ihre Anwendung
  auf Probleme der Entwicklung der Sterne und der Milchstraße.
  Mitteilungen der AG 27, 15

- 1971 Antony Hewish:
  Three years with pulsars.
  Mitteilungen der AG 31, 15

- 1972 Jan H. Oort:
  On the problem of the origin of spiral structure.
  Mitteilungen der AG 32, 15

- 1974 Cornelis de Jager:
  Dynamik von Sternatmosphären.
  Mitteilungen der AG 36, 15

- 1975 Lyman Spitzer, jr.:
  Interstellar matter research with the Copernicus satellite.
  Mitteilungen der AG 38, 27

- 1977 Wilhelm Becker:
  Die galaktische Struktur aus optischen Beobachtungen.
  Mitteilungen der AG 43, 21

- 1978 George B. Field:
  Intergalactic matter and the evolution of galaxies.
  Mitteilungen der AG 47, 7

- 1980 Ludwig Biermann:
  Dreißig Jahre Kometenforschung.
  Mitteilungen der AG 51, 37

- 1981 Bohdan Paczynski:
  Thick accretion disks around black holes.
  Mitteilungen der AG 57, 27

1982 Jean Delhaye:
Die Bewegungen der Sterne
und ihre Bedeutung in der galaktischen Astronomie.
Mitteilungen der AG 57, 123

1983 Donald Lynden-Bell:
Mysterious mass in local group galaxies.
Mitteilungen der AG 60, 23

1984 Daniel M. Popper:
Some problems in the determination
of fundamental stellar parameters from binary stars.
Mitteilungen der AG 62, 19

1985 Edwin E. Salpeter:
Galactic fountains, planetary nebulae, and warm H I.
Mitteilungen der AG 63, 11

1986 Subrahmanyan Chandrasekhar:
The aesthetic base of the general theory of relativity.
Mitteilungen der AG 67, 19

1987 Lodewijk Woltjer:
The future of European astronomy.
Mitteilungen der AG 70, 21

1989 Sir Martin J. Rees:
Is there a massive black hole in every galaxy.
Reviews in Modern Astronomy 2, 1

1990 Eugene N. Parker:
Convection, spontaneous discontinuities,
and stellar winds and X-ray emission.
Reviews in Modern Astronomy 4, 1

1992 Sir Fred Hoyle:
The synthesis of the light elements.
Reviews in Modern Astronomy 6, 1

1993 Raymond Wilson:
Karl Schwarzschild and telescope optics.
Reviews in Modern Astronomy 7, 1

1994 Joachim Trümper:
X-rays from Neutron stars.
Reviews in Modern Astronomy 8, 1

1995 Henk van de Hulst:
Scaling laws in multiple light scattering under very small angles.
Reviews in Modern Astronomy 9, 1

1996 Kip Thorne:
Gravitational Radiation – A New Window Onto the Universe.
Reviews in Modern Astronomy 10, 1

1997 Joseph H. Taylor:
Binary Pulsars and Relativistic Gravity.
not published

| | |
|---|---|
| 1998 | Peter A. Strittmatter:<br>Steps to the LBT – and Beyond.<br>Reviews in Modern Astronomy 12, 1 |
| 1999 | Jeremiah P. Ostriker:<br>Historical Reflections<br>on the Role of Numerical Modeling in Astrophysics.<br>Reviews in Modern Astronomy 13, 1 |
| 2000 | Sir Roger Penrose:<br>The Schwarzschild Singularity:<br>One Clue to Resolving the Quantum Measurement Paradox.<br>Reviews in Modern Astronomy 14, 1 |
| 2001 | Keiichi Kodaira:<br>Macro- and Microscopic Views of Nearby Galaxies.<br>Reviews in Modern Astronomy 15, 1 |
| 2002 | Charles H. Townes:<br>The Behavior of Stars Observed by Infrared Interferometry.<br>Reviews in Modern Astronomy 16, 1 |
| 2003 | Erika Boehm-Vitense:<br>What Hyades F Stars tell us about Heating Mechanisms<br>in the outer Stellar Atmospheres.<br>Reviews in Modern Astronomy 17, 1 |
| 2004 | Riccardo Giacconi:<br>The Dawn of X-Ray Astronomy<br>Reviews in Modern Astronomy 18, 1 |
| 2005 | G. Andreas Tammann:<br>The Ups and Downs of the Hubble Constant<br>Reviews in Modern Astronomy 19, 1 |
| 2007 | Rudolf Kippenhahn:<br>Als die Computer die Astronomie eroberten<br>Reviews in Modern Astronomy 20, 1 |

The **Ludwig Biermann Award** was established in 1988 by the ASTRONOMISCHE GESELLSCHAFT to be awarded in recognition of an outstanding young astronomer. The award consists of financing a scientific stay at an institution of the recipient's choice. Recipients of the Ludwig Biermann Award are

  1989   Dr. Norbert Langer (Göttingen),
  1990   Dr. Reinhard W. Hanuschik (Bochum),
  1992   Dr. Joachim Puls (München),
  1993   Dr. Andreas Burkert (Garching),
  1994   Dr. Christoph W. Keller (Tucson, Arizona, USA),
  1995   Dr. Karl Mannheim (Göttingen),
  1996   Dr. Eva K. Grebel (Würzburg) and
         Dr. Matthias L. Bartelmann (Garching),
  1997   Dr. Ralf Napiwotzki (Bamberg),
  1998   Dr. Ralph Neuhäuser (Garching),
  1999   Dr. Markus Kissler-Patig (Garching),
  2000   Dr. Heino Falcke (Bonn),
  2001   Dr. Stefanie Komossa (Garching),
  2002   Dr. Ralf S. Klessen (Potsdam),
  2003   Dr. Luis R. Bellot Rubio (Freiburg im Breisgau),
  2004   Dr. Falk Herwig (Los Alamos, USA),
  2005   Dr. Philipp Richter (Bonn),
  2007   Dr. Henrik Beuther (Heidelberg) and
         Dr. Ansgar Reiners (Göttingen).

# Contents

*Karl Schwarzschild Lecture:*
Als die Computer die Astronomie eroberten
By Rudolf Kippenhahn (With 10 Figures) .................................... 1

*Ludwig Biermann Award Lecture I:*
Massive Star Formation: The Power of Interferometry
By Henrik Beuther (With 10 Figures) ....................................... 15

*Ludwig Biermann Award Lecture II:*
At the Bottom of the Main Sequence
Activity and Magnetic Fields Beyond the Threshold to Complete Convection
By Ansgar Reiners (With 7 Figures) ......................................... 40

Structure Formation in the Expanding Universe: Dark and Bright Sides
By Anatoly Klypin, Daniel Ceverino & Jeremy Tinker (With 14 Figures) ...... 64

From COBE to Planck
By Matthias Bartelmann (With 6 Figures) .................................... 92

Thirty Years of Research in Cosmology, Particle Physics
and Astrophysics and How Many More to Discover Dark Matter?
By Céline Bœhm (With 1 Figure) ............................................ 107

Gravitational Wave Astronomy
By Konstantinos D. Kokkotas (With 2 Figures) .............................. 140

High-(Energy)-Lights – The Very High Energy Gamma-Ray Sky
By Dieter Horns (With 9 Figures) .......................................... 167

Astronomy with Ultra High-Energy Particles
By Jörg R. Hörandel (With 16 Figures) ..................................... 198

Hydrodynamical Simulations of the Bullet Cluster
By Chiara Mastropietro & Andreas Burkert (With 15 Figures) ............... 228

Pulsar Timing – From Astrophysics to Fundamental Physics
By Michael Kramer (With 10 Figures) ....................................... 255

The Assembly of Present-Day Galaxies as Witnessed by Deep Surveys
By Klaus Meisenheimer (With 18 Figures) ................................... 279

The First Stars
By Volker Bromm (With 9 Figures) .................................... 307

Massive Stars as Tracers for Stellar and Galactochemical Evolution
By Norbert Przybilla (With 22 Figures) ................................ 323

Formation and Evolution of Brown Dwarfs
By Alexander Scholz (With 8 Figures) ................................. 357

Status and Perspectives of Astroparticle Physics in Europe
By Christian Spiering (With 20 Figures) ............................... 375

**Index of Contributors** ............................................... 406

**General Table of Contents** ......................................... 407

*Karl Schwarzschild Lecture*

# Als die Computer die Astronomie eroberten

Rudolf Kippenhahn

Max-Planck-Institut für Astrophysik
Karl-Schwarzschild-Straße 1, 85748 Garching
`rkippen@gwdg.de`

**Abstract**

Around the year 1950, computers became available for scientific institutes. At the Max-Planck-Institut für Physik in Göttingen, lead by Werner Heisenberg, an astrophysics group under Ludwig Biermann used computers, which had been developed in the institute by Heinz Billing and his group. The first machine was the G1. The author was among the young scientists who had the chance to use it for astrophysical calculations. He describes the situation of computing at a time not so much back in the past. But according to the rapid progress in computing today, it appears like it was the stone age.

## 1 Einführung

Als ich vor nahezu 60 Jahren als Mathematiker in die Astronomie ging, fesselte mich der Gedanke, dass man Aussagen über das tiefe Innere der Sterne machen kann, von dort, wohin unser Blick nicht dringt.

Um 1950 waren die physikalischen Gesetze, einschließlich der Energieerzeugung durch Kernprozesse in den Sternen, wohlbekannt. Die Gleichungen, die zu lösen waren, standen längst in den Büchern. Man musste sie nur lösen. Nach meiner ersten Stelle als Astronom in Bamberg kam ich zu Arbeitsgruppe Biermann an dem von Werner Heisenberg geleiteten Max-Planck-Institut in Göttingen. Damals entstanden die ersten elektronischen Rechenmaschinen. Ich hatte das Glück, Zugang zu Ihnen zu bekommen und zu erleben, wie die Computer in die Astrophysik eindrangen. Das geschah nahezu gleichzeitig überall in der Welt. Ich berichte es so, wie ich es bei den Arbeiten in unserer Arbeitsgruppe in Göttingen erlebt habe.

Große und gute Computer allein helfen nicht immer, man muss auch wissen, was man mit ihnen machen muss, wenn man die Vorgänge im Inneren der Sterne auf dem Computer verfolgen will. Ich hatte das Glück, als einer der ersten das entscheidende Rechenverfahren direkt von seinem Erfinder zu lernen.

Das alles aber hätte für mich nicht gereicht, hätte Ulbricht nicht in Berlin die Mauer gebaut, und wäre der Jenenser Astronom Alfred Weigert, der frühmorgens gerne länger schlief, zu dieser Zeit nicht zu Besuch in Westberlin gewesen. Als er an jenem Sonntag aufwachte, war die Mauer schon ziemlich weit hochgezogen, so dass er vor der Frage stand, ob es sich noch lohnt, in den Osten zurückzuklettern. Er blieb, und ich habe über lange Zeit mit ihm gearbeitet. Unsere wichtigsten Veröffentlichungen waren die, die wir gemeinsam geschrieben haben.

**Abbildung 1:** Ludwig Biermann (1907–1986)

## 2 Computer im Eigenbau

Aber ich muss der Reihe nach erzählen. Lassen Sie mich mit den Rechenmaschinen beginnen. Ludwig Biermann leitete Anfang der 50er Jahre die Abteilung Astrophysik in dem Heisenbergschen Max-Planck-Institut für Physik in Göttingen. Die Schwerpunkte der Abteilung und des später daraus entstandenen Instituts für Astrophysik wurden von den jeweiligen Interessen Biermanns bestimmt. Ursprünglich waren es Berechnungen der Elektronenhüllen astrophysikalisch interessanter Atome. Dazu hatte er eine Gruppe von Rechnerinnen und Rechnern angestellt, die auf Tischrechenmaschinen die aufwendigen Rechnungen durchführten. Diese Gruppe wurde von Eleonore Trefftz geleitet.

In den ersten Nachkriegsjahren arbeitete Heinz Billing, ein Physiker aus der Schule von Walther Gerlach, am Göttinger Institut für Instrumentenkunde, das Messgeräte für andere Institute herstellte. Im Jahre 1947 war eine Kommission englischer Wissenschaftler nach Göttingen gekommen, um etwas über die Forschung in Deutschland während des Krieges zu erfahren. Darunter war auch der berühmte Alan Turing. Was die Engländer über die Computer berichteten, weckte Billings Inter-

**Abbildung 2:** Heinz Billing, an einer seiner Rechenmaschinen

esse. Ein Problem in der Computerei ist das Speichern von Befehlsfolgen, die der Reihe nach ausgeführt werden sollen, aber auch die Frage, wie Zwischen- und Endergebnisse gespeichert werden können. Dafür erfand Billing die Trommel. Schon während des Krieges war in Deutschland das Tonbandgerät – damals hieß es noch „Magnetophon" – entwickelt worden. Billing sah die Möglichkeit, eine Trommel mit magnetisierbarer Oberfläche als Speicher zu benutzen. Angeblich beklebte er seine erste Trommel mit Tonbändern. Das war praktisch die Erfindung der Festplatte! Die Trommel führte er Heisenberg und Biermann vor, die davon sehr angetan waren. Doch zu einer damit arbeitenden Rechenmaschine kam es erst einmal nicht, denn damals, kurz nach der Währungsreform, gab es kein Geld. So wäre Heinz Billing fast von der Universität von Sidney abgeworben worden. Doch Heisenberg und Biermann gelang es, ihn aus Australien mit einem Angebot nach Göttingen zurück zu locken, und da begann er, Rechenmaschinen zu bauen. Zu seinen Mitarbeitern gehörte übrigens auch ein Sohn des Astronomen Josef Hopmann, der bis Kriegsende den Lehrstuhl für Astronomie der Leipziger Universität innehatte. Und so entstand die G1.

Es war die erste elektronische Rechenmaschine, mit der ich gearbeitet habe. Dabei trat ein mir bis dahin unbekanntes Phänomen auf: Wenn die Maschine plötzlich stehen bleibt, etwa weil irgendetwas falsch programmiert war, lief die Uhr, welche die zugeteilte Zeit zählte, weiter. Man stand also unter dem Zwang, möglichst schnell das Programm zu ändern, damit die zugeteilte Zeit nicht verstrich. Jeder kam einmal in diese Situation, und wir lernten: vor der stehenden Maschine ist man beliebig dumm. Die Angst, Rechenzeit zu verlieren, lähmt alle Kreativität. Die G1 wurde im Herbst 1952 in Betrieb genommen. Die Eingabe erfolgte durch Lochstreifen. Fünf Lochreihen geben die Möglichkeit, 32 verschiedene Zeichen darzustellen. Die Lesegeräte waren vom Telex-Verkehr der Post. Sie lasen 7 Zeilen pro Sekunde. Die

**Abbildung 3:** Die G1 hatte als Eingabegeräte Lochstreifenleser der Post. Der dritte Leser von links trägt eine Programmschleife. Billing steht an einer der die Röhren und Relais tragenden Wände. Die Ausgabe erfolgte entweder durch den Streifenlocher, rechts neben dem vierten Lesegrät, oder durch die elektrische Schreibmaschine. Rechts hinter der Schreibmaschine am Fußboden die Trommel.

Trommel konnte 26 Dualzahlen mit je 32 Stellen speichern. Heute kann das ein besseres Taschenrechner. Doch sie füllte ein ganzes Zimmer. Ihre 476 Röhren erzeugten so viel Wärme, dass der Raum im Sommer gelüftet und im Winter nicht geheizt werden musste. Über 100 Fernmelderelais klapperten beim Rechnen ununterbrochen. Natürlich gab es immer wieder Probleme, denn Röhren haben eine recht begrenzte Lebensdauer. Die Ergebnisse wurden mit einer alten elektrischen Schreibmaschine ausgedruckt oder als Lochstreifen zur direkten Weiterverwendung ausgegeben. Von Lochstreifen erhielt die Maschine auch die Folge der Befehle, die der Reihe nach auszuführen waren.

Programmschleifen, also Befehlsfolgen, die mehrfach durchlaufen werden, waren zu Schleifen zusammen geklebt. Heute weiß jeder Programmierer, wie Programmschleifen zu schreiben sind. Aber keiner weiß mehr, dass die Schleifen nicht einfach mit Uhu zusammengeklebt werden dürfen, weil sie sonst in den Lesegeräten stecken bleiben. Man muss den Klebstoff mit Aceton verdünnen.

In der Sekunde konnte die Maschine etwa zwei Rechenoperationen ausführen. Die G1 war 10- bis 20-mal schneller als ein geübter Rechner mit einer Tischrechenmaschine der damaligen Zeit.

Im März 1953, die G2, die nächst bessere Maschine. war noch im Bau, tagte in Göttingen die Kommission „Rechenanlagen" der Deutschen Forschungsgemeinschaft. 126 Teilnehmer waren gekommen. Aus Kiel die jungen Astronomen Karl-Heinz Böhm und Volker Weidemann, von der Göttinger Sternwarte Hans Haffner, der Vor-Vorgänger von Herrn Mannheim hier in Würzburg.

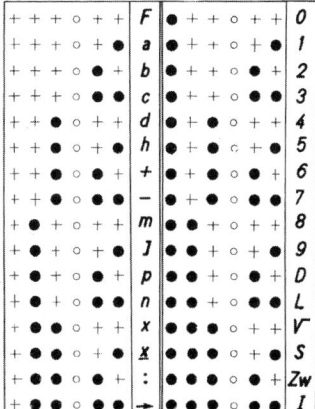

**Abbildung 4:** Die Tabelle der Codezeichen auf dem Lochstreifen. Die kleinen Kreise an der 4. Stelle deuten Transportlöcher an. Entscheidungen, die in heutigen Programmiersprachen mit do while ... enddo oder ähnlich geschrieben werden, erfolgten durch die Zeichen SI ... I. Wenn das Ergebnis der Rechnungen vor diesem Programmteil positiv ist oder 0 wird der hier durch Punkte angedeutete Programmteil übersprungen.

Konrad Zuse, der im Krieg die Z3 gebaut hatte, berichtete über die im Bau befindliche Maschine Z5, auch sie war von Relais gesteuert, wie die G1. Die Göttinger erzählten voller Begeisterung von ihren Erfahrungen mit der G1.

Arnulf Schlüter berichtete über das Rechnen von Bahnen geladener Teilchen, die in das Magnetfeld der Erde gerieten und so genannte Störmerbahnen durchlaufen, ehe sie auf die Erde treffen. Das Programm dazu löste ein System gewöhnlicher Differenzialgleichungen. Reimar Lüst zeigte, wie man Rechenmaschinen das Wurzelziehen beibringt.

## 3 Die G2

Ein Jahr nach dieser denkwürdigen Konferenz, also im Jahre 1954, wurde die G2 in Betrieb genommen. Das Programm wurde zwar noch immer mit Lochstreifen eingegeben, aber die Befehle und Daten waren auf der Trommel gespeichert. Die Maschine erledigte etwa 30 Rechenoperationen in der Sekunde. Erst zwei Jahre später kam der ihr etwa gleichwertige Rechner IBM650 nach Europa. Die G2 war ein Unikat, die IBM-Maschine ging in Serie und wurde ein Welterfolg.

Elektronische Rechenmaschinen wurden bis dahin in der Astronomie noch nicht eingesetzt. Waren sie überhaupt dafür geeignet? Für die G2 kam die Stunde der Wahrheit im Jahre 1955. Die Bestimmung des Abstandes Erde-Sonne, der Grundeinheit für Entfernungen im Weltall, geschieht mit Hilfe nahe vorbeikommender Kleinplaneten. Zwei kleine Planeten bieten sich dafür an. Ihre Namen: Amor und Eros. Der Astronom kann sich also bei der Bestimmung der Astronomischen Einheit

**Abbildung 5:** Hoher Besuch: Bundespräsident Theodor Heuss, Ludwig Biermann, Otto Hahn und Werner Heisenberg betrachten Ausdrucke der G1.

entweder der amourösen oder der erotischen Methode bedienen. Im März 1956 sollte Amor wieder einmal der Erde besonders nahe kommen.

Seit Jahrhunderten hatten die Astronomen Verfahren zur Berechnung der Bewegung eines Himmelskörpers entwickelt, die er unter dem Einfluß des Schwerefeldes der Sonne und der großen Planeten ausführt. Nach der Entwicklung von mechanischen und elektrischen Tischrechenmaschinen konnten die Verfahren diesen Hilfsmitteln angepaßt werden. Die Mitarbeiter des Astronomischen Rechen-Instituts, das damals in Berlin-Babelsberg angesiedelt war, hatten viel Erfahrung in der Berechnung von Planetenephemeriden. Ein Jahr vor der Annäherung von Amor an die Erde schritt man ans Werk. Doch damals arbeitete bereits die G2, und so bot es sich an, auch mit ihr die Bahn des Amor zu berechnen. Da würde man sehen, was die neuen Maschinen, von denen so viel die Rede war, wirklich zu leisten vermochten.

Viele Wissenschaftler blickten damals misstrauisch auf die neu entstehenden Computer. Selbst ein Jahrzehnt später, als Hans-Heinrich Voigt und ich versuchten, in Göttingen ein gemeinsam von der Universität und der Max-Planck-Gesellschaft betriebenes Rechenzentrum zu schaffen, ließen uns Kollegen durchblicken, dass ihrer Meinung nach Computer zwar sehr wichtig sind, dass aber ein guter Mathematiker oder ein guter theoretischer Physiker sie nicht braucht.

Als der Kleine Planet Amor im Anmarsch war, schritten in Göttingen Peter Stumpff, Stefan Temesvary, Arnulf Schlüter und Konrad Jörgens ans Werk.

Sie standen vor einer wohlbekannten mathematischen Aufgabe, der Lösung eines Systems gewöhnlicher Differenzialgleichungen. Was man dazu tun muss, steht in den Lehrbüchern, und wenn es nicht so sehr auf Arbeitszeit ankommt – die Arbeit macht schließlich eine Maschine – muss man nicht die bisher für Tischrechenma-

*Als die Computer die Astronomie eroberten* 7

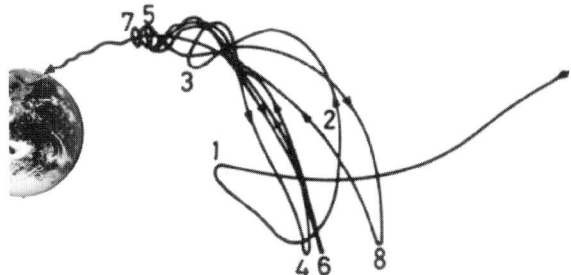

**Abbildung 6:** Die komplizierten Bahnen aus dem Weltall kommender geladener Teilchen im Magnetfeld der Erde zu berechnen, waren eine der Aufgaben der G1.

schinen entwickelten zeitsparenden astronomischen Rechenmethoden verwenden, sondern kann die Differenzialgleichungen nach wohlbekannten Rezepten lösen. Das Göttinger Ergebnis war etwas anders als die von den in Babelsberg nach den klassischen Verfahren berechneten Ephemeriden.

Und dann kam Amor. Wo aber stand er am Himmel? Dort, wo ihm die erfahrenen Babelsberger Rechner einen Platz zugewiesen hatten? Nein, er stand näher an der Stelle, an die ihn die G2 plaziert hatte!

Aber die neuen Maschinen konnten noch mehr! Im März des Jahres 1957 vollendete Ludwig Biermann sein 50. Lebensjahr, und man kam auf die Idee, die G2 ein Portrait des Jubilars drucken zu lassen. Heute wäre das kein Problem: Man nimmt ein Foto, scannt es ein und läßt es ausdrucken. Damals gab es weder Scanner, noch waren Drucker auf dem Markt. Die G2 hatte als Ausgabegerät nur einen Lorenz-Fernschreiber. Mehr als Buchstaben, Ziffern und Interpunktionen konnte er nicht auf das Papier setzen. An Graustufen oder gar Farben war nicht zu denken. Doch zwei junge Doktoranden, Friedrich Meyer und Hermann-Ulrich Schmidt, wußten sich zu helfen. Die verschiedenen Zeichen, über die der Drucker verfügte, benötigten verschieden viel Druckerschwärze. Ein Feld von Leerzeichen war weiß. Felder mit Ziffern oder Buchstaben wirkten aus der Entfernung betrachtet hellgrau. Die 8 auf die 9 geklopft ergibt nahezu Schwarz. Damit war das Problem der Ausgabe von Graustufen gelöst.

Doch wie das Bild in den Computer bringen? Auch da wußten Meyer und Schmidt sich zu helfen. Sie fertigten ein Dia des Bildes an und projizierten es auf eine weiße Wand, auf die viele kleine Quadrate vorgezeichnet waren. Der eine ging zeilenweise das Raster durch und schätzte für jedes Feld die mittlere Helligkeit des projizierten Bildes in diesem Quadrat. Der andere wandelte diese Helligkeit in Zeichen um, die entsprechende Graustufen darstellten. Das alles wurde in einen Lochstreifen gestanzt und ausgedruckt. Auf diese Weise, die ich etwas vereinfacht habe, entstand eines der ersten Computerbilder der Welt.

**Abbildung 7:** Die G2 hatte eine wesentlich größere Trommel, auf der auch das jeweilige Rechenprogramm gespeichert war. Heiz Billing (stehend) mit einem seiner Mitarbeiter. Wurden bei der G1 die Ergebnisse noch auf einer elektrischen Schreibmaschine ausgedruckt, so hatte die G2 bereits einen Fernschreiber der Marke Lorenz, der drucken und Lochstreifen ausgeben konnte.

Am Morgen seines Geburtstages wurde Ludwig Biermann gebeten, sich vor die G2 zu setzen und einen bestimmten Knopf zu drücken. Daraufhin tastete das Lesegerät den vorbereiteten Lochstreifen ab, während Biermann am Fernschreiber zeilenweise sein Bild entstehen sah. Ludwig Biermann, der an der Einführung der Computer in der Wissenschaft so viel Verdienste hatte, hat nie selbst einen Computer angerührt. Nur an jenem Geburtstagsmorgen ließen wir ihn auf einen Knopf drücken. ... und das machte er perfekt.

Stefan Temesvary und ich schrieben damals unser erstes Programm zur Berechnung von Sternmodellen für die G2. Als das Institut im Jahre 1958 nach München übersiedelte, gingen die G1 und die G2 mit, die G1 arbeitete noch einige Zeit an der Münchner Universität und wurde dann verschrottet. Einen Ehrenplatz im Deutschen Museum erhielt sie leider nicht. Die G2 wurde im neuen Institutsgebäude aufgestellt und arbeitete weiter. Damals entstand in der Billingschen Abteilung auch die G3, die etwa 150mal schneller war als die G2. Sie ging mit nach München.

Ich hatte mich in Erlangen habilitiert und hielt damals von München aus dort regelmäßig Vorlesungen. Im WS 1959/60 wagte ich eine Vorlesung über „Programmieren elektronischer Rechenmaschinen". Das war zu einer Zeit, als weder in Erlangen und kaum irgendwo sonst in Deutschland darüber gelehrt wurde, außer vielleicht an der damaliger Münchner TH. Die Informatik kam erst viele Jahre später an die Erlanger Uni.

Da es noch keine allgemeinen Programmiersprachen gab, legte ich den Maschinencode der G2 zugrunde. Gegen Ende des Semesters schrieben die Hörer und ich an der Tafel ein Programm zur Lösung der Emdengleichung, also einer nichtlinea-

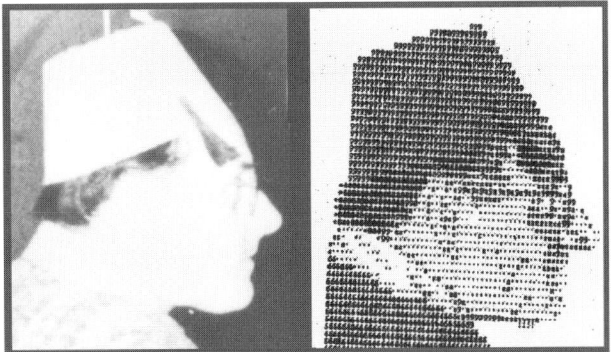

**Abbildung 8:** Links: Ludwig Biermann bei einem Institutsfasching in orientalischem Outfit. Rechts: Die auf der G2 hergestellte Computergrafik.

ren gewöhnlichen Differenzialgleichung mit einer Singularität an einem Ende des Intervalls. Das Programm ließ ich in München auf Lochband stanzen. Der Erlanger Rektor genehmigte meinen Hörern, allzu viele waren es nicht, eine Exkursion nach München. Dort fütterten wir den Lochstreifen in die G2, und es war das erste Mal in meinem Leben, dass ein Computerprogramm, an dem ich beteiligt war, auf Anhieb fehlerlos lief. Es war übrigens auch das letzte Mal!

**Abbildung 9:** Sternentwicklungsrechungen auf der G2 um 1957: Eine Rechnerin, Stefan Temesvary (sitzend), und der Autor (stehend) starren auf die Ergebnisse.

Temesvary und ich setzten unsere Entwicklungsrechnungen für Sterne von etwa einer Sonnenmasse fort, die in ihrem Inneren allen Wasserstoff verbraucht haben und von der Hauptreihe in die Richtung der Roten Riesen marschieren. Sie besitzen einen entarteten Kern aus Helium. Da die Entwicklung durch das nukleare Verbrennen des

Wasserstoffs zu Helium gesteuert wird, ist sie langsam im Vergleich zur Einstellzeit des thermischen und erst recht des hydrostatischen Gleichgewichts. Deshalb sind nur gewöhnliche Differenzialgleichungen zu lösen. Wir begannen die Rechnungen mit Lösungen, die die äußeren Randbedingungen erfüllten und rechneten durch die Hülle nach innen, bis wir auf den Heliumkern trafen. Heliumkerne hatten wir auf Vorrat gerechnet. Natürlich passten Hüllenlösung und Kern nicht aneinander, doch wir lernten schnell, wie wir die noch freien Parameter, mit denen wir die Hüllenrechnungen begannen, abändern musste, um die Anpassung zu verbessern. Wir suchten nach Lösungen der gewöhnlichen Differenzialgleichungen des Sternaufbaus, die an der Sternoberfläche und im Zentrum jeweils zwei Randbedingungen erfüllen mussten. Für jede Rechnung von der Oberfläche zum Kern benötigte die G2 fünf Stunden. Dann erst erfuhren wir, wie gut oder schlecht wir die Anfangswerte für die Hülle gewählt hatten. Eine Reise von der Oberfläche bis fast in das Zentrum eines Sterns in fünf Stunden! Diese Reise ging von der Atmosphäre im Strahlungsgleichgewicht durch die Konvektionszone, in der die aus dem Inneren kommende Energie hauptsächlich durch turbulente Bewegung transportiert wird. Sie führte durch Bereiche, in denen der Strahlungsfluss durch das negative Wasserstoffion oder durch neutralen Wasserstoff kontrolliert wird oder durch Elektronen, die den Atomhüllen abgeschlagen worden sind. Sie führt durch Bereiche, in denen das Helium neutral ist und durch solche, in denen die Heliumatome ein oder zwei Elektronen verloren haben. Schließlich führte die Reise in Gebiete, in denen die Kernenergie frei wird, die den Stern leuchten lässt. Das alles musste das Programm erkennen und zu den entsprechenden Gleichungen umschalten. Es war aufregend, vor der laufenden Maschine den Ausdrucken anzusehen, wie die Rechnung immer tiefer in den Stern eindrang. Und wenn irgendetwas Ungewöhnliches passierte, musste man von Hand eingreifen und den Fehler rasch beseitigen, wenn die Rechenzeit nicht verloren gehen sollte. Beachten Sie bitte, die G2 war eine Festkommamaschine, wenn eine Zahl, auch bei Zwischenrechnungen, größer wurde als 8, blieb die Maschine unweigerlich stehen! Darauf musste man beim Programmieren achten, also zu einer Zeit, zu der man weder die Ergebnisse noch die Zwischenergebnisse kannte!

Für die langen Rechnungen bekamen wir nur nachts Rechenzeit. Deshalb wurden Studenten angestellt, die nachts die Integrationen von der Sternoberfläche in das Innere überwachten. Zur besseren Übersicht hatten wir zu Beginn jedes Integrationsschrittes in einer Zeile Platz für sieben Zeichen gelassen. Dort konnten jeweils ein Stern oder ein Punkt gesetzt werden. Punkt an der ersten Stelle bedeutete zum Beispiel: Die Integration verläuft in einem konvektiven Bereich. War statt dessen an der ersten Stelle ein Stern, bedeutete dies, dass die Rechnung in einem Bereich des Sterns ausgeführt wurde, an der die aus dem inneren kommende Energie durch Strahlung nach außen transportiert wird. Die anderen Stellen zeigten durch Stern oder Punkt an, durch welche Prozesse der Absorptionskoeffizient dominiert wird oder ob die nukleare Energieerzeugung zur Energiebilanz merklich beiträgt. Ich erinnere mich, dass Temesvary einmal mitten in der Nacht angerufen wurde, weil der Nachtrechner den Verdacht hatte, dass etwas schief läuft. „Sind Sie im Bereich der Konvektion oder nicht?" fragte der aus dem Schlaf gerissene Temesvary. „Ich glaube nicht", antwortete es auf der anderen Seite etwas unsicher. „Unsinn, die Rechnung ist mitten in der Konvektionszone, das höre ich doch!", war die Antwort. Tatsächlich konnte man die Anschläge von Punkt und Stern akustisch unterscheiden, da der Stern zwei Anschläge benötigte. Wir haben in dieser Zeit viele Nachtrechner verschlissen. Einige kamen später zu Amt und Würden. Der Mathematiker Bruno Brosowski hatte Lehrstühle in Göttingen und Frankfurt inne. Auch unser Heidelberger Kollege Heinrich Völk zählte zu unseren Nachtrechnern.

Damals kam auch der Physiker Norman Baker als Postdoc nach München. Er und ich begannen auf der G2 pulsierende Sternmodelle zu untersuchen, was wiederum hieß, Eigenwerte gewöhnlicher Differenzialgleichungen zu finden.

Um 1961 war die Situation etwa so: Die Entwicklung von Hauptreihensternen machte keine Probleme, denn diese Sterne sind im mechanischen Gleichgewicht: Schwerkraft und Druck halten einander die Waage. Sie sind auch im thermischen Gleichgewicht: Jedes Gramm Sternmaterie gibt so viel Energie an seine Nachbarschaft ab, wie es erhält. Weder erhitzt es sich mit der Zeit, noch kühlt es ab. Sterne in diesem doppelten Gleichgewicht werden durch ein System gewöhnlicher Differentialgleichungen beschrieben – kein größeres mathematisches Problem. Das wird sofort anders, wenn sich der verfügbare nukleare Brennstoff, der Wasserstoff, im Stern erschöpft oder wenn der Reaktor im Stern instabil wird. Dann verändert sich der Stern bereits im Laufe von Tausenden Jahren, manchmal im Laufe von Minuten. Da müssen die echten partiellen Differenzialgleichungen gelöst werden.

## 4  Ein neues Rechenverfahren

Inzwischen hatte die Industrie zwar leistungsfähigere Computer auf den Markt geworfen, doch die Schwierigkeiten bei den Sternentwicklungsrechnungen konnten sie nicht überwinden. Die lagen nicht in der Hard-, sondern in der Software. Deshalb war es bis 1961 nicht möglich, die Entwicklung von Sternen von merklich mehr Masse als der der Sonne nach dem Erschöpfen des Brennstoffes zu verfolgen, also die Entwicklung von der Hauptreihe weg. Bei Sternen, deren Masse im Bereich der Sonne liegt, ist das zwar möglich, doch in diesen Sternen beginnt nach einiger Zeit die Fusion des Heliums zu Kohlenstoff, Sauerstoff und Neon mit unkontrollierbarer Stärke. Martin Schwarzschild war in Princeton bis dahin gekommen, doch vor diesem so genannten Helium Flash musste er kapitulieren. Damals wusste eben niemand, wie man aus dem Gleichgewicht geratene Sterne am Computer verfolgen kann. Doch einer vielleicht, Louis Henyey, Professor für Astronomie an der Universität von Berkeley. Er hatte eine Arbeit über ein neues Rechenverfahren zur Berechnung von Sternmodellen veröffentlicht. Ganz verstanden hatte die Arbeit wohl niemand, und der Autor hatte auch keine schlagenden Beweise für die Leistungsfähigkeit seiner neuen Methode geliefert.

Wieder einmal hatte ich Glück. Die Internationale Astronomische Union, die alle vier Jahre ihre Generalversammlung abhält, traf sich im August im Jahre 1961 in Berkeley. Ich hatte Gelegenheit, an dieser Konferenz teilzunehmen. Viele angesehene Astronomen waren gekommen. Das wichtigste, das ich danach aus Berkeley mitnahm, waren meine Vortragsnotizen von Henyeys Vorlesung über seine neue Rechenmethode. Verstanden hatte ich sie allerdings nicht. Erst als ich erfuhr, dass Martin Schwarzschild in Princeton nach seinen Notizen von Henyeys Vorlesung ein

**Abbildung 10:** Louis Henyey (1910–1970), der Erfinder der nach ihm benannten Rechenmethode zur Berechnung der zeitlichen Entwicklung der Sterne.

Computerprogramm zum Laufen und seinen Stern durch den Helium-Flash gebracht hatte, sah ich mir meine Notizen des Henyey-Vortrages wieder an – und ich verstand die Methode.

In dieser Zeit bereitete ich nach einem einjährigen Aufenthalt in den USA den Arbeitsplan für die Zeit nach meiner Rückkehr nach München vor. Alfred Weigert hatte an unserem Institut eine Stelle gefunden, außerdem wartete dort eine neue Mitarbeiterin, Emmi Hofmeister, die spätere Emmi Meyer, gelernte Versicherungsmathematikerin, die mit uns arbeiten wollte. Für sie schrieb ich von Pasadena aus eine Kurzfassung der Henyey-Methode.

## 5  Ein Telegramm nach Berkeley

Bald danach kehrte ich nach München zurück. Emmi Hofmeister, Alfred Weigert und ich begannen sofort, ein Henyey-Programm in der Programmiersprache FORTRAN zu schreiben, mit dem wir einen Stern von der Hauptreihe ins Stadium eines Roten Riesen und danach in das Stadium eines pulsierenden Delta-Cephei-Sternes verfolgen wollten. Wir wählten einen Stern von sieben Sonnenmassen. Damals bekam Garching eine IBM7090 und bald spielte es sich ein, dass wir mit einem institutseigenen Kleinbus Kasten mit Lochkarten und Computerausdrucke zwischen Garching und München hin- und herschickten. Einmal änderten wir noch kurz vor der Abfahrt des Busses das Programm und als die neuen Karten gelocht waren, rannte Weigert mit dem Kasten schnell zum Bus, der gerade abfahren wollte. Dabei merkte er nicht, dass auf seinem Weg eine Glastür geschlossen war. Er kam unverletzt durch die Tür und wir erzählten ihm später, dass das Loch im Glas seine Silhouette zeige, einschließlich der Ohren.

Im Januar 1963 lief das Programm und es gelang uns, in Entwicklungsstadien unseres Sterns vorzudringen, in die noch niemand gelangt war. Wir schickten ein Telegramm an Louis Henyey: "The Henyey method is working in Munich!" Der Stern erschöpfte seinen Wasserstoff im Zentrum, wurde zu Roten Riesen. Aus mir heute nicht mehr verständlichen Gründen fürchteten wir, dass in unseren Rechnungen als nächstes nukleares Brennen der Kohlenstoff zündeten würde. Als wir einmal unsere Lochkarten selbst nach Garching brachten, modifizierten wir einen Bayerischen Spruch. „Oh Heiliger Sankt Florian, behüt' das C, zünds Helium an!" Das wirkte. Das Zentrum unseres Modellsterns wurde kurz danach immer dichter und heißer. Schließlich zündete im Zentrum das Helium. Mehrmals führte die Entwicklung den Stern durch Phasen, in denen ihn das Helium seiner Oberflächenschichten zu pulsierenden Schwingungen zwang. Norman Baker, inzwischen Professor an der Columbia-Universität in New York, und ich hatten damit die ersten Sternmodelle für unsere Untersuchungen über die Ursache der Pulsationen. Das lieferte auch die Begründung der berühmten Perioden-Leuchtkraft-Beziehung der Delta-Cephei-Sterne, wie Emmi Hofmeister in ihrer Dissertation zeigte. – In den astrophysikalischen Arbeiten unseres Instituts waren die Computer voll angekommen. Kurt von Sengbusch (mit einer Untersuchung über die Entwicklung der Sonne) und Hans-Christoph Thomas (mit der Entwicklung eines sonnenähnlichen Sterns durch den Helium-Flash) benutzten das Programm für ihre Dissertationen und verbesserten es dabei wesentlich.

## 6 Sterne, die es nicht gibt, und ihre Eigenschaften

Wenn ich jetzt, viele Jahrzehnte danach, sagen soll, was mir im Rückblick als das Wichtigste erscheint. Dann ist es das: Die Computer geben uns die Möglichkeit, viele Eigenschaften der Sterne zu verstehen. Das ist nicht immer einfach. Wer mit Computern arbeitet, weiß es: Man gibt Zahlen ein und bekommt Zahlen heraus. Man gibt andere Zahlen ein und bekommt andere heraus. Aber was bewirkt was? Dazu braucht man mehr als nur die Naturgesetze in ein Programm einzubauen und los zu rechnen. Ich habe einmal so formuliert: Um die Sterne am Himmel zu verstehen, muss man auch die Sterne studieren, die es nicht gibt. Zum Verständnis der Sterne dürfen wir nicht nur die Lösungen der Sternaufbaugleichungen studieren, die Sterne beschreiben, die in der Natur realisiert sind, sondern auch die Mannigfaltigkeit der Lösungen, in die diese Lösungen eingebettet sind. Ich will das an drei Beispielen erläutern:

Vor mehr als 50 Jahren kam die immer wieder diskutierte Frage auf, ob sich die Naturkonstanten nicht mit der Zeit ändern. Damals ließ Schwarzschild einen seiner Schüler die Entwicklung der Sonne unter der Annahme einer künstlichen zeitlich veränderlichen Gravitationskonstanten rechnen. Die Stärke der Gravitation beeinflusst das Alter, Leuchtkraft und Durchmesser der Sonne von heute. Das Ergebnis war eine obere Grenze der Variabilität der Gravitationskonstanten. Hätte die Konstante sich stärker geändert, sähe die Sonne heute anders aus.

Ein zweites Beispiel: Es ist an den Aufbaugleichungen nicht zu erkennen, warum die Sterne Rote Riesensterne werden, wenn sich der Wasserstoff in ihren Zentralgebieten erschöpft. Liegt es daran, dass der Kern eine andere chemische Zusammensetzung hat? Oder daran, dass die Energie jetzt nicht mehr hauptsächlich im Zentrum, sondern an der Oberfläche einer Heliumkugel freigesetzt wird? In seiner Hamburger Zeit setzten Weigert und sein Schüler Höppner eine Punktmasse in das Zentrum eines Hauptreihensterns, die dort die Schwerkraft künstlich erhöht, sonst aber keinerlei Wirkung auf chemischer Zusammensetzung und Erzeugung von Kernenergie hatte. Das Ergebnis: Der Stern, dem das und sonst nichts anderes angetan wurde, verwandelte sich in einen Roten Riesenstern! Was lernen wir daraus? Dass Hauptreihensterne, die während ihrer Entwicklung in ihrem Zentrum eine hohe Dichtekonzentration bilden, eine Art Massenpunkt, genau deshalb zu Roten Riesen werden.

Ein drittes Beispiel: Die Energie, mit der die Sonne leuchtet und uns wärmt, wird bei Kernreaktionen frei. Strahlt die Sonne stärker, wenn die Kernreaktionen effektiver werden? Achim Weiss in Garching hat mit seinen Mitarbeitern Sonnenmodelle mit künstlich bis zu 10fach vergrößerten Wirkungsquerschnitten der Kernreaktionen gerechnet. Das Ergebnis: Die Leuchtkraft des Sterns ändert sich kaum.

Das sind drei Beispiele, wo man bei den Rechnungen der Natur etwas Gewalt angetan hat, um herauszufinden, was was bewirkt. Als wir vor einem halben Jahrhundert Computer einsetzten, wussten wir, dass wir nicht die einzigen waren. Schwarzschild rechnete in Princeton massearme Sterne, in Pasadena hatte Icko Iben etwa gleichzeitig mit uns ein Henyey-Programm geschrieben, das bald Ergebnisse lieferte. Und in Warschau bastelte ein Student an einem Henyey-Programm. Er wurde bald durch seinen Einfallsreichtum weit über die Grenzen Polens bekannt: Bohdan Pacynski, Schwarzschilds späterer Nachfolger in Princeton. Er starb im vergangenen Jahr.

Aber so ist es, viele meiner Kollegen und Freunde von damals sind nicht mehr unter uns: Norman Baker, Ludwig Biermann, Dietmar Lauterborn, Bohdan Pacynski, Martin Schwarzschild, Stefan Temesvary, Alfred Weigert und Marshall Wrubel. Ich habe ihnen allen viel zu verdanken. Jetzt am Ende meiner Vorlesung eine persönliche Bemerkung: Bis auf einen wurde keiner so alt wie ich. Sie haben sich eben weniger geschont.

*Ludwig Biermann Award Lecture I*

# Massive Star Formation: The Power of Interferometry

Henrik Beuther

Max Planck Institute for Astronomy
Königstuhl 17, 69117 Heidelberg, Germany
beuther@mpia.de

### Abstract

*This article presents recent work to constrain the physical and chemical properties in high-mass star formation based largely on interferometric high-spatial-resolution continuum and spectral line studies at (sub)mm wavelengths. After outlining the concepts, potential observational tests, a proposed evolutionary sequence and different possible definitions for massive protostars, four particular topics are highlighted: (a) What are the physical conditions at the onset of massive star formation? (b) What are the characteristics of potential massive accretion disks and what do they tell us about massive star formation in general? (c) How do massive clumps fragment, and what does it imply to high-mass star formation? (d) What do we learn from imaging spectral line surveys with respect to the chemistry itself as well as for utilizing molecules as tools for astrophysical investigations?*

## 1 Introduction

Star formation is a key process in the universe, shaping the structure of entire galaxies and determining the route to planet formation. In particular, the formation of massive stars impacts the dynamical, thermal and chemical structure of the interstellar medium (ISM), and it is almost the only mode of star formation observable in extragalactic systems. During their early formation, massive star-forming regions inject energy to the ISM via their outflows and jets, then during their main sequence evolution, the intense uv-radiation of the massive stars heats up their environment, and at the very end of their life, Supernovae stir up the ISM by their strong explosions. Furthermore, massive stars are the cradles of the heavy elements. Hence, life as we know it today would not exist if massive stars had not formed first the heavy elements in their interiors via nucleosynthesis. In addition to this, almost all stars form in clusters, and within the clusters massive stars dominate the overall luminosities. Isolated low-mass star formation is the exception, and isolated high-mass star formation likely does not exist

*Concepts:* In spite of their importance, many physical processes during the formation of massive stars are not well understood. While there exists a paradigm for low-mass star formation on which large parts of the scientific community agree (e.g., Andre et al. [1], McKee & Ostriker [41]), this is less the case for high-mass star formation (e.g., Beuther et al. [5], Zinnecker & Yorke [71]). The conceptional problem is based on the fact that at least in spherical symmetry the radiation pressure of a centrally ignited star $\geq 8\,M_\odot$ would be large enough to stop any further gas infall and hence inhibit the formation of more massive objects. Over the last decade, two schools have been followed to solve this problem: (a) The turbulent accretion scenario, which is largely an enhancement of the low-mass star formation scenario, forms massive stars in a turbulent core with high accretion rates and a geometry including accretion disks and molecular outflows (e.g., Keto [35], Krumholz et al. [37], McKee & Tan [43], Yorke & Sonnhalter [69]). In contrast to that, (b) the competitive accretion scenario relies on the clustered mode of massive star formation. In this scenario, the accretion rates are determined by the whole cluster potential, and those sources sitting closest to the potential well will competitively accrete most of the mass (e.g., Bonnell et al. [24, 25]).

*Potential observational tests:* How can we observationally discriminate between the different scenarios? For example, the turbulent accretion scenario predicts qualitatively similar outflow and disk properties as known for low-mass stars, however, with quantitatively enhanced parameters like the accretion rates, outflow energies, or disk sizes. In contrast to that, modeling of the proto-cluster evolution in the competitive accretion scenario indicates extremely dynamic movements of all cluster-members throughout their whole evolution. It is unlikely that in such a dynamic environment collimated outflows or large disks could survive at all. Another difference between both scenarios is based on their early fragmentation predictions. While the turbulent accretion scenario draws their initial gas clumps from (gravo-)turbulently fragmented clouds (e.g., Padoan & Nordlund [50]) and these gas clumps do not fragment afterwards much anymore, the competitive accretion scenario predicts that the initial gas clumps fragment down to many clumps all of the order a Jeans-mass ($\sim 0.5\,M_\odot$). Hence, while in the former scenario, the Initial Mass Function (IMF) is determined during the early cloud fragmentation processes, the latter models predict that the IMF only develops during the ongoing cluster formation. Therefore, studying the initial fragmentation and the early core mass functions can give insights in the actual massive star formation processes.

*Evolutionary sequence:* Independent of the formation scenario, there has to be an evolutionary sequence in which the processes take place. For this review, I will follow the evolutionary sequence outlined by Beuther et al. [5]: Massive star-forming regions start as High-Mass Starless Cores (HMSCs), i.e., these are massive gas cores of the order a few 100 to a few 1000 $M_\odot$ without any embedded protostars yet. In the next stage we have High-Mass cores with embedded low- to intermediate-mass protostars below $8\,M_\odot$, which have not started hydrogen burning yet. During that evolutionary phase, their luminosity should still be dominated by accretion luminosity. Following that, the so called High-Mass Protostellar Objects (HMPOs) are still massive gas cores but now they contain embedded massive protostars $>8\,M_\odot$ that have started hydrogen burning which soon dominates the total luminosity of the sources. Hot Molecular Cores (HMCs) and hypercompact HII regions (HCHIIs) are part of that class. The last evolutionary stage then contains the final stars that have stopped accreting. While most ultracompact HII regions (UCHIIs) are likely part of the latter group, some of them may still be in the accretion phase and could then hence still harbor HMPOs.

*Definitions of a massive protostar:* Another debate in massive star formation centers around the exact definition of a "massive protostar". If one followed the low-mass definition which basically means that a protostars is an object that derives most of its luminosity from accretion, then "massive protostars" should not exist or only during a very short period of time because as soon as they are termed "massive" ($>8\,M_\odot$), their luminosity is quickly dominated by hydrogen burning. In this scenario, during the ongoing formation processes, one would then need to talk about "accreting stars". This approach is for example outlined by Zinnecker & Yorke [71]. A different definition for "massive protostars" is advocated, e.g., recently by Beuther et al. [5]: In this picture, a protostar is defined in the sense that each massive object that is still in its accretion phase is called a "massive protostar", independent of the dominating source of luminosity. This definition follows more closely the usual terminology of "proto", meaning objects that are not finished yet.

*Observational challenges:* Whatever physical or chemical processes we are interested in massive star formation, one faces severe observational challenges because of the clustered mode of massive star formation and the on average large distances of a few kiloparsec. Therefore, high spatial resolution is a prerequisite for any such study. Furthermore, the early stages of massive star formation are characterized by on average cold gas and dust temperatures which are best observed at (sub)mm wavelength. Hence, most observations presented in the following are based on (sub)mm interferometer observations of young massive star-forming regions at different evolutionary stages.

The main body of this article is divided into four sections dealing first with the initial conditions that are present prior to or at the onset of massive star formation. The next two sections deal with our current knowledge about the properties of potential massive accretion disks and the fragmentation behavior of massive gas clumps and cores. The following section will then outline the status and future possibilities of astrochemical investigations. Finally, I try to sketch the directions where current and future answers to the questions raised in the Abstract may lead to.

## 2 The earliest stages of massive star formation

What are the initial conditions of massive star formation? Until a few years ago, to address this question observationally in a statistical sense was close to impossible because we had no means to identify large samples of sources prior to or at the onset of massive star formation. The situation has changed significantly since the advent of the space-based near- and mid-infrared missions that surveyed the Galactic plane starting with ISO and MSX, and now conducted with much better sensitivity and spatial resolution by Spitzer. These missions have revealed more than $10^4$ Infrared Dark Clouds (IRDCs), which are cold molecular clouds that are identified as shadows against the Galactic background (e.g., Carey et al. [27], Egan et al. [32], Simon et al. [63]). These clouds are characterized by on average cold temperatures ($\sim$15 K), large masses (a few 100 to a few 1000 $M_\odot$) and average densities of the order $10^4$–$10^5$ cm$^{-1}$ (e.g., Pillai et al. [53], Rathborne et al. [55], Sridharan et al. [64]). Although these clouds appear as dark shadows, they may be starless but they can also harbor embedded forming protostars. In fact, a statistical analysis of the percentage of starless IRDCs versus IRDCs with embedded protostars will be an important step to understand the time-scales important for the earliest evolutionary stages. Currently, the statistical database of in depth IRDC studies is still insufficient for such an estimate (e.g., Beuther & Steinacker [15], Motte et al. [46], Pillai et al. [54], Rathborne et al. [55, 56]), however, it is interesting to note that until now no starless IRDC has been unambiguously identified in the literature, all detailed studies revealed embedded star formation processes. To first order, this triggers the interpretation/speculation that the high-mass starless core phase is likely to be extremely shortlived. Future investigations of larger samples will answer this question more thoroughly.

In a recent spectral line study of a sample of 43 IRDCs (Fig. 1), Beuther & Sridharan [13] detected SiO(2–1) emission from 18 sources. Assuming that SiO is produced solely through sputtering from dust grains, and that this sample is representative for IRDCs in general, it indicates that at least 40% of the IRDCs have ongoing outflow activity. Since the non-detection of SiO does not imply no outflow activity, this number is a lower limit, and even a higher percentage of sources may harbor already ongoing star formation. The range of observed SiO line-widths down to zero intensity varied between 2.2 and 65 km s$^{-1}$. While inclination effects and embedded objects of different mass could account for some of the differences, such effects are unlikely causing the whole velocity spread. Therefore, Beuther & Sridharan [13] speculate whether the varying SiO line-widths are also indicators of their evolutionary stage with the smallest line-width close after the onset of star formation activity. In the same study, Beuther & Sridharan [13] observed $CH_3OH$ and $CH_3CN$. While $CH_3CN$ was detected only toward six sources, $CH_3OH$ was found in approximately 40% of the sample. The derived column densities are low of the order $10^{-10}$ with respect to $H_2$. These values are consistent with chemical models of the earliest evolutionary stages of high-mass star formation (e.g., Nomura & Millar [48]), and the $CH_3OH$ abundances compare well to recently reported values for low-mass starless cores (e.g., Tafalla et al. [66]).

Zooming into selected regions in more detail, we studied one particularly interesting IRDC at high angular resolution with the Plateau de Bure Interferometer and the Spitzer Space Telescope (IRDC 18223-3, see Fig. 1 right panel; Beuther et al. [14], Beuther & Steinacker [15]). Combining the Spitzer mid-infrared data between 3 and 8 μm with the 3.2 mm long-wavelengths observations from the Plateau de Bure Interferometer (PdBI), we did not find any mid-infrared counterpart to the massive gas core detected at 3.2 mm (Fig. 2, Beuther et al. [14]). However, we did detect three faint 4.5 μm features at the edge of the central 3.2 mm continuum core. Since

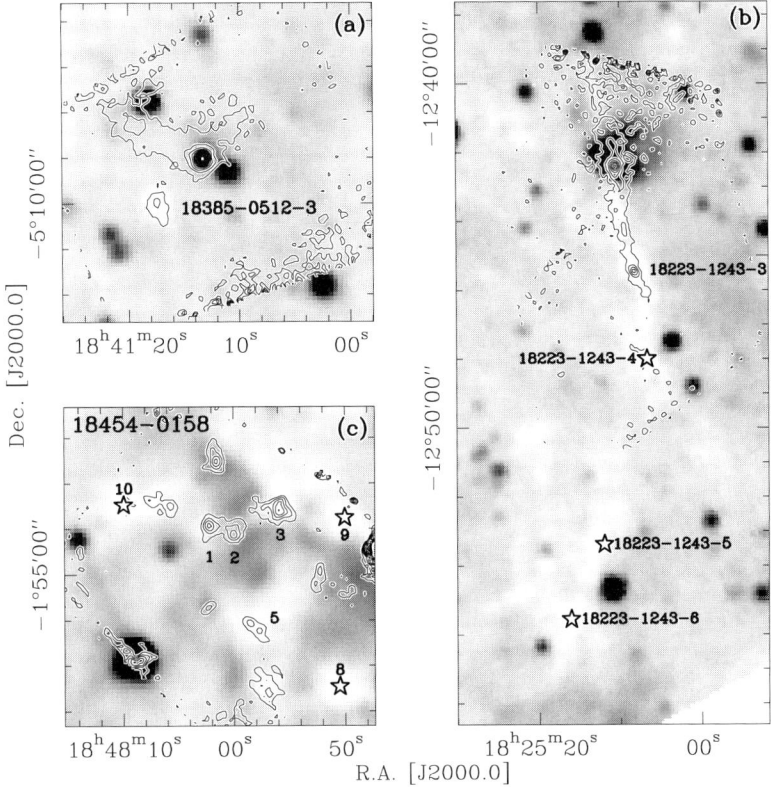

**Figure 1:** Sample IRDCs from Sridharan et al. [64]: MSX A-band (8 μm) images (black is bright) with 1.2 mm emission contours: The first two numbers refer to the corresponding IRAS source and the third number labels the mm sub-sources. The five-pointed stars mark cores lacking good 1.2 mm measurements.

emission features that occur only in the 4.5 μm but no other Spitzer band are usually attributed to shocked $H_2$ emission from molecular outflows (e.g., Noriega-Crespo et al. [49]), we concluded that the region likely hosts a very young protostar that drives a molecular outflow but that is still too deeply embedded by to be detected by the Spitzer IRAC bands. This interpretation found further support by line-wing emission in older CO and CS data. Based on the inferred central source, we predicted that the region should have a strongly rising spectral energy distribution (SED) and hence be detected at longer wavelengths. As soon as the MIPSGAL mid- to far-infrared survey with Spitzer became available, we then could identify the central source at 24 and 70 μm (Fig. 2, Beuther & Steinacker [15]). Combing the available mid-/far-infrared data with the long-wavelengths observations in the mm regime, it is possible to fit the SED with a two component model: one cold component ($\sim$15 K and $\sim$576 M$_\odot$) that contains most of the mass and luminosity, and one warmer component ($\sim$51 K and $\sim$0.01 M$_\odot$) to explain the 24 μm data. The integrated luminosity of $\sim 177\,L_\odot$ can be used to constrain additional parameters of the embedded protostar from the turbulent core accretion model for massive star formation [43]. Following the simulations by Krumholz et al. [37], the data of IRDC 18223-3 are consistent with a massive gas core harboring a low-mass protostellar seed of still less than half a solar mass with high accretion rates of the order $10^{-4}\,M_\odot\,yr^{-1}$ and an age below 1000 yrs. In the framework of this model, the embedded protostar is destined to become a massive star at the end of its formation processes. While this interpretation is attractive, it is not unambiguous, and especially the derived time-scale from this model appears short when comparing with recent outflow data that will be presented in the following section (§3.1).

In summary, these observations indicate that the physical and chemical conditions at the onset of low- and high-mass star formation do not differ significantly (except for largely different initial cloud clump masses and accretion rates), and that the time-scale for massive bound gas clumps to remain starless is likely relatively short.

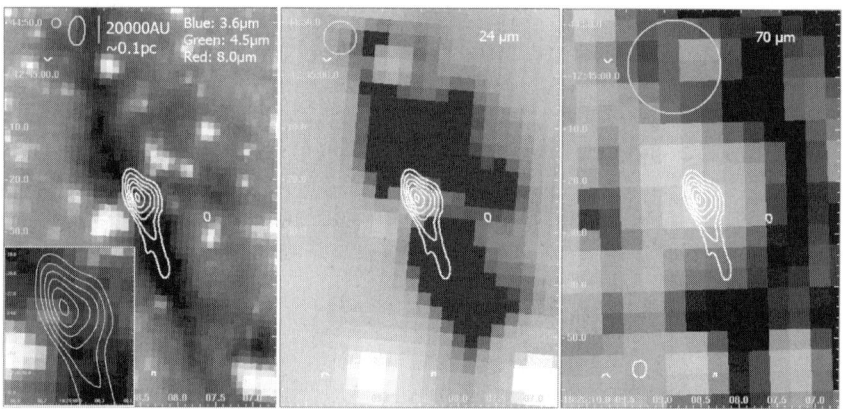

**Figure 2:** IRDC 18223-3 images at different wavelengths from Beuther & Steinacker [15]. The color scales show Spitzer images at various wavelength, and the contours show the 93 GHz (3.2mm) continuum emission observed with the PdBI [14]. The left panel presents a three-color composite with blue 3.6 μm, green 4.5 μm and red 8.0 μm (adapted from Beuther et al. [14]). The inlay zooms into the central core region. The middle and right panel show the Spitzer 24 and 70 μm images, respectively. The circles in each panel present the Spitzer beam sizes and the ellipse in the left panel presents the PdBI 3.2 mm continuum synthesized beam. A size-ruler is also shown in the left panel.

# 3 Massive accretion disks?

As mentioned in the Introduction, molecular outflows and accretion disks can be used to discriminate between the different formation scenarios for massive stars. Massive outflows have been subject to intense studies for more than a decade (e.g., Arce et al. [2], Beuther et al. [11], Shepherd & Churchwell [61], Zhang et al. [70]), and although there is considerable discussion about the details, we find a growing consensus that massive molecular outflows are ubiquitous in high-mass star formation, and that collimated jet-like outflows do exist for massive sources as well, at least during the very early evolutionary stages [12]. The collimation of the outflows is likely to widen with ongoing evolution. Nevertheless, these data are consistent with the turbulent core model for massive star formation, whereas they are less easy to reconcile with the competitive accretion model because the latter is so dynamic that collimated structures likely could not survive very long. Furthermore, the existence of collimated outflows can only be explained by magneto-hydrodynamic acceleration of the jet from an underlying accretion disk. Hence, there is ample indirect evidence for massive accretion disks, however, the physical characterization of disks in massive star formation is still lacking largely the observational basis [29]. The two main reasons for this are that the expected massive accretion disks are still deeply embedded within their natal cores complicating the differentiation of the disk emission from the ambient core, and that the clustered mode of massive star formation combined with the large average distances of the targets makes spatially resolving structures of the order 1000 AU a difficult observational task. In spite of these difficulties, the advent of broad spectral bandpasses allowing us to study several spectral lines simultaneously, as well as the improved spatial resolution of existing and forthcoming interferometers have increased the number of disk studies over the last few years. For a recent review see Cesaroni et al. [29].

Here, I will show three different examples of disk and/or rotation candidates in an evolutionary sense: It starts with a rotation and outflow investigation of the previously discussed IRDC 18223-3 (§2), then I will present recent data from the high-mass disk candidate in the HMPO IRAS 18089-1732, and finally observations from a massive disk candidate at a more evolved evolutionary stage will be discussed.

## 3.1 Rotation in IRDCs: the case of IRDC 18223-3

As a follow-up of the Infrared Dark Cloud study of IRDC 18233-3 discussed in §2, Fallscheer et al. (in prep.) observed the same region with the Submillimeter Array (SMA) in several spectral setups around 230 and 280 GHz covering outflow as well as dense gas tracers. Figure 3 shows a compilation of the CO(2–1) data, one $CH_3OH$ line and the dust continuum emission. The blue- and red-shifted CO(2–1) emission clearly identifies at least one large-scale outflow in the north-west south-east direction. This is consistent with two of the 4.5 μm emission features at the edge of the core (Fig. 2, left panel and inlay). There is another collimated red-shifted feature to the south-west corresponding to the third 4.5 μm feature, however, we do not identify a blue counterpart and refrain from further interpretation of that feature. Since we find for the main north-west south-east outflow blue- and red-shifted emission on both sides of the continuum peak, the orientation of the outflow should be close to the plane of the sky (see, e.g., models by Cabrit & Bertout [26]), and hence the assumed underlying perpendicular rotating structure close to edge-on. Following the approach outlined in Beuther et al. [11], the outflow mass and outflow rate are $13\,M_\odot$ and $3.5 \times 10^{-4}\,M_\odot\,yr^{-1}$, respectively. With the above derived core mass of

$\sim$576 M$_\odot$ (§2), this source fits well into the correlation between outflow rate and core mass previously derived for HMPOs (Fig. 7 in Beuther et al. [11]).

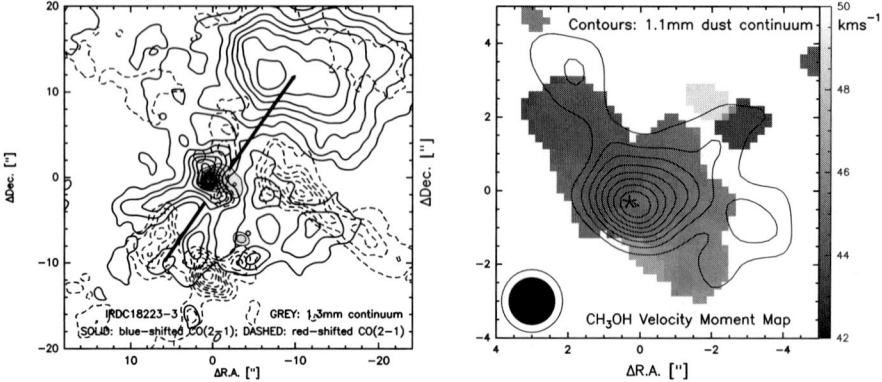

**Figure 3:** SMA observations toward IRDC 18223-3 (Fallscheer et al. in prep.). The left panel shows the blue- and red-shifted CO(2–1) emission as solid and dashed contours overlaid on the grey-scale 1.3 mm dust continuum emission. The central core is the same source as in Fig. 2. The right panel presents in grey-scale a 1st moment map (intensity weighted velocity distribution) of CH$_3$OH overlaid with the 1.1 mm continuum emission. The empty and full circle are the synthesized beams of the line and continuum emission, respectively.

While the outflow rate is consistent with the accretion rate previously derived from the SED (§2), discrepancies arise with respect to the age of the system. Although dynamical timescales are highly uncertain (e.g., Parker et al. [51]), the size of the molecular outflow combined with a low inclination angle allows for at least a timescale estimate for the outflow of the order a few $10^4$ yrs, well in excess of the value $\leq 10^3$ yrs previously derived from the SED applied to models (§2). Notwithstanding the large errors between the different estimates, the discrepancy of more than an order of magnitude appears real. How can we explain that? There is no clear answer to that yet, but a possibility is that the orientation of the disk-outflow system with the disk close to edge-on absorbs a large amount of flux distorting the SED on the Wien-side. If that were the case, the SED-estimated age could underestimate the age of the system. Another possibility to solve the discrepancy is that the initial start of high-mass star formation may proceed slower, i.e., the first low-mass protostar(s) (destined to become massive or not?) form within the massive cores, and they already start driving outflows, but at that stage it is impossible to detect them in the near- to far-infrared because of the large extinction. In this picture at some point the high-mass star formation process would need to accelerate because otherwise the massive stars cannot form in the short time-scales of a few $10^5$ yrs (e.g., McKee & Tan [42]). It is not clear why the whole process should start slow and what could trigger such acceleration later-on. Obviously, more theoretical and observational work is required to explain the different time-scales in more detail.

Figure 3 (right panel) zooms into the central core and shows dust continuum emission as well as the velocity structure of the dense central gas observed in CH$_3$OH ($6_{0,1} - 5_{0,1}$) with a lower level excitation level of $E_{\text{low}} = 34.8$ K. Interestingly, both the continuum and the spectral line emission are elongated in the north-east south-west direction perpendicular to the main molecular outflow. While the continuum emission shows three resolved emission features, $CH_3OH$ exhibits a smooth velocity gradient across the source spanning approximately $3\,km\,s^{-1}$. The $CH_3OH$ line-width FWHM toward the continuum peak is $2.1\,km\,s^{-1}$. The blue-redshifted features in the north-west are likely part of the molecular outflow and one sees even a slight elongation of the continuum emission in that direction. While $CH_3OH$ is a well-known shock tracer and hence regularly found within molecular outflows (e.g., Bachiller et al. [3]), it is more of a surprise that we find it in an elongated structure likely associated with rotation and infall perpendicular to the outflow. The extent of this structure is large with $\sim 6.5''$ corresponding to more than 20000 AU at the given distance of 3.7 kpc. Although we have no methanol isotopologue in the setup to exactly determine the opacity of the line, a low-energy transition like this is likely to be optically thick. Hence, we are tracing some of the outer rotating structures, probably corresponding to a larger scale rotating and potentially infalling/inspiralling toroid (e.g., Cesaroni et al. [31], Keto [35]). The small velocity spread across the structure as well as the relatively narrow $CH_3OH$ line-width toward the core center are also consistent with tracing outer structures because due to momentum conservation rotating structures should have lower velocities further out.

Notwithstanding that we do not exactly know the age of IRDC 18223-3, its non-detection up to $8\,\mu m$ puts it at an early evolutionary phase prior to the better studied HMPOs and Hot Molecular Cores. Our data clearly show that even at such early stages molecular outflows and rotating structures perpendicular to that have been developed, and it is likely that closer toward the core center, one will find a real accretion disk. To investigate the latter in more detail, higher angular resolution observations of an optically thin dense gas tracer are required.

## 3.2 The HMPO disk candidate IRAS 18089-1732

As a more evolved massive disk candidate, we have studied intensely over the last few years the HMPO IRAS 18089-1732. This source at a distance of 3.6 kpc with a luminosity of $10^{4.5}\,L_\odot$ is part of a large sample of HMPOs, it hosts $H_2O$ and Class II $CH_3OH$ maser and has strong molecular line emission indicative of en embedded Hot Molecular Core [10, 65]. During early SMA observations Beuther et al. [6, 20, 21] identified in SiO a molecular outflow in the north-south direction, and perpendicular to that in $HCOOCH_3$ a velocity gradient on scales of a few 1000 AU. Although these data were indicative of rotation and an underlying massive accretion disk, the observations did not allow us to characterize the structure in more detail because of a lack of spatial resolution. Therefore, we observed IRAS 18089-1732 now in high-energy transitions of $NH_3$ at 1.2 cm wavelength with the VLA and the ATCA at a spatial resolution of $0.4''$ [16]. These $NH_3(4,4)$ and $(5,5)$ lines have a two-fold advantage: Their high excitation levels ($> 200\,K$) ensure that we are tracing the warm inner regions and are less confused by the surrounding cold envelope, whereas the cm wavelengths regime is less affected by high optical depth of the dust emission in high column density regions and may hence be particularly well suited for massive disk studies (e.g. Krumholz et al. [36]). Figure 4 presents an integrated image and a 1st moment map (intensity weighted velocity) of the corresponding VLA observations.

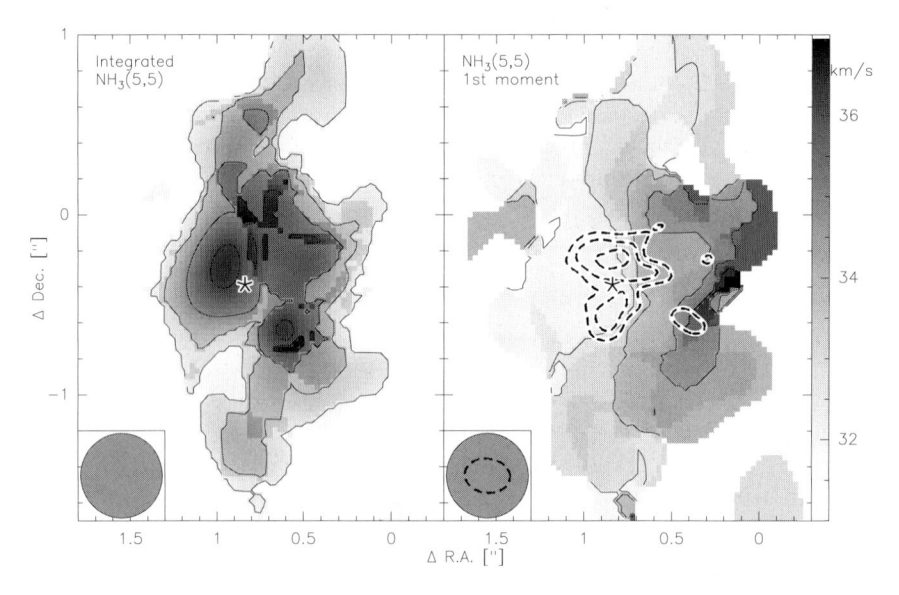

**Figure 4:** The left panel shows the VLA NH$_3$(5,5) emission integrated from 31 to 37 km s$^{-1}$ [16]. The right panel presents the corresponding 1st moment map contoured from 31.5 to 36.5 km s$^{-1}$ (step 1 km s$^{-1}$). The white-black dashed contours show the 1.2 cm continuum emission. The asterisks mark the position of the submm continuum peak [21], and the synthesized beams are shown at the bottom-left (grey NH$_3$ and dashed 1.2 cm emission).

The 1st moment map confirms the previously assessed velocity gradient in east-west direction perpendicular to the molecular outflow. The NH$_3$ line-width FWHM toward the central core is 4.7 km s$^{-1}$, significantly broader than that of IRDC 18223-3 (§3.1). In the simple picture of equilibrium between gravitational and centrifugal force, the rotationally bound mass would be ∼37 M$_\odot$, of the same order as the whole gas mass as well as the mass of the central source (of the order 15 M$_\odot$). Furthermore, the position-velocity diagram is not consistent with Keplerian rotation. It even shows indications of super-Keplerian motion, which is expected for very massive disks where the rotation profile is not only determined by the mass of the central object but also by the disk itself (e.g., Krumholz et al. [37]). Hence, the new VLA and ATCA data clearly confirm the previous assessment of rotation perpendicular to the outflow/jet, however, the kinematic signatures of that rotating structure are not consistent with a Keplerian disk like in low-mass star formation, but they show additional features which can be produced by massive self-gravitating disks as well as by infalling gas that may settle eventually on the disk. In addition to this, the detection of the high-excitation lines in the rotating material indicates high average gas temperatures >100 K for the disk-like structures, well in excess of typical gas temperatures in low-mass disks of the order 30 K (e.g., Piétu et al. [52]). Moreover, we detect double-lobe cm continuum emission close to the core center where the two lobes are oriented in north-south direction parallel to the outflow identified in SiO. With respect to previous data at longer wavelength, we find a spectral index at cm wavelength of 1.9, consistent with an optically thick jet [57].

It will be interesting to further zoom into the innermost regions with future instruments like ALMA and eVLA to asses whether the quantitative deviations from typical low-mass accretion disks continue down to the smallest scales, or whether we will find Keplerian disk structures as known from their low-mass counterparts.

## 3.3 A more evolved massive disk candidate?

Moving along in the evolutionary sequence, we have recently identified a potential disk around a more evolved candidate young stellar object (Quanz et al. in prep.). The source labeled so far mdc1 (massive disk candidate) was identified serendipitously during a near-infrared wide-field imaging campaign on Calar Alto via its K-band cone-like nebulosity and a central dark lane (Fig. 5). First single-dish bolometer and spectral line measurements revealed a 1.2 mm flux of 12 mJy and a velocity of rest of $\sim 51\,\mathrm{km\,s^{-1}}$. The latter value indicates a kinematic distance of $\sim 5\,\mathrm{kpc}$, consistent with distances of a few UCHII regions in the surrounding neighborhood. To investigate this object in more detail, we recently observed it with the SMA at 1.3 mm wavelength mainly in the mm continuum and the $^{12}CO/C^{18}O$ spectral line emission. Figure 5 presents an overlay of the SMA data with the K-band nebulosity, and a few points need to be stressed:

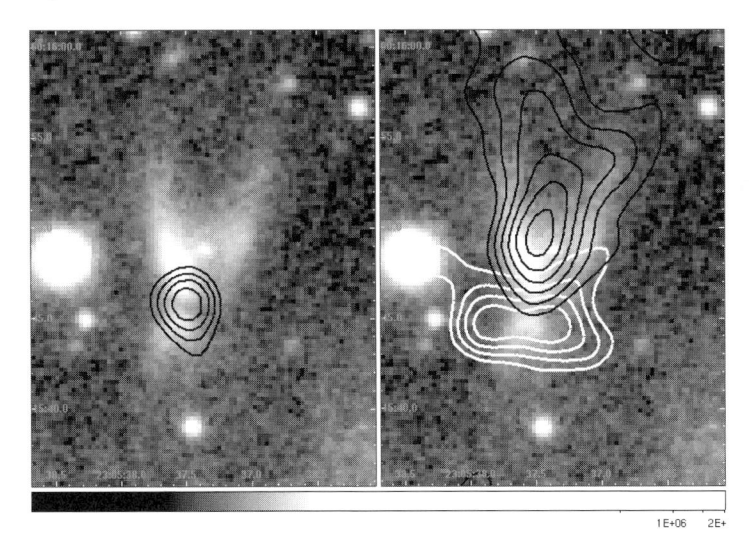

**Figure 5:** The grey-scale in both panels shows the K-band near-infrared nebulosity observed for this massive evolved disk candidate (Quanz et al. in prep.). The contours are the corresponding SMA mm observations where the left panel shows the 1.3 mm continuum emission, and the right panel in black and white contours the blue-shifted $^{12}CO(2-1)$ and the integrated $C^{18}O(2-1)$ emission, respectively.

(a) Although spatially unresolved with a synthesized beam of $\sim 4.0''$ the 1.3 mm continuum peak exactly coincides with the infrared dark lane consistent with the large column densities of the proposed disk-like structure. The flux measured with the SMA is 12 mJy like the previous single-dish measurements. This indicates that we have no surrounding dust/gas envelope but rather an isolated central structure. Assuming optically thin dust emission at 50 K, the approximate gas mass of the central structure is $\sim 5$ $M_\odot$. (b) We detect blue-shifted CO(2–1) spatially well correlated with the K-band nebulosity north of the dark lane. This confirms the initial interpretation of that feature to be due to an outflow. (c) The integrated $C^{18}O(2-1)$ emission is elongated perpendicular to the outflow observed in CO and K-band continuum emission. The line-width FWHM of the $C^{18}O$ emission is narrow with $\sim 0.8$ km s$^{-1}$, however, the spatial extent of this structure is large, of the order $2 \times 10^4$ AU.

While the low gas mass and the missing more massive gas envelope could be interpreted in the framework of a low-mass source, such large disk-structures as indicated by the $C^{18}O$ emission are not known from typical low-mass disk sources (e.g., Simon et al. [62]). Therefore, these observations can also be interpreted as a remnant disk/torus around an intermediate to high-mass (proto)star that has already dispersed much of its envelope. Although $C^{18}O$ is detected only in two channels, these show a clear velocity shift, and the small line-width may be due to the lower rotational motions on large scales assuming momentum conservation in rotating, potentially Keplerian structures.

Synthesizing the three example sources shown here, it is interesting to note that the line-widths are small in the youngest and the supposed to be oldest source, whereas they are large in the HMPO which should be in its main accretion phase. In an evolutionary picture this can be interpreted that at early evolutionary stages infall, turbulence and rotation are not yet that vigorous. Then in the main accretion phase, infall, rotation and outflow processes strongly increase the line-width. And finally, when the accretion stops, the envelope and disk slowly disperse and one observes only a remnant structure with small line-width in the outer regions. This scenario is speculative, however, the number of disk candidates is steadily increasing, and since we start sampling more evolutionary stages, we are getting the chance to address disk evolution questions in high-mass star formation as well.

## 4 Fragmentation in high-mass star formation

How massive gas clumps fragment is one of the key questions if one wants to understand the formation of the Initial Mass Function, and as outlined in §1, the two main massive star formation theories predict differences in the early fragmentation processes. In the following I will present several examples of fragmenting massive cores addressing issues about fragmentation on the cluster-scale, fragmentation of smaller groups, potential proto-trapezia, and the determination of density structures of sub-sources within evolving clusters.

## 4.1 Resolving the massive proto-cluster IRAS 19410+2336

To address fragmentation processes at early evolutionary stages high angular resolution at (sub)mm wavelengths is the tool of choice to resolve the relevant substructures. Beuther & Schilke [8] resolved the young massive star-forming region IRAS 19410+2336 (distance $\sim 2.1$ kpc and luminosity $\sim 10^4\,L_\odot$) at 1.3 mm wavelength with the PdBI at approximately 2000 AU linear resolution into 24 sub-sources. Although from a statistical point of view such numbers cannot compete with the clusters exceeding 100 or even 1000 stars observed at optical and near-infrared wavelength, this is still one of the prime examples of a spatially resolved massive protocluster. Assuming that all emission features are due to cold dust emission from embedded protostars, they were able to derive a core-mass function. With a power-law slope of -2.5, this core mass function is consistent with the Salpeter IMF slope of -2.35 [59]. Therefore, Beuther & Schilke [8] interpreted these observations as support for the turbulent fragmentation put forth by, e.g., Padoan & Nordlund [50].

A few caveats need to be kept in mind: While Beuther & Schilke [8] assumed a uniform gas temperature for all sub-sources, it is more likely that the central peaks are warmer than those further outside. This issue can be addressed by spectral line emission with temperature sensitive molecules (e.g., $H_2CO$) which is an ongoing project by Rodon et al. (in prep.). Furthermore, the assumption that all mm continuum peaks are of pro- or pre-stellar nature is not necessarily always valid, e.g., Gueth et al. [33] or Beuther et al. [9] have shown that mm continuum emission can partly also be caused by collimated jets. However, only the central source is detected at cm wavelength and collimated jets should be detectable at cm wavelengths as well. Therefore, we believe that jets should not affect the analysis much.

Independent of the caveats, it is surprising that IRAS 19410+2336 is still the only young massive star-forming region that is resolved in $>10$ sub-sources in the mm continuum emission. While this can be explained to some degree by the exceptionally good uv-coverage obtained for the given observations, which results in a good sampling of spatial structures, we also need to consider whether different modes of fragmentation may exist. Similar high-spatial-resolution studies of more protoclusters spanning a broad range of luminosities are required to tackle this question in more detail. Another interesting question is associated with the spatial filtering of interferometers and the corresponding large-scale emission: Many interferometric (sub)mm continuum studies of massive star-forming regions filter out of the order 90% of the flux, hence, large amounts of the gas are distributed on larger scales, usually $> 10''$. The question remains whether this gas will eventually participate in the star formation process or not?

## 4.2 Fragmentation of potential proto-trapezia

### 4.2.1 The enigmatic proto-trapezium W3-IRS5

The W3-IRS5 region is one of the prototypical high-mass star-forming regions with $\sim 10^5\,L_\odot$ at a distance of $\sim 1.8$ kpc that shows fragmentation on scales of the order 1000 AU observed at near-infrared as well as cm wavelengths [44, 67]. However, not much was known about the cold dust and gas emission. Therefore, we observed the region with the PdBI at 1.3 and 3.5 mm wavelengths with the new extended baselines resulting in an unprecedented spatial resolution of $\sim 0.37''$ (Rodon et al. in prep.). Figure 6 shows a compilation of the 1.3 mm continuum data and the SiO(5–4) and $SO_2(22_{2,20}-22_{1,21})$ spectral line emission.

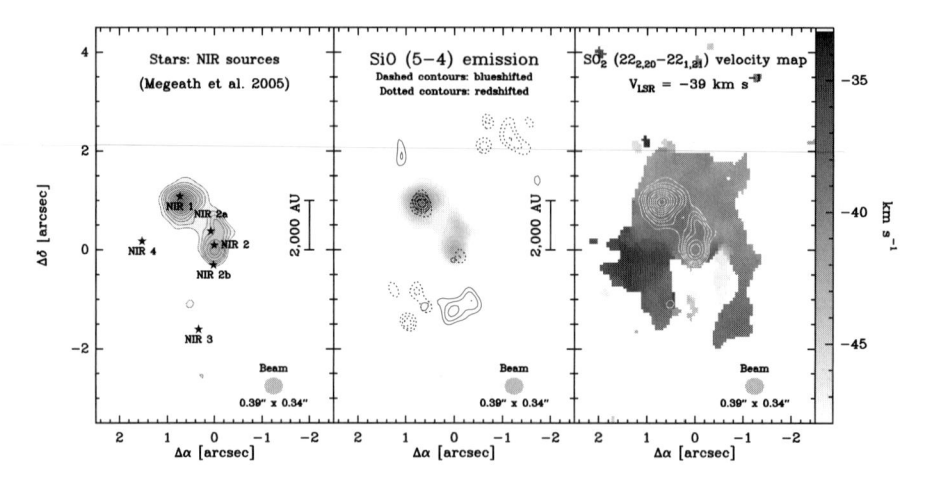

**Figure 6:** PdBI observations of the W3-IRS5 system from Rodon et al. (in prep.). The left panel shows the 1.3 mm continuum emission, and the stars mark near-infrared sources from Megeath et al. [44]. The middle panel presents as solid and dotted contours the blue- and red-shifted SiO(5–4) emission overlaid on the grey-scale 1.3 mm continuum emission. The right panel finally shows the 1st moment map of $SO_2(22_{2,20}-22_{1,21})$ in grey-scale with the 1.3 mm continuum contours.

The mm continuum emission resolves the W3-IRS5 region into five sub-sources, where four of them are coincident with near-infrared and cm emission peaks. Three of the sources are clustered in a very small projected volume of only $\sim$2000 AU. With this high spatial resolution we find extremely large average column densities of the order a few times $10^{24}$ cm$^{-2}$ which corresponds to visual extinctions $A_v$ between $5 \times 10^3$ and $10^4$ averaged over the beam size. Such extinctions should be far too large to allow any detection at near-infrared wavelengths, nevertheless, near-infrared counterparts are detected [44]. This conundrum can likely be explained by the detection of several SiO outflows in the field. In particular, we find very compact blue- and red-shifted SiO emission toward the two main mm peaks, where the blue- and red-shifted emission is barely spatially separated (Fig. 6 middle panel). Since the overall time-scale of the W3-IRS5 outflow system is relatively large (of the order a few times $10^4$ yrs, Ridge & Moore [58]), these compact features are unlikely from very young outflows, but they indicate that the outflows are oriented almost along the line of sight. The opening cones of the outflows are the likely cause that emission from close to the protostars can escape the region and hence make them detectable at near-infrared wavelengths.

The right panel of Fig. 6 shows the 1st moment map of $SO_2$ (intensity weighted velocities) which encompasses the mm continuum peaks. The coherent velocity field over the sub-sources is a strong indicator that the system is a bound structure and not some unbound chance alignment within the field (e.g., Launhardt [39]). In addition to the general velocity gradient from the south-east to the north-west, one tentatively identifies velocity gradients across the two strongest mm continuum peaks. Since we do not know the exact orientation of the outflows with respect to the $SO_2$ rotation axis, it is not yet possible to identify these structures with disk-like components as in §3. Future observations in different tracers may help to shed more light on the rotational structure associated with each sub-source. It should also be noted that the line-width FWHM toward the mm continuum peaks varies between 6.2 and 7 km s$^{-1}$ larger than those reported in §3. While the larger line-width compared with the IRDC and the more evolved source may be explained by the evolutionary sequence sketched at the end of §3, the larger FWHM compared to the HMPO may have different reasons, among them are the larger luminosity of W3-IRS5, its multiplicity compared with the so far unresolved source IRAS 18089-1732, and also the molecular species, because $SO_2$ should be more affected by shocks than the $NH_3$ line used for the IRAS 18089-1732 study. In addition to this, the $SO_2$ moment map exhibits a velocity discontinuity with a velocity jump of the order 4 km s$^{-1}$ south-east of the mm continuum peaks. What is the cause of this discontinuity, is it associated with the original core formation and a shock within converging flows, or is it of different origin?

In summary, the combination of high-spatial-resolution observations of the continuum emission in addition to outflow and dense gas tracers allows us to characterize many physical properties of this proto-trapezium system with respect to its multiple components and their outflow and rotation properties.

### 4.2.2 Fragmentation of the hot core G29

The hot core G29.96 located right next to a well-known cometary H II region comprises another example of several protostellar submm continuum sources within the innermost center of a high-mass star forming region (distance $\sim$6 kpc, luminosity $9 \times 10^4 \, L_\odot$). High-spatial resolution observations with the SMA in its most extended configuration yielded a spatial resolution of $0.36'' \times 0.25''$ in the submm continuum at $\sim$348 GHz, corresponding to a linear resolution of 2000 AU (Fig. 7, Beuther et al. [17], the line data will be discussed in §5.1). The Hot Molecular Core previously identified in a high-excitation $NH_3$ line [30] is resolved by these new data into four sub-sources within a projected diameter of $\sim$6900 AU. Assuming that the emission peaks are of protostellar nature, Beuther et al. [17] estimated a protostellar density of $\sim 2 \times 10^5$ protostars/pc$^{-3}$. This is considered a lower limit since we are limited by spatial resolution, sensitivity and projection effects. Nevertheless, such a protostellar density is about an order of magnitude higher than values usually reported for star-forming regions (e.g., Lada & Lada [38]). Although this value is still about an order of magnitude lower than protostellar densities that would be required in the merging scenario for massive stars (e.g., Bally & Zinnecker [4], Bonnell et al. [25]), it is interesting to note that increasingly higher protostellar densities are reported when going to younger sources and better angular resolution (see also Megeath et al. [44]). This allows us to speculate whether future observations with better spatial resolution and sensitivity toward extremely massive star-forming regions will reveal protostellar densities that may be sufficient to make mergers possible. While such a detection would not be a proof for mergers to exist, it will certainly be important to verify whether the required initial conditions do exist at all.

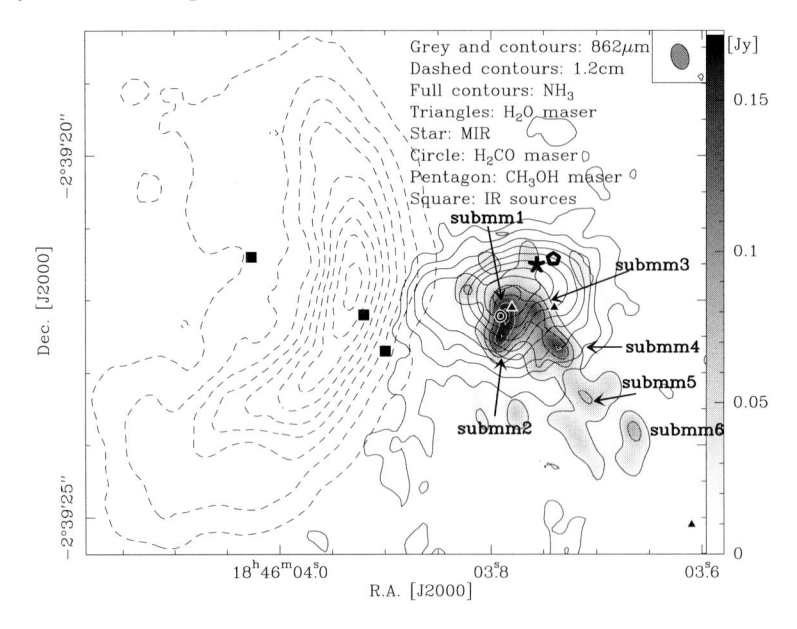

**Figure 7:** Compilation of data toward the UCH$_{II}$/hot core region G29.96 from Beuther et al. [17]. The dashed contours present the cometary UCH$_{II}$ regions whereas the full contours show the older NH$_3$ observation from the hot core [28]. The grey-scale with contours then present the new high-resolution ($0.36'' \times 0.25''$) submm continuum data from the SMA.

## 4.3 Density structure of sub-sources – IRAS 05358+3543

As a final example for the potential of (sub)mm continuum studies, I present the recent multi-wavelength investigation of the HMPO IRAS 05358+3543. This region at a distance of 1.8 kpc with a luminosity of $10^{3.8}$ L$_\odot$ was observed in a combined effort with the PdBI and the SMA at arcsecond resolution in four wavelength bands (3.1 and 1.2 mm, and 875 and 438 µm, Beuther et al. [7], Leurini et al. [40]). While many details about the sub-structure of the forming cluster can be derived, here, I will discuss only two results.

Based on the multi-wavelength data, Beuther et al. [7] fitted the spectral energy distribution on the Rayleigh-Jeans side of the spectrum (Fig. 8). While the main source can well be fitted by a typical protostellar spectrum consisting of free-free emission at long wavelength and a steep flux increase at shorter wavelength due to the dust emission, another sub-source did not fit at all into that picture. In particular the shortest wavelength data-point at 438 µm shows significantly lower fluxes than expected for a typical protostar. The most likely explanation for this effect is that we are dealing with a very cold source and that therefore we are already approaching the peak of the spectral energy distribution. The data allowed us to estimate an upper limit for the dust temperature of $\leq 20$ K. Since we are also not detecting any other line emission from this core (mainly from typical hot core molecules, Leurini et al. [40]), it may well be a starless core right in the vicinity of an already more evolved massive protostar. Further investigations of this sub-source in typical cold gas tracers like $N_2H^+$ or $NH_3$ are required to test this proposal. Independent of whether this source harbors an embedded protostar or not, these observations show the importance of short wavelength data at high spatial resolution if one wants to differentiate between critical core parameters like the dust temperature.

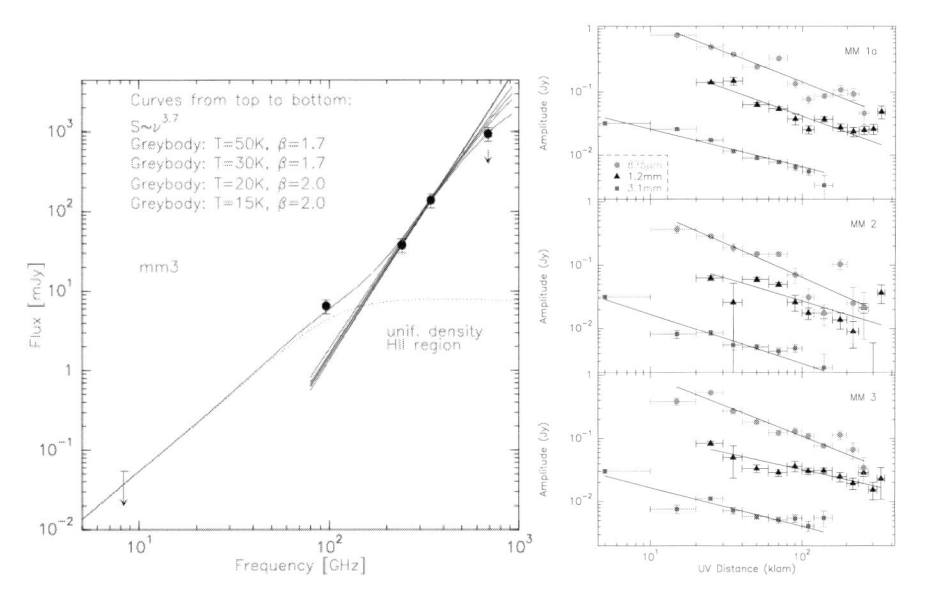

**Figure 8:** The left panel presents the SED toward the coldest sub-source in IRAS 05358+3543 [7]. The parameters of the fits are marked in the figure. The right panel shows intensities averaged in uv-annuli and plotted versus the baseline-length for different sub-sources and wavelengths. Most can be well fitted by power-law distributions.

Another physical parameter which has so far not been observationally constrained for massive star formation, is the density profile of individual sub-sources. While density profiles of low-mass star-forming cores have well been characterized (e.g., Andre et al. [1], Motte et al. [45], Ward-Thompson et al. [68]), in high-mass star formation, density profiles were until now only derived with single-dish observation covering scales of the whole cluster but not individual sub-sources (e.g., Beuther et al. [10], Hatchell & van der Tak [34], Mueller et al. [47]). This is partly due to the technical problem of interferometer observations that filter out large amounts of the flux and hence make density profile determinations from their images extremely unreliable. To overcome this problem, Beuther et al. [7] analyzed the data directly in the uv-domain prior to any fourier transformation. Figure 8 shows the corresponding plots of the observed intensities versus the uv-distance for three sub-sources in three wavelengths bands, respectively. The observations cannot be fitted with Gaussian distributions, but much better fits are achieved with power-law distributions. These power-laws in the uv-domain can directly be converted to the corresponding power-laws of the intensity profiles in the image plane. Assuming furthermore a temperature distribution $T \propto r^{-0.4}$ we can now infer the density profiles of individual sub-sources of the evolving cluster. The derived density profiles $\rho \propto r^{-p}$ have power-law indices $p$ between 1.5 and 2. Although this result is similar to the density profiles previously determined for low-mass cores, to our knowledge this is the first time that they have been observationally constrained for resolved sub-sources in a massive star-forming region. The density structure is an important input parameter for any model of star formation (e.g., McKee & Tan [43]).

## 5 Astrochemistry

### 5.1 Toward a chemical evolutionary sequence

Astrochemistry is a continuously growing field in astronomy. Although line-survey style studies of different sources have existed for quite some time (e.g., Blake et al. [23], Schilke et al. [60]), these studies had usually been performed with single-dish instruments averaging the chemical properties over the whole cluster-forming regions. Since the advent of broadband receivers at interferometers like the SMA, it is now also possible to perform imaging spectral line surveys that allow us to spatially differentiate which molecules are present in which part of the targeted regions, for example, the spatial differentiation between nitrogen- and oxygen-bearing molecules in Orion-KL (e.g., Beuther et al. [18], Blake et al. [22]). In addition to the spatial analysis of individual regions, we are also interested in analyzing how the chemistry evolves in an evolutionary sense. As an early step in this direction we synthesized SMA observations that were observed in the same spectral setup around 862 μm toward four massive star-forming regions over the last few years (Beuther et al. subm.). These four regions comprise a range of luminosities between $10^{3.8}$ L$_\odot$ and $10^5$ L$_\odot$, and they cover different evolutionary stages from young High-Mass Protostellar Objects (HMPOs) to typical Hot Molecular Cores (HMCs): Orion-KL: HMC, $L \sim 10^5$ L$_\odot$, $D \sim 0.45$ kpc [18]; G29.96: HMC, $L \sim 9 \times 10^4$ L$_\odot$, $D \sim 6$ kpc [17]; IRAS 23151, HMPO, $L \sim 10^5$ L$_\odot$, $D \sim 5.7$ kpc [19]; IRAS 05358: HMPO, $L \sim 10^{3.8}$ L$_\odot$, $D \sim 1.8$ kpc [7, 40]. Smoothing all datasets to the same linear spatial resolution of 5700 AU, we are now capable to start comparing these different regions. Figure 9 presents typical spectra extracted toward the HMC G29.96 and the HMPO IRAS 23151.

A detailed comparison between the four sources is given in a forthcoming paper (Beuther et al. subm.), here we just outline a few differences in a qualitative manner.

- The HMCs show far more molecular lines than the HMPOs. Orion-KL and G29.96 appear similar indicating that the nature of the two sources is likely to be comparable as well. Regarding the two HMPOs, the higher luminosity one (IRAS 23151) shows still more lines than the lower-luminosity source (IRAS 05358). Since IRAS 05358 is approximately three times closer to us than

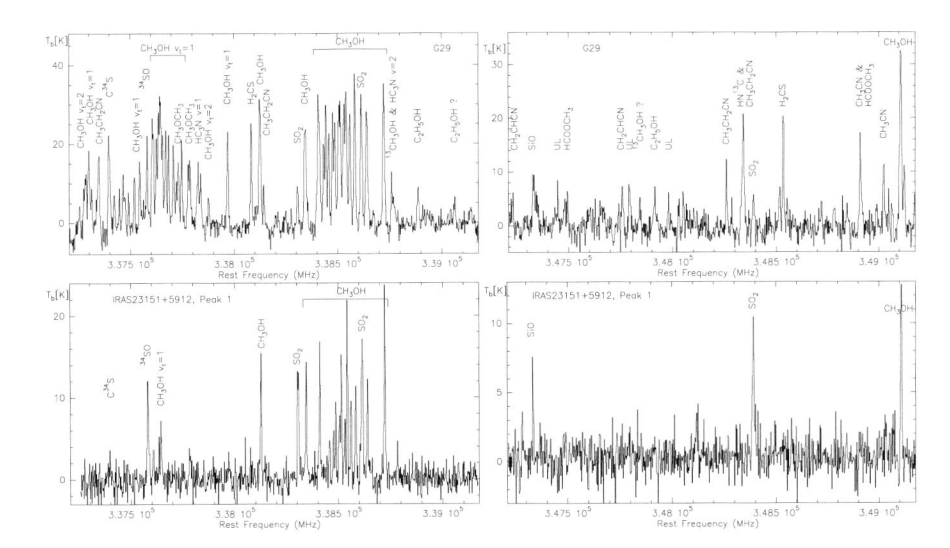

**Figure 9:** SMA spectra extracted toward two massive star-forming regions (G29.96 top row & IRAS 23151+5912 bottom row, Beuther et al., subm.). The spectral resolution in all spectra is 2 km/s. The left and right column show the lower and upper sideband data, respectively.

IRAS 23151, this is not a sensitivity issue but it is likely due to the different luminosity objects forming at the core centers.

- The ground-state $CH_3OH$ lines are detected toward all four sources. However, the vibrational-torsional excited $CH_3OH$ are only strongly detected toward the HMCs Orion-KL and G29.96. Independent of the luminosity, the HMPOs exhibit only one $CH_3OH$ $v_t = 1$ line, which can easily be explained by the lower average temperatures of the HMPOs.

- A more subtle difference can be discerned by comparing the $SO_2$ and the $HN^{13}C$/$CH_3CH_2CN$ line blend near 348.35 GHz (in the upper sideband). While the $SO_2$ line is found toward all four sources, the $HN^{13}C$/$CH_3CH_2CN$ line blend is strongly detected toward the HMCs, but it is not found toward the HMPOs. In the framework of warming up HMCs, this indicates that nitrogen-bearing molecules are either released from the grains only at higher temperatures, or they are daughter molecules which need some time during the warm-up phase to be produced in gas-phase chemical networks. In both cases, such molecules are not expected to be found much prior to the formation of a detectable HMC.

- Comparing the spatial distribution of different molecules, we find, e.g., that $C^{34}S$ is observed mainly at the core edges and not toward the submm continuum peak positions. This difference can be explained by temperature-selective desorption and successive gas-phase chemistry reactions: CS desorbs early from the grains at temperatures of a few 10 K and should peak during the earliest evolutionary phases toward the main continuum sources. Subsequently when the core warms up to $\sim 100$ K, $H_2O$ desorbs and dissociates to OH. The OH then quickly reacts with the sulphur to form SO and $SO_2$ which should then peak toward the main continuum sources. This is what we observe in our data. The fact that the $C^{34}S$ peaks are offset from the submm continuum condensations even toward the younger sources is due to their evolutionary stage where they have already heated up their central regions to more than 100 K. Even younger sources are required to confirm this scenario.

## 5.2 $C_2H$ as a tracer of the earliest evolutionary stages?

In an effort to study a larger source sample with respect to its chemical evolution, we observed 21 massive star-forming regions covering all evolutionary stages from IRDCs via HMPOs/hot cores to UCHII with the APEX telescope at submm wavelengths (Beuther et al. subm.). While most spectral lines were detected mainly toward the HMPO/hot core sources, the ethynyl molecule $C_2H$ is omni-present toward all regions. To get an idea about the spatial structure of ethynyl, we went back to an older SMA data-set targeting the HMPO IRAS 18089-1732 at the same frequency around 349.4 GHz of the $C_2H$ line [21]. Because we were not able to image the spatial distribution of $C_2H$ at that time, we now restricted the data to only the compact configuration allowing us to better image the larger-scale distribution of the gas. Figure 10 presents the resulting molecular line map, and we find that $C_2H$ is distributed in a shell-like fashion around the central protostellar condensation. Comparing this with all other imaged molecules in the original paper, only $C_2H$ exhibits this behavior. To better understand this effect, we ran a set of chemical models in 1D for a cloud of $1200\,M_\odot$, a density power-law $\rho \propto r^{-1.5}$ and different temperature distributions $T \propto r^q$. A snapshot of these models after an evolutionary time of $5 \times 10^4$ yrs is presented in Figure 10. The models reproduce well the central $C_2H$ gap in IRAS 18089-1732 which should have approximately the same age.

While these models reproduce the observations, they give also predictions how the $C_2H$ emission should look like at different evolutionary times. In particular, $C_2H$ forms quickly early on, also at the core center. Since not many molecules exist which do not freeze out and are available to investigate the cold early phases of massive star formation (valuable exceptions are, e.g., $NH_3$ or $N_2H^+$), the detection of $C_2H$ toward the whole sample in combination with the chemical models triggers the prediction that $C_2H$ may well be an excellent molecule to investigate the physical conditions of (massive) star-forming regions at very early evolutionary stages. High-spatial-resolution observations of IRDCs are necessary to investigate this potentially powerful astrophysical tool in more detail.

## 5.3 Employing molecules as astrophysical tools

While the chemical evolution of massive star-forming regions is interesting in itself, one also wants to use the different characteristics of molecular lines to trace various physical processes. In contrast to molecules like SiO and CO that are well-known outflow/jet tracers, the task gets more difficult searching for suitable accretion disk tracers. Investigating our sample and disk claims in the literature, one finds that in

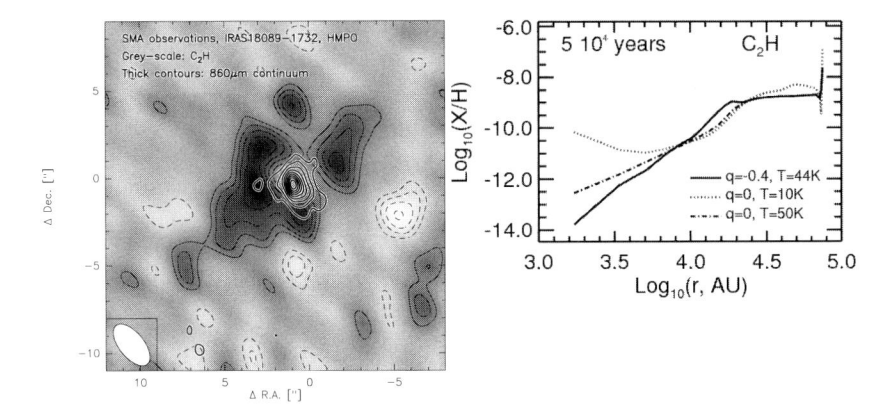

**Figure 10:** The left panel presents in grey-scale the $C_2H$ emission and in thick solid contours the corresponding submm continuum from the SMA toward the HMPO IRAS 18089-1732 (Beuther et al., subm.). The right panel shows a chemical model explaining the decreased emission toward the core center after approximately $5 \times 10^4$ yrs. The parameter $q$ denotes the temperature power-law index, and the $T$ values refer to the temperature at the core edge or to isothermal values ($q = 0$).

many cases exclusively one or the other molecule allows the investigation of rotational motion, whereas most other molecular lines remain without clear signatures. For example, the $HN^{13}C$ line discussed above (§5.1) traces rotation in the hot core G29.96 but it is not even detectable in younger sources. The other way around, $C^{34}S$ traced disk rotation in the young HMPO IRAS 20126 [31], but not anymore toward more evolved sources (§5.1). These differences imply that one will unlikely find a uniquely well suited molecular line allowing the study of large samples of massive accretion disks, but that one has to select for each source or source class the suitable molecule for detailed investigations.

In the following, I give a short table with molecules and their potential usefulness to study different physical processes. This table (1) is restricted to molecules with spectral lines at cm/(sub)mm wavelengths and does not claim any kind of completeness, it should just serve as a rough overview and it only lists the main isotopologues of each species.

## 6 Conclusions and summary

This article tries to outline how far we can currently constrain physical and chemical properties in massive star formation using (sub)mm interferometry. Coming back to the original questions raised in the abstract: (a) What are the physical conditions at the onset of massive star formation? (b) What are the characteristics of potential massive accretion disks and what do they tell us about massive star formation in general? (c) How do massive clumps fragment, and what does it imply to high-mass star formation? (d) What do we learn from imaging spectral line surveys with respect

**Table 1:** A few useful molecules and some of their potential applications.

| | |
|---|---|
| OH | Zeeman effect, magnetic fields, maser signpost of star formation |
| CO | General cloud structure, outflows |
| SiO | Shocks due to jets/outflows |
| $CO^+$ | Far-UV radiation from embedded protostars |
| CS | Dense gas, rotation, also outflows |
| CN | Photodominated regions, Zeeman effect, magnetic fields |
| SO | Shocks, dense gas |
| $H_2O$ | Shocks and hot cores, rotation ($H_2^{18}O$), maser signpost of star formation |
| HDO | Deuterium chemistry |
| $H_2D^+$ | Cold gas, pre-stellar cores, freeze out |
| HCN | Dense cores, also outflows |
| HNC | Dense cores, rotation ($HN^{13}C$) |
| $HCO^+$ | Outflows, infall, cosmic rays, ionization degree, dense gas ($H^{13}CO^+$) |
| $SO_2$ | Shocks, dense gas |
| $C_2H$ | Early evolutionary stages (§5.2) |
| $N_2H^+$ | Early evolutionary stages |
| $N_2D^+$ | Deuteration, freeze out |
| $H_3O^+$ | Cosmic rays |
| $H_2CO$ | Dense gas, temperatures |
| $NH_3$ | Cold and hot cores, rotation, temperatures |
| $CH_3OH$ | Shocks, young rotating structures? (§3.1), temperatures, maser signpost of massive star formation |
| $CH_3CN$ | Hot cores, temperatures, rotation |
| $CH_3CCH$ | Dense gas, temperatures |
| $HCOOCH_3$ | Hot cores, rotation |
| $CH_3CH_3CN$ | Hot Cores |

to the chemistry itself as well as for utilizing molecules as tools for astrophysical investigations?

Can we reasonably answer any of these questions with confidence? There are no clear-cut answers possible yet, however, the observations are paving a way to shedding light on many of the issues, and one can try to give tentative early answers. The following is a rough attempt to outline the directions for current and future answers in these fields:

(a) Massive gas clumps prior or at the onset of high-mass star formation are characterized by cold temperatures of the order 15 K and small line-widths indicative of a low level of turbulence. Their molecular abundances appear comparable to those of low-mass starless cores. Interestingly, the outflow detection rates toward IRDCs are high, and no genuine High-Mass Starless Cores have been reported in the literature yet. Although the statistical basis is not solid enough yet, this allows us to speculate that the high-mass starless phase is likely to be very shortlived.

(b) The detection of a real accretion disk around a massive protostar still remains an open issue. However, we find many rotating structures in the vicinity of young massive star-forming regions all the way from IRDCs to Hot Molecular Cores. These structures are on average large with sizes between $1\times 10^3$ and $2\times 10^4$ AU, and they have masses of the order of the central protostar. Hence, most of them are not Keplerian accretion disks but rather some larger-scale rotating/infalling structures or toroids that may feed more genuine accretion disks in the so far unresolved centers of these regions.

(c) Fragmentation of massive star-forming regions is frequently observed, and the core mass function of one young region is consistent with the Initial Mass Function. However, caveats of unknown temperature distributions or missing flux on larger scales may still affect the results. Furthermore, we find proto-trapezium like structures which show multiple bound sources on small scales of a few 1000 AU implying protostellar densities of the order $10^5$ protostars/pc$^{-3}$. Such densities are still not sufficient to allow coalescence, however, it may be possible to find even higher protostellar densities with the improved observational capabilities of future instruments. Although mergers do not appear necessary to form massive stars in general, they still remain a possibility for the most massive objects.

(d) Astro-chemistry is a young branch in astrophysical research, and we are currently only touching the surface of its potential. The different paths to follow in the coming years are manyfold: With larger source-samples, we will be able to derive a real chemical evolutionary sequence with one of the goals to use chemistry as an astrophysical clock. Furthermore, understanding the chemical differences is important to use the molecular lines as astrophysical tools to investigate the physical processes taking place. Moreover, another current hot topic is planet formation, and in this context astro-biology is a rising subject. In this regard understanding astro-chemistry and detecting new and more complex molecules in space is paving the way for future astro-biological science.

## Acknowledgments

Thanks a lot to Cassie Fallscheer and Javier Rodon for preparing the figures related to the IRDC 18223-3 outflow/disk system and the W3-IRS5 fragmenting core. I further acknowledge financial support by the Emmy-Noether-Program of the Deutsche Forschungsgemeinschaft (DFG, grant BE2578).

# References

[1] Andre, P., Ward-Thompson, D., & Barsony, M. 2000, Protostars and Planets IV, 59
[2] Arce, H. G., Shepherd, D., Gueth, F., et al. 2007, in Protostars and Planets V, ed. B. Reipurth, D. Jewitt, & K. Keil, 245–260
[3] Bachiller, R., Pérez Gutiérrez, M., Kumar, M. S. N., & Tafalla, M. 2001, A&A, 372, 899
[4] Bally, J. & Zinnecker, H. 2005, AJ, 129, 2281
[5] Beuther, H., Churchwell, E. B., McKee, C. F., & Tan, J. C. 2007, in Protostars and Planets V, ed. B. Reipurth, D. Jewitt, & K. Keil, 165–180
[6] Beuther, H., Hunter, T. R., Zhang, Q., et al. 2004, ApJ, 616, L23
[7] Beuther, H., Leurini, S., Schilke, P., et al. 2007, A&A, 466, 1065
[8] Beuther, H. & Schilke, P. 2004, Science, 303, 1167
[9] Beuther, H., Schilke, P., & Gueth, F. 2004, ApJ, 608, 330
[10] Beuther, H., Schilke, P., Menten, K. M., et al. 2002, ApJ, 566, 945
[11] Beuther, H., Schilke, P., Sridharan, T. K., et al. 2002, A&A, 383, 892
[12] Beuther, H. & Shepherd, D. 2005, 105
[13] Beuther, H. & Sridharan, T. K. 2007, ApJ, 668, 348
[14] Beuther, H., Sridharan, T. K., & Saito, M. 2005, ApJ, 634, L185
[15] Beuther, H. & Steinacker, J. 2007, ApJ, 656, L85
[16] Beuther, H. & Walsh, A. 2008, ApJ in press, arXiv:0712.0579
[17] Beuther, H., Zhang, Q., Bergin, E. A., et al. 2007, A&A, 468, 1045
[18] Beuther, H., Zhang, Q., Greenhill, L. J., et al. 2005, ApJ, 632, 355
[19] Beuther, H., Zhang, Q., Hunter, T. R., Sridharan, T. K., & Bergin, E. A. 2007, A&A, 473, 493
[20] Beuther, H., Zhang, Q., Hunter, T. R., et al. 2004, ApJ, 616, L19
[21] Beuther, H., Zhang, Q., Sridharan, T. K., & Chen, Y. 2005, ApJ, 628, 800
[22] Blake, G. A., Mundy, L. G., Carlstrom, J. E., et al. 1996, ApJ, 472, L49
[23] Blake, G. A., Sutton, E. C., Masson, C. R., & Phillips, T. G. 1987, ApJ, 315, 621
[24] Bonnell, I. A., Larson, R. B., & Zinnecker, H. 2007, in Protostars and Planets V, ed. B. Reipurth, D. Jewitt, & K. Keil, 149–164
[25] Bonnell, I. A., Vine, S. G., & Bate, M. R. 2004, MNRAS, 349, 735
[26] Cabrit, S. & Bertout, C. 1990, ApJ, 348, 530
[27] Carey, S. J., Feldman, P. A., Redman, R. O., et al. 2000, ApJ, 543, L157
[28] Cesaroni, R., Churchwell, E., Hofner, P., Walmsley, C. M., & Kurtz, S. 1994, A&A, 288, 903
[29] Cesaroni, R., Galli, D., Lodato, G., Walmsley, C. M., & Zhang, Q. 2007, in Protostars and Planets V, ed. B. Reipurth, D. Jewitt, & K. Keil, 197–212
[30] Cesaroni, R., Hofner, P., Walmsley, C. M., & Churchwell, E. 1998, A&A, 331, 709
[31] Cesaroni, R., Neri, R., Olmi, L., et al. 2005, A&A, 434, 1039
[32] Egan, M. P., Shipman, R. F., Price, S. D., et al. 1998, ApJ, 494, L199
[33] Gueth, F., Bachiller, R., & Tafalla, M. 2003, A&A, 401, L5
[34] Hatchell, J. & van der Tak, F. F. S. 2003, A&A, 409, 589

[35] Keto, E. 2007, ApJ, 666, 976
[36] Krumholz, M. R., Klein, R. I., & McKee, C. F. 2007, ApJ, 665, 478
[37] —. 2007, ApJ, 656, 959
[38] Lada, C. J. & Lada, E. A. 2003, ARA&A, 41, 57
[39] Launhardt, R. 2004, in IAU Symposium, 213
[40] Leurini, S., Beuther, H., Schilke, P., et al. 2007, A&A, 475, 925
[41] McKee, C. F. & Ostriker, E. C. 2007, ARA&A, 45, 565
[42] McKee, C. F. & Tan, J. C. 2002, Nature, 416, 59
[43] —. 2003, ApJ, 585, 850
[44] Megeath, S. T., Wilson, T. L., & Corbin, M. R. 2005, ApJ, 622, L141
[45] Motte, F., Andre, P., & Neri, R. 1998, A&A, 336, 150
[46] Motte, F., Bontemps, S., Schilke, P., et al. 2007, ArXiv e-prints, 0708.2774, 708
[47] Mueller, K. E., Shirley, Y. L., Evans, N. J., & Jacobson, H. R. 2002, ApJS, 143, 469
[48] Nomura, H. & Millar, T. J. 2004, A&A, 414, 409
[49] Noriega-Crespo, A., Morris, P., Marleau, F. R., et al. 2004, ApJS, 154, 352
[50] Padoan, P. & Nordlund, Å. 2002, ApJ, 576, 870
[51] Parker, N. D., Padman, R., & Scott, P. F. 1991, MNRAS, 252, 442
[52] Piétu, V., Dutrey, A., & Guilloteau, S. 2007, A&A, 467, 163
[53] Pillai, T., Wyrowski, F., Carey, S. J., & Menten, K. M. 2006, A&A, 450, 569
[54] Pillai, T., Wyrowski, F., Menten, K. M., & Krügel, E. 2006, A&A, 447, 929
[55] Rathborne, J. M., Jackson, J. M., Chambers, E. T., et al. 2005, ApJ, 630, L181
[56] Rathborne, J. M., Jackson, J. M., & Simon, R. 2006, ApJ, 641, 389
[57] Reynolds, S. P. 1986, ApJ, 304, 713
[58] Ridge, N. A. & Moore, T. J. T. 2001, A&A, 378, 495
[59] Salpeter, E. E. 1955, ApJ, 121, 161
[60] Schilke, P., Groesbeck, T. D., Blake, G. A., & Phillips, T. G. 1997, ApJS, 108, 301
[61] Shepherd, D. S. & Churchwell, E. 1996, ApJ, 457, 267
[62] Simon, M., Dutrey, A., & Guilloteau, S. 2000, ApJ, 545, 1034
[63] Simon, R., Jackson, J. M., Rathborne, J. M., & Chambers, E. T. 2006, ApJ, 639, 227
[64] Sridharan, T. K., Beuther, H., Saito, M., Wyrowski, F., & Schilke, P. 2005, ApJ, 634, L57
[65] Sridharan, T. K., Beuther, H., Schilke, P., Menten, K. M., & Wyrowski, F. 2002, ApJ, 566, 931
[66] Tafalla, M., Santiago-García, J., Myers, P. C., et al. 2006, A&A, 455, 577
[67] van der Tak, F. F. S., Tuthill, P. G., & Danchi, W. C. 2005, A&A, 431, 993
[68] Ward-Thompson, D., Motte, F., & Andre, P. 1999, MNRAS, 305, 143
[69] Yorke, H. W. & Sonnhalter, C. 2002, ApJ, 569, 846
[70] Zhang, Q., Hunter, T. R., Brand, J., et al. 2005, ApJ, 625, 864
[71] Zinnecker, H. & Yorke, H. W. 2007, ARA&A, 45, 481

*Ludwig Biermann Award Lecture II*

# At the Bottom of the Main Sequence
## Activity and Magnetic Fields Beyond the Threshold to Complete Convection

Ansgar Reiners

Georg-August-Universität Göttingen, Institut für Astrophysik
Friedrich-Hund-Platz 1, 37077 Göttingen
Ansgar.Reiners@phys.uni-goettingen.de

### Abstract

*The bottom of the main sequence hosts objects with fundamentally different properties. At masses of about $0.3\,M_\odot$, stars become fully convective and at about $0.08\,M_\odot$ the hydrogen-burning main sequence ends; less massive objects are brown dwarfs. While stars and brown dwarfs experience very different evolutions, their inner structure has relatively little impact on the atmospheres. The generation of magnetic fields and activity is obviously connected to the threshold between partial and complete convection, because dynamo mechanisms involving a layer of shear like the solar $\alpha\Omega$-dynamo must cease. Hence a change in stellar activity can be expected there. Observations of stellar activity do not confirm a rapid break in activity at the convection boundary, but the fraction of active stars and rapid rotators is higher on the fully convective side. I summarize the current picture of stellar activity and magnetic field measurements at the bottom of the main sequence and present recent results on rotational braking beyond.*

## 1 Introduction

Low-mass stars and sub-stellar objects are fascinating astrophysical laboratories affected by various physical processes. The inner structure of low-mass stars undergoes two very interesting phase transitions that seriously influence their structure and evolution. The first transition occurs around $0.35\,M_\odot$ (spectral type $\approx$ M3.5, main-sequence effective temperature $\approx 3500$ K); stars less massive are completely convective while more massive stars possess an inner radiative zone similar to the Sun. Because the solar (interface) dynamo is closely related to the shear layer between the two regions (the tachocline, Ossendrijver, 2003), cooler stars lacking a tachocline cannot maintain such a Sun-like interface dynamo. The second important threshold occurs around 0.08 $M_\odot$, the dividing line between stars and brown dwarfs. Less massive objects do not produce enough heat in their interior to ignite stable hydrogen fusion (e.g., Chabrier & Baraffe, 2000). They keep cooling while contracting which makes them wander through later and later spectral classes getting fainter and fainter.

The spectral appearance of low mass stars is governed by two atmospheric phase transitions. As temperature drops in the atmospheres of cool stars, molecules start to form around effective temperatures of some 4000 K. At about 2500 K, dust grains rain out making the (sub)stellar atmospheres even more complex, and the growing neutrality of the atmosphere entirely changes its physical properties.

Understanding the physics of objects at and beyond the bottom of the main sequence is particularly challenging because these objects are relatively difficult to observe. Although many are close by they are so dim that large telescopes are necessary to uncover their secrets. Most of our knowledge about the physics of stars comes from spectroscopy. Spectra of ultra-cool objects, however, are dominated by molecular absorption, and isolated spectral lines suitable for detailed investigation – a standard in hotter stars – are hardly available.

The physical richness of the bottom at the main sequence and the observational problems are nicely summarized in the review by Liebert & Probst (1987) several years before the first brown dwarf was even discovered. In the 20 years since then large aperture telescopes and much improved observing facilities, sensitive to infrared wavelengths, revealed a huge amount of information on low-mass objects. Not only have we learned that brown dwarfs are actually reality, we have also seen that their formation mechanism as well as their complex atmospheres are indeed as difficult to understand as expected by Liebert & Probst (1987). The history of brown dwarf observations and our knowledge about them is reviewed for example in the articles by Basri (2000), Kirkpatrick (2005), and Chabrier et al. (2005). Here, I will not try to give a comprehensive update on the bottom of the main sequence. I will rather focus on one point: Does the physically important threshold to complete convection have visible effects on low-mass stars?

## 1.1 Convection in the HR-diagram

The presence of convection governs a star's ability to maintain significant magnetic fields that are the reason for the wide range of phenomena summarized as magnetic activity. In the absence of an (outer) convection zone, no dynamo can generate magnetic flux anywhere close to the surface so that fossil fields are the most plausible candidates for the strong magnetic fields observed in hot stars as for example Ap-stars.

In Fig. 1, I show an HR-diagram with pre-main sequence evolutionary tracks taken from Siess et al. (2000). Stars plotted in this diagram are mainly from the Hipparcos catalogue (ESA, 1997) with temperatures from Taylor (1995), Cayrel de Strobel (1997), or converted from $uvby\beta$-colors from Hauck & Mermilliod (1998). Additional low-mass stars are taken from Bessel (1991) and Leggett (1992). For the interesting question of stellar dynamos and their interaction with rotation, the question whether stars have convective envelopes or are fully convective is essential.

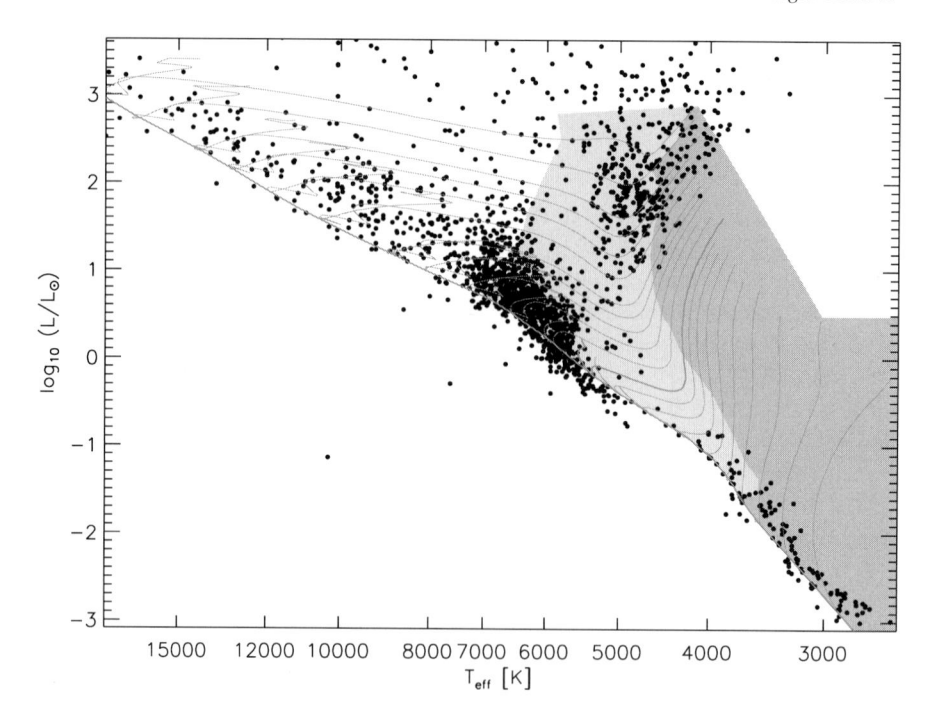

**Figure 1:** HR-diagram with evolutionary tracks from Siess et al. (2000). Regions where stars have outer convective shells are indicated in light grey color, stars occupying the dark grey region are completely convective. The ZAMS is shown with a black line.

I have indicated in Fig. 1 regions in the HR-diagram where stars have outer convective envelopes (light grey) and where they are completely convective (dark grey) during pre-main sequence evolution and on the main sequence (Siess et al. 2000). The temperature region at which stars develop a convective shell is around 5800–6500 K, this is governed by the ionization of hydrogen. On the main sequence, the threshold to complete convection happens around $T_{\rm eff} = 3500$ K but at much higher temperatures in younger stars. Essential information on the nature of Sun-like and fully convective dynamos can be expected by investigating activity close to these two thresholds.

## 1.2 An extended HR-diagram

During the last years, temperature calibrations of low-mass objects of spectral class L and T became available (Dahn et al. 2002; Golimowski et al. 2004). We can use this new information to extend the HR-diagram towards the coolest known objects. Figure 2 shows an HR-diagram covering temperatures well below 1000 K. Additional evolutionary tracks from Baraffe et al. (1998 and 2002) are shown. The temperatures from Dahn et al. (2002) were shifted by 400 K to the hot side in order to achieve consistency (see Golimowski et al 2004). The evolutionary track for

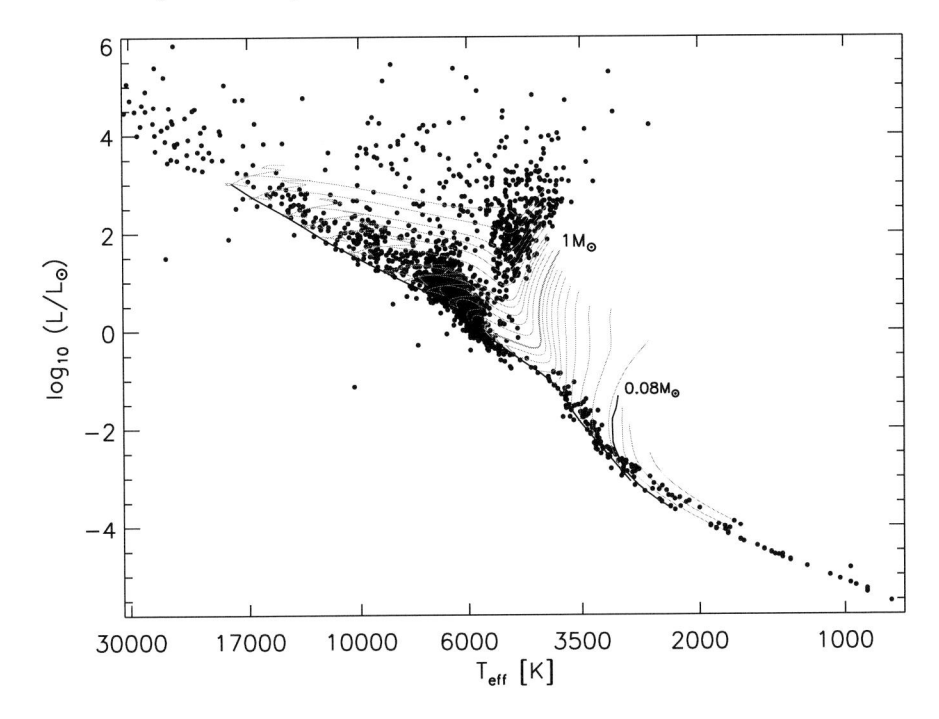

**Figure 2:** Extended HR-diagram including very low-mass objects. Evolutionary tracks from Baraffe et al. (1998 and 2002) are added. The evolution of an 0.08 M. star is shown with a black line.

a $0.08\,M_\odot$ object is plotted as a thick line. It demonstrates a crucial feature in the analysis of brown dwarfs; because hydrogen burning cannot stabilize them, brown dwarfs become cooler and dimmer during their entire lifetime. They never maintain a constant temperature over a (cosmologically) long period. In other words, a certain temperature is not indicative of a certain mass in very low mass objects (VLMs), and because the transition from stars to brown dwarfs is not abrupt, objects of temperatures around 2500–3100 K can either be old stars or young brown dwarfs.

The luminosity of main sequence stars hotter than about 4000 K is declining smoothly. In the brown dwarf regime, i.e. at temperatures lower than $T_{\rm eff} = 2500$ K, the "main sequence" also follows a well-defined but shallower slope. Here it coincides with the line of constant radius, $R \approx 0.1\,R_\odot$; all objects in this regime have approximately the same radius because of electron degeneracy. Between these two regimes, around $T_{\rm eff} = 3500$ K, a step in luminosity appears, which was mentioned, e.g., by Hawley et al. (1996). It occurs close to the temperature of $H_2$-association (Copeland et al. 1970), but theoretical expectations including this effect differ from the observations (e.g., D'Antona & Mazzitelli 1996; Baraffe & Chabrier 1996). The step around $T_{\rm eff} = 3500$ K shows a remarkable coincidence with the onset of complete convection (cp Fig. 1). Clemens et al. (1998) discuss this point and its possible influence on the period distributions of binaries.

## 2 Rotation and magnetic activity in very low-mass objects

The rotation-activity relation has become one of the more solid concepts in stellar astrophysics. In sun-like stars, i.e. stars with an outer convective shell, virtually all tracers of chromospheric and coronal activity correlate with rotation (e.g., Ayres & Linsky, 1980; Noyes et al., 1984; Simon, 2001; Pizzolato, 2003; and references therein). Activity is connected to stellar rotation in the sense that the more rapidly a star rotates, the more emission in tracers of activity is observed until a saturation level is reached beyond which emission does not become stronger anymore (maybe it becomes even less at very high rotation in the "supersaturation" regime; e.g. Randich, 1998). The question whether and how the rotation-activity relation is universal for all types of stars may be subject of debate (e.g., Basri, 1986), but it seems clear that the rotation period is the most important individual parameter for the generation of magnetic activity. The general picture is that of a solar-like dynamo mainly of $\alpha - \Omega$ type. The efficiency of magnetic field generation through this dynamo depends on rotation rate (or Rossby number, Durney & Latour, 1978); more magnetic flux is produced at rapid rotation. Magnetic flux is believed to generate all phenomena of stellar activity in analogy to the solar case.

In order to understand stellar dynamo processes it is always instructive to investigate stars with very different physical conditions for a dynamo to work. A key area is where stars become completely convective. The "classical" $\alpha - \Omega$ dynamo, responsible at least for the cyclic part of solar activity, is believed to be situated at the tachocline between the radiative core and the convective envelope. This dynamo mechanism must cease towards lower mass objects.

### 2.1 Activity

Stellar activity is quantified through the strength of emission observed either in broad wavelength regions (e.g., X-ray) or in specific spectral lines. Which tracer is most suitable for the investigation of stellar activity is a question of the emission produced by the star and a question of the contrast to the spectral energy distribution of the photosphere.

X-rays are well suited indicators in a broad range of stars, because the blackbody emission from the star is virtually free of X-ray emission (i.e. the contrast is very high). X-ray observatories, however, are comparably small (and few in number which reduces the available observing time) so that the detection threshold for X-ray emission is relatively high. It has been found that the ratio of X-ray luminosity $L_X$ to bolometric luminosity $L_{bol}$ even in the most active stars rarely exceeds a ratio of $10^{-3}$ (e.g. Pizzolato et al., 2003, and references therein). This means that X-ray emission from faint (and small) VLMs with low $L_{bol}$ is more difficult to probe than from hotter ones. X-ray measurements in field surveys are available for objects as late as mid-M, only sparse information is available for cooler ones.

In the Sun, X-rays are produced in the corona and we usually assume similar emission processes in other stars. In sun-like stars, coronal emission is closely connected to chromospheric emission. The deep absorption cores of the CaII H&K lines, MgII h&k lines, and other UV-lines are good regions to look for chromospheric emission. In very cool objects, however, H$\alpha$ is the line of choice partly because the contrast to the photosphere is much weaker than in hotter stars; it is rather strong and very easy to observe.

The vast majority of X-ray detections in low mass stars comes from the ROSAT mission (see for example Schmitt, 1995, and references therein). Unfortunately, the sensitivity of ROSAT did not allow to measure X-ray emission in many objects of mid-M spectral type or later. A number of measurements in selected active ultracool objects were carried out with Chandra and XMM (e.g., Stelzer, 2004, and references therein; Robrade & Schmitt, 2005). Normalized X-ray emission between $\log L_{\rm X}/L_{\rm bol} = -2$ and $-4$ was detected during flares in some objects. The overall (quiet) emission level, however, is decreasing (see Fig. 4 in Stelzer et al., 2004) and mostly below the detection level. Emission observed in the H$\alpha$ line is an indicator available for a much larger number of VLMs. Flares in very low mass objects reach a comparable level in normalized H$\alpha$ emission as observed in some X-ray observations. For a first comparison, it is probably not too wrong to assume that observations of H$\alpha$ emission probes similar mechanisms as X-ray observations do.

So far, in order to draw a comprehensive picture of stellar activity from early F stars to brown dwarfs, we have to combine the results from different tracers. Reiners & Basri (2007) showed that X-ray and H$\alpha$ measurements among M-stars exhibit a rough correspondence, i.e. for a first estimate we may assume $L_{\rm X}/L_{\rm bol} \approx L_{\rm H\alpha}/L_{\rm bol}$. In hotter stars, however, Takalo & Nousek (1988) found an offset of about 1.8 dex between these two indicators (in hotter stars there is also a close correspondence between CaII H&K and X-ray emission as found by Sterzik & Schmitt, 1997). Hawley et al. (1996) found a constant offset between $L_{\rm X}$ and $L_{\rm H\alpha}$ of only about half a dex in late-type stars. A census of currently available X-ray and H$\alpha$ measurements is plotted in Fig. 3. Although one has to keep in mind that the two tracers, plotted in grey and black, do not match each other directly (see above), the error introduced is probably not larger than $\sim 0.5$ dex at spectral types where more H$\alpha$ measurements become available and X-ray measurements become rare. In any case, the correspondence between $L_{\rm X}$ and $L_{\rm H\alpha}$ is probably not too bad, and their direct comparison is very instructive.

The current picture of stellar activity may be summarized with the help of Fig. 3: Among the sun-like stars earlier than M0, a clear age-rotation-activity dependence exists. Young (cluster) stars show normalized X-ray luminosities $L_{\rm X}/L_{\rm bol}$ roughly between $-3$ and $-4$, they are rapid rotators in the saturated part of the rotation-activity connection (see Patten & Simon, 1996; Pizzolato et al., 2003). The older field stars have normalized X-ray luminosities an order of magnitude lower than their younger predecessors, i.e., $L_{\rm X}/L_{\rm bol} \leq 10^{-4}$ in the field. The level of activity can be quite low and probably all sun-like stars show some sort of activity that could be measured if the sensitivity was high enough. Activity scales with rotation period or Rossby number, which is rotation period divided by the convective overturn time. Among F–K stars, the convective overturn time does not change dramatically so that relations in rotation period or Rossby number are not too different.

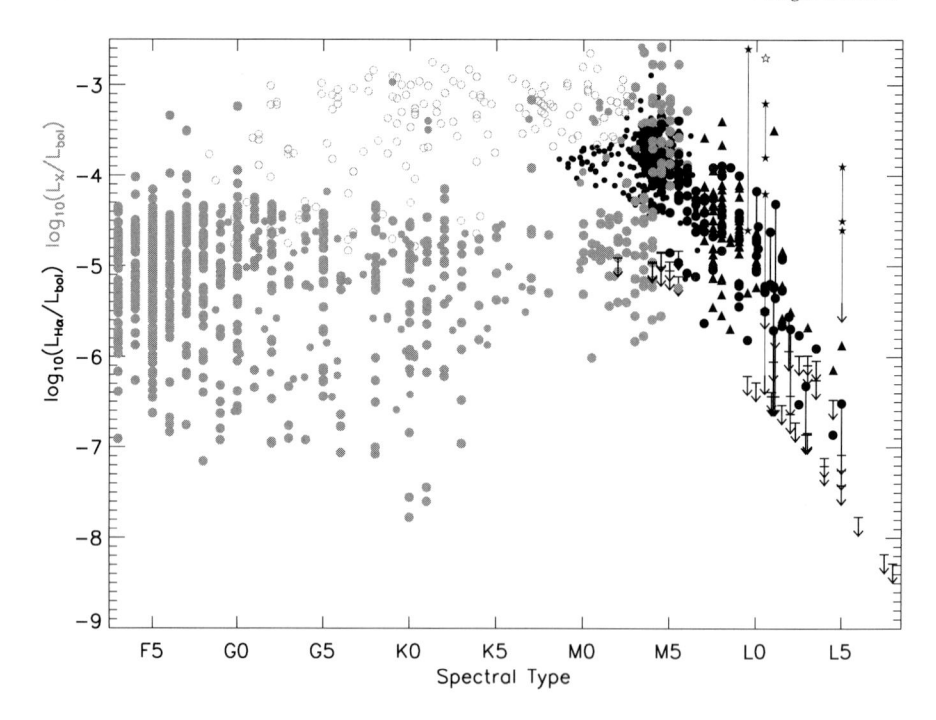

**Figure 3:** Normalized activity vs. spectral type. X-ray and Hα emission are shown on the same scale. Lines connect observations of identical objects. X-ray measurements (grey) are from Voges et al., 1999 (field stars, filled circles) with spectral types from the Hipparcos catalogue (ESA, 1997). Open circles are cluster stars from Pizzolato et al. 2003, field stars from the same publication are plotted as filled circles. Hα (black): small circles from Reid et al., 1995 and Hawley et al., 1996; large circles from Mohanty & Basri, 2003, Reiners & Basri, 2007, Reiners & Basri, in prep.; triangles from Schmidt et al., 2007; stars from Burgasser et al., 2002 and Liebert et al., 2003.

A change in normalized X-ray activity among the field stars is visible in Fig. 3 between spectral types M0 and M5 (both in X-ray and in Hα). Around spectral type M5, normalized X-ray luminosity ramps up and some mid-M stars exhibit values as high as $\log L_X/L_{bol} > -3$. Stars in the spectral region M0–M7 from the volume-limited sample of Delfosse et al. (1998) are shown separately in Fig. 4. This plot suggests a rise in maximum normalized X-ray luminosity between spectral types M2 and M4 (a similar behavior is visible in the Hα data in Fig. 2 of Hawley et al., 1996). This spectral range coincides with the mass range where stars become completely convective. The rise in X-ray activity is probably not the consequence of an entirely different process of magnetic field generation. If that was the case, it would imply that stars in the completely convective regime are even *more* efficiently generating magnetic fields, which is difficult to believe. As shown in Section 2.2, the reason for the rise in activity is more likely weaker rotational braking in completely convective stars. Fields stars beyond spectral type M4 are generally more rapidly

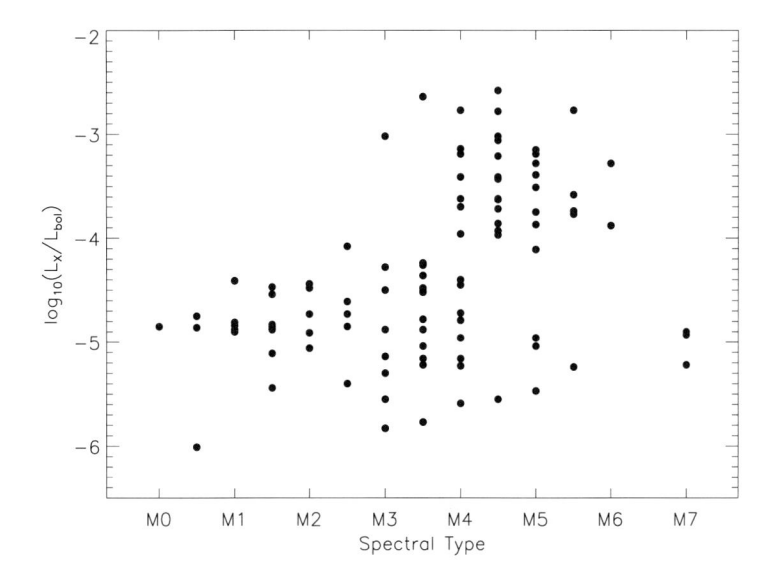

**Figure 4:** Normalized X-ray activity as a function of spectral class for M-type stars. Data are taken from Delfosse et al., 1998.

rotating, and there is ample evidence that the rotation activity connection still applies in fully convective mid-M stars (Mohanty & Basri, 2003; Reiners & Basri, 2007). Thus, more rapid rotation leads to higher X-ray emission. I will further discuss what happens at the threshold to complete convection in Section 3.

From the growing number of H$\alpha$ observations in VLMs, it is clear that around spectral type M9, normalized H$\alpha$ emission gradually decreases with spectral type, i.e. with temperature. This effect is probably not directly related to a lack of magnetic flux, but can rather be explained by the growing neutrality of the cold atmospheres (Meyer & Meyer-Hofmeister, 1999; Fleming et al., 2000; Mohanty et al., 2002).

Another way to search for stellar activity is looking for radio emission. Güdel & Benz (1993) showed that in stars between spectral class F and early-M, X-ray and radio emission are intimately related. A probable explanation is gyrosynchrotron emission of mildly relativistic electrons. Plasma heating and particle acceleration probably occur in the same process. Berger (2006) searched for radio emission in a sample of low mass objects. Although the sensitivity in terms of normalized radio luminosity $\log L_{\rm rad}/L_{\rm bol}$ would not allow the detection of radio emission according to the scaling found by Güdel & Benz (1993), Berger (2006) finds much stronger radio emission in some very low mass objects. Hallinan et al. (2006) argue that the strong and modulating radio emission observed in some very low mass objects can be explained by coherent radio emission (in contrast to incoherent emission in more massive stars). The magnetic fields they derive for very low mass objects are on the order of kilo-Gauss which is consistent with the observations by Reiners & Basri

(2007) and higher than the estimates from incoherent emission (see Berger, 2006, 2007).

## 2.2 Rotation

Rotation is generally believed to be intimately connected to stellar activity through a dynamo process depending on rotation rate. The more rapidly a star rotates the more magnetic flux is produced leading to enhanced activity. At a certain rotation rate, activity reaches a saturation level beyond which activity remains at the same level independent of the rotation rate. At very high velocities, activity may even "supersaturate", i.e. fall below the saturation limit again (Patten & Simon, 1996; Randich, 1998). The level of the saturation velocity is thought to scale with convective overturn time, i.e. saturation sets in at constant Rossby number. Combined with the smaller radius of lower mass stars, this results in a drastic divergence of saturation velocities at the stellar surface. In early G-stars, this velocity is on the order of 25–30 km s$^{-1}$ while in mid-M dwarfs it is less than 5 km s$^{-1}$ (e.g., Reiners, 2007). Such small rotation velocities are difficult to measure particularly in M dwarfs so that the unsaturated part of the rotation activity connection is not very well sampled (but it still holds, see Reiners, 2007).

Rotation rates or surface velocities can be measured through rotational line broadening. This method, however, is only sensitive to the projected rotation velocity $v \sin i$ and is limited by the resolving power of the spectrograph (typically to $v \sin i \approx 3$ km s$^{-1}$ in M dwarfs). Rotation periods can directly be detected if stars have (temperature) spots stable enough to persist longer than the rotation period so that the brightness modulation induced can be followed for some rotations. In very cool stars, this is a difficult task because the temperature contrast is probably very low so that the amplitude of the brightness modulation is very small. Detecting rotation periods in M-stars turned out to be a very frustrating business with only a few successful attempts so far (e.g., Pettersen, 1983; Torres et al., 1983; Benedict et al., 1998; Kiraga & Stepień, 2007). One explanation for this is that lifetimes of spots in very low mass stars are probably short while rotation periods are relatively long. Such spots will not show the same configuration after one full rotation causing non-periodic brightness variations. Satellite missions like *COROT* and *Kepler* will provide a fresh look into rotation periods of low mass stars.

Very active M-stars, dMe stars, show strong brightness variations on very short timescales. Such "flares" are observed in outbursts of emission lasting minutes to hours. Very active dMe stars show flaring activity every some hours. These stars are generally rapid rotators with rotation velocities on the order of a few km s$^{-1}$, and rotational line broadening can be relatively easy to measure. They occupy the saturated part of the rotation-activity connection.

Less active M dwarfs are rotating very slowly. Their surface rotation is on the order of one km s$^{-1}$ and hence very difficult to measure spectroscopically (Reiners, 2007), because a resolving power of more than $10^5$ is required. Rotation periods are on the order of days to weeks, which in the absence of strong brightness modulation and probably short lived spots (if any) is even more difficult to detect. For the reasons given above, measurements of rotation rates in M dwarfs are almost exclusively

*The Bottom of the Main Sequence* 49

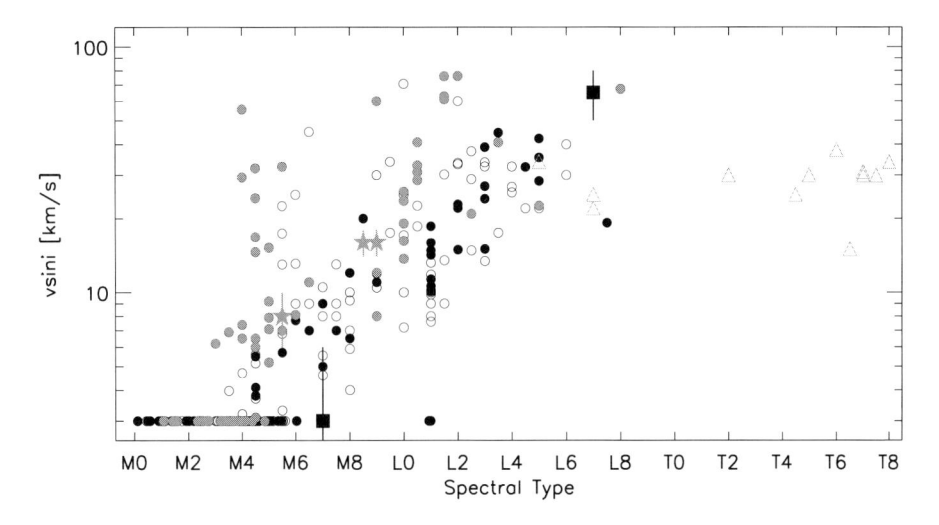

**Figure 5:** Projected rotation velocities in very low mass objects. Circles: Field M- and L-dwarfs from Reiners & Basri, 2007; Mohanty & Basri, 2003; Delfosse et al., 1998; Reiners & Basri, in prep. Filled grey and black circles show objects that are probably young and old, respectively. Open symbols are objects with no age information available. Squares: Subdwarfs from Reiners & Basri, 2006a. Stars: The three components of LHS 1070 from Reiners et al., 2007. Open triangles: Field L- and T-dwarfs from Zapatero Osorio et al., 2006.

available for "rapid" rotators, not much is known about the distribution of rotation velocities below $v \sin i \approx 3\,\mathrm{km\,s^{-1}}$.

In Fig. 5, I show a collection of measured projected rotation velocities, $v \sin i$, in field objects (for references see caption to Fig. 5). The distribution of $v \sin i$ among VLMs shows two remarkable features: (1) A sudden rise in rotation velocities at spectral type M3.5/M4; and (2) a rising lower envelope of minimum rotation velocities in the mid-M and L dwarf regime. I discuss the two features in the following sections.

### 2.2.1 The M dwarf spin-down puzzle

Delfosse et al. (1998) obtained projected rotation velocities in a volume-limited sample of roughly 100 field M dwarfs. Although they come to the conclusion that "the present data show no obvious feature in the rotational velocity distribution at this type [around spectral type M3], or elsewhere within the M0–M6 range", their Fig. 3 shows quite a remarkable feature: The old disk and Halo population exhibit very low rotation velocities (with possibly a gradual increase at spectral types later than M5). With only one exception (at spectral type M8), all old objects including the latest spectral types have $v \sin i < 10\,\mathrm{km\,s^{-1}}$. On the other hand, the young disk population of field M dwarfs exhibits a break in the distribution of rotation velocities. All young population members earlier than spectral type M3 show low rotation rates (projected rotation velocities on the order of the detection limit), but several rapid rotators with $v \sin i > 20\,\mathrm{km\,s^{-1}}$ are found at spectral types M4 and later. Delfosse et al. conclude that the spin-down timescale "is of the order of a few Gyrs at spectral type M3–M4, and of the order of 10 Gyr at spectral type M6".

The sample of Delfosse et al. (1998) is included in Fig. 5. The break in the $v \sin i$ distribution around spectral type M3.5 is clearly visible: In this compilation, all stars earlier than spectral type M3 rotate slower than the detection limits (usually around $v \sin i = 3\,\mathrm{km\,s^{-1}}$). Stars of spectral type M4 or later exhibit rotation speeds up to $60\,\mathrm{km\,s^{-1}}$ and more in some exceptional cases. In general, the rapid rotators belong to the young population (grey symbols), i.e., rotational braking is still functioning at mid-M spectral class, but it is obviously much less efficient than in hotter stars.

The "standard" theory of angular momentum evolution assumes that stars accelerate during the first million years because of gravitational contraction. Once on the main sequence, angular momentum loss by a magneto-thermal wind brakes the star. Chaboyer et al. (1995) and Sills et al. (2000) provide a prescription for angular momentum loss during stellar evolution. Angular momentum loss is proportional to some power of the angular velocity $\omega$ (Mestel, 1984; Kawaler, 1988). The power law itself depends on the magnetic field geometry with very strong braking in the presence of a radial field and lower braking if the field is dipolar. Skumanich-type rotational braking with $v \propto t^{-1/2}$ can be achieved by a magnetic topology between these two cases. In this model, magnetic braking is assumed to be proportional to some power of the angular velocity $\omega$ as long as $\omega$ is small. At a critical angular velocity, $\omega_{\mathrm{crit}}$, the relation between angular momentum loss and $\omega$ changes (see, e.g., Sills et al., 2000). One choice of the scaling of $\omega_{\mathrm{crit}}$ is given in Krishnamurti et al. (1997). They assume that $\omega_{\mathrm{crit}}$ is inversely proportional to the convective overturn timescale (implying that saturation sets in at constant Rossby number).

The model of angular momentum evolution is very successful in sun-like stars of very different ages and a variety of spectral classes (e.g., Barnes, 2007). The main parameters governing rotational braking in this model are the strength of the magnetic field and its geometry. If a similar braking law applies to completely convective stars – and there is no reason to believe the opposite – one of the two parameters must dramatically change at least in the young population of fully convective stars. As mentioned above, there is no reason to believe that the magnetic field strength has a discontinuity at spectral type M3.5 (maybe later it does), and I will discuss direct observations of magnetic fields in completely convective objects in Section 2.3.

### 2.2.2 L-dwarf rotation and the age of the galaxy

The second striking feature of the velocity distribution (Fig. 5) is the rising lower envelope of rotation velocities at spectral types M7–L8. At spectral type M7, some stars still show very low rotation velocities on the order of the detection limit (about $3\,\mathrm{km\,s^{-1}}$). In cooler objects, however, the lowest rotation velocities grow to about $10\,\mathrm{km\,s^{-1}}$ around spectral type L2 up to the order of several ten $\mathrm{km\,s^{-1}}$ at spectral type L6–L8. In the L-dwarf data shown in Fig. 5 (mostly from Mohanty & Basri, 2003; Reiners & Basri, in prep.), this lower envelope is rather well defined with only two objects falling below a virtual line from M7 (zero rotation) up to L8 (about $40\,\mathrm{km\,s^{-1}}$). The two "outliers" at spectral types L1 and L7.5 may be seen under small inclination angles. Zapatero Osorio et al. (2006) measured rotation velocities in a sample of brown dwarfs finding two L7 dwarfs at comparably low rotation velocities. Furthermore, they measured rotation velocities between 15 and 30 km s$^{-1}$ in a couple of T dwarfs. To what extent systematic effects due to the different techniques used affect the results should not be discussed here but has to be kept in mind for interpretating these results. Nevertheless, there is agreement that rotation velocities in VLMs are much higher than in hotter stars, which implies that rotational braking is probably much weaker.

Is the rise of minimum rotational velocities with later spectral type indeed a physical effect, or could it be due to an observational bias? Brown dwarfs do not establish a stable configuration like stars do. Young brown dwarfs are much brighter than old brown dwarfs so that a brightness limited survey (as here) in general favors young brown dwarfs. The observed rise in minimum rotational velocities could mean that, for example, at spectral type L6 we are only observing young (bright) objects that are still not efficiently braked, while at earlier spectral classes we can already reach older objects that suffered braking for a much longer time. In this picture a lack of slowly rotating late-L and T dwarfs would simply mean that such old objects are not contained in our sample because they are too faint.

The fact that brown dwarfs cool during their entire lifetime makes the interpretation of the brown dwarf rotational velocity distribution much more complex than the stellar one. There, evolutionary tracks in the $v \sin i$ / spectral-type diagram essentially are vertical lines on which the stars rise and fall during phases of acceleration and braking. Brown dwarfs, on the other hand, migrate through the spectral classes, they start somewhere between mid- and late-M spectral class and eventually cool down to T-type or even later spectral class. Thus, the distribution of rotation velocities with spectral class among brown dwarfs is a function of rotational evolution, coolings tracks, and of formation rates.

In Fig. 5, age information is included for a number of brown dwarfs. Although sparse, the age information strongly suggests that the rise of the lower envelope of minimum rotation velocities is not an observational bias but rather a consequence of rotational braking. If the brightness limit was the reason for the observed distribution of rotation velocities, we would expect to see only young objects among the mid- to late-L spectral classes (those would actually be of planetary mass). Old objects would only be visible at earlier spectral types. Instead, the age information of our sample clearly shows that the entire lower envelope of rotation velocities is occupied mainly by objects of the old disk or Halo populations, while stars rotating more rapidly are generally younger at each spectral type. This indicates that tracks of rotational evolution go from the upper left to the lower right in Fig. 5. Thus, brown dwarfs are indeed being braked during their evolution.

This interpretation of Fig. 5 is also supported by the five individual objects contained. The three objects plotted as stars are the three members of LHS 1070 (GJ 2005, Reiners et al., 2007b). LHS 1070 B and C have very similar spectral types (around M9) while the A-component is a little more massive (M5.5). There are good arguments that the three components are of same age, and that they are seen under comparable inclination angles. The two cooler objects show similar projected rotation velocities of $v \sin i \approx 16$ km s$^{-1}$ while the earlier one is rotating at half that pace. This result also supports the idea that isochrones in the rotation/spectral-type diagram run from the lower left to the upper right. A possible explanation for this behavior is mass-dependent rotational braking. The two objects plotted as filled squares in Fig. 5 are subdwarfs that are probably among the oldest objects in our galaxy (Reiners & Basri, 2006a). While the earlier subdwarf at spectral type sdM7 exhibits very low rotation, the sdL7 is rotating at the remarkably high speed of $v \sin i \approx 65\,\mathrm{km\,s^{-1}}$. After the long lifetime of probably several Gyrs, the sub-L dwarf has not significantly slowed down, which means that rotational braking in this object must be virtually non-existing.

It is important to realize that, if the lower envelope of minimum rotational velocities is occupied by the oldest brown dwarfs, these did not have enough time to decelerate any further. Thus, the lower envelope of rotational velocities is directly connected to the age of the galaxy, and it should extend to lower $v \sin i$ in older populations.

Zapatero Osorio et al. (2006) show the rotational evolution of brown dwarfs through the spectral classes M, L, and T in the absence of braking. Starting at velocities of $v \sin i = 10\,\mathrm{km\,s^{-1}}$, their test objects spin up due to gravitational contraction to the final rotation speed of some ten $\mathrm{km\,s^{-1}}$. This picture is in qualitative agreement with the overall distribution of rotation velocities among the latest brown (T-)dwarfs, but not with the rotation/age distribution of L dwarfs. Nevertheless, the simulations of Zapatero Osorio et al. show that the rotation of T dwarfs may be explained by gravitational contraction in the absence of any braking. The initial rotation velocities between the L and T dwarfs populations must then be different, because this model cannot explain rotation velocities as high as observed in the mid-M/early-L type objects.

## 2.3 Magnetic Fields

In order to understand the physical processes behind stellar magnetic activity, it is obviously desirable to directly measure magnetic fields. Unfortunately, the direct measurement is much more difficult than measuring most of the indirect activity tracers. The direct measurement of stellar magnetic fields usually means to measure Zeeman splitting in magnetically sensitive lines, i.e. lines with a high Landé-g factor (e.g., Robinson, 1980; Marcy et al. 1989; Saar, 1996; Saar, 2001; Solanki, 1991 and references therein). This is usually achieved by comparing the profiles of magnetically sensitive and insensitive absorption lines between observations and model spectra. An alternate method that relies on the change in line equivalent widths has been developed by Basri et al. (1992). Modeling the Zeeman effect on spectral lines in both cases requires the use of a polarized radiative transfer code and knowledge of the Zeeman shift for each Zeeman component in the magnetic field. Furthermore, it requires the observed lines to be isolated and that they can be measured against a well-defined continuum. The latter becomes more and more difficult in cooler stars since atomic lines vanish in the low-excitation atmospheres and among the ubiquitous molecular lines that appear in the spectra of cool stars. Measurements of stellar magnetic fields carried out through detailed calculations of polarized radiative transfer so far extend to stars as late as M4.5 (Johns-Krull & Valenti, 1996; 2000; Saar,

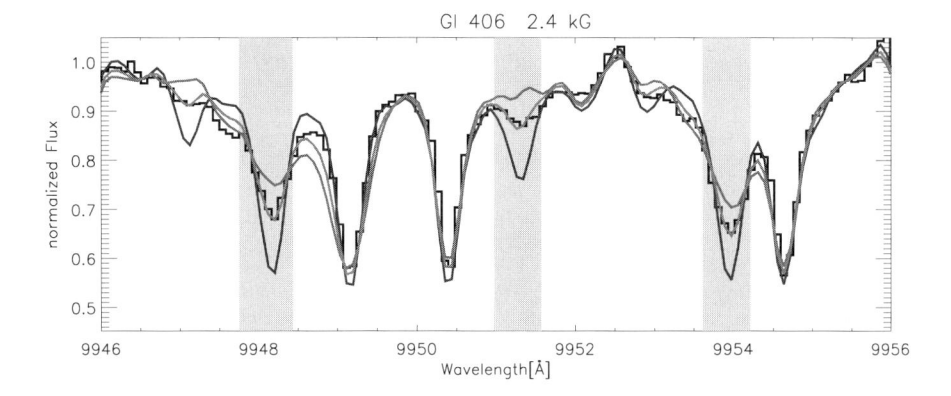

**Figure 6:** Measurements of magnetic flux in the dMe star Gl 406 (CN Leo). Scaled template spectra of a non-magnetic star and a magnetic star ($Bf \sim 3.9\,\text{kG}$) are shown. The fit to the data is the interpolation between the two template spectra that best fits the data (see Reiners & Basri, 2007).

2001). In cooler objects, atomic lines could not be used for the above-mentioned reasons, and because suitable lines become increasingly rare.

Some magnetic field measurements in sun-like stars and early-M stars are compiled in the articles by Saar (1996 and 2001), results are available for spectral types G, K, and early M. Magnetic field strengths and filling factors, i.e. the fraction of the star that is filled with magnetic fields, apparently grow with later spectral type. To my knowledge, no detection exists so far in stars of spectral type F. At early-M spectral classes, filling factors in the (very active) dMe stars approach unity with field strengths of several kilo-Gauss.

In later stars and brown dwarfs, one alternative to atomic lines are molecular bands with well separated individual lines so that they can be distinguished from each other. Valenti et al. (2001) suggested that FeH would be a useful molecular diagnostic for measuring magnetic fields on ultra-cool dwarfs, but they point out that improved laboratory or theoretical line data are required in order to model the spectra directly. Reiners & Basri (2006b) investigated the possibility of detecting (and measuring) magnetic fields in FeH lines of VLMs through comparison between the spectrum of a star with unknown magnetic field strength and a spectrum of a star in which the magnetic field strength is calibrated in atomic lines (early M dwarfs). Although the Zeeman splitting in lines of molecular FeH is not theoretically understood (but see Afram et al., 2007), the effect of kilo-Gauss magnetic fields on the sensitive lines of FeH are easily visible, and magnetic flux differences on the order of a kilo-Gauss can be differentiated. This method was employed to measure the magnetic flux in M-type objects down to spectral type M9 by Reiners & Basri (2007) and Reiners et al. (2007a, 2007b) discovering strong magnetic fields in many ultra-cool dwarfs. An example of the detection of a magnetic field in Gl 406 (CN Leo) is shown in Fig. 6. A list of measurements of magnetic flux in M-type objects is compiled in Table 1. Magnetic flux $Bf$ is plotted as a function of spectral type in Fig. 7.

**Table 1:** Measurements of magnetic flux among M dwarfs. Young objects are shown in the lower part of the table.

| Name | Spectral Type | $v \sin i$ [km s$^{-1}$] | $Bf$ [kG] | Ref |
|---|---|---|---|---|
| Gl 70 | M2.0 | $\leq 3$ | 0.0 | a |
| Gl 729 | M3.5 | 4 | 2.2 | a,b |
| Gl 873 | M3.5 | $\leq 3$ | 3.9 | a,b |
| AD Leo | M3.5 | $\approx 3$ | 2.9 | a,b |
| Gl 876 | M4.0 | $\leq 3$ | 0.0 | a |
| GJ 1005A | M4.0 | $\leq 3$ | 0.0 | a |
| GJ 299 | M4.5 | $\leq 3$ | 0.5 | a |
| GJ 1227 | M4.5 | $\leq 3$ | 0.0 | a |
| GJ 1224 | M4.5 | $\leq 3$ | 2.7 | a |
| YZ Cmi | M4.5 | 5 | $> 3.9$ | a,b |
| Gl 905 | M5.0 | $\leq 3$ | 0.0 | a |
| GJ 1057 | M5.0 | $\leq 3$ | 0.0 | a |
| LHS 1070 A | M5.5 | 8 | 2.0 | c |
| GJ 1245B | M5.5 | 7 | 1.7 | a |
| GJ 1286 | M5.5 | $\leq 3$ | 0.4 | a |
| GJ 1002 | M5.5 | $\leq 3$ | 0.0 | a |
| Gl 406 | M5.5 | 3 | 2.1–2.4 | a,d |
| GJ 1111 | M6.0 | 13 | 1.7 | a |
| VB 8 | M7.0 | 5 | 2.3 | a |
| LHS 3003 | M7.0 | 6 | 1.5 | a |
| LHS 2645 | M7.5 | 8 | 2.1 | a |
| LP 412–31 | M8.0 | 9 | $> 3.9$ | a |
| VB 10 | M8.0 | 6 | 1.3 | a |
| LHS 1070 B | M8.5 | 16 | 4.0 | c |
| LHS 1070 C | M9.0 | 16 | 2.0 | c |
| LHS 2924 | M9.0 | 10 | 1.6 | a |
| LHS 2065 | M9.0 | 12 | $> 3.9$ | a |
| CY Tau | M1.0 | 11$^e$ | 1.2 | f |
| DF Tau | M1.0 | 19$^g$ | 2.9 | f |
| DN Tau | M1.0 | 10$^g$ | 2.0 | f |
| DH Tau | M1.5 | 8$^g$ | 2.7 | f |
| DE Tau | M2.0 | 7$^g$ | 1.1 | f |
| AU Mic | M2.0 | 8$^h$ | 2.3 | i |
| 2MASS1207 | M8.0 | 13 | $< 0.8$ | j |

[a] Reiners & Basri, 2007 (anchored at measurements from Johns-Krull & Valenti, 2000)
[b] Johns-Krull & Valenti, 2000
[c] Reiners et al., 2007b
[d] Reiners et al., 2007a
[e] Hartmann et al., 1986
[f] Johns-Krull, 2007
[g] Johns-Krull & Valenti, 2001
[h] Scholz et al., 2007
[i] Saar, 1994
[j] Reiners & Basri, submitted

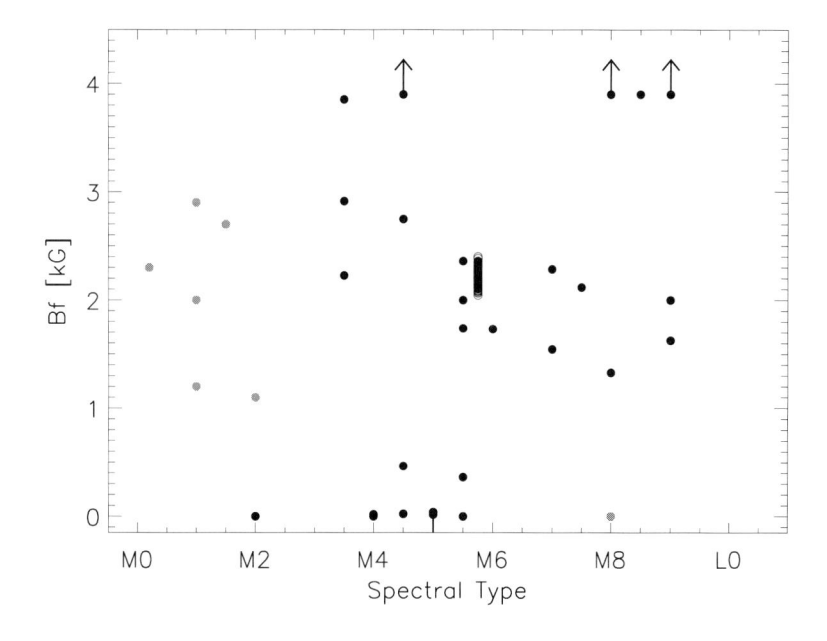

**Figure 7:** Measurements of magnetic flux in M dwarfs. Grey circles are young objects, black circles are old objects. Uncertainties are usually on the order of several hundred Gauss.

The detections of strong magnetic flux in objects as late as spectral type M9 show that magnetic field generation is very efficient in completely convective objects, too. The idea of vanishing dynamo action at the threshold to complete convection is certainly invalid hence a lack of magnetic flux cannot be the reason for the weak magnetic braking discussed above. Looking only at the currently available measurements of magnetic flux, the opposite seems to be true: Integrated flux grows with later spectral type, i.e. with deeper convection zones (e.g. Saar, 1996). This impression, however, is probably due to an observational bias. It is known that activity (hence magnetic field generation) scales with rotation period or Rossby number. At a given Rossby number (or period), hotter (and larger) stars have higher surface rotation velocities than cooler objects. The measurement of magnetic Zeeman splitting requires narrow spectral lines in order to discriminate between Zeeman splitting and other broadening mechanisms. Thus, the strong fields that are probably generated in more rapidly rotating, earlier stars cannot be detected by current observational strategies.

Nevertheless, the result from the growing amount of magnetic flux measurements in low mass stars is that a fully convective dynamo can easily generate mean fields of kG-strength. Reiners & Basri (2007) showed that H$\alpha$ activity in VLMs scales with magnetic flux just as X-ray emission does in hotter stars (e.g. Saar, 2001). From this perspective, no change appears in the generation of magnetic activity at the convection boundary.

Fully convective objects occupy a large region in the HR-diagram (Fig. 1). Objects later than spectral type ∼M3.5 are always completely convective, and more massive stars develop a radiative core within the first ∼10 Myr with the exact time depending on their mass. Fully convective dynamos can also be studied in pre-main sequence stars during their fully convective phase. It is interesting to note that with one exception (the M2 dwarf Gl 70), probably all objects shown in Fig. 7 are fully convective. Among the field stars, however, only the slowly rotating ones have no strong (and large) magnetic field. The rotation-activity connection seems to be intact at least in the sense that rapid rotation involves the generation of a strong magnetic field (Mohanty & Basri, 2003; Reiners & Basri, 2007). It is somewhat disturbing that the only object later than M6 without a strong magnetic field is the young accreting brown dwarf 2MASS 1207334–393254 (hereafter 2MASS 1207), which was found to drive a jet (Whelan et al., 2007). The surface rotation velocity of 2MASS 1207 is at the high end of velocities compared to its older (but more massive) counterparts at similar spectral type ($v \sin i \approx 5\text{--}12 \,\mathrm{km\,s^{-1}}$), which is a good argument for a strong magnetic field. The lack of a strong magnetic field on 2MASS 1207 may indicate that magnetic field generation is also a function of age. Reiners & Basri (submitted) estimate that the magnetic field required for magnetospheric accretion in 2MASS 1207 is only about 200 G, i.e. fully within the uncertainties of the magnetic flux measurement. They speculate that during the accretion phase the magnetic field may be governed by the accretion process rather than by the internal generation through a convective dynamo, as is probably the case in older objects.

The puzzle of magnetic field generation in fully convective stars is currently receiving tremendous attention. Complementary strategies to investigate the strength and topology of magnetic fields in fully convective objects are applied very successfully. With the Zeeman Doppler Imaging technique, Donati et al. (2006) showed that the rapidly rotating fully convective M4 star V379 Peg exhibits a large scale mostly axisymmetric magnetic topology. This finding is apparently contradicting the idea that fully convective objects only generate small scale fields. Browning (2007), performed simulations of dynamo action in fully convective stars demonstrating that kG-strength magnetic fields with a significant mean (axisymmetric) component can be generated without the aid of a tachocline (see also Durney et al., 1993; Küker & Rüdiger, 1999; Chabrier & Küker, 2006; Dobler et al., 2006). From radio observations, Berger (2006) concluded that fully convective objects must have strong magnetic fields, and he finds that in very low mass stars the ratio of radio to X-ray emission is larger than in sun-like stars. Hallinan et al. (2006) found rotational modulation of radio emission from a rapidly rotating M9 dwarf. They conclude that the radio emission is difficult to reconcile with incoherent gyrosynchrotron radiation, and that a more likely source is coherent, electron maser emission from above the magnetic poles. This suggestion, motivated by independent observations, also requires the magnetic dipole (or multipole) to take the form of a large-scale field with kG-strength. And recently, Berger et al. (2007) showed in a multi-wavelength observation in an M9 object that X-ray, Hα, and radio observations not necessarily correlate in time. This raises the interesting question whether heating mechanisms differ between sun-like stars and VLMs.

Another interesting aspect of magnetic fields in VLMs is the question whether magnetism can effectively suppress convective heat transport. Stassun et al. (2006, 2007) discovered the eclipsing binary brown dwarf 2MASS J05352184–0546085 (2MASS 0535). They found that the more massive primary surprisingly is cooler than the less massive secondary. A possible explanation for this temperature reversal is a strong magnetic field on the primary inhibiting convection (Chabrier et al., 2007). Reiners et al. (2007c) discovered that H$\alpha$ emission in the primary is at least a factor of 7 stronger than in the secondary. This supports the idea of a strong magnetic field on the primary of 2MASS 0535. Effective cooling due to the presence of magnetic fields on low-mass stars would have impact on the mass-luminosity relation (see, e.g., Stauffer & Hartmann, 1984; Hawley et al., 1996; López-Morales, 2007).

## 3 What happens at the threshold to complete convection?

There is no doubt that around spectral type M3.5 a change in the interior of main-sequence stars occurs, although the exact locus of the convection boundary may shift towards later spectral types in the presence of strong magnetic fields (Mullan & McDonald, 2001). Stars earlier than the convection boundary develop a radiative core and a tachocline of shear at which a sun-like interface dynamo can work. Stars cooler than the convection threshold do not harbor a tachocline because no radiative core is developed. From the Sun we know that at least the cyclic part of dynamo action is due to a dynamo operating at the tachocline. This sort of dynamo can certainly not work in completely convective objects, and if the interface dynamo was the only one able to produce strong magnetic fields, objects later than M3.5 simply could not produce magnetic flux and magnetic activity at all.

Activity measurements across the convection boundary and the direct detection of kG-strength magnetic fields in completely convective objects rule out the possibility that the interface dynamo is the only functioning type of dynamo in the stellar context. Obviously, completely convective objects manage to generate large-scale fields and probably also strong axisymmetric components. Furthermore, the rapid rise of the fraction of objects exhibiting H$\alpha$ emission at mid-M spectral class (West et al., 2004) could even be interpreted by dynamo efficiency that is higher in the fully convective regime. However, the clue to an understanding of dynamo activity in completely convective objects is probably rotation and rotational braking. The rising fraction of active stars at mid-M spectral type is possibly due to the fact that fully convective stars in general are more rapidly rotating than early-M dwarfs with radiative cores, and that the rotation-activity relation is still working.

The rotational velocities of field M dwarfs show a remarkable rise around spectral type M3.5 – exactly where they are believed to become fully convective. Braking timescales on either side of the convective boundary differ by about an order of magnitude. The change in rotational braking could be sufficient to explain the rising fraction of active M dwarfs. Thus, the key question is: What is the reason for the weak rotational braking in fully convective objects?

The strength of rotational braking depends on two main parameters: (I) the strength of the magnetic field, and (II) the geometry of the magnetic field. Because the strength of the magnetic field does not seem to differ between partially and fully convective stars (the latter may even have stronger fields), it is probably the magnetic field geometry that differs between the two regimes. A sudden break in magnetic field geometry at spectral type M3.5 could explain why no rapid rotators are found at spectral types M0–M3, but are found at later spectral types. A predominantly small-scale magnetic field would lead to less braking than a large-scale field would. Indirect observations of the magnetic field topology, however, suggest that large scale axisymmetric fields do exist at least in some of them. Theory also seems not to be in conflict with the generation of such fields through a fully convective dynamo. However, in order to decide whether the magnetic topology differs on either side of the convection boundary, we must convince ourselves that our observational methods are sensitive to the aspects in question. In the case of fully convective stars, this means that we must look for small scale fields, which cannot be entirely excluded by the recent results of Zeeman Doppler Imaging, because it is not sensitive to the very small scales. One possibility is that a high density of closed loops near the stellar surface prevents the outflow of a wind which normally would brake the object. Measurements of radio emission due to axisymmetric fields also cannot exclude the presence of small scale fields on the surface of fully convective objects.

Mass-dependent rotational braking can in principle explain the slope of rotational velocities in L-dwarfs. Whether an abrupt change is required in the magnetic properties of stars around the convection boundary in order to explain the distribution of their rotation rates, is not known. In order to answer this question, a larger, statistically well defined sample is needed covering stars of different ages. With the current instrumentation, this task can be addressed within the near future.

## 4 Summary

During the last decade our picture of and beyond the bottom of the main sequence has tremendously sharpened. In the ESO symposium on "The Bottom of the Main Sequence – and beyond" in 1994, the last spectral bin in an M-dwarf sample consisted of objects of spectral class M5.5+ (Hawley, 1995). Since then, our view of VLMs rapidly reached out much further thanks to several surveys and large telescopes capable of collecting high quality data of faint stars and brown dwarfs. We have followed the main sequence down to the faintest stars and beyond it to brown dwarfs cooler than 1000 K. In particular, we are now scrutinizing the two regions of major changes in the physics of low mass objects; the threshold to complete convection and the threshold to brown dwarfs. From an observational point, there is still not much to say about the latter; it does not seem to have very large impact on the surface properties of low mass objects (and its activity, Mokler & Stelzer, 2002). Discriminating between low mass stars and heavy brown dwarfs is still a challenge.

The threshold to complete convection is expected to influence the observable properties of low mass stars. Observations at either side of the convection boundary can be sorted into two groups, those without evidence for a change in magnetic field generation and those with evidence for it:

**Parameters not affected by the threshold to complete convection:**

a) The mean levels of activity measurements in X-ray, H$\alpha$, and other tracers do not provide significant evidence for a change in magnetic field generation. In particular, mean chromospheric and coronal emission is not observed to diminish around spectral type M3.5. At later spectral class (around M9), activity decreases with lower temperature. The reason for this is probably not less dynamo efficiency but the growing neutrality of the cooler atmosphere.

b) Direct measurements of magnetic fields show that kG-strength magnetic fields are still generated in fully convective stars. Such fields are detected in objects as late as M9, which is far beyond the convection boundary. Fully convective stars apparently harbor strong fields that occupy a large fraction of their surface.

c) Zeeman Doppler Imaging and measurements of persistent periodic radio emission suggest that fully convective objects still harbor large scale axisymmetric magnetic field components.

**Parameters indicating a break at the convection boundary:**

a) Around spectral type M3.5, a sudden break is observed in the distribution of rotational velocities $v \sin i$. On its cool side, hardly a single field M dwarf is known with detected rotational broadening; members of the young and the old disk populations are rotating at a very low rate. On the cool side of the boundary, however, many young disk objects are known with detected rotation velocities, some of them rotating at several ten km s$^{-1}$. At the threshold to complete convection, the braking timescale changes by about an order of magnitude.

b) The highest normalized emission found in X-ray or H$\alpha$ emission of field dwarfs is about an order of magnitude larger at spectral type M4 than it is around M2.

c) The fraction of active stars exhibits a steep rise from below 10 % at spectral class M3 to more than 50 % at M5.

The two last points in favor of a change at the convection boundary may be explained by the higher fraction of rapid rotators among completely convective objects (point 1). A possible scenario is that in fully convective objects weaker braking leads to higher rotation velocities in objects observed in the field. With the rotation activity connection still working, this implies a higher fraction of active stars. Higher rotation velocities also lead to higher maximum emission levels. If this scenario is true, the key question is: "Why is rotational braking less efficient in fully convective objects?" One possible answer is a change in the magnetic topology due to a different dynamo process, and this possibility is currently receiving high attention.

Related open questions are the following:

- In stars with radiative cores, what is the fraction of the magnetic flux that is generated within the convection zone, and what is the fraction generated in the interface dynamo?
- Does (and to what extent does) the rotation activity connection hold in fully convective objects?
- Does the fully convective dynamo depend on differential rotation (and how does the interface dynamo)?
- What is the magnetic topology of fully convective objects (and what is the topology of rapidly rotating sun-like stars)?
- How does magnetic field generation evolve with age?
- Are strong magnetic fields generated in brown dwarfs, too?

The investigation of the properties of VLMs and of stars around the convection boundary has long been hampered by technical difficulties obtaining suitable data (and a sufficient amount of them). Today, our observational equipment is certainly suited to tackle some of these questions. Theoretical considerations together with numerical calculations are pushing and questioning our understanding of low mass star and brown dwarf physics. Within the near future, we can certainly expect answers to some of the open questions at the bottom of the main sequence and beyond it.

## Acknowledgements

I thank the Astronomische Gesellschaft for the *Ludwig-Biermann Prize* 2007, and I acknowledge research funding from the DFG through an Emmy Noether Fellowship (RE 1664/4-1).

## References

Afram, N., Berdyugina, S.V., Fluri, D.M., Semel, M., Bianda, M., Ramelli, R., 2007, A&A, 473, L1

Baraffe, I., Chabrier, G., Allard, F., & Hauschildt, P.H., 1998, A&A, 337, 403

Baraffe, I., Chabrier, G., Allard, F., & Hauschildt, P.H., 2002, A&A, 382, 563

Barnes, S.A., 2007, arXiv:0704.3068

Basri, G., 2000, ARA&A, 38, 485

Basri, G., Marcy, G.W., & Valenti, J.A., 1992, ApJ, 390, 622

Benedict, F., et al., 1998, AJ, 116, 429

Berger, E., 2006, ApJ, 648, 629

Berger, E., 2007, arXiv:0708.1511

Bessell, M.S., 1991, AAS 83, 357

Browning, M.K., 2007, ApJ in press, arXiv:0712.1603

Burgasser, A.J.Liebert, J., Kirkpatrick, J.D., Gizis, J.E., 2002, AJ, 123, 2744

Cayrel de Strobel G., Soubiran C., Friel E.D., Ralite N., Francois P., 1997, A&AS, 124, 299

Chabrier, G., Baraffe, I., & Plez, B., 1996, ApJ, 459, L91

Chabrier, G., & Baraffe, I., 2000, ARAA, 38, 337

Chabrier, G., Baraffe, I., Allard, F., & Hauschildt, P.H., 2005, ASP Conf. Ser., arXiv:astro-ph/0509798

Chabrier, G., & Küker, M., 2006, A&A, 446, 1027

Chabrier, G., Gallardo, J., & Baraffe, I., 2007, A&A, 472, L17

Chaboyer, B., Demarque, P., & Pinsonneault, M.H., 1995, ApJ, 441, 876

Clemens, J.C., Reid, I.N., Gizis, J.E., & O'Brien, M.S., 1998, ApJ, 496, 352

Copeland, H., Jensen, J.O, & Jorgensen, H.E., 1970, A&A, 5, 12

Dahn, C.C., et al., 2002, AJ, 124, 1170

Dobler, W., & Stix, M., & Brandenburg, A., 2006, ApJ, 638, 336

Donati, J.-F., Forveille, T., Collier Cameron, A., Barnes, J.R., Delfosse, X., Jardina, M.M., & Valenti, J.A., 2006, Science, 311, 633

D'Antona, F., & Mazzitelli, I., 1996, ApJ, 456, 329

Delfosse, X., Forveille, T., Perrier, C., & Mayor, M., 1998, A&A, 331, 581

Durney, B.R., & Latour, J., 1978, Geophys. Astrophys. Fluid Dynamics, 1978, 9, 241

Durney, B.R., De Young, B.S., & Roxburgh, I.W., 1993, Sol.Phys.,145, 207

ESA, 1997, The Hipparcos & Tycho Catalogues, ESA-SP 1200

Fleming, T.A., Giampapa, M.S., & Schmitt, J.H.M.M., 2000, ApJ, 533, 372

Golimowski, D.A., Leggett, S.K., Marley, M.S., et al., 2004, AJ, 127, 3516

Hallinan, G., Antonova, A., Doyle, J.G., Bourke, S., Brisken, W.F., & Golden, A., 2006, ApJ, 653, 690

Hartmann, L., Hewett, R., Stahler, S., Mathieu, R.D., 1986, ApJ, 309, 275

Hawley S.L., Gizis J.E., Reid I.N., AJ, 1996, 112, 2799

Hauck, B. & Mermilliod, M., 1998, A&AS, 129, 431

Johns-Krull, C., & Valenti, J. A. 1996, ApJ, 459, L95

Johns-Krull, C., & Valenti, J.A., 2000, ASPC, 198, 371

Johns-Krull, C., & Valenti, J. A. 2001, ApJ, 561, 1060

Johns-Krull, C., 2007, ApJ, 664, 975

Kawaler, S.D., 1988, ApJ, 333, 236

Kiraga, M., & Stepień, 2007, arXiv:0707.2577

Kirkpatrick, J.D., 2005, ARA&A, 43, 195

Krishnamurti, A., Pinsonneault, M.H., Barnes, S., & Sofia, S., 1997, ApJ, 480, 303

Küker, M., & Rüdiger, G., 1999, A&A, 346, 922

Leggett, S.K., 1992, ApJS 82, 351

Liebert, J., Lirkpatrick, J.D., Cruz, K.L., Reid, I.N., Burgasser, A., Tinney, C.G., & Gizis, J.E., 2003, AJ, 125,343

Liebert, J., & Probst, R.G., 1987, ARAA, 25, 473

López-Morales, M., 2007, ApJ, 660, 732

Marcy, G.W., & Basri, G., 1989, ApJ, 345, 480

Mestel, 1984, in Third Cambridge Workshop on Cool Stars, Stellar Systems, and the Sun, ed. S.L. Baliunas & L. Hartmann (New York: Springer), 49

Meyer, F., & Meyer-Hofmeister, E., 1999, A&A, 341, L23

Mohanty, S., Basri, G., Shu, F., Allard, F., Chabrier, G., 2002, ApJ, 571, 469

Mohanty, S., & Basri, G., 2003, ApJ, 583, 451

Mohanty, S., Jayawardhana, R., & Basri, G., 2005, ApJ, 626, 498

Mokler, F., & Stelzer, B., 2002, A&A, 391, 1025

Noyes, R.W., Hartmann, L.W., Baliunas, S.L., Duncan, D.K., & Vaughan, A.H., 1984, ApJ, 279, 763

Ossendrijver, M., 2003, A&AR, 11, 287

Patten, B.M., & Simon, T., 1996, ApJSS, 106, 489

Pettersen, B.R., 1983, in Byre P.B., Rodono M., eds., Activity in Red-dwarf Stars, Reidel, Dordrecht, p.17

Pizzolato, N., Maggio, A., Micela, G., Sciortino, S., & Ventura, P., 2003, A&A, 397, 147

Randich, S., 1998, ASP Conf. Ser., 154, 501

Reid I.N., Hawley S.L., Gizis J.E., AJ, 1995, 110, 1838

Reiners, A., 2007, A&A, 467, 259

Reiners, A., & Basri, G., 2006a, AJ, 131, 1806

Reiners, A., & Basri, G., 2006b, ApJ, 644, 497

Reiners, A., & Basri, G., 2007, ApJ, 656, 1121

Reiners, A., Schmitt, J.H.M.M., & Liefke, C., 2007a, A&A 466, L13

Reiners, A., Seifahrt, A., Siebenmorgen, R., Käufl, H.U., & Smette, Al., 2007b, A&A, 471, L5

Reiners, A., Seifahrt, A., Stassun, K.G., Melo, C., & Mathieu, R.D., 2007c, ApJL, in press, `arXiv:0711.0536`

Robinson, R.D., 1980, ApJ, 239, 961

Robrade, J., & Schmitt, J.H.M.M., 2005, A&A, 435, 1077

Saar, S.H., 1994, IAU Symp. 154, Infrared Solar Physics, eds. D.M. Rabin et al., Kluwer, 493

Saar, S.H., 1996, in IAU Symp. 176, Stellar Surface Structure, eds., Strassmeier, K., J.L. Linsky, Kluwer, 237

Saar, S.H., 2001, ASPCS 223, 292

Scholz, A., Coffey, J., Brandeker, A., & Jayawardhana, R., 2007, ApJ, 662, 1254

Schmidt, S.J., Cruz, K.L., Bongiorno, B.J., Liebert, J., & Reid, I.N., 2007, AJ, 133,2258

Schmitt, J.H.M.M., 1995, ApJ, 450, 392

Siess, L., Dufour, E., & Forestini, M., 2000, A&A, 358, 593

Simon, T., 2001, ASP Conf. Ser., 223, 235

Solanki, S.K., 1991, RvMA, 4, 208

Stauffer, J.R., & Hartmann, L.W., 1986, ApJS, 61, 531

Stassun, K.G., Mathieu, R.D., & Valenti, J.A., 2006, Nature, 440, 311

Stassun, K.G., Mathieu, R.D., & Valenti, J.A., 2007, ApJ, 664, 1154

Sterzik, M.F., & Schmitt, J.H.M.M, 1997, AJ, 114, 167

Takalo, L.O., & Nousek, J.A., 1988, ApJ, 326, 779

Taylor, B.J., 1995, PASP, 107, 734

Torres, C.A.O., Busko, I.C., & Quast, G.R., 1983, in Byre P.B., Rodono M., eds., Activity in Red-dwarf Stars, Reidel, Dordrecht, p.175

Valenti, J.A., Johns-Krull, C.M., & Piskunov, N.P., 2001, ASP 223, 1579

Voges, W., et al., 1999, A&A, 349, 389

West, A.A., et al., 2004, AJ, 128, 426

Whelan, E.T., Ray, T.P, Randich, S., Bacciotti, F., Jayawardhana, R., Testi, L., Natta, A., & Mohanty, S., 2007, ApJ, 659, L45

Zapatero Osorio, M.R., Martín, E.L., Bouy, H., Tata, R., Deshpande, R., & Wainscoat, R.J., 2006, ApJ, 647, 1405

# Structure Formation in the Expanding Universe: Dark and Bright Sides

Anatoly Klypin & Daniel Ceverino,

New Mexico State University, Las Cruces, NM 88001

aklypin@nmsu.edu, ceverino@nmsu.edu http://astronomy.nmsu.edu/aklypin

Jeremy Tinker,

Kavli Institute for Cosmological Physics,

University of Chicago, Chicago, IL,60637, U.S.A.

tinker@kicp.uchicago.edu

## Abstract

*We give a brief overview of the current state of that part of cosmology, which deals with the formation of different structures in the hierarchical model. Modern cosmology covers a vast range of physical phenomena and scales ranging from large scales, which are dominated by gravitational dynamics to interior of galaxies, where the physics of baryons play an important role. Driven by both observations and by theoretical effort of echelons of cosmologists, the gravitational dynamics has made remarkable progress, giving rise to the notion of precision cosmology. We discuss such characteristics as the power spectrum, mass function, and halo concentration, which are predicted by the theory with the accuracy of few percent. The situation is different once the physics of baryons becomes important. Predictions are much less certain even on large scales, and on the scales of galaxies they become absolutely exciting but qualitatively uncertain. We argue that a significant effort – comparable to what cosmology has done with the gravitational dynamics – must be devoted to the theory of galaxy formation.*

## 1 Introduction

The current cosmological paradigm, the $\Lambda$CDM Universe, is remarkably successful on scales larger than $\sim$Megaparsec. It predicted the amplitude and spectrum of fluctuations in the cosmic microwave background (CMB) and the spectrum of large-scale fluctuations in the distribution of galaxies (Bardeen *et al.*, 1987; Holtzman, 1989). Those predictions were later verified by observational measurements

(Netterfield et al., 2002; Tegmark et al., 2004; Cole et al., 2005; Spergel et al., 2007). The recent success of theoretical cosmology is impressive, but one should not forget that the current model is the result of a long process of selection and perfection of competing theoretical models. Many models did not make it and were rejected. Historically, the $\Lambda$CDM model had not been a favored model. The main objection to the model involves an ad-hoc component – the cosmological constant or the dark energy. The $\Lambda$CDM model started to win well before the CMB measurements became available. A combination of the measurements of the Hubble constant (thus, the age of the Universe) and the abundance of high redshift objects (clusters of galaxies in particular) left a choice of either the $\Lambda$CDM model (a flat universe with low density of the dark matter) on an open CDM model (a model with negative curvature). Then the first measurements of the position of the first acoustic peak in the CMB spectrum (de Bernardis, 2000; Bond et al. 2000) eliminated the open CDM model.

To a large degree, the theory is successful because it is based on well understood and simple physics: the growth of linear perturbations. We should not underestimate the effort put into making the physics "well understood": it took more than 30 years (e.g., Lifshitz & Khalatnikov, 1963; Eisenstein & Hu, 1999). Still, if we compare this part of cosmology with other fields of astronomy such as star formation or stellar evolution, it is clear that linear perturbations in cosmology deal with a simple physics. This is why theoretical predictions can be very accurate.

The field of structure formation (quasi-linear and strongly non-linear fluctuations) is split into two almost independent parts: physics of clustering of the dark matter and the theory of the galaxy formation. Gravitational dynamics is a complicated field in comparison with the growth of linear perturbations. Yet it is much simpler than galaxy formation. Thus statistics and properties of structures in the dark matter can be predicted with stunning accuracy. The gravitational dynamics of the dark matter is actually a complicated field. The structures in general are not spherical nor stationary. There is substantial coupling of different modes (scales) with long waves substantially affecting small-scale dynamics. The range of scales is astounding. For example, in modern computer simulations it is not unusual to have a dynamic range of 100,000 or even more. Reducing this dynamical range may result in incorrect results. These are the reasons why it is very difficult to make analytical predictions in this field. Indeed, there are very few useful analytical solutions.

There are two factors that dramatically simplify the theory. (1) In cosmology we deal with collisionless dynamics. Mathematically this means solving the Vlasov-Poisson system of equations for which the phase-space density is preserved along trajectory of each particle. Physically this means that we do not have complicated processes such as two-body scattering or the gravothermal catastrophe. (2) Long waves are only weakly affected by the short-scales: what happens on small scales does not significantly affect perturbations on large scales. Because of this weak coupling, one can simulate a relatively small volume and obtain reasonably accurate results. How small? This depends on required accuracy and particular properties studied. There is no a priori rule for how large the volume should be. This is found only by comparing results from different volumes.

The theory of galaxy formation, which involves complicated physics of gas and star formation, is still in a very preliminary stage. Here the situation is very different compared to the dynamics of the dark matter. The accuracy of the results is not much of an issue: one does not worry about the accuracy, when it is not clear whether all the basic physics is included. So, the main thrust is to identify and include the most important physical processes. The complexity of the problem is coming from the fact that small-scale phenomena affect large scales in a very profound way. For example, motion and metallicity of gas in a galactic halo ($\sim$ 100 kpc scales) is affected by energy release and gas motion in regions of active star formation in the galaxy itself (few parsecs scales). In turn, the physics on parsec scales is affected by the stars formation precesses and by their environment scales well below a parsec. In other words, the minimum scale is not clear, but it must be included in analysis to produce the correct treatment of gas and stars on large scales. The situation may not be as hopeless as it looks: there should be a minimum scale, which is sufficient for accurate description of the main physical processes responsible for formation of galaxies and for the large-scale dynamics of baryons inside the galaxies.

Because of the complexities and uncertainties in galaxy formation, it is highly unlikely that predictions of the galaxy formation will be used to constrain the parameters of cosmological models. Fluctuations in the CMB, geometrical tests (such as distances to SNI), and results from dark matter clustering are the main tools for testing the global parameters of cosmological models.

The theory of non-linear structures has different components.

(1) *Numerical simulations* play an important role. Sophisticated codes were developed to simulate different aspects of the theory (e.g., ART – Kravtsov *et al.*, 1997; GADGET – Springel *et al.*, 2001; GASOLINE – Wadsley *et al.*, 2004; Enzo – O'Shea *et al.*, 2004). To large degree, the progress in the field was driven by numerical simulations. In turn, numerical effort is driven by observational needs and by computer hardware. Numerical simulations have one limiting factor: they do not give explanations. For example, simulations predict that the density profiles of dark matter halos are reasonably well fit by the Navarro-Frenk-White (NFW, 1997) approximation. Yet, the simulations do not tell us why this is the case.

(2) *Analytics* is important because they can provide understanding of what occurs in simulations. Unfortunately, analytical models are very difficult to make. There are two special cases: the halo mass mass function and the quasi-linear regime of clustering. There has been a dramatic improvement in both statistics from traditional-style approximations. In the case of quasi-linear clustering, one expands the fluctuations using perturbation theory and collects leading terms. The calculations are tedious but tractable. Predictions were tested against N-body simulations with impressive results (Crocce & Scoccimarro, 2007). In fact, some simulations were not accurate enough and were re-done. The mass function is estimated using a different type of approximation. Press and Schechter (1974) derived the scaling law for the mass function, which is based on Gaussian statistics. The scaling is now used with additional fitting factors to approximate the halo mass function. Results depend on particular choice (and ideas) for the extra fitting factors (Sheth & Tormen, 1999; Jenkins *et al.*, 2001).

(3) There is another type of prediction in cosmology, which is called *semi-analytics* (e.g., Kauffmann, Guiderdoni, & White, 1994; Somerville & Primack, 1999; Benson *et al.*, 2003). It is not an approximation in a usual sense: it does not make an analytical approximation of any law of physics or any equations. It is a set of prescriptions, which guess the outcome of complicated (and often poorly understood) precesses. For example, it assumes that the collision of two spiral galaxies of comparable mass (major merger) results in the formation of an elliptical galaxy. In reality this may not be true, but there is no way of testing it within the framework of the semi-analytical models. Semi-analytics has another limitation. They cannot be used to learn about new physical phenomena. If the rules include only spiral and elliptical galaxies, the results cannot tell you that there are other types of galaxies (e.g., lenticular or barred galaxies). Semi-analytics is often criticized for having too many free parameters: this is not right. It is an ambitious prospect to explain the whole universe, and an extra 200 free parameters is a small price to pay. Semi-analytics can be a power tool, if used cautiously and in conjunction with numerical simulations. It is a great tool for making mock catalogs for testing statistical effects or for designing observational strategies.

Results of numerical simulations is the main goal of this review. Numerical simulations in cosmology have a long history and numerous important applications. It began in the 60s (Aarseth, 1963) and 70s (Peebles, 1970) with simple problems solved using N-body codes with few hundred particles. Peebles (1970) studied collapse of a cloud of 300 particles as a model of cluster formation. After the collapse and virialization the system resembled a cluster of galaxies. Those early simulations of cluster formation, though they produced cluster-like objects, signaled the first problem – a simple model of an initially isolated cloud (top-hat model) results in a density profile which is too steep (power-law slope -4) as compared with real galaxy clusters (slope -3). The problem was addressed by Gunn & Gott (1972), who introduced a notion of secondary infall in an effort to solve the problem. Another keystone work of those times is the paper by White (1976), who studied the collapse of 700 particles with different masses. It was shown that if one distributes the mass of a cluster to individual galaxies, two-body scattering will result in mass segregation not compatible with observed clusters. This was another manifestation of the dark matter in clusters, demonstrating that inside a cluster the dark matter can not reside inside individual galaxies.

Survival of substructures in galaxy clusters was another problem addressed by White (1976). It was found that lumps of dark matter, which in real life may represent galaxies, do not survive in dense environment of galaxy clusters. White & Rees (1978) argued that the real galaxies survive inside clusters because of energy dissipation by the baryonic component. That point of view was accepted for almost 20 years. However, recently it was shown that energy dissipation does not play a dominant role in the survival of galaxies and that dark matter halos are not destroyed by tidal stripping and galaxy-galaxy collisions inside clusters (Klypin et al. 1999; Moore et al. 1999). The reason early simulations came to the wrong result was pure numerical: they did not have high enough resolution and the integration of trajectories was not accurate enough. But 20 years ago it was physically impossible to make a simulation with sufficient resolution; even if at that time we had present-day codes, it would have taken about 600 years to make one run.

Starting in the mid 1980s the field of numerical simulations began to bloom: new numerical techniques were invented, old ones were perfected. The number of publications based on numerical modeling sky-rocketed. To large extent, this has changed our way of doing cosmology. Instead of questionable assumptions and waving-hands arguments, we have tools of testing our hypotheses and models. As an example, we mention two analytical approximations which were validated by numerical simulations. The importance of both approximations is difficult to overestimate. The first is the Zeldovich approximation, which paved the way for understanding the large-scale structure of the galaxy distribution. The second is the Press & Schechter (1974) approximation, which gives the number of objects formed at different scales at different epochs. Both approximations cannot be formally proved. The Zeldovich approximation is not applicable for hierarchical clustering because the Zeldovich approximation works only with smooth perturbations which have a truncated spectrum. Nevertheless, numerical simulations have shown that even for hierarchical clustering the approximation can be used with appropriate filtering of the initial spectrum. The Press-Schechter approximation and its siblings are also difficult to justify without numerical simulations. The approximations utilize the initial spectrum and linear theory, but then (a very long jump) it predicts the number of objects at a very nonlinear stage. Because it is not based on any realistic theory of nonlinear evolution, its justification is based solely on numerical simulations.

## 2 Dark Matter

### 2.1 Evolution of the perturbations

Because numerical simulations play a vital role in cosmology, testing the codes and the accessing the accuracy of simulations is very important. There are different ways of testing codes: (1) testing a code against known analytical solution (e.g., Navarro & White, 1993; Kravtsov *et al.* 1997; Heitmann *et al.*, 2005), (2) comparing results of different codes (Frenk *et al.*, 1999; Heitmann *et al.*, 2005), or (3) testing convergence of a given code. All three methods are widely used in cosmology.

Figure 1 shows the evolution of the power spectrum of perturbations in a large simulation of the $\Lambda$CDM model: 1Gpc box with $1024^3$ particles. Longest waves in the simulation have small amplitude and must grow according to the linear theory. This indeed is the case, as seen in the bottom panel. Note, that a small dip $\sim -2\%$ at $k = 0.05$–$0.07 h^{-1}$Mpc is what the quasi-linear theory of perturbations predicts (Crocce & Scoccimarro, 2007). The plot also shows the main tendency: in the nonlinear stage the perturbations at the beginning grow faster than the predictions of the linear theory (we neglect a possible small negative growth extensively discussed by Crocce & Scoccimarro (2007)). At later stages the growth slows down, which is seen as bending down of $P(k)$ at high frequencies (Peacock & Dodds, 1996).

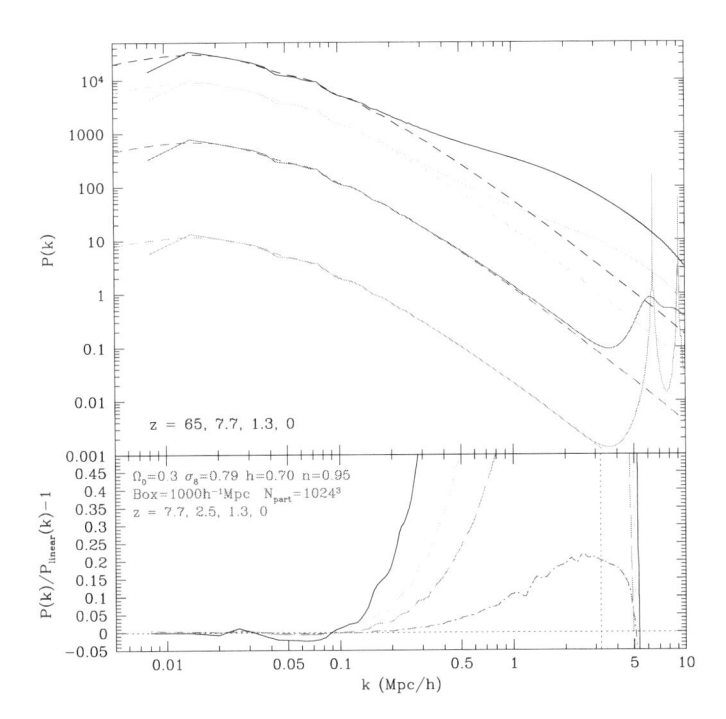

**Figure 1:** Growth of perturbations in the ΛCDM model. The top panel shows the evolution of the power spectrum $P(k)$ in the simulation (full curves) as compared with the linear theory (dashed curves). The simulation used a realization of the spectrum for $1024^3$ particles. The deviation in the first harmonic (the smallest $k$) is due to small statistics of longest waves. Two strong spikes at large $k$'s are above the Nyquist frequency: the N-body code does not "see" them. Fluctuations on large scales (small $k$'s) grow according to the linear theory (bottom panel).

Figure 2 shows the distribution of the dark matter in a $\sim 10h^{-1}$Mpc slice of a $1024^3$ particles simulation. The simulation used a special technique – constrained simulations – to set initail conditions, when long waves are taken for the distribution of matter and velocities of galaxies in real Universe, but small-scale perturbations are from a realization of the ΛCDM model. Details can be found in Klypin *et al.* (2003). Because the long wave are taken from the observations, the simulation has large-scale features in right places. In this figure our Galaxy (or a candidate for MW) is in the middle in a relatively weak filament. Just above it at a distance of about $10h^{-1}$Mpc there is "Virgo". cluster. Other objects are also marked in the plot. Most of the attention in the field is paid to dark matter halos, which in this figure are seen as tight bright round knots. Yet, the whole distribution is a complex web of filaments, voids, and halos.

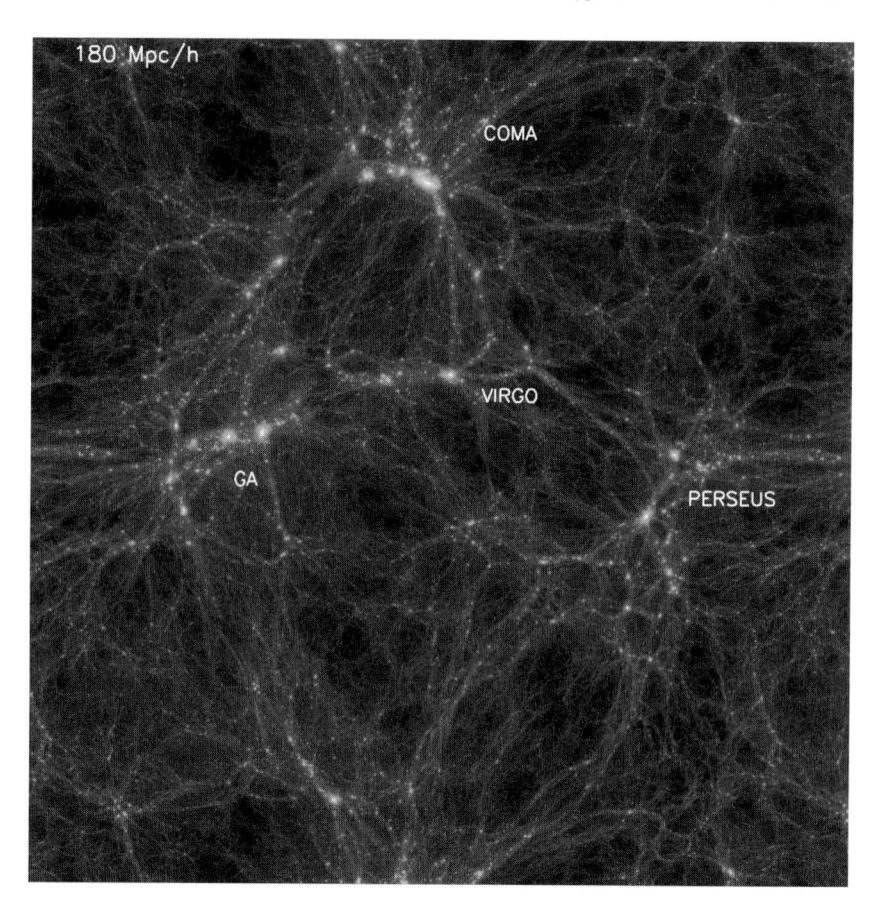

**Figure 2:** Distribution of the dark matter in simulations constrained by the distribution and motion of galaxies in the real Universe. An analog of our Galaxy is a tight small halo in the center of the plot. The distribution of dark matter is a hierarchy of filaments, voids, and halos

## 2.2 Halo mass function and halo concentration

For a given set of cosmological parameter the number of dark matter halos at different redshifts is defined by the shape and amplitude of the initial density field, making observational measurements of the mass function a powerful constraint on cosmological parameters, primarily the matter density $\Omega_m$ and the amplitude of the power spectrum $\sigma_8$ defined at scale the $8h^{-1}$Mpc (scale of galaxy clusters). The evolution of cluster statistics with time is a primary method for measuring the equation of state of dark energy $w$, a leading problem in observational cosmology and fundamental physics. Precise theoretical understanding of the mass function and its dependence on cosmology and redshift are required by upcoming programs that seek to use the abundance of clusters to constrain cosmology. In addition to constraining cosmological parameters, dark matter halos are also key to our understanding of galaxies. Precise interpretation of observational data requires precise theoretical predictions for the properties of dark matter halos: their abundance, their bias, and their density profiles.

The standard for precision determination of the mass function from simulations was set by Jenkins et al. (2001), who improved on earlier analytic predictions by Press & Schechter (1974), Lee & Shandarin (1998), and Sheth & Tormen (1999). The primary result of the Jenkins study was a fitting function accurate to $\sim (10-20)\%$. The authors also claimed that this function was universal for variations in cosmology and redshift. Although encouraging, efforts to constrain dark energy from galaxy clusters require theoretical uncertainty of order 5%, a level of precision achieved by Warren et al. (2006) for a fixed cosmology at $z = 0$. These studies are augmented by other studies investigating the universality of the mass function out to high redshift (Reed et al. 2003; Lukic et al. 2007; Cohn & White 2007). The majority of these studies emphasize results based on halos identified by the friends-of-friends algorithm. Although computationally efficient and straightforward to implement, the relation between friends-of-friends (FOF) halos and objects detected in the real universe is murky at best. Another choice for identifying halos in simulations is the spherical overdensity (SO) approach. In the SO method halos are defined as the total mass within a sphere of radius $R_\Delta$ such that $\Delta = 3M/4\pi R_\Delta^3 \Omega_m \rho_{\rm crit}$, where $\Omega_m \rho_{\rm crit}$ is the average density of the dark matter. The SO definition has significant benefits over FOF both theoretically and observationally. Analytic halo models deal with spherically averaged halos, and the statistics derived are sensitive to the exact halo definition. Many of the observable quantities for galaxy clusters are defined on a spherical aperture: FOF algorithm is not used by observers. Instead of the mass of halos $M$, it is convenient to express results in terms of the overdensity $\sigma(M)$, which is the rms fluctuation of mass as estimated by the filtering of the linear power spectrum with the top-hat filter with mass $M$.

Tinker et al. (2008a) use a large set of cosmological N-body simulations to estimate the halo mass function. The simulation set spans six orders of magnitude in volume and has different cosmologies and redshifts in the analysis. Comparison of results presented in Figure 3 and obtained with different resolutions and with different codes indicate that the halo mass function is now predicted with accuracy better than 5% over a mass range of five orders of magnitude. The results also clear indicate that the halo mass function does not depend only on the amplitude of perturbation at given scale $\sigma(M)$. If it were true, the mass function at redshift $z = 0$ $f(\sigma; z = 0)$ would provide the best fit for the mass function at redshift $z = 1.25$. This is not the case: $f(\sigma; z = 0)$ overpredicts the mass function at $z = 1.25$ by 20%.

Concentration is another important property of a dark matter halo. It is defined as the ratio of the virial radius $r_{\rm vir}$ to the radius $r_s$, where the slope of the density profiles is equal to $-2$ (core radius in the NFW approximation): $c_{\rm vir} = r_{\rm vir}/r_s$. There are ambiguities in defining the virial radius. A radius of constant overdensity $\Delta = 200$ is often used as a substitute for the virial radius. This gives another concentration $c_{200} = R_\Delta/r_s$. The reference point for the concentrations is the paper by Bullock et al. (2001), who parametrized the halo concentration as a function of mass and redshift. Simulations used in Bullock et al. (2001) had a small size of

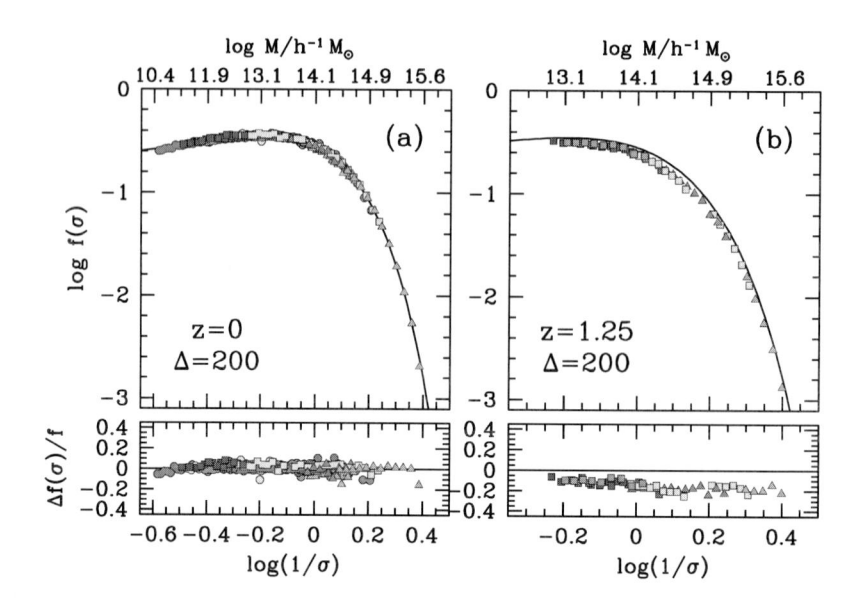

**Figure 3:** Halo mass function for different redshifts and cosmological models. Panel (a): The measured halo mass function $f(\sigma(M))$ from a large set of N-body simulations. Results are presented at $z = 0$ and for the overdensity $\Delta = 200$. The solid line is the best fit function. The lower window shows the percentage residuals with respect to the fitting function. Panel (b): The measured $f(\sigma)$ at $z = 1.25$. The solid line is the same as in panel (a), which was calibrated at $z = 0$. The lower window shows that the $z = 1.25$ mass function is offset by $\sim 20\%$ with respect to the results at z = 0.

the computation box ($60h^{-1}$Mpc) and small number of particles ($256^3$), and, thus, are very limited in mass resolution: $M_{\rm vir} = 3 \times 10^{11} h^{-1} M_\odot - 5 \times 10^{13} h^{-1} M_\odot$. Still, within the mass range the results are remarkably accurate and are very close (within the errors) to those of much larger Millenium simulation with 8 billion particles (Neto et al. (2007)). The main problem with the Bullock et al. approximation is that it was used well outside it original range of masses. When extrapolated to large mass, it significantly underpredicts the concentration. Remarkably, Bullock et al. give another set of parameters in a footnote ($F = 0.001$, $K = 3.0$), which provided an equally good fit for their results and gives a much better fit for current data.

Figure 4 presents results of Tinker et al. (2008b) for halos in a large set of simulations and resolutions. The analytical fits in the plot use almost the same set of parameters as in the footnote of Bullock et al. (2001): $K = 2.9$, $F = 0.001$ (see also Wechsler et al. (2006)). At small masses the data scale nearly as the power-law:

$$c_{200} = 5.2(M/10^{14}h^{-1}M_\odot)^{-0.12}(1+z)^{-0.6}. \tag{1}$$

Here we define concentration relative to overdensity $200\rho_{\rm crit}$. This approximation also provides an excellent fit to data presented by Neto et al. (2007). At large masses

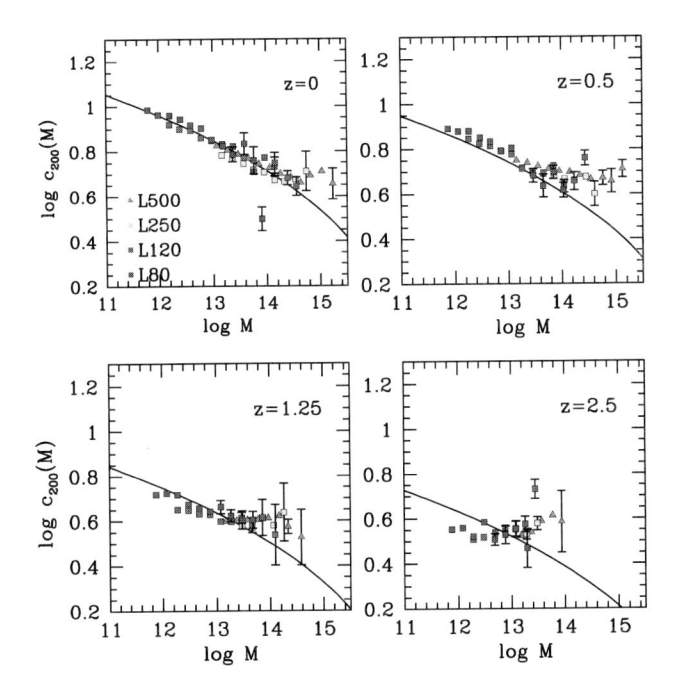

**Figure 4:** Halo concentration as a function of mass at different redshifts. The full curves are analytical approximation of footnote of Bullock et al. (2001) scaled with redshift as $(1+z)^{-0.6}$. At small masses the results are equally well fit by a power-law with the slope $-0.12$. The halo concentration stops declining at large masses. In order to make easy comparison with other results, we define concentration relative to overdensity $200\rho_{\rm crit}$.

the data indicate that the concentration does not decline with mass: a simple power-law fit badly fails, which is clearly seen at high redshifts.

## 2.3 Angular momentum

It is convenient to characterize rotation of a dark matter halo using a dimensionless spin parameter $\lambda$:

$$\lambda = \frac{J|E|^{1/2}}{GM^{5/2}}, \qquad (2)$$

here $J$ is the angular momentum, $E$ is the total energy, and $M$ is the mass of a halo. The value of the spin parameter roughly corresponds to the ratio of the angular momentum of an object to that needed for rotational support. Typical values of the spin parameter of individual halos in simulations are in the range 0.02 to 0.11 (e.g. Barnes & Efstathiou, 1987). The distribution of spin parameters in N-body

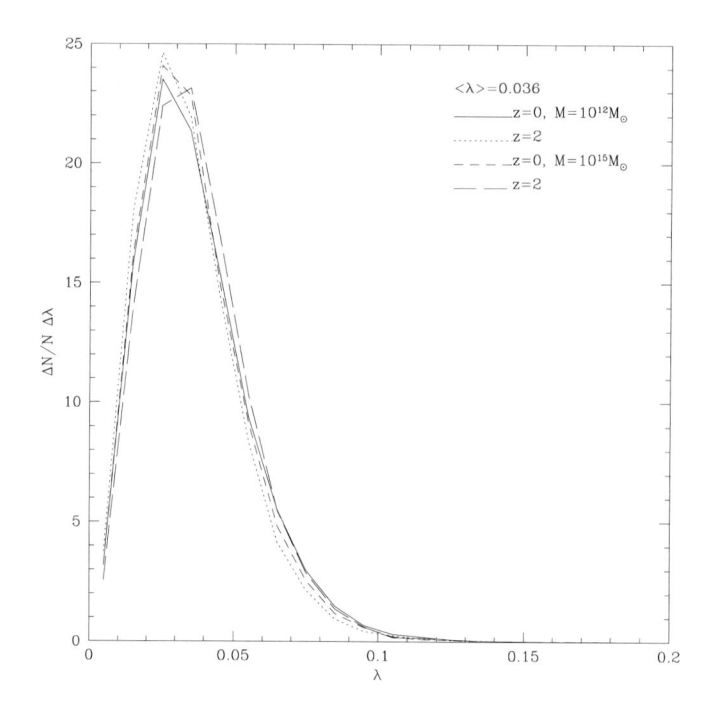

**Figure 5:** The distribution of spin parameters of halos at $z = 0$ with masses $10^{12} M_\odot$ and $10^{15} M_\odot$ and their progenitors at $z = 3$. The spin distribution has rather weak dependence on both mass and redshift.

simulations is well described by the log-normal distribution:

$$p(\lambda)d\lambda = \frac{1}{\sigma_\lambda \sqrt{2\pi}} \exp\left(-\frac{\ln^2(\lambda/\bar{\lambda})}{2\sigma_\lambda^2}\right) \frac{d\lambda}{\lambda}, \qquad (3)$$

where $\bar{\lambda} = 0.037$ and $\sigma_\lambda = 0.51$ (e.g., Bett *et al.*, 2007). There are indications that the log-normal distribution underestimates the number of halos with small spin parameters (Bett *et al.*, 2007). Figure 5 shows the distribution of spin parameters for halos with different mass at $z = 0$ and their progenitors at $z = 3$ (Vitvistska *et al.*, 2002). It is clear that the distribution of the spin parameter is very stable: it does not depend on halo mass and on redshift (see also Hahn *et al.*, 2007). Spins of neighboring halos do not correlate (e.g., Hahn *et al.*, 2007): the source of the angular momentum (be it the tidal torques or satellite accretion) does not have a long-distance memory. There are some correlations of the halo spin. The direction of the spin tend to be parallel to the minor axis of halo mass distribution (e.g., Bett *et al.*, 2007), though the distribution of dis-alignment angles is quite broad. Halos with larger spin are more clustered ( Bett *et al.*, 2007).

In spite of the fact that the overall distribution of spins does not change with time, each individual halo changes its angular momentum quite substantially (Vitvitska et al., 2002; Hetznecker & Burkert, 2006; D'Onghia & Navarro, 2007). Figure 6 shows the evolution of spin parameters and masses for three halos with mass $\sim 10^{12} h^{-1} M_\odot$ at $z = 0$. The spin tends to increase to very large values of $\lambda = 0.05$–0.1 when a large satellite is accreted. Figure 3 in D'Onghia & Navarro (2007) gives two more examples of this process: $\lambda$ increased by factor of three, when either a major merger (mass ratio 1:3 of satellite to primary) or a 1:15 merger occurred. Interpretation of the subsequent evolution is still controversial. Vitvitska et al. argued that the decline of the spin after merger is caused by accretion of satellites with small net angular momentum resulting in dilution of the large initial angular momentum over bigger mass. D'Onghia & Navarro argue that the decline "is due to the internal redistribution of mass and angular momentum that occur during virialization". This is not easy to arrange. If a system does not grow and does not interact with its environment, then its total energy, mass, and angular momentum are preserved. Because the spin parameter depends only on those quantities, no internal process (including virialization) can change it. The angular momentum can be lost to immediate environment of a halo through tidal interactions. Whatever is the explanation of the subsequent evolution of the angular momentum, all the results agree that the angular momentum dramatically increases during the merging process and it stays in the system for a long time. This may have an effect on the angular momentum of gas during this merging process: if the gas has large angular momentum, it will change the final angular momentum of the galaxy forming from the gas. If we further assume that large mergers produce ellipticals, than ellipticals should have larger angular momentum.

## 2.4 Structure of Dark matter halos

There are numerous aspects regarding the internal structure of dark matter halos. Spherically averaged density profile is probably the most well studied statistics. The density profile is reasonably accurately approximated by the NFW profile: $\rho(r) = \rho_o/x(1+x)^2$, where $x = r/r_s$. Here $\rho_0$ and $r_s$ are two free parameters, which can be replaced by, say, virial mass and concentration. The central slope of the density profile is a very much debated characteristics. For the NFW it is $\gamma - 1$. Recent simulations of Diemand et al. (2007) with $\sim 1/4$ billion partices give $\gamma = -1$ at 0.34% od $r_{\rm vir}$. The problem with the NFW is that on average it overestimates by $\sim 20\%$ the density in the region $r \approx r_s$. Figure 7 shows the average profile for halos with mass $< M > = 3 \times 10^{11} h^{-1} M_\odot$ (Prado et al. 2006). The NFW provides are reasonable fit over a wide range of radii. Still, it is not accurate even on average. A more accurate fit is provided by the three parameter Einasto profile (Navarro et al., 2004; Prada et al., 2006; Gao et al., 2007):

$$\rho = \rho_s \exp\left(-\frac{2}{\alpha}\left[\left(\frac{r}{r_s}\right)^\alpha - 1\right]\right). \tag{4}$$

On average the parameter $\alpha$ correlates with halo mass: the larger is the mass the larger is $\alpha$. Combining results presented by Prada et al. (2006) and by Gao et al.

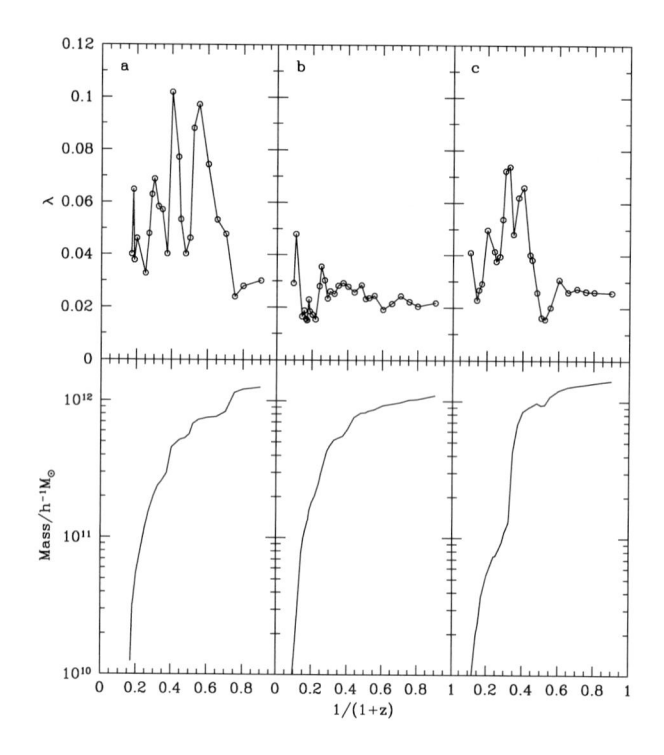

**Figure 6:** Three examples of evolution tracks of Milky-Way galaxy-size halos in N-body simulations. All halos show fast mass growth at high redshifts. At that epoch their spin parameters changes very violently, but subsequently they mostly decline as the halo masses the growth slows down.

(2007) we get the following approximation for $\alpha(M)$ for halos at $z = 0$:

$$\alpha = 0.155 + 0.015 \log(M/10^{12} h^{-1} M_\odot). \tag{5}$$

The steep increase in the density of the dark matter results in significant tensions with observations of rotation curves of dwarf irregular galaxies (e.g., de Blok *et al.* (2002)). There are some cases where observed rotation curves measured by motion of neutral hydrogen, or molecular gas (Simon *et al.*, 2005) or H$\alpha$ emission (de Blok *et al.*, 2002) fall below theoretical predictions and, if naively interpreted, predict constant density core for the dark matter. This is the so called cores vis. cusps problem. These contradictions with the theoretical predictions are not universal: in a large fraction of measured galaxies the NFW profile gives a match to observations (Hayashi *et al.*, 2004, Simon *et al.*, 2005).

So far the main contenders for the resolution of the problem are different effects related with the motion of gas in the central regions of galaxies. Valenzuela *et al.* (2006) argue that cold gas in the central regions of dwarf spheroidals moves slower than what would be naively expected. Additional support for the gas is provided by a combination of few factors: small-scale non-circular motions and pressure sup-

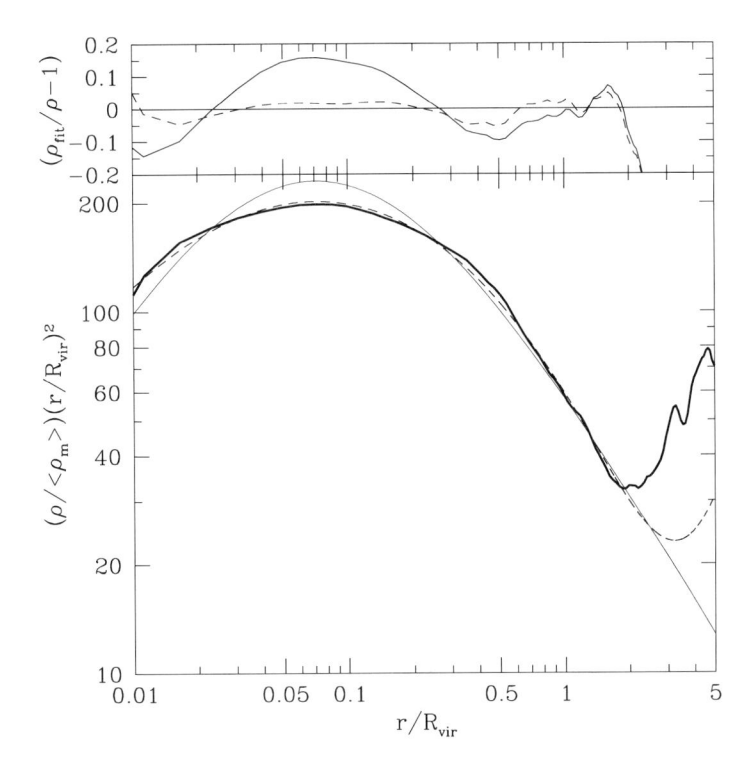

**Figure 7:** Comparison of the NFW and the Einasto fits. The thick full curve in the bottom panel shows the average density for halos with mass $<M> = 3 \times 10^{11} h^{-1} M_\odot$. The Einasto fit is the dashed curve. The NFW fit is presented by the thin full curve. The top panel shows the errors of the fits. The full curve is for the NFW approximation, and the dashed curve is for the Einasto fit. In outer regions $r = (0.3 - 2) R_{\rm vir}$ both fits have practically the same accuracy. Both fits start to fail at larger distances. Overall, the Einasto approximation provides remarkable accurate fit.

port due to low-level star formation (measured in these galaxies), and large-scale distortions in the gravitational potential produced either by weak stellar bars or non-spherical dark matter halos. Recently, Maschenko et al. (2007) blamed violent motions in the gas during early stages of galaxy formation for flattening the dark matter cusp. It is a possibility, though it is not clear how efficient is the process and how realistic are the processes lading to such violent gas motions.

Interestingly enough, it seems that dwarf spheroidal galaxies, which are typically smaller than dwarf irregulars, do not have much of the problem. Recent extensive measurements of random velocities of stars in many dwarf spheroidals in the Local Group are coming along with the theoretical expectations of cuspy dark matter profiles (Walker et al., 2007)

Figure 8 illustrates the overall dynamical structure of dark matter halos (see Cuesta et al., 2007 for details). All the curves were normalized to the virial values for each halo. So, we are looking to relative differences. The plots show that massive halos are less concentrated than small halos (bottom panels). The velocity dispersions (top panels) indicate that larger halos dynamically dominate their environment: rms velocities in the environment of clusters are much smaller than in their interiors. Dwarf halos are different in that respect: just outside their formal radii the velocities start to increase.

Results in this figure give us an important lesson: There is nothing very special about the virial radius. There are no discontinuities, there is no "truncation" or a boundary of a halo, and halo properties smoothly change when we cross the formal virial radius. Prado et al. (2006) and Cuesta et al. (2007) made a detailed analysis of outer regions of halos with a conclusion that halos of small mass ($\sim 5 \times 10^{12} h^{-1} M_\odot$ and below) may extend 2–3 times beyond their formal virial radius. They measure the average radial velocity of dark matter as a function of distance. Figure 9 shows the profiles for halos of different mass. The cluster-size halos behave as expected: just outside their virial radius there is a broad region where matter falls into the central halo as indicated by negative radial velocities. Dwarf halos gave a surprise: there is no infall region and on average the matter does not fall.

It should be reminded that the formal virial radius is not estimated using any obvious virial relations. One may think of using the virial ratio $2K/U - 1$, but it appears not very useful in practice. So, to large degree, th "virial radius" in cosmology is mostly convention. This also explains why there are many definitions of halo radii. Typically one uses some constant overdensity radius. Early simulations indicated that the radius of overdensity 178 is close to the virial radius. Thus, the radius of overdensity 200 ($\approx 178$) became the virial radius. The reason why this was a good approximation is simply coincidental: the early simulations were mostly done for cluster-size haloes and, indeed, for those masses the virial radius is close to the radius of overdensity 200. The early models were flat models without the cosmological constant. Models with the cosmological constant have produced significant confusion in the community. The top-hat model must be modified to incorporate the changes due to the different rate of expansion and due to the different rate of growth of perturbations. That path produced the so called virial radius, which for the standard cosmological model gives the radius of overdensity relative to matter of about 340. Still, a large group of cosmologists uses the old overdensity 200 relative to the critical density even for the models with the cosmological constant.

Cuesta et al. (2007) argue that the radius inside which the radial velocity is zero – the static radius – is a better approximation for the virial radius. In the sense of dynamics the static radius is the virial radius. Here the system is stationary (may not be relaxed) and stationary systems obey the virial theorem. The notion of the static radius should be important for dynamics of environment of isolated halo. For example, motion of satellites at 2–3 formal virial radii reflects the mass distribution of the central object. For other tests, such as abundance of clusters of galaxies at different redshifts, the distinction between true and convention virial radii does not seem to be an issue. There is only one condition: We need to follow the same definition in the theory and in observations.

There is another interesting result which follows from the analysis of out regions of halos. It appears that halos of small mass (such as our Milky Way) may not have

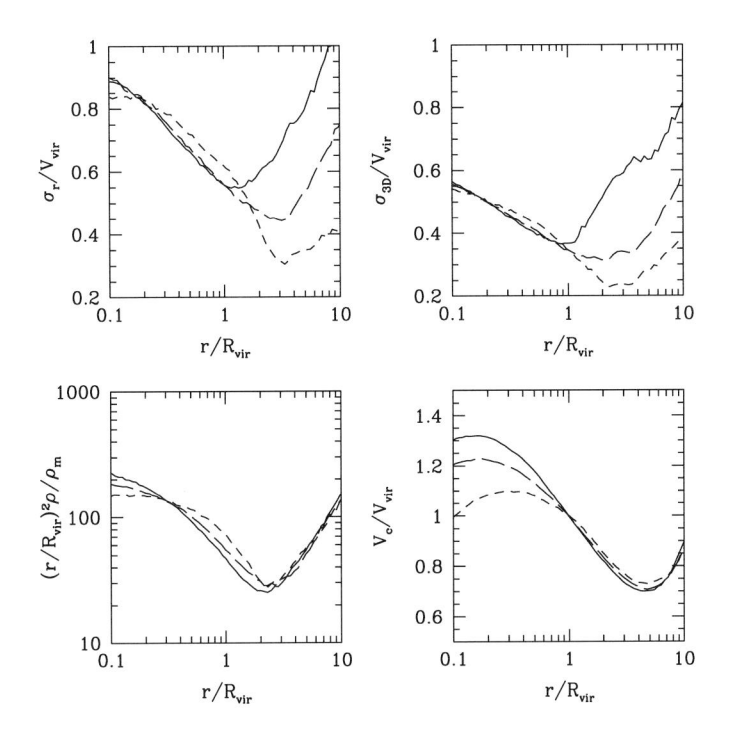

**Figure 8:** Profiles for halos with different mass. The profiles were obtained by averaging over hundreds of distinct haloes on each mass bin. Top left panel: radial velocity dispersion. Top right: 3D velocity dispersion. Bottom left: density profile. Bottom right: Circular velocity profile. Full curves are for dwarf with mass $\sim 10^{10} h^{-1} M_\odot$; long dash curves are for Milky Way-type halos with $\sim 10^{12} h^{-1} M_\odot$. Short dash curves are for cluster-size halos with $\sim 10^{14} h^{-1} M_\odot$.

changed much their density distribution since $z = 1\text{--}2$. Figure 10 shows the density profile of the halos in proper (physical) units. Within 400 kpc the density of the halos was nearly the same at $z = 1$ as it is now.

## 3 Galaxy formation

The classical picture of galaxy formation was outlined several decades ago. These early models showed the importance of gas-dynamical dissipative processes in galaxy formation (Rees & Ostriker, 1977; Silk 1977). Gas falls into dark matter halos and it is shock heated at the virial radius. Then the gas looses its energy through radiative cooling and collapses inside galaxies within dark matter halos (White & Rees 1978). The angular momentum is preserved in this collapse. Thus, cold gas settles in a thin galactic disk (Fall & Efstathiou 1980). Inside the disk, the gas achieves high densities and forms stars. Finally, these stars modify the surrounding

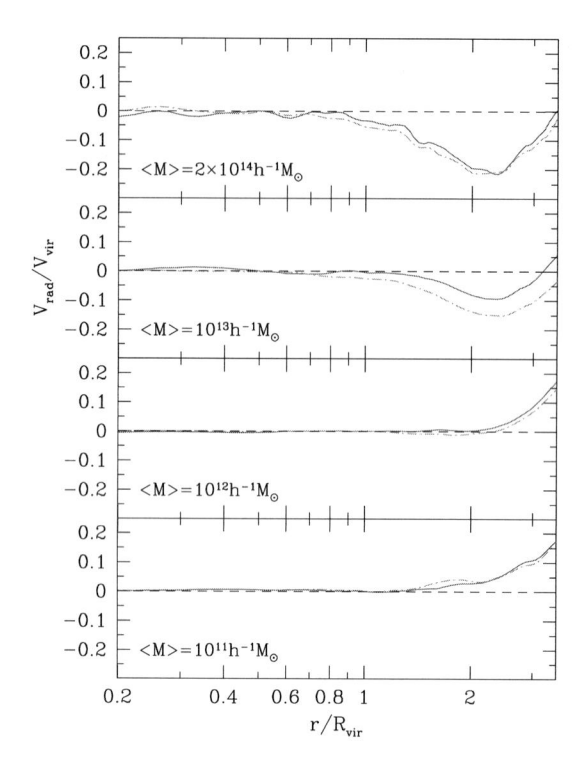

**Figure 9:** Average radial velocities for halos with different virial masses. The full curves show results for isolated halos and the dot-dashed curves are for distinct halos. The velocities are practically zero within $(2-3)R_{\rm vir}$ for halos with mass smaller than $10^{12}h^{-1}M_\odot$. The situation is different for group- and cluster- sized halos (two top panels). For these large halos there are large in- fall velocities, which amplitude increases with halo mass.

gas through different mechanisms of feedback, such as supernova explosions (Dekel & Silk 1986).

This classical picture shows that radiative cooling, star formation and stellar feedback are key processes in the assembly of baryons in galaxies. However, there is a complex interplay between these processes and large-scale cosmological processes. Two examples can illustrate this interconnection. For example, baryons can flow from cosmological distances to the center of dark matter halos along cold flows, which are not being shock heated at the virial radius (Birnboim & Dekel 2003; Keres et al. 2005; Dekel & Birnboim 2006). Another example is the role of stellar feedback in galaxy formation. Stellar feedback injects energy at parsecs scales but it results in galactic outflows, that reach very large scales (Ceverino & Klypin 2007; Springel & Hernquist 2003; Oppenheimer & Davé 2006). These outflows are routinely detected in high redshift star-forming galaxies (Law et al. 2007).

In short, the current picture of galaxy formation is more complicated than the classical paradigm outlined above. An interconnected set of processes with very

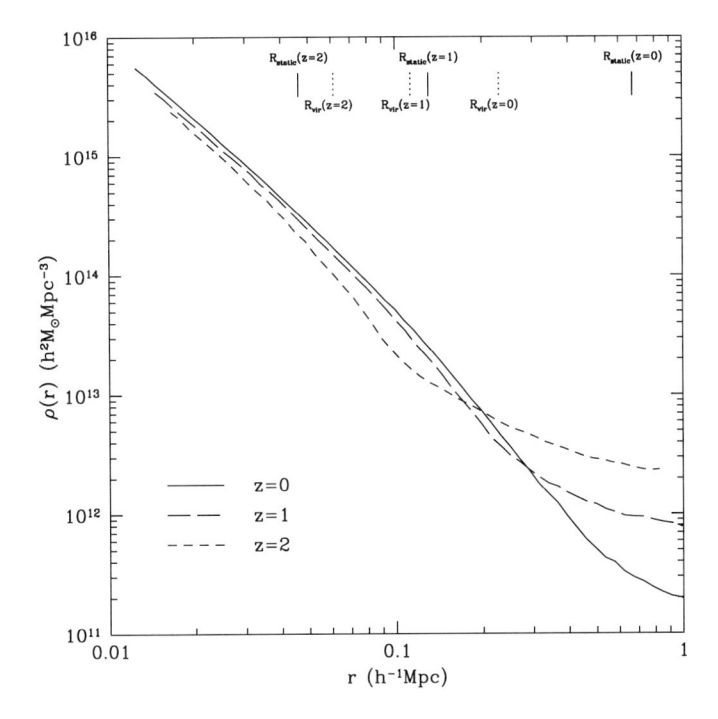

**Figure 10:** Evolution of dark matter density profile. The evolution of the density profile from $z = 2$ to $z = 0$ for galactic-size haloes. There has been very little change in this profile from $z = 1$ up to the present epoch. The vertical dotted lines marks the virial radii at $z = 2$ (left), $z = 1$ (middle) and $z = 0$ (right). Static radii are shown with vertical solid lines. The growth of the static radius in Milky-Way size haloes from $z = 1$ to $z = 0$ is not due to accretion but virialization of the surrounding regions of the halo.

different spatial scales control the galaxy assembly. Cosmological gasdynamical simulations have become ideal tools to study these wide variety of processes. However, current simulations face two main problems: the overcooling problem and the angular momentum catastrophe.

## 4 The overcooling and the angular momentum problems

Because the radiative cooling is very efficient in low-mass halos at high redshift, the overcooling problem is an issue in the hierarchical clustering scenario of galaxy formation (White & Rees 1978; White & Frenk 1991). If most of the gas inside these galactic halos cools, collapses, and forms stars, almost all baryons inside galactic halos should have been locked into stars well before the assembly of present-day galaxies. However, the fraction of baryons found in galaxies at low redshift is significant smaller than the universal baryonic fraction (Fukugita & Peebles, 2004). In addition, a large fraction of baryons (about 1/2), which is naively expected to be inside galaxies such as our Milky Way is not accounted for (Klypin et al., 2002; Conroy et al. 2007) – they are not in stars and not in the observed gas). It is clear that in order to produce realistic models, the theory should include mechanisms, which prevent catastrophic cooling of gas and its conversion into stars.

Early cosmological simulations of galaxy formation reported a catastrophic loss of angular momentum during the galaxy assembly, resulting in too small and concentrated galactic disks and very massive spheroids (Navarro & Benz 1991; Navarro, Frenk & White 1995; Navarro & Steinmetz 1997, 2000; Sommer-Larsen; Gelato & Vedel 1999). This angular momentum problem has a double origin. A part of the problem is related with the spatial resolution of the simulations. Sommer-Larsen et al. (2003), Governato et al. (2004) and Kaufmann et al. (2006) show that a sub-kpc resolution is necessary to minimize an artificial loss of angular momentum. Recent improvements in resolution have resulted in simulations with extended galactic disks (Governato et al 2004; Robertson et al. 2004; Brook et al. 2004; Okamoto et al 2005; D'Onghia et al. 2006; Governato et al 2007). However, a clear convergence in resolution has not been reached (Governato et al. 2008; Naab at al. 2007). Thus, simulations with a better resolution are still needed to check for any artificial loss of angular momentum.

Even if there is no numerical loss of angular momentum, the overcooling problem causes a loss of angular momentum during the accretion of protogalaxies into larger halos and during merger episodes. Clumps of too cold and dense gas lose all angular momentum due to dynamical friction during the infall or during a merger (Navarro & Benz 1991; D'Onghia et al. 2006).

Current state-of-the-art cosmological simulations still suffer from an angular momentum problem, although it is not as severe as it was before. Two main issues remain unsolved. First, current simulations are unable to produce bulgeless galaxies. Second, simulated galaxies does not have a rotation curve consistent with observations. They raise too fast at the center and tend to have a declining shape, inconsistent with the typical almost flat rotation curve of observed disk galaxies. The bottom line is that current simulations still have too much material with low angular momentum (van den Bosch et al. 2001).

A detailed discussion of these issues can be found in recent review paper by Mayer *et al.* (2008).

## 5 Stellar feedback and galaxy formation

It was quickly realized that stellar feedback can potentially solve the problems of galaxy formation described in the previous sections (White & Frenk 1991; Navarro & Benz 1991). However, the modeling of stellar feedback and its efficiency in cosmological simulations turned to be a difficult task. Cosmological simulations lack the necessary resolution to follow correctly the effect of supernova explosions in the ISM. Because of this lack of resolution, the modeling of stellar feedback has relied on ad-hoc assumptions about the effect of stellar feedback on scales unresolved by simulations. Early attempts to introduce stellar feedback into simulations found the obstacle of a strong radiative cooling. The energy deposited by supernova explosions was quickly radiated away without any effect in the ISM (Katz 1992). Several shortcuts have been proposed to get around this problem.

The most common method is to artificially stop cooling when the stellar energy is deposited (Gerritsen & Icke 1997; Thacker & Couchman 20000; Sommer-Larsen et al. 2003; Kereš et al. 2005; Governato et al. 2007) This approach prolongs the adiabatic phase of supernova explosion (the Sedov solution) to about 30 Myr. The motivation behind this assumption is that the combination of blastwaves from different supernova explosions and turbulent motions produces hot bubbles much larger than individual supernova remnants and last longer.

Another method is to introduce a sub-resolution model in which the energy from supernova explosions is stored in an unresolved hot phase, which does not cool and looses energy through the evaporation of cold clouds (Yepes et al. 1997; Springel & Hernquist 2003). In this model, the only effect of stellar feedback is to regulate the star formation: the hot gas is coupled with the cold phase through cloud formation and evaporation. As a result, this high entropy gas is artificially trapped within the galactic disk. Thus, galactic winds are introduced in a simplified way in order to reproduce other natural effects of stellar feedback, such as galactic outflows.

An alternative approach assumes kinetic feedback instead of thermal feedback (Navarro & White 1993). In that case, the energy from supernova explosions or stellar winds is transferred to the kinetic energy of the surrounding medium. This energy is not dissipated directly by radiative cooling. However, in order to resolve this effect accurately, simulations should be able to resolve the expansion of individual supernova explosions or the stellar winds from individual stars. Currently, this is not possible. At larger scales, the picture is more complicated. Different blastwaves from different supernova explosions can collide, dissipating their kinetic energy. The same dissipation of energy happens in collisions of stellar winds in stellar clusters. So, it is commonly assumed that the kinetic energy from stellar feedback is dissipated into thermal energy at the smallest scales resolved by simulations. Nevertheless, this feedback-heated gas can expand. As a result, thermal energy can be transferred to kinetic energy. The net result is formation of flows at large scales powered by the thermal feedback. However, feedback heating should dominates over radiative cooling: only in this case those flows are produced.

These improvements in the modeling of stellar feedback have produced simulations with smaller loses of angular momentum during mergers. The result is that a merger of two gas-rich galaxies can produce a disk galaxy, rather than a elliptical galaxy (Springel & Hernquist 2005; Cox et al. 2006; Governato et al. 2007). The orbital angular momentum of the merger is transfered to the spin of the remnant. Thus, a rotation-supported disk can be formed from the leftover gas after the merger.

## 5.1 A multiphase ISM

We develop a model of stellar feedback which minimizes ad hoc assumptions about stellar feedback (Ceverino & Klypin 2007). We start with developing high resolution simulations of the ISM and formulate conditions required for its realistic functionality: formation of multi-phase medium with hot chimneys, super-bubbles, cold molecular phase, and very slow consumption of gas. We find that this can be achieved only by doing what the real Universe does: formation of dense ($> 10$ atoms cm$^{-3}$), cold ($T \approx 100$ K) molecular phase, where the star formation happens, and which is disrupted by young stars. Another important ingredient is the runaway stars: massive binary stars ejected from molecular clouds when one of the companions becomes a supernova. Those stars can move to 10–100 parsecs away from molecular clouds before exploding themselves as supernovae. This greatly facilitates the feedback.

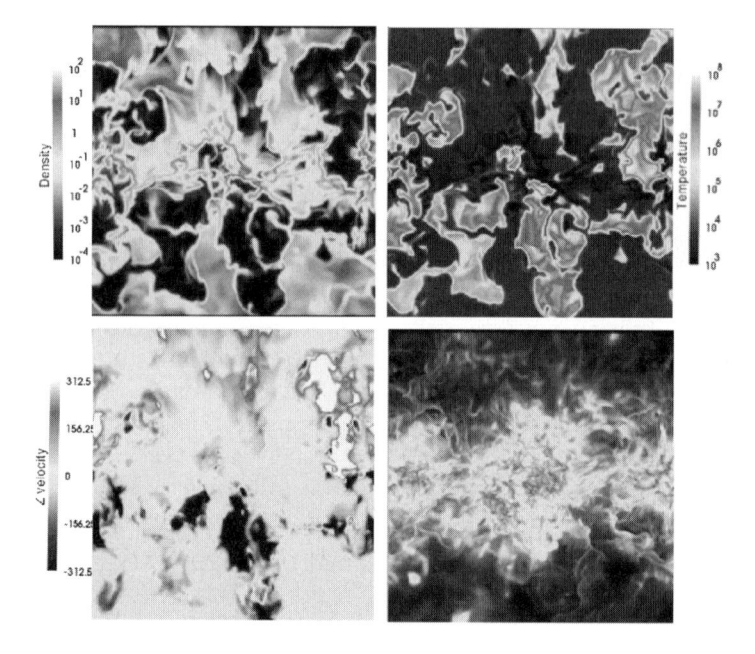

**Figure 11:** Slices perpendicular to a 4-Kpc piece of a galactic disk, showing the density in hydrogen atoms per cm$^3$ (top-left), temperature in Kelvin (top-right), gas velocity in the z-direction (bottom-left), and surface density in hydrogen atoms per cm$^2$ (bottom-right). Panels show an edge-on view perpendicular to that plane. The three phases of the ISM are clearly visible: cold and dense clouds, warm and diffuse medium and hot bubbles with very low densities. Velocities exceeding 300 km s$^{-1}$ can be seen in hot outflows at both sides of the galactic plane. This multi-phase ISM is driven by stellar feedback from young stellar clusters formed from cold clouds

A ISM-scale simulation of a $4 \times 4 \times 4$ Kpc$^3$ piece of a galactic disk with 8–16 pc resolution was used as a benchmark for the effect of stellar feedback at galactic scales. Figure 11 shows representative slices perpendicular to the simulated galactic disk. The medium is very inhomogeneous at different scales. Large bubbles of low density coexist with long filamentary structures of dense clouds. Overall, the medium covers more than 6 orders of magnitude in density and temperature. The cold phase forms dense and cold clouds near the galactic plane. The warm phase fills old cooled bubbles and low-density clouds. Finally, the hot phase is present in form of hot bubbles of few hundred pc wide and Kpc-scale chimneys. The gas in these chimneys is flowing away from the plane with velocities exceeding $\pm 300$ km s$^{-1}$. These bubbles even break the dense plane in hot spots surrounded by cold and dense shells. All this phenomenology associated with the hot phase is driven by stellar feedback. As a result, the effect of the stellar feedback is to sustain a three-phase ISM and produce galactic-scale outflows. These simulations tell us what are the necessary ingredients to reproduce the truly effect of stellar feedback at the resolution that we can afford in cosmological simulations of galaxy formation.

## 5.2 The result of a multi-phase ISM in cosmological simulations: Galactic outflows

After the ISM-scale models, we study the effect of stellar feedback in galaxy formation. We use the same feedback models in cosmological hydrodynamics simulations with a resolution of 35–70 pc. The simulation follows the formation of a MW-type galaxy starting from primordial density fluctuations in a 10 $h^{-1}$ Mpc comoving box. Our model develops naturally a multiphase ISM: molecular phase with temperatures close to 100 K and hot bubbles with temperatures above $10^6$ K.

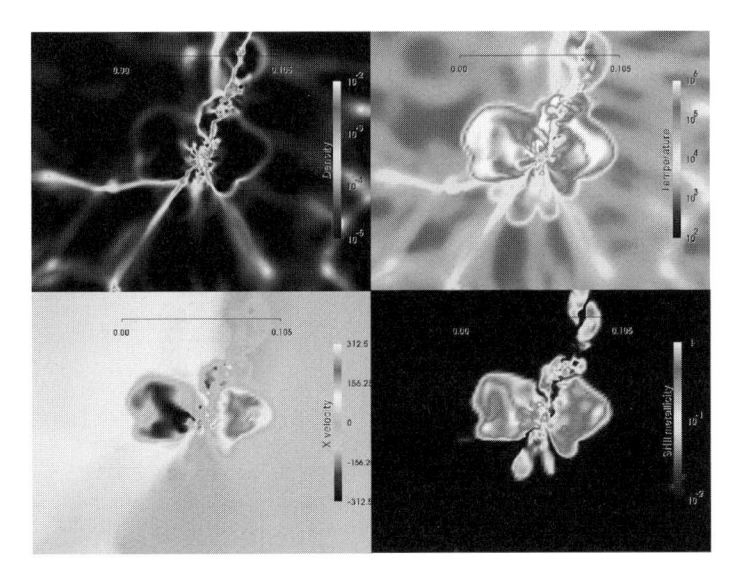

**Figure 12:** Protogalaxy in a cosmological simulation with a resolution of 45 pc at redshift 3.4. The slices of 600 Kpc on a side show the gas density (top left), temperature (top right), velocity in the horizontal direction (bottom left), and metallicity (bottom right). There are inflows of low-metallicity gas in cold filaments, as well as outflows of hot, metal-rich gas produced by chimneys in a multi-phase interstellar medium. Outflow velocities exceeds 300 km s$^{-1}$. The virial radius is 70 Kpc and the total virial mass is $10^{11}$ solar masses.

This model produces galactic chimneys that combine in a galactic wind. As a result, galactic winds are the natural outcome from stellar feedback. These galactic-scale outflows can be seen in Figure 12. It shows a slice of the simulation through the main MW-progenitor at redshift $z = 3.4$. At that redshift, its virial radius is 70 Kpc and the total virial mass is $10^{11}$ M$_\odot$. The gas density panel shows the galaxy embedded in a cosmological web of filaments. The galaxy at the center is blowing a galactic wind of hot and dilute gas with outflows velocities exceeding 300 km s$^{-1}$. The wind is rich in $\alpha$-elements and other products of the ejecta of core-collapse supernova. These metal-rich outflows can contribute to the enrichment of the halo and the inter-galactic medium. These outflows can reach even higher velocities and can escape the galactic halo and enrich the inter-galactic medium. The galactic wind is produced by the combination of different galactic chimneys anchored in the multi-phase ISM of the galaxy. The multi-phase nature of the ISM can be seen in a close view of the ISM of the proto-galaxy shown in Figure 13.

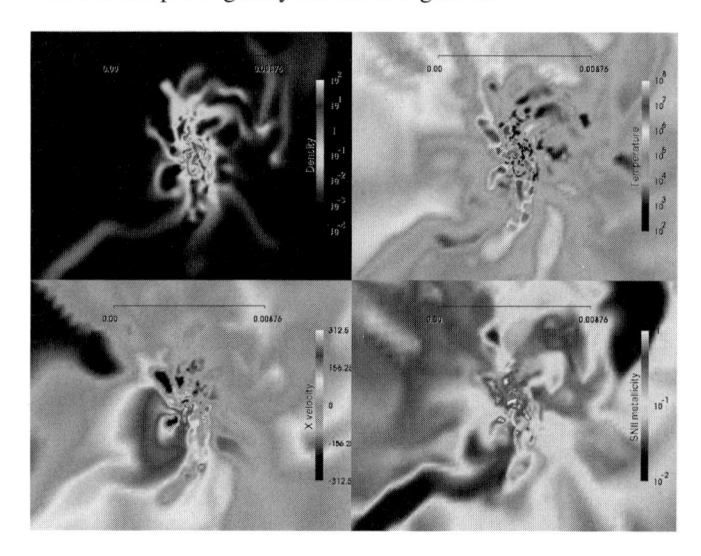

**Figure 13:** The same as in figure 2, but now the size of the images is 50 Kpc. It shows a multi-phase ISM of cold and dense clouds surrounded by bubbles of hot and dilute gas. Inflow and outflows velocities can reach 300 km s$^{-1}$. The outflows are galactic chimneys powered by core-collapse supernova. Therefore, they are rich in $\alpha$-elements. In contrast, the inflow of gas has almost primordial composition.

## 5.3 Overcooling model versus multi-phase model of galaxy formation

We can see now the effect of this multiphase medium in the galaxy assembly by comparing two cosmological simulations with the same spatial resolution but with different models of feedback. In the overcooling model, the effect of stellar feedback is very small because the local efficiency of star formation is small. This model

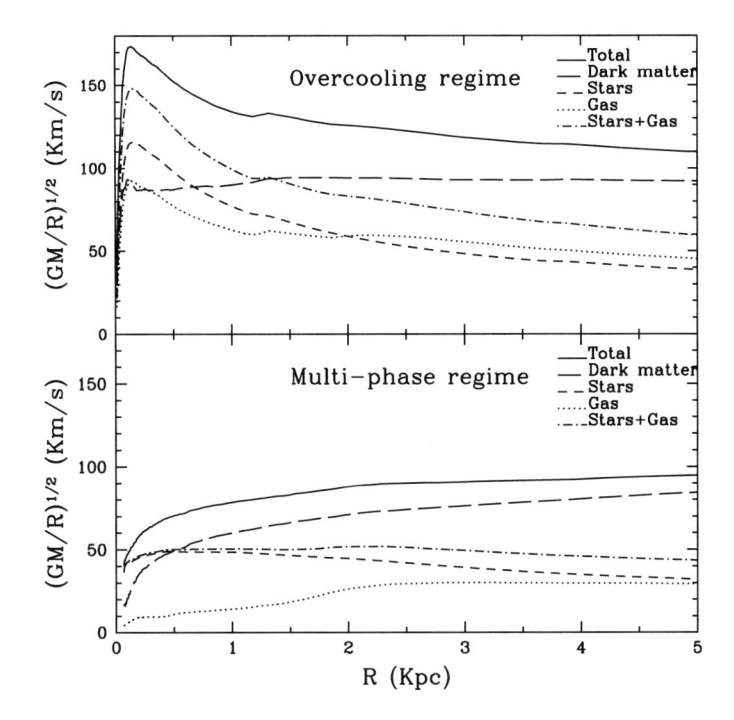

*Structure Formation in the Expanding Universe* 87

**Figure 14:** Circular velocity profile of the main progenitor of a MW-type galaxy at redshift 5. The top panel shows the results of an inefficient stellar feedback. The galaxy is too concentrated and has a too massive spheroidal component. By contrast, the bottom panel shows a regime in which stellar feedback is more efficient and it can regulate the growth of the galaxy. The resolution in both cases is 70 pc.

suffers from the overcooling problem and the ISM is close to a isothermal medium with a temperature close to $10^4$ K. In more realistic model discussed above the local star formation rate is higher. This results in much lower global star formation rate and in formation of healthy multiphase ISM. This higher efficiency of utilization of the stellar energy drastically affect the distribution of gas in the central region of the forming galaxy.

This can be tested by studying the profile of circular velocity as a proxy of the mass distribution, $V_c = \sqrt{GM/R}$, where $G$ is the gravitational constant and $M$ is the mass inside a radius $R$. Figure 14 shows the profile of the circular velocity for the same galaxy in the two cases. The simulation with the overcooling problem shows a strong peak in the baryonic component of the circular velocity. Both gas and stars are very concentrated in the first Kpc. In contrast, the simulation with multiphase medium has a more shallow circular velocity profile. This indicates a less concentrated galaxy. As a result, by solving the overcooling problem, the angular momentum problem is greatly minimized.

# References

Aarseth, S. J. 1963, MNRAS, 126, 223

Bardeen, J. M., Bond, J. R., Efstathiou, G., 1987, ApJ, 321, 28

Barnes, J., & Efstathiou, G. 1987, ApJ, 319, 575

Benson, A. J., Frenk, C. S., Baugh, C. M., Cole, S., & Lacey, C. G. 2003, MNRAS, 343, 679

Bett, P., Eke, V., Frenk, C. S., Jenkins, A., Helly, J., & Navarro, J. 2007, MNRAS, 376, 215

Bond, J. R., et al. 2000, arXiv:astro-ph/0011378

Birnboim, Y., & Dekel, A. 2003, MNRAS, 345, 349

Brook, C. B., Kawata, D., Gibson, B. K., & Freeman, K. C. 2004, ApJ, 612, 894

Bullock, J. S., *et al.*, 2001, MNRAS, 321, 559

Ceverino, D., Klypin, A. ArXiv e-prints, 712, arXiv:0712.3285

Cohn, J. D., & White, M. 2007, ArXiv e-prints, 706, arXiv:0706.0208

Cole, S., *et al.*, 2005, MNRAS, 362, 505

Crocce, M., & Scoccimarro, R. 2007, ArXiv e-prints, 704, arXiv:0704.2783

Conroy, C., et al., 2007, ApJ, 654, 153

Cox, T. J., Jonsson, P., Primack, J. R., & Somerville, R. S. 2006, MNRAS, 373, 1013

de Bernardis, P., *et al.,* 2000, Nature, 404, 955

de Blok, W. J. G., McGaugh, S. S., Rubin, V. C., 2001, AJ, 122, 2396

Dekel A., Silk J., ApJ, 303, 39

Dekel, A., & Birnboim, Y. 2006, MNRAS, 368, 2

Diemand, J., Kuhlen, M., & Madau, P. 2007, ApJ, 657, 262

D'Onghia, E., Burkert, A., Murante, G., & Khochfar, S. 2006, MNRAS, 372, 1525

D'Onghia, E., & Navarro, J. F. 2007, MNRAS, 380, L58

Eisenstein, D. J., & Hu, W. 1999, ApJ, 511, 5

Fall S.M., Efstathiou G., 1980, MNRAS, 193, 189

Flores, R. A., & Primack, J. R., 1994, ApJ, 427, L1

Frenk, C. S., et al. 1999, ApJ, 525, 554

Fukugita, M., & Peebles, P. J. E. 2004, ApJ, 616, 643

Gao, L., Navarro, J. F., Cole, S., Frenk, C., White, S. D. M., Springel, V., Jenkins, A., & Neto, A. F. 2007, ArXiv e-prints, 711, arXiv:0711.0746

Gerritsen, J. P. E., & Icke, V. 1997, AAP, 325, 972

Governato, F., *et al.*, 2004, ApJ, 607, 688

Governato, F. *et al.* 2007, MNRAS, 374, 1479

Governato, F., Mayer, L., & Brook, C. 2008, ArXiv e-prints, 801, arXiv:0801.1707

Gunn, J. E., & Gott, J. R. I. 1972, ApJ, 176, 1

Hahn, O., Porciani, C., Carollo, C. M., & Dekel, A. 2007, MNRAS, 375, 489

Hayashi, E., et al. 2004, MNRAS, 355, 794

Hayashi, E., & Navarro, J. F., 2006, MNRAS, 373, 1117

Hetznecker, H., & Burkert, A. 2006, MNRAS, 370, 1905

Heitmann, K., Ricker, P. M., Warren, M. S., & Habib, S. 2005, ApJS, 160, 28

Holtzman, J., 1989, ApJS, 71, 1

Jenkins, A., *et al.*, 2001, MNRAS, 321, 372

Katz, N. 1992, ApJ, 391, 502

Kauffmann G., Guiderdoni B., White S. D. M., 1994, MNRAS, 267, 981

Kaufmann, T., Mayer, L., Wadsley, J., Stadel, J., & Moore, B. 2006, MNRAS, 370, 16

Kereš, D., Katz, N., Weinberg, D. H., & Davé, R. 2005, MNRAS, 363, 2

Klypin, A., Hoffman, Y., Kravtsov, A. V., & Gottlöber, S. 2003, ApJ, 596, 19

Klypin, A., Kravtsov, A. V., Valenzuela, O., & Prada, F. 1999, ApJ, 522, 82

Klypin, A., Zhao, H., & Somerville, R. S. 2002, ApJ, 573, 597

Kravtsov, A. V., Klypin, A. A., & Khokhlov, A. M. 1997, ApJS, 111, 73

Law, D. R. *et al.* 2007, ApJ, 669, 929

Lee, J., & Shandarin, S. F. 1998, ApJ, 500, 14

Lifshitz, E.M., Khalatnikov, I.M., 1963, Adv. Phys., 12, 465

Lukić, Z., Heitmann, K., Habib, S., Bashinsky, S., & Ricker, P. M. 2007, ApJ, 671, 1160

Mashchenko, S., Wadsley, J., & Couchman, H. M. P. 2007, ArXiv e-prints, 711, arXiv:0711.4803

Mayer, L., Governato, F., & Kaufmann, T., 2008, Advanced Science Letters, in press.

Moore, B., 1994, Nature, 370, 629

Moore, B., Ghigna, S., Governato, F., Lake, G., Quinn, T., Stadel, J., & Tozzi, P. 1999, ApJ, 524, L19

Naab, T., Johansson, P. H., Ostriker, J. P., & Efstathiou, G. 2007, ApJ, 658, 710

Navarro, J. F., & Benz, W. 1991, ApJ, 380, 320

Navarro, J. F., & White, S. D. M. 1993, MNRAS, 265, 271

Navarro, J. F., Frenk, C. S., & White, S. D. M. 1995, MNRAS, 275, 56

Navarro, J. F., Frenk, C. S., & White, S. D. M. 1997, ApJ, 490, 493

Navarro, J. F., & Steinmetz, M. 1997, ApJ, 478, 13

Navarro, J. F., & Steinmetz, M. 2000, ApJ, 538, 477

Navarro, J. F., & White, S. D. M. 1993, MNRAS, 265, 271

Navarro, J. F., et al. 2004, MNRAS, 349, 1039

Netterfield, C. B., *et al.*, 2002, ApJ, 571, 604

Neto, A. F., et al. 2007, MNRAS, 381, 1450

O'Shea, B. W. *et al.* 2004, arXiv:astro-ph/0403044

Okamoto, T., Eke, V. R., Frenk, C. S., & Jenkins, A. 2005, MNRAS, 363, 1299

Oppenheimer, B. D., & Davé, R. 2006, MNRAS, 373, 1265

Peacock, J. A., & Dodds, S. J. 1996, MNRAS, 280, L19

Peebles, P. J. E. 1970, AJ, 75, 13

Prada, F., it et al., 2006, ApJ, 645, 1001

Reed, D., Gardner, J., Quinn, T., Stadel, J., Fardal, M., Lake, G., & Governato, F. 2003, MNRAS, 346, 565

Rees M.J., Ostriker J.P., 1977, MNRAS, 179, 541

Robertson, B., Yoshida, N., Springel, V., & Hernquist, L. 2004, ApJ, 606, 32

Sheth, R. K. & Tormen, G., 1999, MNRAS, 308, 119

Silk J., 1977, ApJ, 211, 638

Simon, J. D., Bolatto, A. D., Leroy, A., Blitz, L., & Gates, E. L. 2005, ApJ, 621, 757

Sommer-Larsen, J., Gelato, S., & Vedel, H. 1999, ApJ, 519, 501

Sommer-Larsen, J., Götz, M., & Portinari, L. 2003, ApJ, 596, 47

Somerville, R. S., & Primack, J. R. 1999, MNRAS, 310, 1087

Spergel, D. N., et al. 2007, ApJS, 170, 377

Springel, V., & Hernquist, L. 2003, MNRAS, 339, 289

Springel, V., & Hernquist, L. 2005, ApJL, 622, L9

Springel, V., Yoshida, N., & White, S. D. M. 2001, New Astronomy, 6, 79

Tegmark, M., *et al.*, 2004, ApJ, 606, 702

Thacker, R. J., & Couchman, H. M. P. 2000, ApJ, 545, 728

Tinker, J. *et al.*, 2008a, in preparation

Tinker, J. *et al.*, 2008b, in preparation

van den Bosch, F. C., Burkert, A., & Swaters, R. A. 2001, MNRAS, 326, 1205

Valenzuela, O., Rhee, G., Klypin, A., Governato, F., Stinson, G., Quinn, T., & Wadsley, J., 2007, ApJ, 657, 773

Vitvitska, M., Klypin, A. A., Kravtsov, A. V., Wechsler, R. H., Primack, J. R., & Bullock, J. S. 2002, ApJ, 581, 799

Wadsley, J. W., Stadel, J., & Quinn, T. 2004, New Astronomy, 9, 137

Walker, M. G., *et al.* 2007, ApJ, 667, L53

Warren, M. S., Abazajian, K., Holz, D. E., & Teodoro, L. 2006, ApJ, 646, 881

Wechsler, R. H., Zentner, A. R., Bullock, J. S., Kravtsov, A. V., & Allgood, B. 2006, ApJ, 652, 71

White, S. D. M. 1976, MNRAS, 177, 717

White, S. D. M., & Rees, M. J. 1978, MNRAS, 183, 341

White, S. D. M., & Frenk, C. S. 1991, ApJ, 379, 52

White S.D.M, Rees M.J., 1978, MNRAS, 183, 341

Yepes, G., Kates, R., Khokhlov, A., & Klypin, A. 1997, MNRAS, 284, 235

# From COBE to Planck

Matthias Bartelmann

Zentrum für Astronomie der Universität Heidelberg
Institut für Theoretische Astrophysik, Albert-Überle-Str. 2, 69120 Heidelberg
mbartelmann@ita.uni-heidelberg.de, http://www.ita.uni-heidelberg.de/~msb/

### Abstract

*The results of the WMAP satellite mark the as yet most spectacular success of CMB observations and their comparison with theoretical models. They are forming a cornerstone of the cosmological standard model which can now incorporate virtually all cosmological observations. After a brief historical review of CMB theory and observations, I summarise what can be expected from the next CMB satellite, Planck. Among the most important improvements will be more precise cosmological parameters, better control of systematic effects, detailed polarisation measurements, numerous astrophysical foreground detections and possible signatures of inflationary or new physics.*

## 1 The light elements and the CMB

The history of CMB research may begin with Gamow's search for an origin of the light elements in the Universe. We could begin earlier with Lemaître's keen speculation about the origin of the Universe in a fireball, but the question where the light elements might have come from led Gamow, Alpher and Herman to the first prediction not only of the CMB, but even of its temperature. The argument is ingeniously simple. How can it be that $\sim 25\%$ of the cosmic gas, by mass, is $^4$He rather than hydrogen? Why so much, and why not more? It can quickly be estimated that stellar fusion is insufficient to produce that amount of $^4$He, even ignoring the problem how the $^4$He produced in stars could have escaped from them afterwards. Gamow realised that the problem could be solved assuming that the entire early Universe may have served as a fusion reactor. If so, it must have gone through a phase comparably hot to the interior of stars. If so, and if thermodynamic equilibrium was maintained, there must have been a thermal radiation background which should still exist, cooled by the expansion of the Universe [12, 13][1].

But Gamow, Alpher and Herman went further. It is unlikely to produce $^4$He from protons and neutrons directly because that would require four particles to meet.

---

[1] It is virtually impossible to cite all papers relevant for CMB physics. The references given here should be seen as examples for pioneering work or review articles.

Much more likely is the fusion from deuterium through two-body encounters. But then, it is remarkable that neither 0 nor 100% of the hydrogen was fused to $^4$He, beacuse it implies that deuterium had to be produced in just the right quantity. Had there been much more, all hydrogen would have been turned into helium, and none had there been much less.

Conditions favourable for $^4$He production end a few minutes after the Big Bang. Then, radiation dominated all other forms of energy and thus controlled the cosmic expansion rate. High radiation density implies fast expansion and fast cooling, leaving little time for fusion. Knowledge of the fusion cross section for deuterium allowed calculations how long fusion could have lasted to produce the "just right" amount of deuterium and $^4$He, and thus estimates of the present radiation temperature assuming a blackbody spectrum. Gamov found $\sim 5$ K. The search for the origin of $^4$He thus led to two predictions: there should be a radiation background left over from the hot early phase of the Universe, and its present temperature should be a few Kelvin [1].

Thus predicted in 1949, it was serendipitously found by Penzias and Wilson in 1964–65 calibrating a horn antenna for the AT&T-Bell company. They published their result as a Letter to the Astrophysical Journal [36], in which they referred to "a possible explanation for the observed excess noise temperature" given by Dicke, Peebles, Roll and Wilkinson in the same issue of the ApJ Letters [8]. Penzias and Wilson received the Nobel Prize for their discovery.

## 2  Structures in the CMB

Five years later, in 1970, two papers appeared predicting structures in the CMB [35, 47]. Since the universe is not completely homogeneous, structures are likely to have formed by gravitational collapse from small fluctuations seeded in the early Universe, and should thus have left their imprint on the CMB. In linear theory, structures grow approximately by the same amount as the Universe expands. Since the Universe expanded by a factor of $\sim 1000$ since the CMB was released, and linear structures have grown to a density contrast near unity by today, temperature fluctuations at the milli-K level should be seen in the CMB. The pioneering papers in 1970 by Peebles & Yu and by Sunyaev & Zel'dovich explained for the first time how matter density fluctuations should have produced temperature fluctuations in the CMB.

Three effects are mainly responsible (Fig. 1, left panel). First, density fluctuations cause fluctuations of the gravitational potential, giving rise to a gravitational red- or blueshift. Photons escaping from hills or valleys in the potential landscape gain or lose energy, afterwards appearing slightly warmer or cooler than the mean. This is the Sachs-Wolfe effect [39]. Driven by potential fluctuations, it only appears on the largest scales because the potential is much smoother than the density.

Second, as the excess gravity tends to contract matter overdensities, the pressure of the enclosed gas and radiation counter-acts. This gives rise to acoustic oscillations of the primordial plasma as long as two conditions are satisfied: the density fluctuations must be small enough for sound waves to propagate through them, and the photons must be tightly coupled to the gas. Both conditions set an upper limit to

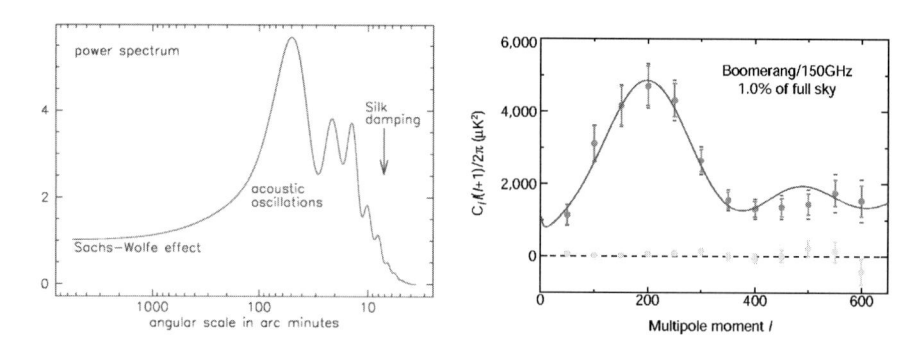

**Figure 1:** *Left panel*: Schematic drawing of the CMB power spectrum illustrating the three main effects characterising its shape. *Right panel*: Power spectrum measured by the Boomerang-1 balloon experiment.

the size of fluctuations that could gave undergone acoustic oscillations. It is straightforward to calculate that the Universe was approximately 400,000 years old when it had cooled down sufficiently to release the CMB. Due to the high photon density in the primordial plasma, the sound speed was $\sim c/\sqrt{3}$, almost 60% of the speed of light. Thus, acoustic oscillations could only have occurred on scales smaller than $\sim 230000$ light years, or $\sim 70$ kpc, the so-called sound horizon.

Third, two damping effects occur on the smallest scales. The mean free path of the photons was small, but finite, and grew rapidly as the primordial plasma recombined. Structures smaller than the mean free path were wiped out by the freely-streaming photons. This is called Silk damping [43]. In addition, recombination was not instantaneous, but lasted $\sim 40,000$ years. Observing the CMB, we thus look into a shell of finite width rather than at a surface. Structures smaller than the width of that shell are averaged over and appear suppressed.

Structures in the CMB are conveniently quantified by their *rms* amplitude as a function of angular scale, i.e. by their angular power spectrum. Since it has to be measured on the sphere, spherical-harmonic transforms must be used instead of Fourier transforms. The three effects just described create the characteristic shape of the expected CMB power spectrum: a flat tail caused by the Sachs-Wolfe effect at the largest scales, the onset of the acoustic oscillations on scales below the sound horizon, and the Silk damping causing a rapid decrease towards the smallest scales. Qualitatively, power spectra like this appear in the pioneering papers by Peebles & Yu and Sunyaev & Zel'dovich. They were quantitatively refined by Bond & Efstathiou, Hu & Sugiyama, and Seljak & Zaldarriaga, to name some of the most important few [4, 5, 19, 42].

## 3 COBE

The search for fluctuations in the CMB led cosmology into a crisis because fluctuations at the milli-K level were not found when experiments reached the required sensitivity level. After subtracting foreground signals, the CMB remained perfectly

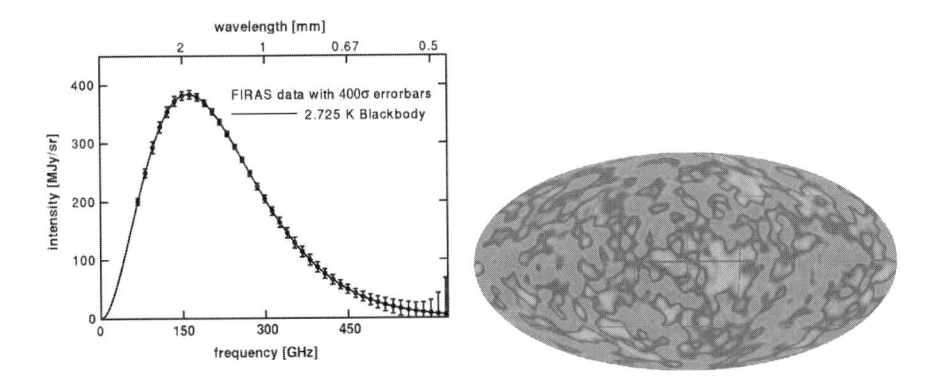

**Figure 2:** *Left panel:* CMB spectrum obtained by the COBE-FIRAS instrument, confirming its perfect Planckian shape. *Right panel:* CMB temperature fluctuation map obtained by COBE after four years of operation.

isotropic (see [52] for an example). The paradigm of cosmological structure formation by gravitational collapse, and with it the cosmological model as a whole, was shaking.

A convincing, if speculative, way out was proposed by Peebles [34]. Suppose matter in the Universe is dominated not by the ordinary baryons, but by some form of dark matter not interacting electromagnetically. Then, this form of matter could not have caused CMB temperature fluctuations directly, but only indirectly through its gravitational effect on the electromagnetically-interacting matter. CMB temperature fluctuations at the micro- rather than the milli-K level could then be reconciled with the present-day cosmic structures. This is still by far the most powerful argument for a form of "dark matter" incapable of interacting with light. Peebles suggested that this postulated dark matter should be cold because otherwise the formation of small cosmic structures was hard or impossible to understand. The subsequent simulations of structure formation in a universe dominated by cold dark matter, CDM, by Davis, Efstathiou, Frenk & White quickly showed the virtues of this new paradigm [7, 11, 54, 55].

The COBE satellite achieved two ground-breaking results. It had two instruments, the far infrared spectrometer FIRAS and the differential microwave radiometer DMR. Within a very short time after launch, FIRAS showed that the electromagnetic spectrum of the CMB had a perfect black-body shape measured to a precision never reached in the laboratory (Fig. 2, left panel). A temperature of $2.725$ K could be derived from it [30]. This confirmed the origin of the CMB from thermal equilibrium in the hot early Universe.

The DMR instrument was designed to detect the temperature fluctuations at the micro-K level compatible with the idea of cosmological structure growth through gravitational collapse under the assumption that the dominant form of matter is cold and dark. It ended the profound cosmological crisis when it found the fluctuations at the predicted level [44] (Fig. 2, right panel). COBE's angular resolution was too poor to see more than the far Sachs-Wolfe tail of the CMB power spectrum, but it was enough to convincingly support the concept of a universe filled with CDM undergoing gravitational collapse. These two discoveries were awarded by the Nobel Prize for John Mather and George Smoot, the scientists responsible for FIRAS and DMR, respectively.

## 4 From COBE to WMAP

In the years following COBE's discoveries, CMB research focussed on measuring the CMB temperature fluctuations at higher angular resolution to verify the onset of the acoustic oscillations. Microwave observations must be carried out in dry and cold environments to reach the required sensitivity and to avoid the absorption by water vapour in the atmosphere. Of the many experiments carried out, perhaps the most important were the Boomerang and Maxima balloon experiments, and the measurements with the ground-based Dasi interferometer.

Boomerang and Maxima had bolometer detectors carried into the stratosphere by balloons. Both produced maps of small sections of the sky at an angular resolution much better than COBE's. These maps demonstrated two things. First, they confirmed COBE's measurements in the common regions of the sky, and second, they revealed the existence of at least the first and possibly the second acoustic peak in the CMB power spectrum (Fig. 1, right panel). As described above, acoustic oscillations are confined to scales smaller than the sound horizon. The angular scale of the first acoustic peak thus reflects the angular size of the sound horizon. How large a fixed physical scale at a known distance appears on the observer's sky depends on the curvature of space. It could be immediately inferred from the location of the first peak that space must be flat or nearly flat. This implies that the total energy density of the Universe must equal the critical value allowing flat space, which is considerably more than the amount of matter in the Universe inferred from other observations. Even allowing for dark matter, most of the known cosmic energy density must thus be in some entirely unknown form of "dark energy" [14, 20, 38].

The release of the CMB was controlled by Thomson scattering. It happened when the photon energy had dropped to $0.3\,\mathrm{meV}$, so much below the rest mass of the electron that Compton scattering can very safely be reduced to its low-energy limit. The anisotropy of Thomson scattering must have created linear polarisation at the level of $\sim 10\%$ from the quadrupole anisotropy of the CMB at the time of recombination. Thus, the detection of this degree of linear polarisation, with fluctuations correlated with those of the temperature, was an important, and perhaps the last, step in the experimental confirmation of the origin of the CMB in a hot primordial plasma perturbed by mild gravity fluctuations. It was achieved with the DASI interferometer [23, 27].

## 5 WMAP

In many respects, the Wilkinson Microwave Anisotropy Probe [2], or WMAP, is the successor of COBE as a satellite experiment (Fig. 3, left panel). WMAP is a purely

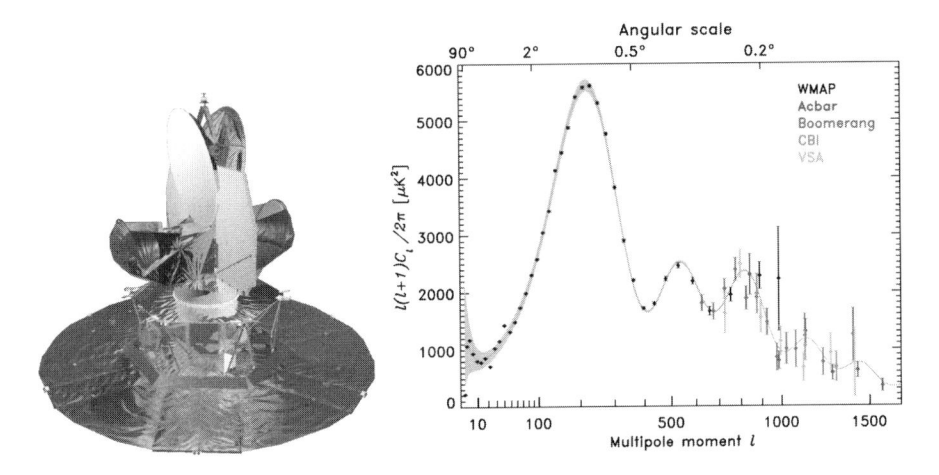

**Figure 3:** *Left panel*: The WMAP satellite. *Right panel*: Power spectrum of the CMB temperature fluctuations obtained from the 3-year data release of WMAP, and from other experiments on small angular scales, as marked in the figure.

differential radiometer measuring microwave radiation in five frequency bands at 22, 30, 40, 60, and 94 GHz, or between $\sim 7\,\mathrm{mm}$ and $\sim 3\,\mathrm{cm}$ wavelength. WMAP has three main advantages compared to the preceding balloon-borne and ground-based experiments. It covers the full sky and thus reduces error bars considerably, it reaches higher sensitivity and better angular resolution, and it is capable of measuring polarisation. Unlike COBE, it has no spectrometer on-board and focuses exclusively on the measurement of CMB intensity fluctuations.

Perhaps it is fair to say that the first-year data release of WMAP did not go much beyond what had already been inferred from the preceding experiments, albeit with much lower accuracy [17, 22, 33, 46]. The first and also the second acoustic peak in the CMB were now clearly confirmed at the locations and amplitudes found especially by the balloon-borne experiments (Fig. 3, right panel). Fitting theoretical models to the CMB temperature power spectrum, cosmological parameters could be derived with unprecedented accuracy. The location of the first acoustic peak confirmed the spatial flatness of the universe, and the depth of the first minimum together with the heights of the first and the second acoustic peaks constrained the baryon density and the ratio of the dark and baryonic matter densities. Matter was found to contribute $\sim 25\%$ to the critical density required for spatial flatness, and $\sim 16\%$ of the matter was found to be baryonic. Ascribing the rest of the cosmic energy density to a cosmological constant allows a determination of the Hubble constant, confirming the result of the HST Key Project [10].

A potential problem was the so-called optical depth. We know e.g. from the Gunn-Peterson effect in QSO spectra that the Universe was re-ionised by the first energetic light sources some time after recombination. Thereafter, the CMB photons had to travel through much diluted, but ionised material. Its optical depth due to Thomson scattering damped the CMB fluctuations to a degree which is degenerate with the primordial fluctuation amplitude, thus the optical depth due to reionisation cannot be inferred from the CMB temperature fluctuations alone. However, just like at recombination, Thomson scattering added secondary polarisation to the CMB at angular scales reflecting the size of the cosmic horizon at the time of reionisation. In that way, the time of recombination can be recovered from the polarisation signal. In turn, this allows a determination of the optical depth after reionisation, which then allows breaking the degeneracy between the primordial fluctuation amplitude and the optical depth. The redshift of reionisation, inferred to fall between 12 and 17, was unexpectedly high and at odds with constraints from QSO spectra.

This has changed with the third-year release of the WMAP data [16, 32, 45]. Improved measurements of the cross-correlation between temperature and polarisation, and of the polarisation auto-correlation itself, allowed better constraints on the epoch of reionisation and thus also of the amplitude of primordial matter fluctuations. The reionisation redshift was lowered to $\sim 10$, and the fluctuation amplitude, expressed by the conventional $\sigma_8$ parameter, was found to be as low as $\sigma_8 = 0.74$ with a relative error of 6%.

A very important finding from the 3rd-year data of WMAP was that the large-scale behaviour of the density-fluctuation power spectrum turned out to be exactly as expected from the inflationary theory of structure formation. Inflation asserts that cosmic structures originated from quantum fluctuations prior to inflation which were frozen in when driven out of the horizon in the course of inflation. Inflation predicts that the power spectrum of density fluctuations should be very close to a power law $\propto k^n$ in wave number $k$, with a power-law exponent, the so-called spectral index, of $n \approx 1$. Eternal inflation would have produced at strictly scale-invariant spectrum with $n = 1$, but the finite duration of inflation must have caused a slight deviation from that, $n \lesssim 1$. It is an important if indirect support for the inflationary paradigm that WMAP has found $n = 0.951$, more than $3\sigma$ below unity.

The success of WMAP is impressive. Cosmological parameters are now known to a degree of accuracy which seemed out of reach even a decade ago, in particular when CMB data are combined with other cosmological probes, in particular galaxy clustering, gravitational lensing or supernovae of type Ia. However, some doubts remain. One of the key issues is that the polarisation measurement with WMAP has revealed a surprisingly high degree of polarisation in the foreground radiation components emitted in particular by dust in the Milky Way. Since the polarisation signal is an order of magnitude fainter than the temperature signal, accurate knowledge and precise subtraction of the polarised foreground is indispensable for further reliable conclusions. Thus, it remains to be seen whether the determination of the reionisation epoch, and the determinations of the optical depth due to reionisation and the fluctuation amplitude hinging on it, will persist as the data improve. Another potential remaining problem is the fact that the multipole of order 5 appears to be aligned with the Galactic plane. While this is of course statistically possible, it is unlikely, and thus raises further questions regarding the reliability of the foreground subtraction.

*From COBE to Planck*  99

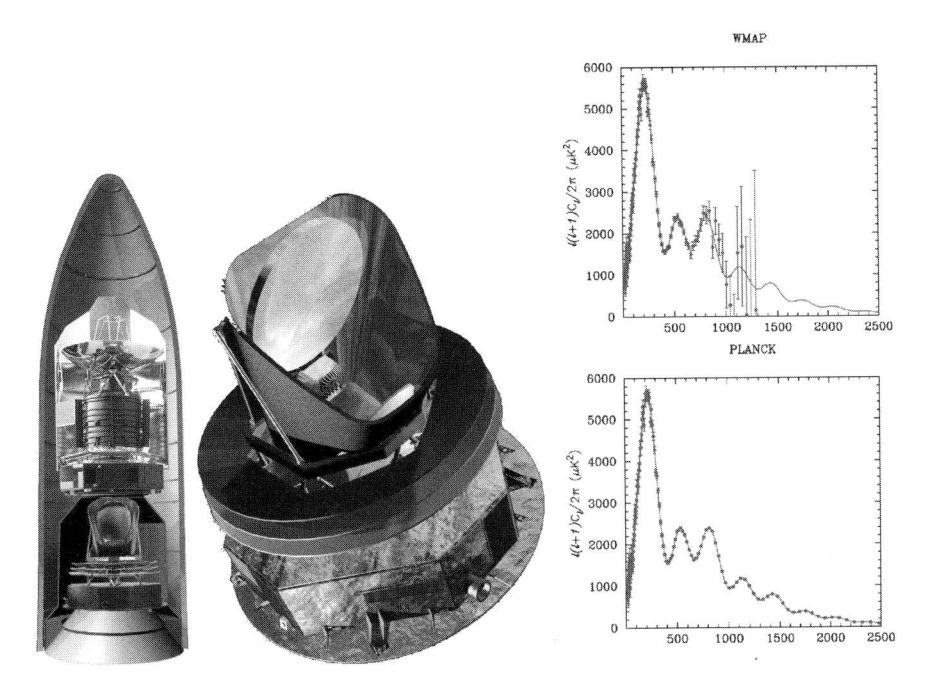

**Figure 4:** *Left panel*: Planck and Herschel in the cargo bay of an Ariane-5 rocket. *Middle panel*: Sketch of the Planck satellite, showing the telescope mirror in its shielding on top of the spacecraft. *Right panel*: Expected improvement on the CMB temperature-fluctuation power spectrum from WMAP (top) to Planck (bottom).

## 6 Planck

The ESA satellite Planck [3, 49, 50] is scheduled for launch by an Ariane-5 rocket (Fig. 4, left and middle panels) in August, 2008. It will improve the sensitivity of WMAP by approximately an order of magnitude, and its angular resolution by about a factor of three. Besides, it will cover the frequency range between 30 and 857 GHz with nine channels at 30, 44, 70, 100, 143, 217, 353, 545, and 857 GHz. This is possible because of two extremely sensitive instruments, the low-frequency instrument LFI [29, 41, 53] operating between 30 and 70 GHz, and the high-frequency instrument HFI [24–26] operating between 100 and 857 GHz (Fig. 5). The LFI uses high electron-mobility transistors (HEMTs) for directly measuring amplitudes and phases of the incoming waves. The HFI consists of bolometers which need to be cooled to 0.1 K and part of which are polarisation-sensitive due to their feed horns. In many aspects, the Planck instrumentation is pioneering. Part of the detectors reach limits set by quantum physics, and detectors cooled to 0.1 K have never been operated in space before. Tests carried out so far have shown that the Planck instruments operate at least to specification, and mostly better.

Like WMAP, Planck will not orbit the Earth, but the outer Lagrangian point (L2) of the Sun-Earth system. There, 1.5 million kilometres from Earth, the satellite can always turn its back to the Sun, the Earth and the Moon at the same time, lowering

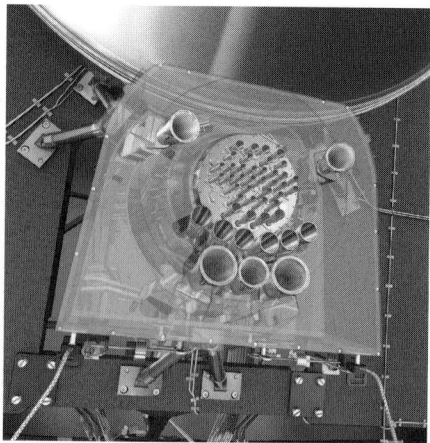

**Figure 5:** *Left panel*: Planck's mirror illuminating the focal-plane instruments. *Right panel*: The feed-horns of the two instruments in Planck's focal plane: HFI (centre) and LFI (surrounding ring).

the level of systematic noise. Planck's paramount design concept is the control of systematic errors. Its broad frequency range, its much enhanced polarisation sensitivity, and absolute radiometry by means of an on-board reference source will allow a substantial improvement in the subtraction of foregrounds.

Planck's primary goal is the further increase of the accuracy of cosmological parameters. With its angular resolution reaching $5'$, Planck will allow the measurement of the CMB power spectrum well into the Silk damping tail and thus obtain virtually all information contained in the CMB temperature fluctuations (Fig. 4, right panel). The substantially more accurate determination of the polarisation will remove much of the uncertainty due to the polarised foregrounds, thus enable a more reliable determination of the epoch of reionisation and the primordial amplitude of density fluctuations.

Although the nature of the dark energy leaves very little information in the primordial CMB itself, its possible modifications of the cosmic expansion rate at late times cause the gravitational potential to change in characteristic ways along the lines-of-sight towards the CMB. This leads to the so-called integrated Sachs-Wolfe effect [39], whose amplitude is determined by the line-of-sight integral of the time derivative of the gravitational potential. It can be detected by cross-correlating the CMB with the distribution of distant galaxies and contains information possibly useful for constraining the dark energy (cf. [37]).

If nature cooperates, it may be possible for Planck to constrain the energy scale of cosmological inflation. Besides the scalar fluctuations seen in density perturbations, tensor perturbations should also have been excited by inflation [21]. Such primordial gravitational waves would imprint a so-called $B$-mode, or odd-parity, polarisation pattern which cannot be created by scalar perturbations. Such a direct confirmation of the inflationary paradigm would be a triumph for Planck, even if we must be lucky for it to come within reach.

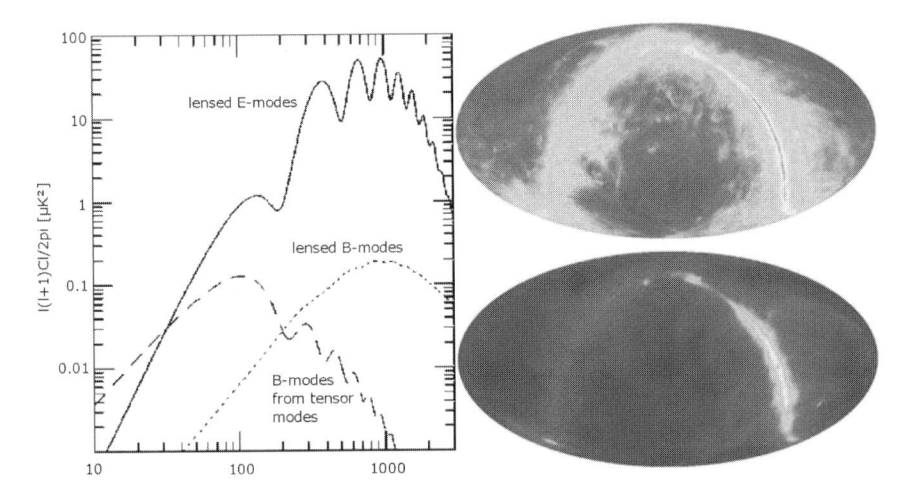

**Figure 6:** *Left panel*: Power spectra for the gravitationally-lensed $E$- and $B$-mode polarisation signals and a primordial $B$-mode signal as it could arise from gravitational waves excited during inflation. *Right panel*: Galactic dust (top) and synchrotron emission (bottom) at 100 GHz, in ecliptic coordinates.

Planck will also detect a large variety of foreground effect which will make its data extremely useful for other branches of astrophysics. Galaxy clusters, distant galaxies, gravitational lensing and microwave emission by the Milky Way are just four prominent examples.

Galaxy clusters appear on the microwave sky because of the Sunyaev-Zel'dovich effect [48]. CMB photons Thomson-scattered off electrons in the hot intracluster plasma gain energy and are thus transported from low to high energies. They are then missing at low and reappearing at high frequencies, with a redshift-independent boundary at 217 GHz. Galaxy clusters thus cast shadows on the CMB below, appear as microwave sources above, and disappear at 217 GHz. This very peculiar spectral feature allows the straightforward detection of galaxy clusters in the Planck data. Estimates show that Planck may find up to $\sim 10^4$ galaxy clusters, approximately $\sim 1\%$ of which may be at high redshift, $z \gtrsim 0.8$. Due to the limited angular resolution of Planck, most clusters will appear point-like. Another class of point sources are distant galaxies, whose infrared emission due to starbursts will appear redshifted into the sub-mm regime [51]. It is almost guaranteed that the Planck data will reveal many surprises regarding these distant and luminous galaxies.

The CMB is inevitably gravitationally lensed because it has to travel through almost the entire Universe on its way to us [28, 42]. Gravitational lensing has two main effects on the CMB. First, it mildly damps the CMB fluctuations to a degree which is invisible at the lowest-order acoustic peaks and increases to a level perceptible for Planck at the higher-order peaks. This damping, which is due to the fact that lensing by large-scale structures resembles a diffusion process, also transports CMB fluctuations into the Silk-damping tail, where they cause a substantial increase in the fluctuation level [31].

If this was the only effect, we had no chance of disentangling it from the primordial fluctuations. However, lensing imprints a characteristic pattern on the CMB which mixes temperature-fluctuation modes of different wavelengths. Since this is impossibly intrinsic in Gaussian random fields, the gravitational deflection can be recovered from the degree of mode-mixing [18]. While the primordial CMB polarisation caused by density fluctuations conserves parity and thus consists purely of $E$-modes, the misalignment of lensing density fluctuations with primordial CMB fluctuations moves some of the $E$-mode power to $B$-modes [56]. Thus, another signature of gravitational lensing in the CMB possibly detectable with Planck is the $B$-mode polarisation on small scales, which needs to be distinguished from the $B$-mode polarisation possibly caused by primordial gravitational waves on larger scales (Fig. 6, left panel).

Finally, the Planck data will also be an important resource for studies of our Galaxy (Fig. 6, right panel). At least three foreground components are contributed by the Milky Way: synchrotron radiation at the lowest frequencies due to relativistic electrons gyrating in the Galactic magnetic field [15], emission increasing towards high frequencies due to warm dust and molecular clouds [6, 40], and the free-free emission from thermal electrons in ionised hydrogen clouds [9]. The synchrotron and dust emission components are also polarised. The study of all these components will add important information to our understanding of the Milky Way.

Of course, Planck is first and foremost a cosmological mission which will measure the CMB temperature-fluctuation power spectrum with definitive accuracy and the CMB polarisation power spectra at much increased precision. In addition, however, Planck will also close the largest remaining gap in the astronomically observed electromagnetic spectrum. Its data set will thus also be an immensely valuable astrophysical resource.

# References

[1] R. A. Alpher and R. C. Herman. Remarks on the Evolution of the Expanding Universe. *Physical Review*, 75:1089–1095, Apr. 1949.

[2] C. L. Bennett, M. Bay, M. Halpern, G. Hinshaw, C. Jackson, N. Jarosik, A. Kogut, M. Limon, S. S. Meyer, L. Page, D. N. Spergel, G. S. Tucker, D. T. Wilkinson, E. Wollack, and E. L. Wright. The Microwave Anisotropy Probe Mission. *ApJ*, 583:1–23, Jan. 2003.

[3] M. Bersanelli, N. Mandolesi, C. Cesarsky, L. Danese, G. Efstathiou, M. Griffin, J. M. Lamarre, H. U. Norgaard-Nielsen, O. Pace, J. L. Puget, A. Raisanen, G. F. Smoot, J. Tauber, and S. Volonte. COBRAS/SAMBA: the European space mission to map the CBR anisotropy. *Astrophysical Letters Communications*, 33:19–24, Feb. 1996.

[4] J. R. Bond and G. Efstathiou. Cosmic background radiation anisotropies in universes dominated by nonbaryonic dark matter. *ApJL*, 285:L45–L48, Oct. 1984.

[5] J. R. Bond and G. Efstathiou. The statistics of cosmic background radiation fluctuations. *MNRAS*, 226:655–687, June 1987.

[6] T. M. Dame, D. Hartmann, and P. Thaddeus. The Milky Way in Molecular Clouds: A New Complete CO Survey. *ApJ*, 547:792–813, Feb. 2001.

[7] M. Davis, G. Efstathiou, C. S. Frenk, and S. D. M. White. The evolution of large-scale structure in a universe dominated by cold dark matter. *ApJ*, 292:371–394, May 1985.

[8] R. H. Dicke, P. J. E. Peebles, P. G. Roll, and D. T. Wilkinson. Cosmic Black-Body Radiation. *ApJ*, 142:414–419, July 1965.

[9] D. P. Finkbeiner. A Full-Sky H$\alpha$ Template for Microwave Foreground Prediction. *ApJS*, 146:407–415, June 2003.

[10] W. L. Freedman, B. F. Madore, B. K. Gibson, L. Ferrarese, D. D. Kelson, S. Sakai, J. R. Mould, R. C. Kennicutt, Jr., H. C. Ford, J. A. Graham, J. P. Huchra, S. M. G. Hughes, G. D. Illingworth, L. M. Macri, and P. B. Stetson. Final Results from the Hubble Space Telescope Key Project to Measure the Hubble Constant. *ApJ*, 553:47–72, May 2001.

[11] C. S. Frenk, S. D. M. White, M. Davis, and G. Efstathiou. The formation of dark halos in a universe dominated by cold dark matter. *ApJ*, 327:507–525, Apr. 1988.

[12] G. Gamow. Expanding Universe and the Origin of Elements. *Physical Review*, 70:572–573, Oct. 1946.

[13] G. Gamow. The Origin of Elements and the Separation of Galaxies. *Physical Review*, 74:505–506, Aug. 1948.

[14] S. Hanany, P. Ade, A. Balbi, J. Bock, J. Borrill, A. Boscaleri, P. de Bernardis, P. G. Ferreira, V. V. Hristov, A. H. Jaffe, A. E. Lange, A. T. Lee, P. D. Mauskopf, C. B. Netterfield, S. Oh, E. Pascale, B. Rabii, P. L. Richards, G. F. Smoot, R. Stompor, C. D. Winant, and J. H. P. Wu. MAXIMA-1: A Measurement of the Cosmic Microwave Background Anisotropy on Angular Scales of 10'-5°. *ApJL*, 545:L5–L9, Dec. 2000.

[15] C. G. T. Haslam, C. J. Salter, H. Stoffel, and W. E. Wilson. A 408 MHz all-sky continuum survey. II - The atlas of contour maps. *A&AS*, 47:1–+, Jan. 1982.

[16] G. Hinshaw, M. R. Nolta, C. L. Bennett, R. Bean, O. Doré, M. R. Greason, M. Halpern, R. S. Hill, N. Jarosik, A. Kogut, E. Komatsu, M. Limon, N. Odegard, S. S. Meyer, L. Page, H. V. Peiris, D. N. Spergel, G. S. Tucker, L. Verde, J. L. Weiland, E. Wollack, and E. L. Wright. Three-Year Wilkinson Microwave Anisotropy Probe (WMAP) Observations: Temperature Analysis. *ApJs*, 170:288–334, June 2007.

[17] G. Hinshaw, D. N. Spergel, L. Verde, R. S. Hill, S. S. Meyer, C. Barnes, C. L. Bennett, M. Halpern, N. Jarosik, A. Kogut, E. Komatsu, M. Limon, L. Page, G. S. Tucker, J. L. Weiland, E. Wollack, and E. L. Wright. First-Year Wilkinson Microwave Anisotropy Probe (WMAP) Observations: The Angular Power Spectrum. *ApJs*, 148:135–159, Sept. 2003.

[18] W. Hu. Mapping the Dark Matter through the Cosmic Microwave Background Damping Tail. *ApJL*, 557:L79–L83, Aug. 2001.

[19] W. Hu and N. Sugiyama. Anisotropies in the cosmic microwave background: an analytic approach. *ApJ*, 444:489–506, May 1995.

[20] A. H. Jaffe, P. A. Ade, A. Balbi, J. J. Bock, J. R. Bond, J. Borrill, A. Boscaleri, K. Coble, B. P. Crill, P. de Bernardis, P. Farese, P. G. Ferreira, K. Ganga, M. Gi-

acometti, S. Hanany, E. Hivon, V. V. Hristov, A. Iacoangeli, A. E. Lange, A. T. Lee, L. Martinis, S. Masi, P. D. Mauskopf, A. Melchiorri, T. Montroy, C. B. Netterfield, S. Oh, E. Pascale, F. Piacentini, D. Pogosyan, S. Prunet, B. Rabii, S. Rao, P. L. Richards, G. Romeo, J. E. Ruhl, F. Scaramuzzi, D. Sforna, G. F. Smoot, R. Stompor, C. D. Winant, and J. H. Wu. Cosmology from MAXIMA-1, BOOMERANG, and COBE DMR Cosmic Microwave Background Observations. *Physical Review Letters*, 86:3475–3479, Apr. 2001.

[21] M. Kamionkowski, A. Kosowsky, and A. Stebbins. Statistics of cosmic microwave background polarization. *PRD*, 55:7368–7388, June 1997.

[22] A. Kogut, D. N. Spergel, C. Barnes, C. L. Bennett, M. Halpern, G. Hinshaw, N. Jarosik, M. Limon, S. S. Meyer, L. Page, G. S. Tucker, E. Wollack, and E. L. Wright. First-Year Wilkinson Microwave Anisotropy Probe (WMAP) Observations: Temperature-Polarization Correlation. *ApJs*, 148:161–173, Sept. 2003.

[23] J. M. Kovac, E. M. Leitch, C. Pryke, J. E. Carlstrom, N. W. Halverson, and W. L. Holzapfel. Detection of polarization in the cosmic microwave background using DASI. *Nat*, 420:772–787, Dec. 2002.

[24] J. M. Lamarre, P. R. Ade, A. Benoît, P. de Bernardis, J. Bock, F. Bouchet, T. Bradshaw, J. Charra, S. Church, F. Couchot, J. Delabrouille, G. Efstathiou, M. Giard, Y. Giraud-Héraud, R. Gispert, M. Griffin, A. Lange, A. Murphy, F. Pajot, J. L. Puget, and I. Ristorcelli. The High Frequency Instrument of Planck: Design and Performances. *Astrophysical Letters Communications*, 37:161–+, 2000.

[25] J. M. Lamarre, M. Piat, P. A. R. Ade, J. Bock, P. de Bernardis, M. Giard, A. Lange, A. Murphy, J. P. Torre, A. Benoit, R. Bhatia, F. R. Bouchet, B. Maffei, J. L. Puget, R. Sudiwala, and V. Yourchenko. Use of high sensitivity bolometers for astronomy: Planck high frequency instrument. *Low Temperature Detectors*, 605:571–576, Feb. 2002.

[26] J. M. Lamarre, J. L. Puget, F. Bouchet, P. A. R. Ade, A. Benoit, J. P. Bernard, J. Bock, P. de Bernardis, J. Charra, F. Couchot, J. Delabrouille, G. Efstathiou, M. Giard, G. Guyot, A. Lange, B. Maffei, A. Murphy, F. Pajot, M. Piat, I. Ristorcelli, D. Santos, R. Sudiwala, J. F. Sygnet, J. P. Torre, V. Yurchenko, and D. Yvon. The Planck High Frequency Instrument, a third generation CMB experiment, and a full sky submillimeter survey. *New Astronomy Review*, 47:1017–1024, Dec. 2003.

[27] E. M. Leitch, J. M. Kovac, C. Pryke, J. E. Carlstrom, N. W. Halverson, W. L. Holzapfel, M. Dragovan, B. Reddall, and E. S. Sandberg. Measurement of polarization with the Degree Angular Scale Interferometer. *Nat*, 420:763–771, Dec. 2002.

[28] A. Lewis and A. Challinor. Weak gravitational lensing of the CMB. *Phys. Rep.*, 429:1–65, June 2006.

[29] N. Mandolesi, M. Bersanelli, C. Burigana, and F. Villa. The Planck Low Frequency Instrument. *Astrophysical Letters Communications*, 37:151–+, 2000.

[30] J. C. Mather, E. S. Cheng, D. A. Cottingham, R. E. Eplee, Jr., D. J. Fixsen, T. Hewagama, R. B. Isaacman, K. A. Jensen, S. S. Meyer, P. D. Noerdlinger, S. M. Read, L. P. Rosen, R. A. Shafer, E. L. Wright, C. L. Bennett, N. W.

Boggess, M. G. Hauser, T. Kelsall, S. H. Moseley, Jr., R. F. Silverberg, G. F. Smoot, R. Weiss, and D. T. Wilkinson. Measurement of the cosmic microwave background spectrum by the COBE FIRAS instrument. *ApJ*, 420:439–444, Jan. 1994.

[31] R. B. Metcalf and J. Silk. Gravitational Magnification of the Cosmic Microwave Background. *ApJ*, 489:1–+, Nov. 1997.

[32] L. Page, G. Hinshaw, E. Komatsu, M. R. Nolta, D. N. Spergel, C. L. Bennett, C. Barnes, R. Bean, O. Doré, J. Dunkley, M. Halpern, R. S. Hill, N. Jarosik, A. Kogut, M. Limon, S. S. Meyer, N. Odegard, H. V. Peiris, G. S. Tucker, L. Verde, J. L. Weiland, E. Wollack, and E. L. Wright. Three-Year Wilkinson Microwave Anisotropy Probe (WMAP) Observations: Polarization Analysis. *ApJs*, 170:335–376, June 2007.

[33] L. Page, M. R. Nolta, C. Barnes, C. L. Bennett, M. Halpern, G. Hinshaw, N. Jarosik, A. Kogut, M. Limon, S. S. Meyer, H. V. Peiris, D. N. Spergel, G. S. Tucker, E. Wollack, and E. L. Wright. First-Year Wilkinson Microwave Anisotropy Probe (WMAP) Observations: Interpretation of the TT and TE Angular Power Spectrum Peaks. *ApJs*, 148:233–241, Sept. 2003.

[34] P. J. E. Peebles. Large-scale background temperature and mass fluctuations due to scale-invariant primeval perturbations. *ApJL*, 263:L1–L5, Dec. 1982.

[35] P. J. E. Peebles and J. T. Yu. Primeval Adiabatic Perturbation in an Expanding Universe. *ApJ*, 162:815–+, Dec. 1970.

[36] A. A. Penzias and R. W. Wilson. A Measurement of Excess Antenna Temperature at 4080 Mc/s. *ApJ*, 142:419–421, July 1965.

[37] A. Rassat, K. Land, O. Lahav, and F. B. Abdalla. Cross-correlation of 2MASS and WMAP 3: implications for the integrated Sachs-Wolfe effect. *MNRAS*, 377:1085–1094, May 2007.

[38] J. E. Ruhl, P. A. R. Ade, J. J. Bock, J. R. Bond, J. Borrill, A. Boscaleri, C. R. Contaldi, B. P. Crill, P. de Bernardis, G. De Troia, K. Ganga, M. Giacometti, E. Hivon, V. V. Hristov, A. Iacoangeli, A. H. Jaffe, W. C. Jones, A. E. Lange, S. Masi, P. Mason, P. D. Mauskopf, A. Melchiorri, T. Montroy, C. B. Netterfield, E. Pascale, F. Piacentini, D. Pogosyan, G. Polenta, S. Prunet, and G. Romeo. Improved Measurement of the Angular Power Spectrum of Temperature Anisotropy in the Cosmic Microwave Background from Two New Analyses of BOOMERANG Observations. *ApJ*, 599:786–805, Dec. 2003.

[39] R. K. Sachs and A. M. Wolfe. Perturbations of a Cosmological Model and Angular Variations of the Microwave Background. *ApJ*, 147:73–+, Jan. 1967.

[40] D. J. Schlegel, D. P. Finkbeiner, and M. Davis. Maps of Dust Infrared Emission for Use in Estimation of Reddening and Cosmic Microwave Background Radiation Foregrounds. *ApJ*, 500:525–+, June 1998.

[41] M. Seiffert, A. Mennella, C. Burigana, N. Mandolesi, M. Bersanelli, P. Meinhold, and P. Lubin. 1/f noise and other systematic effects in the Planck-LFI radiometers. *A&A*, 391:1185–1197, Sept. 2002.

[42] U. Seljak and M. Zaldarriaga. A Line-of-Sight Integration Approach to Cosmic Microwave Background Anisotropies. *ApJ*, 469:437–+, Oct. 1996.

[43] J. Silk. Cosmic Black-Body Radiation and Galaxy Formation. *ApJ*, 151:459–+, Feb. 1968.

[44] G. F. Smoot, C. L. Bennett, A. Kogut, E. L. Wright, J. Aymon, N. W. Boggess, E. S. Cheng, G. de Amici, S. Gulkis, M. G. Hauser, G. Hinshaw, P. D. Jackson, M. Janssen, E. Kaita, T. Kelsall, P. Keegstra, C. Lineweaver, K. Loewenstein, P. Lubin, J. Mather, S. S. Meyer, S. H. Moseley, T. Murdock, L. Rokke, R. F. Silverberg, L. Tenorio, R. Weiss, and D. T. Wilkinson. Structure in the COBE differential microwave radiometer first-year maps. *ApJL*, 396:L1–L5, Sept. 1992.

[45] D. N. Spergel, R. Bean, O. Doré, M. R. Nolta, C. L. Bennett, J. Dunkley, G. Hinshaw, N. Jarosik, E. Komatsu, L. Page, H. V. Peiris, L. Verde, M. Halpern, R. S. Hill, A. Kogut, M. Limon, S. S. Meyer, N. Odegard, G. S. Tucker, J. L. Weiland, E. Wollack, and E. L. Wright. Three-Year Wilkinson Microwave Anisotropy Probe (WMAP) Observations: Implications for Cosmology. *ApJs*, 170:377–408, June 2007.

[46] D. N. Spergel, L. Verde, H. V. Peiris, E. Komatsu, M. R. Nolta, C. L. Bennett, M. Halpern, G. Hinshaw, N. Jarosik, A. Kogut, M. Limon, S. S. Meyer, L. Page, G. S. Tucker, J. L. Weiland, E. Wollack, and E. L. Wright. First-Year Wilkinson Microwave Anisotropy Probe (WMAP) Observations: Determination of Cosmological Parameters. *ApJs*, 148:175–194, Sept. 2003.

[47] R. A. Sunyaev and Y. B. Zeldovich. Small-Scale Fluctuations of Relic Radiation. *ApSS*, 7:3–19, Apr. 1970.

[48] R. A. Sunyaev and Y. B. Zeldovich. The Observations of Relic Radiation as a Test of the Nature of X-Ray Radiation from the Clusters of Galaxies. *Comments on Astrophysics and Space Physics*, 4:173–+, Nov. 1972.

[49] J. A. Tauber. The Planck Mission: Overview and Current Status. *Astrophysical Letters Communications*, 37:145–+, 2000.

[50] J. A. Tauber. The Planck mission. *Advances in Space Research*, 34:491–496, 2004.

[51] L. Toffolatti, F. Argueso Gomez, G. de Zotti, P. Mazzei, A. Franceschini, L. Danese, and C. Burigana. Extragalactic source counts and contributions to the anisotropies of the cosmic microwave background: predictions for the Planck Surveyor mission. *MNRAS*, 297:117–127, June 1998.

[52] J. M. Uson and D. T. Wilkinson. Improved limits on small-scale anisotropy in cosmic microwave background. *Nat*, 312:427–429, Nov. 1984.

[53] L. Valenziano, M. Sandri, G. Morgante, C. Burigana, M. Bersanelli, R. C. Butler, F. Cuttaia, F. Finelli, E. Franceschi, M. Galaverni, A. Gruppuso, M. Malaspina, N. Mandolesi, A. Mennella, F. Paci, L. Popa, P. Procopio, L. Stringhetti, L. Terenzi, M. Tomasi, F. Villa, and J. Zuccarelli. The low frequency instrument on-board the Planck satellite: Characteristics and performance. *New Astronomy Review*, 51:287–297, Mar. 2007.

[54] S. D. M. White, M. Davis, G. Efstathiou, and C. S. Frenk. Galaxy distribution in a cold dark matter universe. *Nat*, 330:451–453, Dec. 1987.

[55] S. D. M. White, C. S. Frenk, M. Davis, and G. Efstathiou. Clusters, filaments, and voids in a universe dominated by cold dark matter. *ApJ*, 313:505–516, Feb. 1987.

[56] M. Zaldarriaga and U. Seljak. Gravitational lensing effect on cosmic microwave background polarization. *PRD*, 58(2):023003–+, July 1998.

# 30 Years of Research in Cosmology, Particle Physics and Astrophysics and How Many More to Discover Dark Matter?

Céline Bœhm

LAPTH, UMR 5108
9 chemin de Bellevue, 74 941 Annecy-Le-Vieux
France
celine.boehm@cern.ch,

### Abstract

*The nature of dark matter is a long standing problem. The latest observation in cosmology now indicate that dark matter represents about 23 % of the total content of the Universe and 80% of its matter component. Many candidates have been proposed. Most of them are spin-1/2 fermionic particles with a mass greater than a few GeV. Yet spin-0 and spin-1 bosons have received a lot of attention during the last decade. The lack of strong indication in favor of any of these candidates nevertheless incites to question the particle hypothesis. Yet modifying gravity in a consistent way with observations is not an easy task. So is dark matter a fluke, an impossible issue to solve, a soon answered problem or "simply" a frustrating, challenging, yet fascinating question of physics? To try to assess these questions, we review the progress which have been made in both cosmology, particle physics and astrophysics during the last thirty years regarding the nature of dark matter.*

## 1 Disclaimer

Many astrophysicists, cosmologists and particle physicists have worked on various theoretical and observational aspects of dark matter during the last 30 years. It is impossible to give a short but exhaustive review of all the contributions in this field. I will therefore only mention the literature which, in my opinion, have contributed to change (or set) the direction in the field or that I am aware of. I deeply apologize to the authors that I may have forgotten.

## 2 Introduction

Although the presence of invisible matter in the universe now seems a well established concept, it took about 40 years of research to admit/convince the community that this hypothesis was actually interesting and perhaps, even, the correct answer to several puzzling observations. It all started in 1933, when F. Zwicky noticed that the velocity in the COMA cluster was too large to maintain its cohesion [1, 2]. This puzzling observation lead him to postulate the existence of a new invisible substance in the Universe, a daring hypothesis when one remembers that both the neutron and neutrino were discovered in the 1930s (1932 and 1930–1933 respectively), well after the electron (1897) and the proton (1919).

Although Zwicky's idea did not appeal to any of his colleagues, evidence in the same direction continued to be accumulated: S. Smith thus obtained a similar conclusion than Zwicky in 1936 by studying the Virgo cluster [3] while, in 1939, both Babcock and J. Oort noticed a too large rotation curve in the Andromeda and NGC 3115 galaxies respectively which was the sign that there was also a missing piece in the understanding of galaxy dynamics [4, 5].

These conclusions were confirmed later on with the use of both optical measurements of galactic rotation curves [6], neutral hydrogen rotational velocity [7–9] and other elements velocity. However, although early measurements of the rotation curve of M31 by Rubin and Ford in 1970 [10, 11], M101 by Rogstad and Shostak in 1972 [12] and M81 by Roberts and Rots in 1973 [13] suggested that the total mass of the galaxy was increasing linearly with the radius $r$ (the distance to the center of the galaxy), it is only in 1978 – with the advent of better optical instrumentation– that the evidence of flat rotation curves of galaxies have been firmly established. Indeed, this very year, two groups published the rotation curves of 10 spiral galaxies (Sa through Sc) and 6 spiral galaxies respectively which are now often presented as the ultimate evidence in favor of the dark matter existence ([14] and [15, 16]).

Interestingly enough, though, there were already several cosmological indications of the dark matter existence before 1978. For example, the authors of Ref. [17] noticed that the early mass measurements of the M31,81,101 galaxies could be translated into a total energy density which largely exceeds the estimate derived from primordial big bang nucleosynthesis observations (see [18]). Also, as we will see, J. Silk proposed an argument in 1968 which, in my opinion, was certainly the first theoretical argument in favor of the existence of dark matter [19].

As explained, the flat behavior of rotation curves suggested that the mass profile of the galaxy increases linearly with the distance to the center. The Virial theorem, $V^2(r) = GM(r)/r$, indeed indicates that at large radius $M(r)$ must be proportional to the radius $r$ to flatten the velocity $V(r)$. Yet, for a galaxy of a given size, the luminous matter mass distribution is expected to decrease beyond a maximal radius $r_{max}$. So how can these two behaviors be reconciled? The current, commonly accepted, solution is that the total mass of the galaxy $M(r)$ is in fact the sum of two different contributions. One is visible $M_{visible}(r)$ and can be inferred by optical observations (it vanishes at large radii). The other one $M_{invisible}(r)$ is meant to be related to an invisible substance which would extend well beyond the optical radius of a given galaxy, as initially postulated by F. Zwicky about 40 years earlier.

One may find surprising that dark matter does not have the same mass distribution as that of the luminous matter. However this is due to the absence of electromagnetic interactions which prevents dark matter to collapse and form a disk, like the baryons do.

The existence of dark matter has been also confirmed via

- lensing measurements and, in particular, the existence of rings around cluster of galaxies which are due to strong lensing effects owing to the presence of dark matter or simply image distortion which corresponds to weak lensing effects.

- Primordial Big Bang Nucleosynthesis (BBN) which indicates that there is only 3–5% of baryonic matter in our Universe [20–23]. Note that in cosmology, the word "baryonic" describes the baryons, mesons and leptons altogether. The idea of primordial nucleosynthesis was first proposed in 1938 by Weizsacker, Chandrasekhar and Henrich in 1942 [24] and Gamow in 1948 [25]. However this mechanism started to be studied in detail only since 1966 after that J. Peebles pointed out that the blackbody radiation was an indication that the Universe did pass through a very hot phase during which light elements (hydrogen, deuterium, helium, tritium) could be formed [26].

- the absence of Massive Compact Halo Object (MACHOs). Two experiments of gravitational microlensing, named EROS and MACHO, have indeed confirmed that MACHOs could not fill the galactic halo, therefore suggesting that our halo is made of new particles [27].

## 3 Progress made in cosmology

In this section we summarize the works in cosmology which lead to the introduction of dark matter and the determination of its characteristics.

### 3.1 Primordial matter density fluctuations: the very first step towards the characteristics of dark matter particles

As mentioned in the introduction the existence of dark matter could be gathered from many puzzling observations. However there is one theoretical argument that is, in fact, very rarely mentioned in dark matter reviews but which is, in my view, among the most important evidence in favor of dark matter. This point is certainly the hardest to overcome when one tries to build alternatives to the dark matter hypothesis and is called the Silk damping.

While it became clear in the early 1960s that the Universe was quite homogenous and isotropic, the formation of galaxies appeared quite problematic. Indeed, how come galaxies and clusters of galaxies form in a homogeneous Universe? The first author who answered this question was Gamow in 1948, explaining that as soon as the Universe becomes matter dominated the "homogeneous gas starts to break up into separate clouds which are later pulled apart" due to the Jeans gravitational instabilities [28]. In 1965, Peebles realized that the Jeans instability principle alone was not sufficient to explain structure formation [29]. One has to postulate initial conditions. This lead him to introduce the concept of initial density perturbations (i.e. regions of matter that are overdense and others that are underdense) and even derive the spectrum of primordial fluctuations that is necessary to form galaxies and clusters of galaxies.

Peebles however neglected the possibility that the initial spectrum could be damped due to viscosity effects; a critical point that was quickly noticed by C. Misner in 1967 [30]. In particular, he pointed out that neutrino-electron interactions erased the primordial spectrum below a critical mass of the order of $10^{-4} M_\odot$. Although Misner's damping effect certainly exists, the corresponding scale appeared however too small to be cosmologically relevant, thus making Peebles' assumption perfectly valid. However, during the same year, J. Silk realized that charged Standard Model particles have electromagnetic interactions which erase the primordial fluctuations with a size smaller than 1 Mpc. Unfortunately, observations at that time were not sufficient to tell whether primordial fluctuations of the required size to form galaxies could actually survive until the epoch were galaxy formation was possible. Nevertheless, it became clear a few years later that there were not. Thus the conclusion that can be drawn a posteriori from J. Silk work is that, in a purely baryonic matter-dominated Universe, small galaxies cannot form as efficiently as observations indicate. Given the now known contradiction with observation, one can interpret the Silk damping as the first theoretical argument (on top of BBN) in favour of the existence of dark matter.

J. Silk's argument could have been overcame nevertheless with the presence of massive neutrinos. For example, Szalay and Marx noticed that primordial fluctuations in the neutrino gas may initiate the formation of clusters of galaxies providing that neutrinos were lighter than 15 eV [31]. Such a limit (based on the age of the Universe and the deceleration parameter $q < 2$) was compatible with the result from R. Cowsik and J. McClelland (who were using the same argument in 1972 [32]) and Gershtein and Zel'dovich in 1966 [33] who obtained $m_{\nu_\mu} < 400 eV$ by imposing that the neutrino energy density does not exceed the critical energy density of the Universe. Yet as demonstrated by Pryor et al in 1980 [34, 35] as well as Bond, Efstathiou, Silk in 1980 [36] neutral (almost) relativistic particles cannot form Milky-Way size galaxies efficiently. Indeed these particles free-stream and can never cluster in "small"-size objects. These latest results were actually confirmed by the Tremaine-Gunn bound derived in 1979 [37] which indicate that fermionic particles with a mass lower than 1 MeV cannot form the observed halos due to phase-space argument.

Hence, from the late 1970s, it became obvious that neither charged baryons nor neutrinos would lead to a satisfactory mechanism to explain the formation of Milky Way-size galaxies. So here is the question which was then posed: what kind of substance can exist in the Universe and simultaneously explain the formation of objects such as galaxies?

Following up the suggestion by Shvartsman in 1969 [38] that there could be several massless particle species in the Universe, Steigman et al constrained in 1977 [39] the number of relativistic degrees of freedom in the Universe. This seems to have inspired P. Hut [40] who then set a more general limit on the mass and number of neutral weakly interacting particles the same year. In particular, he obtained that the Universe could not be fill with weakly interacting particles with a mass smaller than 3 GeV and greater than 120 eV. Interestingly enough, a very similar result was obtained from M.I Vysotskii, A.D. Dolgov, I.B. Zeldovich [41] while Peebles in

1982 [42] argued (without specifying a kind of interactions as done by previous authors) that any massive particle with weak interactions could explain the formation of the structures we know if it is heavier than 1 keV. This limit has been slightly improved since but remains of the same order of magnitude.

Since the formation of "small" galaxies could not be explained neither with ordinary matter nor new neutral (fermionic) weakly interacting particles lighter than 3 GeV, the most promising candidate to explain structure formation was the existence of very massive ($m > 3$ GeV) neutral weakly interacting particles.

Such (fermionic) candidates were proposed in 1977 by Lee-Weinberg [43] who also derived an upper bound on their mass. In particular, they showed that these particles could not be heavier than 1 TeV. We will detail about this in the next section.

## 3.2 Large-scale-structure formation and the dark matter scenarios

Altogether the works we have summarized gave birth to the scenario of Cold Dark Matter (CDM). Assuming that dark matter was once in thermal and chemical equilibrium, this means that the dark matter particles first become non-relativistic, then leave the chemical equilibrium and finally leave their thermal equilibrium. The last stage typically occurs in the MeV range while the second happens at $m/x$ where $x$ is a factor ranging from 10–20 for annihilating particles. For comparison, relativistic neutrinos are referred as to Hot Dark Matter (HDM) particles since they thermally decouple well before becoming non-relativistic. There is also a third scenario, which was first proposed in 1988 by Schaeffer and Silk [44] and studied in detail by Ostriker, bode and Turok [45], and many other authors (see e.g. [46]) namely collisionless Warm Dark Matter (WDM). This initially describes the case of particles with a mass of about a few keV and with negligible interactions. Such particles would have a free-streaming damping scale that is about the size of Lyman-$\alpha$ forest. Hence, all the scales in their matter power spectrum are cut below 10 kpc. Nowadays, collisionless WDM particles refer to particles for which the damping scale matches the smallest observed scales [47]. For example, sterile neutrinos which are typical WDM examples, are still currently discussed in the literature [48–50]. The associated transfer function $T_X$ is exponentially suppressed and proportional to the dark matter mass while the CDM transfer function does not have any cut-off. The HDM power spectrum is exponentially suppressed at the cluster scale. $T_X$ can therefore be expressed as

$$P_X = T_X^2 \cdot P_{CDM}, \qquad (1)$$

with $P_{CDM}$ the corresponding Cold Dark Matter power and $P_X$ the real dark matter power spectrum. For collisionless WDM particles, a Boltzmann code calculation gave [45]

$$T_{WDM} = [1 + (\alpha k)^{2\nu}]^{-5/\nu}, \qquad (2)$$

with

$$\alpha = 0.048 \text{Mpc} \left(\frac{m_{dm}}{\text{keV}}\right)^{-1.15} \left(\frac{\Omega_{dm}}{0.4}\right)^{0.15} \left(\frac{h}{0.65}\right)^{1.3} \left(\frac{g_{dm}}{1.5}\right)^{0.29},$$

with $\nu = 1.2$, and $g_{dm} = 1.5$ for a neutrino-like dark matter candidate.

The three dark matter scenarios (CDM, HDM, WDM) that we have mentioned so far are certainly the most popular. They are all based on the hypothesis that dark matter has negligible interactions. Yet we know that dark matter has interactions since it must be able to annihilate or decay so that the relic density matches observations. Hence the question which may be posed is: can realistic dark matter interactions be neglected in numerical simulations indeed? This was specifically answered in [47, 51].

These two references showed that dark matter can suffer from large damping effects which in some specific cases can even be of cosmological interest. Indeed, any dark matter particles with (even weak) interactions, experience both collisional damping and free-streaming damping. The problem is to determine whether the damping scale is cosmologically relevant or not. In order to be as most general as possible, we therefore classified all dark matter candidates that have been once in thermal equilibrium and then derived their damping scale. For the purpose of the classification, we used the quantities that are most relevant to compute damping effects, i.e. the thermal decoupling time ($t_{dec}$), the non-relativistic transition time ($t_{nr}$) and finally we matter-radiation equality epoch ($t_{eq}$). Those can be translated into associated scale factors ($a_{dec}$, $a_{nr}$, $a_{eq}$) respectively. From the combination of these three characteristic times, we obtained six types of dark matter candidates for which it is easy to compute the free-streaming and self-damping scales, see Fig. 1.

Region I corresponds to HDM and WDM, region II to CDM, region III to candidates with very large interaction rates such as self-interacting dark matter [52]. Region IV, V, VI are generally never considered in the literature but their free-streaming and self-damping scales are anyway too large. Perhaps one of the most important points that is stressed in this figure is that CDM and HDM/WDM particles do not have the same free-streaming length. Since CDM particles in general become non-relativistic before they thermally decouple, their free-streaming length depends on both the dark matter mass and the total dark matter interaction rate at decoupling:

$$l_{CDM} = 330 \text{ kpc } \mathcal{A} \left(\frac{m_{dm}\kappa_{dm}}{\text{MeV}}\right)^{-1/2} \left(\frac{\tilde{\Gamma}}{6\,10^{-24}\text{s}^{-1}}\right)^{1/2}$$

while for WDM, the free-streaming length only depends on the dark matter mass and is given by:

$$l_{WDM} = 100 \text{ kpc } \mathcal{B} \left(\frac{m_{dm}}{\text{keV}}\right)^{-4/3}.$$

Here $\mathcal{A} = f\, g_\star'^{3/4}(T_{dec(dm)})$ and $\mathcal{B} = 0.9 \left(\frac{\epsilon g_{dm}}{2}\right)^{-2/3} g'^{-8/3} \left(\frac{\Omega_{dm}h^2}{0.3}\right)^{2/3}$. All the definitions of the parameters can be found in Ref. [47].

A little bit before the publication of Ref. [51], the free-streaming damping length of neutralino dark matter was published in Ref. [53] (under specific assumptions on the mass and relevant couplings). Given that neutralinos are a specific case of CDM, the estimate of Ref. [53] should match that from the expression of $l_{CDM}$ given above. However the comparison showed some discrepancies which have never been clarified. In addition several groups [54, 55] *numerically* computed the neutralino

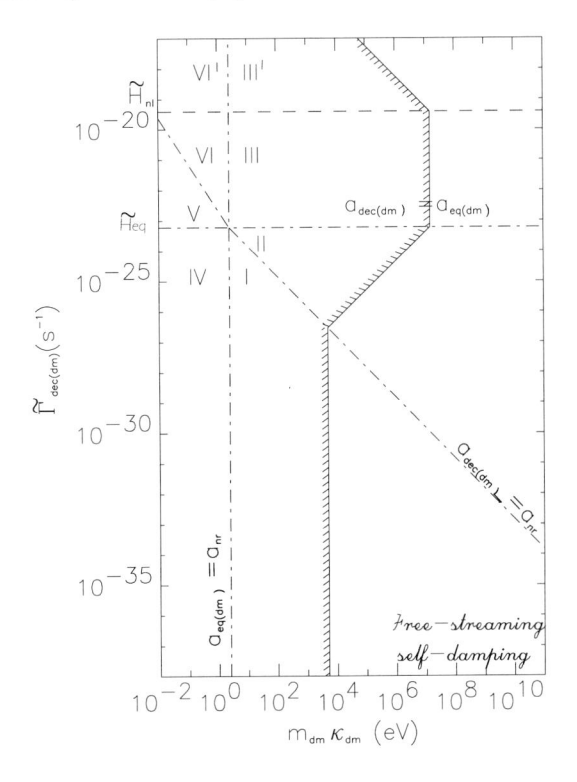

**Figure 1:** Free-streaming scales

damping length and found a larger value than in Ref. [53]. To my knowledge, the value of the neutralino damping length is still quite controversial.

In the same vein, I computed the free-streaming scale of light dark matter candidates (as well as the other collisional damping scale) and found that all the scales below a critical mass of $0.1 M_\odot$ would be erased but the authors of Ref. [56] found that it would be rather $10^7 M_\odot$. This discrepancy means that my estimate is about a factor $\sim 500$ smaller than the damping length of Ref. [56]. Indeed the damping mass is given by $M_{damping} \sim l_{damping}^3$, where $l_{damping}$ is the damping length.

For the candidates with a reasonable free-streaming damping scale, it is possible to estimate the collisional damping scale ($l_{cd}$) due to dark matter interactions with ordinary particles. The expressions of the dissipative coefficients (i.e. shear viscosity $\eta$, bulk viscosity $\xi$ and heat conduction $\lambda T$) were first given by Weinberg in 1971 [57] and were generalized to the case of dark matter in Ref. [47]. This translates into a damping scale that is given by:

$$l_{cd}^2 = \pi^2 \int^{t_{dec(dm)}} \frac{\zeta + \frac{4}{3}\eta + \lambda T \frac{\rho_m^2}{4\rho\rho_r}}{\rho a^2} dt. \qquad (3)$$

Here $\rho = \sum_i \rho_i$. Hence the mass-scale associated with the length $l_{cd}$ is given by $M_{cd} = (4\pi/3)\rho_m l_{cd}^3$. The integral runs over all the period during which Dark Matter is collisional. For any reasonable energy-dependence of the cross-sections, the integral in Eq. 3 is dominated by late times, so that the value of $l_{cd}$ is determined by the Dark Matter decoupling. Replacing the coefficients by their expression, one obtains:

$$l_{cd}^2 = l_{sd}^2 + \sum_{i \neq dm} l_{id}^2 , \qquad (4)$$

with

$$l_{sd}^2 = \frac{2\pi^2}{3} \int_0^{t_{dec}(dm)} \frac{\rho_{dm} v_{dm}^2 t}{\not{p} a^2 \Gamma_{dm}} (1 + \Theta_{dm}) \, \frac{dt}{t} , \qquad (5)$$

and for each species $i \neq dm$:

$$l_{id}^2 = \frac{2\pi^2}{3} \int_0^{t_{dec}(dm-i)} \frac{\rho_i v_i^2 t}{\not{p} a^2 \Gamma_i} (1 + \Theta_i) \, \frac{dt}{t} . \qquad (6)$$

In Ref. [47, 51], $l_{sd}$ was named the self-damping scale since it is a function of the dark matter characteristics only and $l_{id}$ the induced damping scale since it describes the damping that is experienced by a species $i$ and communicated to the dark matter via dark matter-$i$ collisions.

The self-damping scale is quite similar to the dark matter free-streaming scale and is quite easy to compute. The induced-damping scale $l_{id}$ is a little bit more tricky. Since $l_{id}^2$ is proportional to the energy density and velocity of the species $i$, the most interesting case is when dark matter interacts with photons or neutrinos. Indeed the radiation energy density never get as suppressed as the baryonic matter energy density (since baryons annihilate along the way) and its velocity is maximal.

In the case of dark matter interactions with photons, the shear viscosity term is inversely proportional to the photon interaction rate which is the sum of the photon-baryon interaction rate ($\sigma v_{\gamma-b} \times n_b$) and the photon-dark matter interaction rate ($\sigma v_{\gamma-dm} \times n_{dm}$). To avoid the Silk damping, dark matter must decouples from the photon before recombination. This implies $\sigma v_{\gamma-b} \times n_b > \sigma v_{\gamma-dm} \times n_{dm}$, i.e. $\Gamma_\gamma \sim \sigma v_{\gamma-b} \times n_b$. Hence, the collisional damping scale reduces to the following expression:

$$l_{\gamma d} = 8.2 \text{ Mpc } r_\gamma \, g_*'^{-1} \left(\frac{\Omega_b h_{70}^2}{0.05}\right)^{-\frac{1}{2}} \left(\frac{\tilde{\Gamma}_{dm-\gamma}}{6 \, 10^{-24} \text{ s}^{-1}}\right)^{\frac{3}{2}}$$

using $\tilde{\Gamma}_{dm-\gamma} \equiv \Gamma_{dm-\gamma} a^3$ Imposing that this damping scale be smaller than the smallest scale of the matter power spectrum ever been observed, one obtains:

$$\sigma_{DM-\gamma} v < 1.1 \, 10^{-23} \text{ cm}^3\text{s}^{-1} \, \mathcal{C} \left(\frac{m_{dm} \kappa_{dm}}{1 \text{MeV}}\right) \left(\frac{l_{struct}}{100 \text{kpc}}\right)^{\frac{4}{3}}$$

with $\mathcal{C} = r_\gamma^{-\frac{4}{3}} g_*'^{\frac{5}{6}}(T) \left(\frac{\Omega_b h_{70}^2}{0.05}\right)^{\frac{2}{3}} \kappa_{dm}^{-1}$ for a dark matter candidate that first becomes non-relativistic, then decouples from photons and finally passes the equality epoch.

This result was confirmed numerically in [58]. However, in this very reference, an additional effect was pointed out which constrain the dark matter-photon cross section to be:

$$\sigma_{DM-\gamma} v < 2.\ 10^{-24}\ \text{cm}^3\text{s}^{-1} \left(\frac{m_{dm}\kappa_{dm}}{1\text{MeV}}\right)$$

in order to reproduce structures with a characteristic size $l > 150$ kpc. The important point is that it is the maximum dark matter-photon cross section that is permitted by structure formation. Surprisingly enough, for heavy particles, it is far from being a weak-strength cross section!

Ref. [47, 51] also pointed out that dark matter-neutrino interactions could give birth to a new damping mechanism, namely the *mixed damping effect*, if the three following criteria were satisfied:

- the dark matter-neutrino interactions must be the dominant dark matter interactions. This sets a lower limit on the dark matter-neutrino cross section,

- the dark matter-neutrino interaction rate must be large enough so that the dark matter thermal decoupling occur at a temperature below a few MeV (i.e. below the neutrino-electron thermal decoupling),

- the dark matter-neutrino interaction rate must be such that the neutrino thermal decoupling remains fixed by the neutrino-electron thermal decoupling. I.e., below a few MeV, neutrinos must be free-streaming, as expected in the standard case. This actually sets an upper limit on the dark matter-neutrino cross section,

This is the case in particular for MeV particles having weak-strength interactions. The neutrino free-streaming is then "communicated" to the dark matter via the dark matter-neutrino interactions, erasing the dark matter primordial fluctuations below a characteristic scale [47, 51]. This particular situation gave birth to the light dark matter scenario studied in detail in [59–61].

Although the outcome of Refs. [47, 51] was the discovery of a new damping effect, named mixed damping, the main purpose of these two references was to stress the importance of dissipative effects for dark matter fluctuations whatever the dark matter candidate is and to estimate of the associated damping length. We found that unless dark matter is of a few MeV and has weak strength interactions (in contrast with super weak), it is indeed quite correct to neglect the dark matter interactions in numerical simulations, at least until simulations become sensitive to the physics below $10^3$–$10^6 M_\odot$.

Incidentally, this means that large cut-off in the matter power spectrum rather refers to collisionless WDM, i.e. dark matter particles with weak or super weak-strength interaction rates and a mass of a few keV.

In [62], it was nevertheless pointed out that – with current simulations– simulations of a WDM scenario with a cut-off in the matter power spectrum $P(k)$ at scales $l < 100$ kpc gave a very similar result as a CDM scenario. Indeed one can observe a small-scale-regeneration mechanism. This might be an artefact due to the way the numerical simulations have been performed. However if it reveals to be true, this would mean that although the WDM initial matter power spectrum (i.e. linear matter power spectrum) is truncated, observing nowadays a CDM-like matter power spectrum cannot be unambiguously interpreted as the manifestation of CDM particles. To distinguish between CDM and WDM scenarios, one would thus have to measure the $P(k)$ at different values of the redshift.

CDM and HDM numerical simulations were already performed in the 1980s but one had to wait until 1990s for the confirmation of the existence of primordial matter density fluctuations in the Universe. The COBE [63, 64] experiment was indeed the first experiment to detect temperature fluctuations in the Cosmological Microwave Background (CMB), thereby confirming Peebles hypothesis and indirectly validating J. Silk calculations. With observations of the matter power spectrum at small galactic size, it then became obvious that our Universe could not be dominated by ordinary matter: dark matter seems really a necessity to explain the existence of small-scale objects in the Universe.

Although COBE discovery was certainly a major progress in the field I always found quite ironical that the latest CMB experiments [65–68] together with the Hubble constant, high redshift supernovae, primordial nucleosynthesis and the matter power spectrum measurements, now indicates that dark matter is not in fact the main component of the Universe. It now seems that it only represents about 23% of the Universe, the rest being mostly dominated by the so-called Dark energy [69–71].

## 3.3 The dark matter power spectrum

As explained before the presence dark matter is more or less a necessity to explain why there are so many small-scale-structures in our Universe. However analytically, the only reliable prediction that can be made is the shape of the matter power spectrum in the linear regime, before structures start to form (i.e. at redshift $z > 30$). At smaller redshift, the effect gravity become very important and fluctuations enter a non-linear phase in which their growth is accelerated. Due to this effect, the number of small-scale structures (with a size smaller than 100 kpc) cannot be predicted analytically and one has to perform numerical simulations to estimate the number (and the properties) of the structures of a given size that should exist nowadays.

Many authors have contributed to the developments of numerical simulations of large-scale-structure formation, the determination of the dark matter halo profile and galaxy properties (for a non-exhaustive list, see e.g.[72–93]). The resolution of the simulations have been considerably improved with the advent of very powerful computers [94–96]. Until very recently, most of these simulations were aiming at characterizing the dark matter power spectrum (for CDM, HDM, WDM) as accurately as possible. However the tendency is now to include the baryonic gas and perform the so-called hydrodynamical simulations in order to study more realistic situations and determining the effects of the dynamics of the disc on the dark matter halo.

In dark matter N-body simulations, the Universe is simulated as a periodic box. The size of this box depends on the simulations. As an example, in Millennium simulations, the size is about $500h^{-1}$ Mpc. Consequently, the individual particles which are simulated have a size about e.g. $\sim 10^9 M_\odot$ [96]. This precision is insufficient to study dark matter weak interactions but is enough to confirm that the presence of dark matter overcomes the Silk damping.

They are also sufficient to outline interesting issues. For example, numerical simulations based on the existence of a collisionless fluid do reproduce well the number of objects as small as Lyman-$\alpha$ forest but seem to overpredict the number of very small size objects. This is called the missing satellite problem. Perhaps the reason for this discrepancy is that those satellites are very hard to see. They are indeed expected to be dark matter dominated and therefore quite invisible. Lensing experiments will be crucial to solve this issue. Also CDM simulations predict quite cuspy dark matter halo profile. This was a problem in particular for dwarf galaxies since observations seemed to prefer a rather flat dark matter halo profile. We will discuss this point in the next subsection.

At last let us mention a point that is not generally presented as an issue but is, however, quite peculiar: the necessity to include a cosmological constant (or dark energy) in order to fit observation within the CDM scenario. This is the so-called $\Lambda$(C)DM model. Although there are some data to support this hypothesis, one may nevertheless wonder whether a model based on two unknown substance to fit observation does not simply indicate how gravity behaves at different characteristic scales?

Perhaps the answer will come from better measurements of the matter power spectrum at very small scales. The latest measurements (see e.g. 2dF [97] and SDDS surveys [98]) have already shown that the matter power spectrum was consistent with the Cold Dark Matter scenario on scales larger than 100 kpc (i.e. $10^9 M_\odot$). Neglecting the small-scale regeneration problem [62], these observations constrain the ratio of the dark matter cross section to mass and set stringent constraints on collisionless WDM models such as sterile neutrinos [50, 99].

## 3.4 The dark matter halo profile

One of the most surprising prediction associated with the dark matter particle hypothesis is the existence of a halo of dark matter particles surrounding the galactic disk. Numerical simulations have demonstrated that the dark matter halos were tri-axial (neglecting astrophysical process and the physics associated with the formation of the disk) but could be well-approximated by a sphere.

Within this hypothesis, it is easy to derive the dark matter particles energy density inside the halo using the rotation curves of galaxies and the relation

$$\rho(r) = \frac{1}{4\pi r^2} \frac{dM}{dr}.$$

One obtains a behavior that is proportional to $r^{-3}$ (with the $r$ the distance from the center of the galaxy) for distances greater than a characteristic scale $r_c$, $r^{-2}$ at intermediate distances and $r^{-1}$ for $r < r_c$.

The behavior at small radii is given by the numerical simulations. However they are not exactly consistent with each others. Some indeed give a profile that is inversely proportional to the distance $r^{-1}$ in the inner part of the galaxy while others predict $r^{-1.5}$. In addition if one adds astrophysical processes such as adiabatic contraction, one obtains a mush steeper slope. Of course, theoretical prediction should be by passed by observations. However there is no data yet for distances smaller than the inner kpc.

Despite these uncertainties, it was found that the dark matter halo profile of galaxies, dwarf galaxies and even cluster of galaxies were well-approximated by an universal energy density profile: the so-called Navarro-Frenk-White (NFW) profile:

$$\rho = \frac{\rho_0}{\left(\frac{r}{r_c}\right)^\gamma \left(1 + \left(\frac{r}{r_c}\right)^\alpha\right)^{(\beta-\gamma)/\alpha}}$$

with $\rho_0$ a normalization, $r_c$ the core radius and $\alpha, \beta, \gamma$ parameters predicted by the numerical simulations. For example, typical parameters to describe the Milky Way halo are:

|     | $\alpha$ | $\beta$ | $\gamma$ | $r_s$ kpc |
| --- | --- | --- | --- | --- |
| NFW | 1 | 3 | 1 | 25 |
| KRA | 2 | 3 | 0.2 | 11 |
| ISO | 2 | 2 | 0 | 4 |
| BE  | 1 | 3 | 0.3 | 4 |

and for the Virgo and Coma clusters:

|     | $\alpha$ | $\beta$ | $\gamma$ | $r_s$ | $D$ | $\rho_0$ |
| --- | --- | --- | --- | --- | --- | --- |
| C-NFW | 1 | 3 | 1 | $0.25/h$ | $70/h$ | $0.090h^2$ |
| C-$\beta$-pr. | 2 | 2.25 | 0 | $0.2/h$ | $70/h$ | $0.13h^2$ |
| V-NFW | 1 | 3 | 1 | 0.56 | 15 | 0.012 |
| V-$\beta$-pr. | 2 | 1.41 | 0 | 0.015 | 15 | 0.76 |

Such a profile is in fact very particular. It means that dark matter:

- is distributed in a sphere, surrounding the galaxy. The radius of this sphere is actually greater than the visible radius of the galaxy,

- has an energy density that is much greater in the inner part of the galaxy than at large radii. Given that the energy density is the product of the number density times the dark matter mass, this means that the dark matter number density (i.e. the ratio of the number of dark matter particles to the volume) is very suppressed in the outer part of the galaxy.

The behavior of the dark matter halo profile in the inner part of galaxies remains controversial since several years. For example, observations of dwarf and low surface brightness galaxies suggested that their inner profile was shallower than predicted by CDM. This puzzling discrepancy is often referred as the so-called CDM crisis (altogether with the missing satellite problem). It may be solved once the effect of inclination, non circular orbits and triaxiality are taken into account [100]. However some observations remain puzzling [101, 102].

Interestingly enough, the inner energy density profile of the Milky-Way is also unknown. Observations of other spiral (Milky-Way size) galaxies tend to indicate that they have a rather flat profile. This is also in contradiction with N-body dark matter simulations which predict cuspy profiles. One could nevertheless argue that astrophysical processes and the dynamics of the disk flatten the dark matter halo profile. However adiabatic contraction effects (that describe the impact of the contraction of the gas on the dark matter halo) seems to worsen the discrepancy. Indeed the dark matter halo profile gets steeper when this process is added [86, 103].

## 3.5 The key pending questions

Despite the many efforts in cosmology to prove the nature of dark matter, there is still no evidence in favor of the dark matter particle hypothesis. The only very strong argument in favor of dark matter particles is the fact that it is extremely difficult to avoid the Silk damping effect in modifying gravity scenarios.

Another worrying aspect which was summarized in the previous subsections is that structure formation is not by itself very constraining. It simply indicates that:

- the dark matter mass must be greater than a few keV.

- the dark matter interactions with photons must be smaller than [58]

$$\sigma_{dm-\gamma} v < 2.\,10^{-24} \text{ cm}^3\text{s}^{-1} \left(\frac{m_{dm}\kappa_{dm}}{1\text{MeV}}\right).$$

- the dark matter interactions with baryons must be smaller than [104]

$$\sigma_{dm-\gamma} v < 10^{-15} \text{ cm}^3\text{s}^{-1} \left(\frac{m_{dm}\kappa_{dm}}{1\text{MeV}}\right)$$

- The dark matter self-interactions must be smaller than:

$$\sigma_{dm-dm} v < 10^{-13} \text{ cm}^3\text{s}^{-1} \left(\frac{m_{dm}\kappa_{dm}}{1\text{MeV}}\right)$$

to not modify the DM halo profile inside clusters of galaxies [105, 106].

Recently there has been a claim supporting the existence of dark matter particle. Observations of the so-called bullet cluster (which describes the collision of two clusters) indicate that the gas and the dark matter behave quite differently. While the two clusters collide, the gas and the dark matter are pulled apart. This was confirmed from X-ray observations and lensing measurements. In modifying gravity scenarios, the invisible mass should match the position of the baryonic gas. Since this feature was not observed, the conclusion is that dark matter is made of particles which have been "ejected" during the collision [107, 108]. This also indicates that the dark matter-baryon interaction strength is smaller than that of electromagnetic interactions [109]. However the observation of a dark core in Abell 520 [110] may challenge the collisionless hypothesis and the conclusions drawn in Ref. [108].

Also, while many researchers are actually convinced that dark matter is made of very heavy particles (we will discuss this point in the next section), the constant dispersion velocity of dwarf galaxies may question the range of the dark matter mass that is actually relevant. There is now a lot of interest regarding the impact of dark matter on star formation. A proper modeling may nevertheless requires to know the dark matter mass and interaction rate with precision. Alternatively the comparison of the theoretical prediction of the star formation for a given dark matter scenario with observations may actually constrain the dark matter mass and interaction rates. At last, the distinction between (collisionless or collisional) WDM and CDM scenarios should tell us about the nature of the dark matter particles, if they indeed exist. However, a few issues associated with numerical simulations remain to be solved to make firm predictions. Is there indeed a small-scale regeneration mechanism that makes the non-linear power spectrum of a WDM scenario looking alike that of CDM nowadays? Are the dark matter collisional and free-streaming damping scales detectable in the dark matter power spectrum? Is the dark matter halo profile in galaxies compatible with observations when one adds also the baryonic physics?

To answer these questions, astronomers and observational as well as theoretical cosmologists will probably have to work closely together in the next few years.

## 4 Modifying gravity

Many alternatives to the dark matter hypothesis based on modifying gravity scenarios have been proposed to explain the rotation curves of galaxies. The simplest and most famous of these alternatives is called the Modified Newtonian Dynamics (MOND) program and was introduced by Milgrom in 1983 [111]. In MOND, the Newton's second law in a gravitational field is modified so that the Newtonian theory is recovered for large value of the acceleration $|\vec{a}|$ and modified at small values:

$$\mu(|\vec{a}|/a_0)\vec{a} = -\nabla\Phi$$

where $\Phi$ is the Newtonian potential and $\mu(x)$ is a function with a scale $x \simeq 1$. For $|\vec{a}| > a_0$, $\mu(x) = 1$ while for $|\vec{a}| < a_0$, $\mu(x) \simeq x$. Milgrom's theory has been extremely successful in explaining a number of observational properties of galaxies [112] but it is not generally covariant and hence cannot be studied in a general setting. In addition it faces immediate difficulties. Since MOND is based on a purely baryonic universe, there is no additional substance (apart from massive neutrinos) that could counteract the Silk damping.

A few years ago, J. Bekenstein proposed a new theory that explains lensing measurements, solve the covariance problem and might also avoid the Silk damping effect. This theory is the result of several previous attempts to modify gravity and is indeed a relativistic version of MOND [113]. In Brans-Dicke theory, the tensor field in Einstein's theory of gravity is replaced by a scalar and a tensor field. In Bekenstein's theory, the tensor is now replaced by a scalar, a vector and a tensor which interact in such a way to give MOND in the weak-field non-relativistic limit. This is called the Tensor-Vector-Scalar theory (TeVeS).

Bekenstein's theory has two metrics. One of the metrics, $\tilde{g}_{\mu\nu}$ has its dynamics governed by the Einstein–Hilbert action,

$$S_g = \frac{1}{16\pi G} \int d^4x \sqrt{-\tilde{g}} \tilde{R},$$

where $G$ is Newton's constant and $\tilde{R}$ is the scalar curvature of $\tilde{g}_{\mu\nu}$. The second metric, $g_{\mu\nu}$ is minimally coupled to all the matter fields in the Universe. The two metrics are related through

$$g_{\mu\nu} = e^{-2\phi}(\tilde{g}_{\mu\nu} + A_\mu A_\nu) - e^{2\phi} A_\mu A_\nu.$$

Two fields are required to connect the two metrics. The scalar field, $\phi$ has dynamics given by the action

$$S_s = -\frac{1}{16\pi G} \int d^4x \sqrt{-\tilde{g}} \left[\mu \left(\tilde{g}^{\mu\nu} - A^\mu A^\nu\right) \phi_{,\mu}\phi_{,\nu} + V(\mu)\right],$$

where $\mu$ is a non-dynamical field and $V$ is a free function which can be chosen to give the correct nonrelativistic MOND limit and depends on two free parameters, $\ell_B$ and $\mu_0$ (related to $\kappa$ in [113] as $\mu_0 = 8\pi/\kappa$). The unit timelike vector field, $A_\mu$ has dynamics given by the action

$$S_v = -\frac{1}{32\pi G} \int d^4x \sqrt{-\tilde{g}} \left[K F^{\alpha\beta} F_{\alpha\beta} - 2\lambda(A^\mu A_\mu + 1)\right],$$

where $F_{\mu\nu} = A_{\mu,\nu} - A_{\nu,\mu}$, indices are raised with $\tilde{g}$ and where $K$ is the third parameter in this theory. The Lagrange multiplier $\lambda$ is completely fixed by variation of the action.

Surprisingly enough, TeVeS theory does fit WMAP data perfectly up to the second peak and large-scale surveys despite a competition between overproducing large scale power in the CMB but also overcoming damping on small scale [114] (see [115] for more details). This is due, notably, to the large number of free parameters (five in fact) and the introduction of massive (mass degenerated) neutrinos; in this scenario each neutrino would have a mass of about 2 eV.

The fit to CMB data is not perfect though. In a purely baryonic Universe, the magnitude of the third peak is expected to be suppressed due to the Silk damping effect. In a dark matter-dominated Universe, the third peak is almost at the same level of the second one because the dark matter counteracts the Silk damping effect. Observations now favor a dark matter dominated Universe. Hence, to be successful, TeVeS must mimic the effect of dark matter on the $C_l$ and therefore predicts a non-suppressed third peak. We did not find any combination of the free parameters that could reproduce the magnitude of the third peak. Note also that the neutrino mass that enables us to reproduce WMAP data up to the second peak are is marginally compatible with present experiments and may become problematic once the neutrino mass splitting will be known accurately. Nonetheless, one should stress that the present results are based on a scalar potential $V$ which could perhaps be modified so that the neutrino masses play a less crucial role.

## 5 Progress made in particle physics

Cosmology and astrophysics have actually started the dark matter problematic but so far they fail to answer what is dark matter. Unfortunately, it is difficult to predict whether they will be successful in this task in a short amount of time. In contrast particle and astroparticle physics could turn out to be much more powerful tools to solve the dark matter issue. Indeed they may either demonstrate that dark matter is made of particles and determine their characteristics (i.e. mass and interaction rates) or simply exclude a certain number of candidates, thus reducing the number of viable possibilities to explain all the cosmological observations.

### 5.1 Thermal particles versus non-thermal particles

An important question regarding the nature of dark matter is whether the particles have been produced thermally or not.

Particles which have been once in thermal equilibrium have a very nice property: their number density (i.e. the number of particles per unit of volume) is proportional to the temperature of the photon to the cube: $n_{dm} \propto T^3$. This property is valid whether they are fermionic or bosonic. Hence their energy density, $\rho_{dm} \propto E/V$ with $V \propto T^{-3}$ the volume, can be rewritten as: $\rho_{dm} \propto E \times n_{dm}$. When the temperature of the photons in the Universe is greater than the mass threshold of the dark matter particles, $\rho_{dm} \propto T^4$ while $\rho_{dm} \propto m_{dm} \times T^3$ in the opposite case. One can thus trace the evolution of the energy density of the dark matter particles with the expansion of the Universe and predict the value of this energy density nowadays (see next section) quite accurately. The confrontation of this "relic" energy density with data then gives an information on the dark matter mass and the type of interactions dark matter had during the period where it was in thermal and chemical equilibrium. Examples of thermal candidates are supersymmetric and Kaluza-klein particles.

In contrast, in the case of non-thermal particles, one cannot do such predictions. The energy density has to be fine-tuned so that it matches observations. Typical non-thermal candidates are axions, gravitinos, sterile neutrinos, WIMPzillas. Pseudoscalar particles, such as axions, have been searched for in many experiments. The range of their mass and coupling to two photons have been constrained by astrophysical observations. Yet, a few years ago, the PVLAS experiment announced evidence for such particles [116]. Although their result was inconsistent with the CAST experiments [117], several groups tried to reconcile the various constraints (see for example e.g. [118, 119]). However another "light shining through a wall" experiment in Toulouse and PVLAS itself [120, 121] finally excluded the parameter space initially emphasized by PVLAS (see also [122]).

In the following we will only consider thermal particles and describe how their present energy density constrain their characteristics.

## 5.2 The relic density argument

As discussed previously dark matter is meant to represent only 23% of the matter content of the Universe. Hence any dark matter candidate must actually explain or, at least be compatible with, this fraction.

Yet, as we explained also, the present energy density (also referred to as relic density) of a massive particle with $m > \text{keV}$ is given by $\rho_{dm}^0 \propto m_{dm} T_0^3$ with $T_0 = 2.73 K \sim 3.\,10^{-4}$ eV the temperature of the photon nowadays. The energy density of photons nowadays is given by: $\rho_\gamma^0 \propto T_0^4$. Hence the ratio of the relic dark matter energy density to that of photons is given by:

$$\frac{\rho_{dm}^0}{\rho_\gamma^0} \propto \frac{m_{dm}}{T_0}.$$

Since $m_{dm} \gg T_0$, one obtains that

$$\rho_{dm}^0 \sim 10^6 \, \frac{m_{dm}}{\text{keV}} \, \rho_\gamma^0.$$

WMAP precision measurements however indicates $\rho_{dm}^0 \sim 500 \, \rho_\gamma^0$. Hence a very massive particle failed in principle to explain the observation.

P. Hut, Vysotskii et al, and Lee-Weinberg in 1977 were the first to realize the importance of annihilations in the attainment of the correct (thermal) dark matter relic density. When the temperature of the Universe drops below the dark matter mass threshold, the dark matter number density becomes exponentially suppressed $n_{dm} \propto \left(\frac{mT}{2\pi}\right)^{3/2} e^{-\frac{m}{T}}$ until the dark matter particles leave the chemical equilibrium. This moment can be determined from the resolution of the Boltzman equation:

$$\frac{dn_{dm}}{dt} = -3H\, n_{dm} - \sigma v \, (n_{dm}^2 - n_{dm,0}^2)$$

where $\sigma v$ is the dark matter total annihilation cross section and $H$ the Hubble constant ($H \propto T^2$ in the radiation dominated era).

### 5.2.1 Chemical equilibrium and freeze-out

Altogether the product $\sigma v \, n_{dm}$ defines the rate $\Gamma_{dm}$ at which the dark matter particles annihilate. The annihilation cross section can be seen as the probability for a dark matter particle and its antiparticle to disappear. In the case of a Majorana fermion or real scalar, the annihilation involve two identical dark matter particles instead of involving a particle and antiparticle. The larger the annihilation cross section, the smaller the relic density. In contrast, the smaller the annihilation cross section, the larger the relic density.

This can be seen analytically (and in a quite good approximation) as follows:

The evolution of the dark matter number density with the time, which is described by the Boltzman equation, indicates that there is a competition between the expansion rate (the so-called Hubble rate) and the annihilation rate. Since the expansion rate only dilutes the dark matter number density with the volume, the comoving number density (i.e. the number density times the volume) remains almost constant as soon as the Hubble rate dominates the annihilation rate. Hence the particle leave the chemical equilibrium around a time fixed by the condition:

$$H^{dec} = \sigma v \, n_{dm}^{dec}$$

which using the relations $n_{dm,0} = (\Omega_{dm}\rho_c)/m_{dm}$ and $n_{dm} \times a^3 = n_{dm,0} \times a_0^3$, with $a$ the scale factor and $a_0 = 1$, translates into a temperature $T_{dec}$ given by:

$$x_{dec} = \frac{m_{dm}}{T_{dec}} \simeq 17.2 + \ln(g/\sqrt{g_\star}) + \ln(m_{dm}/\text{GeV}) + \ln\sqrt{x_F}$$
$$\in [12-22]$$

for particles in the MeV-100 GeV range.

The moment at which the dark matter particles leave the chemical equilibrium is called the freeze-out. It is almost independent of the dark matter mass. One then obtains that:

$$\Omega_{dm} h^2 = \frac{T_0}{H_r} \frac{1}{\rho_c/h^2} \frac{x_{dec}}{\sigma v}$$

with $H_r = 2.\,10^{-20} \text{s}^{-1}$ which confirms the naive intuition that the larger the cross section, the smaller the relic density.

The amazing point is that this formula has only constant terms. Hence the annihilation cross section that is required to explain the observed dark matter energy density is the same whatever the candidate:

$$\sigma v_r|_{RD} = 3.\,10^{-27} \text{cm}^3/\text{s} \times \left(\frac{\Omega_{dm} h^2}{0.1}\right). \tag{7}$$

Incidentally this means that any candidate having such a cross section can be the dark matter! Said differently, although the relic density criterion is a rather strong constraint, it does not select one type of candidates but instead validate all models for which the dark matter total annihilation cross section is of $3.\,10^{-27} \text{cm}^3/\text{s}$.

### 5.2.2 Constraints on the dark matter mass

To use this constraints, one has to postulate a candidate and specify its interactions with Standard Model particles. Once this is done, one must obtains the theoretical expression of the total dark matter pair annihilation cross section ($\sigma v_{total}$) and compare it with eq. 7. This procedure is meant to constrain both the mass of the dark matter particles and their couplings.

In the case of massive neutrinos, as mentioned by Hut or Lee and Weinberg, the couplings are known but the mass is unknown. In this case imposing that the total massive neutrino pair annihilation cross section matches the value given in eq. 7 simply constrains the value of the mass for which massive neutrinos can be a good dark matter candidate.

For example, using the condition $\Omega_{dm} < 1$, Hut obtained a lower limit of 3 GeV on the mass of the massive neutrino. This was actually confirmed by Lee and

Weinberg who found 2 GeV. However, by also requiring $\Omega_{dm} > 0$, they found an upper limit of a few TeV on the mass of these massive neutrinos.

In the same way, many authors have constrained the sneutrino and neutralino properties. The most popular extension of the Standard Model, namely supersymmetry, predicts indeed the existence of a neutral massive weakly interacting particles. In the simplest version of supersymmetric realization, the Lightest Supersymmetric Particle (LSP) is often a neutralino (previously called a photino), that is a linear combination of the supersymmetric partner of the gauge and Higgs bosons. In some cases, the LSP can also be the supersymmetric partner of the neutrino, the so-called sneutrino, see Ref. [123, 124]. If the symmetry called R-parity [125] is conserved, the LSP is stable. Indeed, in this case, a supersymmetric particle can only decay into another supersymmetric particle plus a Standard Model particle. A neutral and stable LSP is then a perfect dark matter candidate providing that its relic density matches observation (there is a very long literature on the subject; among the first references, one can cite e.g. [126–131]). Note that neutralinos are Majorana particles, therefore they are their own anti particle. Sneutrinos are complex scalars and therefore not their own antiparticle.

Given that the $N = 1$ supersymmetric parameter space is quite broad, the use of the relic density criterion used to let a quite large region in which the neutralino could be a good dark matter candidate. In particular, if the neutralino was Bino-like, the main annihilation channel was into a pair fermion-anti fermion. Nevertheless with the LEP2 experiment, other laboratory experiments constraints, and the latest measurements of the dark matter relic density, the so-called bulk region in the CMSSM reduced to a very narrow strip [132].

### 5.2.3 Some complications: coannihilations, focus point and Higgs pole

Other mechanisms than ordinary annihilations have been proposed to explain the neutralino relic density. They somewhat became a necessity given that direct dark matter detection experiments now set very stringent constraints on the dark matter elastic-scattering cross section with protons, indicating that the annihilation cross section is less likely to match eq. 7 and more likely to have a smaller value.

One of them is called the coannihilation mechanism (see for example among the first references on the subject Ref. [133–138]). The term "coannihilation" means that the neutralino (or more generally the LSP) annihilates with another particle instead of annihilating with itself. Since the neutralino is assumed to be the LSP, all the other supersymmetric particles are meant to be heavier than neutralinos, and thus decay well before that neutralinos become non-relativistic. One condition for coannihilations to happen is therefore that the Next to LSP (NLSP) is light enough to decay at the same epoch as the neutralino annihilations. In this case, the neutralino number density decreases more efficiently. There is nevertheless an exception to this condition. When the NLSP has strong interactions, the mass splitting between the LSP and the NLSP can be larger, as shown in [137].

Another possibility to explain why neutralinos have the correct relic density is that the neutralino mass is half that of the pseudo-Higgs mass. The annihilation cross section is then increased due to a resonance. This is called the Higgs-pole region.

At last another case, very specific to supersymmetry, where the neutralino relic density can match observations is when the neutralinos and charginos (the superpartners of the charged Higgs) are mass degenerated. This happens when the neutralino is Higgsino-like. In this case neutralino annihilations into a pair of $W$ bosons are efficient because the particle that is mediated during the annihilations is a chargino with a mass very similar to that of the neutralino, i.e. there is only one suppression scale.

Handling all the supersymmetric parameters is not an easy task and codes have been established to simultaneously perform scans on the parameter space, compute the relic density, check the possibility of direct and indirect detection and the compatibility with accelerator or collider constraints. The two codes of references are micrOMEGAs [139] and DarKSUSY [140] presently used by many authors.

### 5.2.4 When the spin of the particle matters

In the previous sections, we have illustrated the relic density argument using massive neutrinos and discussed the limits that can be derived on the DM mass from this criterion in the case of the neutralino. All these particles are fermionic with a spin-1/2. However there is no reason why dark matter should be a fermion. It could well be a scalar particle or a gauge boson.

Scalar particles, such as e.g. the sneutrino in supersymmetry, can annihilate into a pair of Standard Model particles via the exchange of a fermion $F$ or a gauge (vector) boson. For example, sneutrinos can annihilate into electron-positron via the exchange of a chargino (the linear combination of the charged Higgsino) or into a neutrino pair (or neutrino-anti neutrino pair) via the exchange of a neutralino and a $Z$-boson. One can generalize this example to a non-supersymmetric situation where a pair of scalar dark matter particle (real or complex) annihilates via the exchange of a new heavy spin-1/2 particle $F$ via the t and u-channels.

The generalized annihilation cross section (with S- and P-waves) is given in Ref. [60]. However it can be simply rewritten as

$$\sigma v = A \frac{C_l^2 C_r^2}{m_F^2}$$

with $A$ a coefficient equal to $1/(4\pi)$ or $1/\pi$ depending on whether the particle is a complex or real scalar respectively and $C_{l,r}$ the left and right-handed couplings of the scalar particle to the fermion $F$ and the Standard Model particle.

This cross section is independent of the dark matter mass. Hence, if one imposes that it matches the value $\sigma v_{ann} = 10^{-26} \text{cm}^3/\text{s}$, one constrains the mass of the particle that is exchanged ($m_F$) and the couplings $C_{l,r}$ but not on the dark matter mass itself. Thanks to this unique property of scalar particles, one can conclude that scalar dark matter can be as light as a few MeV (the MeV lower limit is set by the BBN requirement) and, yet, have the correct relic density. Spin-0 particles are therefore counter-examples of Hut and Lee-Weinberg conclusions [59, 60]. Such light scalar (annihilating) dark matter particles (decaying and annihilating) have later been studied in detail as they might explain the 511 keV line observed in the centre of the galaxy [61, 141].

Gauge boson dark matter particles are predicted for example in models of Universal Extra dimensions (UED). It was shown in particular that the first levels of the Kaluza-Klein (KK) modes of the neutral gauge boson could be a viable dark matter candidate [142, 143]. This case requires coannihilations between the lightest KK modes because the annihilation cross section is proportional to the square of the dark matter mass. Since the 2000s, many new dark matter candidates have been proposed. However, to my knowledge, they do not involve any new mechanism other than those already discussed in this review.

## 5.3 Direct detection

Although many cosmological evidence tend to suggest that dark matter is made of particles, modifying gravity might finally be the answer to the dark matter issue. A dark matter signature in a laboratory experiment would remove any ambiguity. Besides it will help to discriminate between different kinds of candidates (spin-0, spin-1/2 or spin-1 particles). In 1985 Goodman and Witten [144] have noticed that one could directly test the WIMP hypothesis by using a detector in a low background environment. Dark matter being in a halo around the galactic disk and the orbits being unperfect, it is meant to fall on earth and eventually interact with the nuclei in the detector. The expected signatures are ionization charge, scintillation light, heat (thermal phonons) nuclear recoil. A few exceptions are e.g. PICASSO which is based on the formation of superheated droplets and the acoustic signal generated when they explode and DAMA which aim at measuring the annual modulation associated with the presence of dark matter. Target detector material are e.g. Germanium or Sapphire crystals [145–148], Xenon [149], Fluor [150], nitrogen, NaI [151] and Argon [152] liquids.

The main characteristic of these detectors is their sensitivity to spin-dependent and/or spin-independent elastic scattering dark matter cross sections. The spin-dependent cross section is sensitive to axial couplings (i.e. to the dark matter couplings to ordinary matter with a $\gamma_5$ matrix) while the spin-independent cross section is sensitive to vectorial couplings. The spin-independent cross sections are much more constrained. The latest results from XENON 10 confirm that particles with a mass in the range 10GeV-TeV and a WIMP-nucleon cross section greater than $10^{-42}$cm$^2$ are excluded. In addition they showed that particles with a mass of 30–40 GeV and a cross section as small as $[6-7]\ 10^{-44}$cm$^2$ are excluded. Supersymmetry used to predict larger cross sections but do to the coannihilation mechanism, the Higgs pole and the focus point region, a big part of the parameter space lies between WIMP-nucleon cross section of $10^{-44} - 10^{-42}$cm$^2$ [149].

So far the lack of evidence help in constraining the long list of dark matter candidates. In models where the dark matter elastic scattering cross section is related to the annihilation cross section, these experiments set very stringent constraints on the annihilation cross section. In fact they often exclude models in which these two cross sections are strongly related, favoring scenarios where they are independent. It is nevertheless important to remember that direct detection experiments are for the moment insensitive to particles lighter than a few GeV and much heavier than a few TeVs.

## 5.4 Indirect detection

So far I mentioned how to detect dark matter in a laboratory experiment but there is also a way to detect dark matter indirectly thanks to the secondary particles that are produced during the dark matter decay or annihilations. For example, as was noticed in 1978 by Stecker [153] and later by [154, 155], dark matter annihilations can produce a continuum of photons or a monoenergetic gamma ray line. Also it could produce anti positrons [156, 157] as well positrons and neutrinos [158]. These annihilations can either take place in the galactic centre or in the sun [159]. Hence there is a hope to detect a dark matter signature by either searching for monoenergetic photons (i.e. gamma ray line) which trace the dark matter mass, an excess of gamma ray or neutrino continuum, gamma ray or neutrino lines and finally anti protons and/or anti electrons.

Since neutrinos are weakly interacting, one way to detect neutrinos is to use their ability to convert into muons or electrons. This is expected to happen in very rich medium such as the earth or the sun. The electrons and muons thus produced can then be detected in a water or ice detector through the Cherenkov light they emit when they interact with the nuclei of the detector. This factor of conversion can be estimated quite accurately. It is about $10^{-9}$. I.e. the flux of neutrinos that can be detected is about $10^{-9}$ times the flux that is emitted. They are two big experiments dedicated to neutrino astronomy: Antares (in the Mediterranean sea) and ICECUBE (in antartic). Satellite experiments are used on the other hand to detect gamma rays [160].

# 6 Progress and key pending questions

Despite the tremendous efforts made in experimental particle and astroparticle physics to directly or indirectly detect dark matter particles, there has been no conclusive evidence yet in favour of dark matter particles. Until dark matter particles are found (if they exist), scenarios of modifying gravity will remain a possible candidate to explain observations.

Thus, if dark matter is indeed made of particles, their spin, mass and interaction rates with Standard Model particles still remain to be determined. In the likely case where it is made of particles, it is also unclear whether dark matter is made of one or several neutral stable kind of particles. Most of the theoretical works assume, in general, that it is made of one species. Not only this greatly simplifies the calculations but it also a very useful assumption to make predictions.

Another open question is whether dark matter is thermal or not. The chance of detection if dark matter is non-thermal are quite smaller than the thermal case.

# 7 Conclusions

The dark matter field is "officially" born about thirty years ago with the precise measurement of the rotation curves of spiral galaxies. However, despite the fact that cosmologists, astrophysicists and particle physicists have worked very hard to gather information about the nature of dark matter, one may have the very unpleasant feeling that observations (or the lack of evidence) always bring us back to the naive picture that emerged in the early 1980s. I.e. dark matter is likely to be made of WIMPs whose nature remains to be discovered.

Yet we learnt a lot. For example, thanks to the discovery of the primordial density fluctuations and the confirmation of the Silk damping effect, we now know that dark matter plays a fundamental role. It is indeed the substance that explains the number of galaxies, dwarf galaxies and even smaller objects. Besides it prevents these objects to be pulled apart.

We also learnt that dark matter indeed represents the largest fraction of the matter component of the Universe (about 80%) and somehow progressed in our understanding of Einstein's gravity. While for many years, there has been a controversy about the existence of the so-called cosmological constant, latest developments in cosmology now strongly suggest that, on top of dark matter, a second unknown substance (referred to as Dark Energy and which looks alike the cosmological constant[1]) which "fills" about 70% of the Universe. Unfortunately, its discovery makes dark matter a more puzzling topic. Are dark matter and dark energy two different aspects of a same problem or are they totally disconnected as often considered when one computes the relic density of a specific dark matter candidate? Would a connection between these two substances drastically change our understanding of the law of physics? Wouldn't it suggest a modification of gravity?

Although this is currently a hot topic-, cosmologists are presently gathering information that may solve these big questions. For example, with the precise measurement of the matter and angular power spectra ($P(k)$ and $C_l$), we start to know the number of small-size cosmological objects (say down to scales comparable to Lyman-alpha clouds). These measurements are extremely important since they rule out many modifying gravity scenarios. However to some extent they also point out the deficiency of the CDM scenario (this is the so-called missing satellite problem). Hence, as long as dark matter has not been discovered directly, these measurements simply teach us how gravity behaves at the different scales that we test.

Among the questions that are still unsolved, one could cite structure formation. Although numerical simulations unambiguously determine how large scale structures form in the CDM scenario, there is presently a big debate about the WDM scenario. Collisionless or collisional WDM predicts a cut-off in the linear matter power spectrum that depends on the dark matter mass and/or its interactions. However, this cut-off may not show-off (or not in the same way) in the non-linear matter power spectrum [62]. Also, so far, observations are not precise enough to efficiently constrain this scenario. Hence, it is too premature to conclude that dark matter particles must belong to the CDM scenario. Let us stress in addition that CDM faced/faces difficulties (e.g. cuspy profiles, missing satellites, correct description of dwarf galaxies). However, there are many progress to do regarding the modeling of astrophysical process in galaxies, known to potentially have an impact on halo formation and change the CDM predictions based on dark matter simulations only. Will these dif-

---

[1] The determination of the equation of state will be crucial. The relation $p = -\rho$ would suggest that the so-called dark energy is Einstein's cosmological constant.

ficulties remain and encourage physicists to consider other scenarios than CDM? At present, this is difficult to say.

From the particle physics side, many challenging experiments have been proposed and performed. Yet dark matter particles still remain invisible. If we do not discover any evidence of new physics at LHC nor detect some WIMP signal in direct detection and/or indirect detection experiments, we will be able to exclude a large class of dark matter models. Would this mean that dark matter is lighter than conventional WIMPs as proposed in supersymmetry or heavier than a few TeV? Would this lead to reconsider modifying gravity scenarios? Or would this be synonym of very weak elastic scattering cross sections and a velocity-dependence annihilation cross section for example? In my opinion there is a non-negligible chance that dark matter remains invisible for many more years despite all the experimental efforts that are being done. The situation may drastically change nevertheless if ATLAS and CMS experiments at CERN/LHC discover new heavy charged particles, sign of new physics beyond the Standard Model.

Although the lack of experimental evidence would prevent theoreticians to converge towards a unique dark matter candidate (or at least a unique class of candidates), progress in astrophysics might help physicists and cosmologists to determine the relevant mass and interaction scale for dark matter. For example, WARM or COLD scenarios predict different pattern for structure formation and star formation. The annihilation/decay rate could also modify star formation. Hence new information might be gathered from the study of astrophysical process.

In the 2000s, the advent of neutrino and gamma ray telescopes for detecting evidence of dark matter have lead particle physicists to discover new astrophysical sources. Since cosmology is now getting closer to astrophysics for modeling galaxy formation, the dark matter halo, reionization sources, and understanding CMB physics at very small scales, one may wonder if the natural direction of the three fields (e.g. particle physics, cosmology, and astrophysics) is not to merge into one single field centered around the "dark" thematic.

> "What does she do? Is she an archeologist?"
> "Oh ...she's a physicist. She studies dark matter," said Lyra [...]
> "Dark matter?" he was saying. "How fascinating! I saw something about this in the *Times* this morning. The universe is full of this mysterious stuff, and nobody knows what it is!"
>
> *The Subtle Knife, Philip Pullman*

# References

[1] F. Zwicky. Spectral displacement of extra galactic nebulae. *Helv. Phys. Acta*, 6:110–127, 1933.
[2] F. Zwicky. Die Rotverschiebung von extragalaktischen Nebeln. *Helvetica Physica Acta*, 6:110–127, 1933.
[3] S. Smith. The Mass of the Virgo Cluster. *APJ*, 83:23–+, January 1936.

[4] H. W. Babcock. *On the Rotation of the Andromenda Nebula.* PhD thesis, AA(UNIVERSITY OF CALIFORNIA, BERKELEY.), 1938.

[5] J. H. Oort. Some Problems Concerning the Structure and Dynamics of the Galactic System and the Elliptical Nebulae NGC 3115 and 4494. *APJ*, 91:273–+, April 1940.

[6] E. M. Burbidge and G. R. Burbidge. *The Masses of Galaxies*, pages 81–+. Galaxies and the Universe, January 1975.

[7] H. I. Ewen and E. M. Purcell. Observation of a Line in the Galactic Radio Spectrum: Radiation from Galactic Hydrogen at 1,420 Mc./sec. *Nature*, 168:356–+, September 1951.

[8] C. A. Muller and J. H. Oort. Observation of a Line in the Galactic Radio Spectrum: The Interstellar Hydrogen Line at 1,420 Mc./sec., and an Estimate of Galactic Rotation. *Nature*, 168:357–358, September 1951.

[9] M. S. Roberts. *Radio Observations of Neutral Hydrogen in Galaxies*, pages 309–+. Galaxies and the Universe, January 1975.

[10] V. C. Rubin and W. K. Ford, Jr. A Comparison of Dynamical Models of the Andromeda Nebula and the Galaxy. In W. Becker and G. I. Kontopoulos, editors, *The Spiral Structure of our Galaxy*, volume 38 of *IAU Symposium*, pages 61–+, 1970.

[11] V. C. Rubin and W. K. J. Ford. Rotation of the Andromeda Nebula from a Spectroscopic Survey of Emission Regions. *APJ*, 159:379–+, February 1970.

[12] D. H. Rogstad and G. S. Shostak. Gross Properties of Five Scd Galaxies as Determined from 21-CENTIMETER Observations. *APJ*, 176:315–+, September 1972.

[13] M. S. Roberts and A. H. Rots. Comparison of Rotation Curves of Different Galaxy Types. *AAP*, 26:483–+, August 1973.

[14] V. C. Rubin, N. Thonnard, and W. K. Ford, Jr. Extended rotation curves of high-luminosity spiral galaxies. IV - Systematic dynamical properties, SA through SC. *APJ Letters*, 225:L107–L111, November 1978.

[15] P. C. van der Kruit and A. Bosma. Optical Surface Photometry of the Barred Spiral Galaxy NGC 5383. *AAP*, 70:63–+, November 1978.

[16] A. Bosma and P. C. van der Kruit. The local mass-to-light ratio in spiral galaxies. *AAP*, 79:281–286, November 1979.

[17] J. P. Ostriker, P. J. E. Peebles, and A. Yahil. The size and mass of galaxies, and the mass of the universe. *APJ Letters*, 193:L1–L4, October 1974.

[18] R. V. Wagoner. Big-Bang Nucleosynthesis Revisited. *APJ*, 179:343–360, January 1973.

[19] J. Silk. Fluctuations in the Primordial Fireball. *Nature*, 215:1155–+, September 1967.

[20] J. Yang, D. N. Schramm, G. Steigman, and R. T. Rood. Constraints on cosmology and neutrino physics from big bang nucleosynthesis. *APJ*, 227:697–704, January 1979.

[21] S. Sarkar. Big bang nucleosynthesis and physics beyond the standard model. *Reports of Progress in Physics*, 59:1493–1609, 1996.

[22] D. Romano, M. Tosi, F. Matteucci, and C. Chiappini. Light element evolution resulting from WMAP data. *MNRAS*, 346:295–303, November 2003.

[23] O. Pisanti, A. Cirillo, S. Esposito, F. Iocco, G. Mangano, G. Miele, and P. D. Serpico. PArthENoPE: Public Algorithm Evaluating the Nucleosynthesis of Primordial Elements. *ArXiv e-prints*, 705, May 2007.

[24] S. Chandrasekhar and L. R. Henrich. An Attempt to Interpret the Relative Abundances of the Elements and Their Isotopes. *APJ*, 95:288–+, March 1942.

[25] G. Gamow. The origin of elements and the separation of galaxies. *Phys. Rev.*, 74(4):505–506, Aug 1948.

[26] P. J. E. Peebles. Primordial Helium Abundance and the Primordial Fireball. II. *APJ*, 146:542–+, November 1966.

[27] C. Alcock, R. A. Allsman, D. Alves, R. Ansari, E. Aubourg, T. S. Axelrod, P. Bareyre, J.-P. Beaulieu, A. C. Becker, D. P. Bennett, S. Brehin, F. Cavalier, S. Char, K. H. Cook, R. Ferlet, J. Fernandez, K. C. Freeman, K. Griest, P. Grison, M. Gros, C. Gry, J. Guibert, M. Lachieze-Rey, B. Laurent, M. J. Lehner, E. Lesquoy, C. Magneville, S. L. Marshall, E. Maurice, A. Milsztajn, D. Minniti, M. Moniez, O. Moreau, L. Moscoso, N. Palanque-Delabrouille, B. A. Peterson, M. R. Pratt, L. Prevot, F. Queinnec, P. J. Quinn, C. Renault, J. Rich, M. Spiro, C. W. Stubbs, W. Sutherland, A. Tomaney, T. Vandehei, A. Vidal-Madjar, L. Vigroux, and S. Zylberajch. EROS and MACHO Combined Limits on Planetary-Mass Dark Matter in the Galactic Halo. *APJ Letters*, 499:L9+, May 1998.

[28] J. H. Jeans. *Astronomy and cosmogony*. Cambridge [Eng.] The University press, 1928., 1928.

[29] P. J. E. Peebles. The Black-Body Radiation Content of the Universe and the Formation of Galaxies. *APJ*, 142:1317–+, November 1965.

[30] C. W. Misner. Transport Processes in the Primordial Fireball. *NATURE*, 214:40–+, 1967.

[31] A. S. Szalay and G. Marx. Neutrino rest mass from cosmology. *AAP*, 49:437–441, June 1976.

[32] R. Cowsik and J. McClelland. An upper limit on the neutrino rest mass. *Phys. Rev. Lett.*, 29(10):669–670, Sep 1972.

[33] S.S. Gershtein and Y.B. Zeldovich. Rest Mass of Muonic Neutrino and Cosmology. *Pisma Zh. Eksp. Teor. Fiz.*, 4, 1966.

[34] C. Pryor, M. Davis, M. Lecar, and E. Witten. Galaxy Formation With Massive Neutrinos. In *Bulletin of the American Astronomical Society*, volume 12 of *Bulletin of the American Astronomical Society*, pages 861–+, September 1980.

[35] M. Davis, M. Lecar, C. Pryor, and E. Witten. The formation of galaxies from massive neutrinos. *APJ*, 250:423–431, November 1981.

[36] J. R. Bond, G. Efstathiou, and J. Silk. Massive neutrinos and the large-scale structure of the universe. *Physical Review Letters*, 45:1980–1984, December 1980.

[37] S. Tremaine and J. E. Gunn. Dynamical role of light neutral leptons in cosmology. *Physical Review Letters*, 42:407–410, February 1979.

[38] V. F. Shvartsman. Density of relict particles with zero rest mass in the universe. *Soviet Journal of Experimental and Theoretical Physics Letters*, 9:184–186, 1969.

[39] G. Steigman, K. A. Olive, and D. N. Schramm. Cosmological constraints on superweak particles. *Physical Review Letters*, 43:239–242, July 1979.

[40] P. Hut. Limits on masses and number of neutral weakly interacting particles. *Physics Letters A*, 69:85–88, July 1977.

[41] M. I. Vysotskij, A. D. Dolgov, and Y. B. Zel'Dovich. Cosmological restrictions of the masses of neutral leptons. *Pis ma Zhurnal Eksperimental noi i Teoreticheskoi Fiziki*, 26:200–202, 1977.

[42] P. J. E. Peebles. Large-scale background temperature and mass fluctuations due to scale-invariant primeval perturbations. *APJ letters*, 263:L1–L5, December 1982.

[43] B. W. Lee and S. Weinberg. Cosmological lower bound on heavy-neutrino masses. *Physical Review Letters*, 39:165–168, July 1977.

[44] R. Schaeffer and J. Silk. Cold, warm, or hot dark matter - Biased galaxy formation and pancakes. *APJ*, 332:1–16, September 1988.

[45] P. Bode, J. P. Ostriker, and N. Turok. Halo Formation in Warm Dark Matter Models. *APJ*, 556:93–107, July 2001.

[46] A. Knebe, J. E. G. Devriendt, A. Mahmood, and J. Silk. Merger histories in warm dark matter structure formation scenarios. *MNRAS*, 329:813–828, February 2002.

[47] C. Boehm and R. Schaeffer. Constraints on Dark Matter interactions from structure formation: damping lengths. *AAP*, 438:419–442, August 2005.

[48] M. Shaposhnikov. How to find sterile neutrinos ? *ArXiv e-prints*, 706, June 2007.

[49] A. D. Dolgov and S. H. Hansen. Massive sterile neutrinos as warm dark matter. *Astroparticle Physics*, 16:339–344, January 2002.

[50] M. Viel, J. Lesgourgues, M. G. Haehnelt, S. Matarrese, and A. Riotto. Can Sterile Neutrinos Be Ruled Out as Warm Dark Matter Candidates? *Physical Review Letters*, 97(7):071301–+, August 2006.

[51] C. Bœhm, P. Fayet, and R. Schaeffer. Constraining dark matter candidates from structure formation. *Physics Letters B*, 518:8–14, October 2001.

[52] D. N. Spergel and P. J. Steinhardt. Observational Evidence for Self-Interacting Cold Dark Matter. *Physical Review Letters*, 84:3760–3763, April 2000.

[53] S. Hofmann, D. J. Schwarz, and H. Stöcker. Damping scales of neutralino cold dark matter. *PRD*, 64(8):083507–+, October 2001.

[54] J. Diemand, B. Moore, and J. Stadel. Earth-mass dark-matter haloes as the first structures in the early Universe. *NATURE*, 433:389–391, January 2005.

[55] A. Loeb and M. Zaldarriaga. Small-scale power spectrum of cold dark matter. *PRD*, 71(10):103520–+, May 2005.

[56] D. Hooper, M. Kaplinghat, L. E. Strigari, and K. M. Zurek. MeV dark matter and small scale structure. *PRD*, 76(10):103515–+, November 2007.

[57] S. Weinberg. Entropy Generation and the Survival of Protogalaxies in an Expanding Universe. *APJ*, 168:175–+, September 1971.

[58] C. Bœhm, A. Riazuelo, S. H. Hansen, and R. Schaeffer. Interacting dark matter disguised as warm dark matter. *PRD*, 66(8):083505–+, October 2002.

[59] C. Boehm, T. A. Ensslin, and J. Silk. Are light annihilating dark matter particles possible? *J. Phys.*, G30:279–286, 2004.

[60] C. Boehm and P. Fayet. Scalar dark matter candidates. *Nucl. Phys.*, B683:219–263, 2004.

[61] Celine Boehm, Dan Hooper, Joseph Silk, Michel Casse, and Jacques Paul. Mev dark matter: Has it been detected? *Phys. Rev. Lett.*, 92:101301, 2004.

[62] C. Boehm, H. Mathis, J. Devriendt, and J. Silk. Wimp matter power spectra and small scale power generation. 2003.

[63] J. C. Mather, E. S. Cheng, R. E. Eplee, Jr., R. B. Isaacman, S. S. Meyer, R. A. Shafer, R. Weiss, E. L. Wright, C. L. Bennett, N. W. Boggess, E. Dwek, S. Gulkis, M. G. Hauser, M. Janssen, T. Kelsall, P. M. Lubin, S. H. Moseley, Jr., T. L. Murdock, R. F. Silverberg, G. F. Smoot, and D. T. Wilkinson. A preliminary measurement of the cosmic microwave background spectrum by the Cosmic Background Explorer (COBE) satellite. *APJ Letters*, 354:L37–L40, May 1990.

[64] C. L. Bennett, G. F. Smoot, G. Hinshaw, E. L. Wright, A. Kogut, G. de Amici, S. S. Meyer, R. Weiss, D. T. Wilkinson, S. Gulkis, M. Janssen, N. W. Boggess, E. S. Cheng, M. G. Hauser, T. Kelsall, J. C. Mather, S. H. Moseley, Jr., T. L. Murdock, and R. F. Silverberg. Preliminary separation of galactic and cosmic microwave emission for the COBE Differential Microwave Radiometer. *APJ Letters*, 396:L7–L12, September 1992.

[65] C. B. Netterfield et al. A measurement by boomerang of multiple peaks in the angular power spectrum of the cosmic microwave background. *Astrophys. J.*, 571:604–614, 2002.

[66] Andrew H. Jaffe et al. Cosmology from maxima-1, boomerang and cobe/dmr cmb observations. *Phys. Rev. Lett.*, 86:3475–3479, 2001.

[67] C. L. Bennett et al. First year wilkinson microwave anisotropy probe (wmap) observations: Preliminary maps and basic results. *Astrophys. J. Suppl.*, 148:1, 2003.

[68] D. N. Spergel et al. Wilkinson microwave anisotropy probe (wmap) three year results: Implications for cosmology. *Astrophys. J. Suppl.*, 170:377, 2007.

[69] Adam G. Riess et al. Observational evidence from supernovae for an accelerating universe and a cosmological constant. *Astron. J.*, 116:1009–1038, 1998.

[70] S. Perlmutter et al. Measurements of omega and lambda from 42 high-redshift supernovae. *Astrophys. J.*, 517:565–586, 1999.

[71] Martin Kunz. Why we need to see the dark matter to understand the dark energy. 2007.

[72] Edward R. Harrison. Fluctuations at the threshold of classical cosmology. *Phys. Rev.*, D1:2726–2730, 1970.

[73] Simon D. M. White and M. J. Rees. Core condensation in heavy halos: A two stage theory for galaxy formation and clusters. *Mon. Not. Roy. Astron. Soc.*, 183:341–358, 1978.

[74] Jeremiah P. Ostriker and Lennox L. Cowie. Galaxy formation in an intergalactic medium dominated by explosions. *Astrophys. J.*, 243:L127–L131, 1981.

[75] Joel R. Primack. Galaxy and cluster formation in a universe dominated by cold dark matter. Invited talk given at 8th Johns Hopkins Workshop, Particles and Gravity, Baltimore, MD, Jun 20-22, 1984.

[76] S. D. M. White. Angular momentum growth in protogalaxies. *Astrophys. J.*, 286:38–41, 1984.
[77] Simon D. M. White, C. S. Frenk, and M. Davis. Clustering in a neutrino-dominated universe. *Astrophys. J.*, 274:L1–L5, 1983.
[78] G. Efstathiou and J. W. Eastwood. On the clustering of particles in an expanding universe. *Mon. Not. Roy. Astron. Soc.*, 194:503–525, February 1981.
[79] J. A. Peacock and S. J. Dodds. Nonlinear evolution of cosmological power spectra. *Mon. Not. Roy. Astron. Soc.*, 280:L19, 1996.
[80] G. L. Bryan and M. L. Norman. Statistical properties of x-ray clusters: Analytic and numerical comparisons. *Astrophys. J.*, 495:80, 1998.
[81] Tom Abel, Greg L. Bryan, and Michael L. Norman. The formation of the first star in the universe. *Science*, 295:93, 2002.
[82] Eric Hayashi, Julio F. Navarro, James E. Taylor, Joachim Stadel, and Thomas Quinn. The structural evolution of substructure. *Astrophys. J.*, 584:541–558, 2003.
[83] Julio F. Navarro et al. The inner structure of lambdacdm halos iii: Universality and asymptotic slopes. *Mon. Not. Roy. Astron. Soc.*, 349:1039, 2004.
[84] Yago Ascasibar, G. Yepes, S. Gottlober, and V. Muller. On the physical origin of dark matter density profiles. *Mon. Not. Roy. Astron. Soc.*, 352:1109, 2004.
[85] Jeremy Blaizot et al. Galics iii: Predicted properties for lyman break galaxies at redshift 3. *Mon. Not. Roy. Astron. Soc.*, 352:571, 2004.
[86] Oleg Y. Gnedin, Andrey V. Kravtsov, Anatoly A. Klypin, and Daisuke Nagai. Response of dark matter halos to condensation of baryons: cosmological simulations and improved adiabatic contraction model. *Astrophys. J.*, 616:16–26, 2004.
[87] Gabriella De Lucia, Volker Springel, Simon D. M. White, Darren Croton, and Guinevere Kauffmann. The formation history of elliptical galaxies. *Mon. Not. Roy. Astron. Soc.*, 366:499–509, 2006.
[88] Volker Springel. The cosmological simulation code gadget-2. *Mon. Not. Roy. Astron. Soc.*, 364:1105–1134, 2005.
[89] Kentaro Nagamine, Renyue Cen, Lars Hernquist, Jeremiah P. Ostriker, and Volker Springel. Massive galaxies and eros at $z = 1 - 3$ in cosmological hydrodynamic simulations: near-ir properties. *Astrophys. J.*, 627:608–620, 2005.
[90] Liang Gao et al. The redshift dependence of the structure of massive lcdm halos. 2007.
[91] T. Sousbie, C. Pichon, S. Colombi, D. Novikov, and D. Pogosyan. The three dimensional skeleton: tracing the filamentary structure of the universe. 2007.
[92] David H. Weinberg, Stephane Colombi, Romeel Dave, and Neal Katz. Baryon dynamics, dark matter substructure, and galaxies. 2006.
[93] Roya Mohayaee, Hugues Mathis, Stephane Colombi, and Joseph Silk. Reconstruction of primordial density fields. *Mon. Not. Roy. Astron. Soc.*, 365:939–959, 2006.
[94] Marcel Zemp, Ben Moore, Joachim Stadel, C. Marcella Carollo, and Piero Madau. Multi-mass spherical structure models for n-body simulations. 2007.

[95] Philip Bett et al. The spin and shape of dark matter haloes in the millennium simulation of a lambdacdm universe. *Mon. Not. Roy. Astron. Soc.*, 376:215–232, 2007.
[96] Volker Springel et al. Simulating the joint evolution of quasars, galaxies and their large-scale distribution. *Nature*, 435:629–636, 2005.
[97] Shaun Cole et al. The 2df galaxy redshift survey: Power-spectrum analysis of the final dataset and cosmological implications. *Mon. Not. Roy. Astron. Soc.*, 362:505–534, 2005.
[98] Max Tegmark et al. Cosmological constraints from the sdss luminous red galaxies. *Phys. Rev.*, D74:123507, 2006.
[99] Steen H. Hansen, Julien Lesgourgues, Sergio Pastor, and Joseph Silk. Closing the window on warm dark matter. *Mon. Not. Roy. Astron. Soc.*, 333:544–546, 2002.
[100] E. Hayashi, J. F. Navarro, A. Jenkins, C. S. Frenk, C. Power, S. D. M. White, V. Springel, J. Stadel, T. Quinn, and J. Wadsley. Disk Galaxy Rotation Curves in Triaxial CDM Halos. *ArXiv Astrophysics e-prints*, August 2004.
[101] G. Gentile, P. Salucci, U. Klein, D. Vergani, and P. Kalberla. The cored distribution of dark matter in spiral galaxies. *MNRAS*, 351:903–922, July 2004.
[102] W. J. G. de Blok. Halo Mass Profiles and Low Surface Brightness Galaxy Rotation Curves. *APJ*, 634:227–238, November 2005.
[103] G. R. Blumenthal, S. M. Faber, R. Flores, and J. R. Primack. Contraction of dark matter galactic halos due to baryonic infall. *APJ*, 301:27–34, February 1986.
[104] Xue-lei Chen, Steen Hannestad, and Robert J. Scherrer. Cosmic microwave background and large scale structure limits on the interaction between dark matter and baryons. *Phys. Rev.*, D65:123515, 2002.
[105] Romeel Dave, David N. Spergel, Paul J. Steinhardt, and Benjamin D. Wandelt. Halo properties in cosmological simulations of self- interacting cold dark matter. *Astrophys. J.*, 547:574–589, 2001.
[106] Haakon Dahle, Steen Hannestad, and Jesper Sommer-Larsen. The density profile of cluster-scale dark matter halos. *Astrophys. J.*, 588:L73, 2003.
[107] Douglas Clowe, Anthony Gonzalez, and Maxim Markevitch. Weak lensing mass reconstruction of the interacting cluster 1e0657-558: Direct evidence for the existence of dark matter. *Astrophys. J.*, 604:596–603, 2004.
[108] Douglas Clowe et al. A direct empirical proof of the existence of dark matter. *Astrophys. J.*, 648:L109–L113, 2006.
[109] Maxim Markevitch et al. Direct constraints on the dark matter self-interaction cross-section from the merging cluster 1e0657-56. *Astrophys. J.*, 606:819–824, 2004.
[110] A. Mahdavi, H. y Hoekstra, A. y Babul, D. y Balam, and P. Capak. A dark core in abell 520. 2007.
[111] M. Milgrom. A modification of the newtonian dynamics as a possible alternative to the hidden mass hypothesis. *Astrophys. J.*, 270:365–370, 1983.
[112] R. H. Sanders. Mond and cosmology. 2005.
[113] Jacob D. Bekenstein. Relativistic gravitation theory for the mond paradigm. *Phys. Rev.*, D70:083509, 2004.

[114] Constantinos Skordis, D. F. Mota, P. G. Ferreira, and C. Boehm. Large scale structure in bekenstein's theory of relativistic mond. *Phys. Rev. Lett.*, 96:011301, 2006.

[115] Constantinos Skordis. Teves cosmology: Covariant formalism for the background evolution and linear perturbation theory. *Phys. Rev.*, D74:103513, 2006.

[116] E. Zavattini et al. Experimental observation of optical rotation generated in vacuum by a magnetic field. *Phys. Rev. Lett.*, 96:110406, 2006.

[117] K. Zioutas et al. First results from the cern axion solar telescope (cast). *Phys. Rev. Lett.*, 94:121301, 2005.

[118] Eduard Masso and Javier Redondo. Compatibility of cast search with axion-like interpretation of pvlas results. *Phys. Rev. Lett.*, 97:151802, 2006.

[119] R. N. Mohapatra and Salah Nasri. Reconciling the cast and pvlas results. *Phys. Rev. Lett.*, 98:050402, 2007.

[120] Cecile Robilliard et al. No light shining through a wall. *Phys. Rev. Lett.*, 99:190403, 2007.

[121] E. Zavattini et al. New pvlas results and limits on magnetically induced optical rotation and ellipticity in vacuum. 2007.

[122] A. S. Chou et al. Search for axion-like particles using a variable baseline photon regeneration technique. 2007.

[123] Toby Falk, Keith A. Olive, and Mark Srednicki. Heavy sneutrinos as dark matter. *Phys. Lett.*, B339:248–251, 1994.

[124] John S. Hagelin, Gordon L. Kane, and S. Raby. Perhaps scalar neutrinos are the lightest supersymmetric partners. *Nucl. Phys.*, B241:638, 1984.

[125] Glennys R. Farrar and Pierre Fayet. Phenomenology of the production, decay, and detection of new hadronic states associated with supersymmetry. *Phys. Lett.*, B76:575–579, 1978.

[126] H. Goldberg. Constraint on the photino mass from cosmology. *Phys. Rev. Lett.*, 50:1419, 1983.

[127] John R. Ellis, J. S. Hagelin, Dimitri V. Nanopoulos, Keith A. Olive, and M. Srednicki. Supersymmetric relics from the big bang. *Nucl. Phys.*, B238:453–476, 1984.

[128] K. Griest. Cross sections, relic abundance, and detection rates for neutralino dark matter. *PRD*, 38:2357–2375, October 1988.

[129] Manuel Drees and Mihoko M. Nojiri. The neutralino relic density in minimal n=1 supergravity. *Phys. Rev.*, D47:376–408, 1993.

[130] G. Jungman, M. Kamionkowski, and K. Griest. Supersymmetric dark matter. *Physics REport*, 267:195–373, March 1996.

[131] L. Roszkowski. Lower bound on the neutralino mass. *Physics Letters B*, 252:471–475, December 1990.

[132] John R. Ellis, Keith A. Olive, Yudi Santoso, and Vassilis C. Spanos. Supersymmetric dark matter in light of wmap. *Phys. Lett.*, B565:176–182, 2003.

[133] P. Binetruy, G. Girardi, and P. Salati. Constraints on a system of two neutral fermions from cosmology. *Nucl. Phys.*, B237:285, 1984.

[134] Kim Griest and David Seckel. Three exceptions in the calculation of relic abundances. *Phys. Rev.*, D43:3191–3203, 1991.

[135] Satoshi Mizuta and Masahiro Yamaguchi. Coannihilation effects and relic abundance of higgsino dominant $lsp_s$. *Phys. Lett.*, B298:120–126, 1993.

[136] John R. Ellis, Toby Falk, and Keith A. Olive. Neutralino stau coannihilation and the cosmological upper limit on the mass of the lightest supersymmetric particle. *Phys. Lett.*, B444:367–372, 1998.

[137] Celine Boehm, Abdelhak Djouadi, and Manuel Drees. Light scalar top quarks and supersymmetric dark matter. *Phys. Rev.*, D62:035012, 2000.

[138] M. E. Gomez, G. Lazarides, and C. Pallis. Supersymmetric cold dark matter with yukawa unification. *Phys. Rev.*, D61:123512, 2000.

[139] G. Belanger, F. Boudjema, A. Pukhov, and A. Semenov. micromegas2.0: A program to calculate the relic density of dark matter in a generic model. *Comput. Phys. Commun.*, 176:367–382, 2007.

[140] P. Gondolo et al. Darksusy 4.00 neutralino dark matter made easy. *New Astron. Rev.*, 49:149–151, 2005.

[141] Yago Ascasibar, P. Jean, C. Boehm, and J. Knoedlseder. Constraints on dark matter and the shape of the milky way dark halo from the 511-kev line. *Mon. Not. Roy. Astron. Soc.*, 368:1695–1705, 2006.

[142] Geraldine Servant and Tim M. P. Tait. Is the lightest kaluza-klein particle a viable dark matter candidate? *Nucl. Phys.*, B650:391–419, 2003.

[143] Hsin-Chia Cheng, Jonathan L. Feng, and Konstantin T. Matchev. Kaluza-klein dark matter. *Phys. Rev. Lett.*, 89:211301, 2002.

[144] Mark W. Goodman and Edward Witten. Detectability of certain dark-matter candidates. *Phys. Rev.*, D31:3059, 1985.

[145] Jonghee Yoo. Cdms experiment: current status and future. FERMILAB-CONF-07-606-E.

[146] R. Lemrani. Search for dark matter with edelweiss: Status and future. *Phys. Atom. Nucl.*, 69:1967–1969, 2006.

[147] W. Westphal et al. Dark-matter search with cresst. *Czech. J. Phys.*, 56:535–542, 2006.

[148] H. Kraus et al. Eureca: The european future of cryogenic dark matter searches. *J. Phys. Conf. Ser.*, 39:139–141, 2006.

[149] J. Angle et al. First results from the xenon10 dark matter experiment at the gran sasso national laboratory. 2007.

[150] M. Barnabe-Heider et al. Improved spin dependent limits from the picasso dark matter search experiment. *Phys. Lett.*, B624:186–194, 2005.

[151] R. Bernabei et al. Results from dama/nai and perspectives for dama/libra. 2003.

[152] M. Messina and A. Rubbia. Status report of ardm project: A new direct detection experiment, based on liquid argon, for the search of dark matter. Prepared for 9th ICATPP Conference on Astroparticle, Particle, Space Physics, Detectors and Medical Physics Applications, Villa Erba, Como, Italy, 17-21 Oct 2005.

[153] F. W. Stecker. The cosmic gamma-ray background from the annihilation of primordial stable neutral heavy leptons. *Astrophys. J.*, 223:1032–1036, 1978.

[154] L. Bergström and H. Snellman. Observable monochromatic photons from cosmic photino annihilation. *PRD*, 37:3737–3741, June 1988.

[155] Alain Bouquet, Pierre Salati, and Joseph Silk. gamma-ray lines as a probe for a cold dark matter halo. *Phys. Rev.*, D40:3168, 1989.
[156] Joseph Silk and Mark Srednicki. Cosmic-ray antiprotons as a probe of a photino-dominated universe. *Phys. Rev. Lett.*, 53:624, 1984.
[157] S. Rudaz and F. W. Stecker. Cosmic ray anti-protons, positrons and gamma-rays from halo dark matter annihilation. *Astrophys. J.*, 325:16, 1988.
[158] Joseph Silk, Keith A. Olive, and Mark Srednicki. The photino, the sun, and high-energy neutrinos. *Phys. Rev. Lett.*, 55:257–259, 1985.
[159] William H. Press and David N. Spergel. Capture by the sun of a galactic population of weakly interacting, massive particles. *Astrophys. J.*, 296:679–684, 1985.
[160] Y. Edmonds et al. Estimate for glast lat milky way dark matter wimp line sensitivity. *AIP Conf. Proc.*, 921:514–515, 2007.

# Gravitational Wave Astronomy

Konstantinos D. Kokkotas

Theoretical Astrophysics, Auf der Morgenstelle 10,
Eberhard Karls University of Tübingen, Tübingen 72076, Germany
kostas.kokkotas@uni-tuebingen.de,
http://www.tat.physik.uni-tuebingen.de/~kokkotas

### Abstract

*As several large scale interferometers are beginning to take data at sensitivities where astrophysical sources are predicted, the direct detection of gravitational waves may well be imminent. This would open the gravitational-wave window to our Universe, and should lead to a much improved understanding of the most violent processes imaginable; the formation of black holes and neutron stars following core collapse supernovae and the merger of compact objects at the end of binary inspiral.*

## 1 Introduction

Gravitational waves are ripples of spacetime generated as masses are accelerated. It is one of the central predictions of Einsteins' general theory of relativity but despite decades of effort these ripples in spacetime have still not been observed directly. Yet we have strong indirect evidence for their existence from the excellent agreement between the observed inspiral rate of the binary pulsar PSR1913+16 and the theoretical prediction (better than 1% in the phase evolution). This provides confidence in the theory and suggests that "gravitational-wave astronomy" should be viewed as a serious proposition. This new window onto universe will complement our view of the cosmos and will help us unveil the fabric of spacetime around black-holes, observe directly the formation of black holes or the merging of binary systems consisting of black holes or neutron stars, search for rapidly spinning neutron stars, dig deep into the very early moments of the origin of the universe, and look at the very center of the galaxies where supermassive black holes weighting millions of solar masses are hidden. Second, detecting gravitational waves is important for our understanding of the fundamental laws of physics; the proof that gravitational waves exist will verify a fundamental 90-year-old prediction of general relativity. Also, by comparing the arrival times of light and gravitational waves, from, e.g., supernovae, Einstein's prediction that light and gravitational waves travel at the same speed could be checked. Finally, we could verify that they have the polarization predicted by general relativity.

These expectations follow from the comparison between gravitational and electromagnetic waves. That is : a) While electromagnetic waves are radiated by individual particles, gravitational waves are due to non-spherical bulk motion of matter. In essence, this means that the information carried by electromagnetic waves is stochastic in nature, while the gravitational waves provide insights into coherent mass currents. b) The electromagnetic waves will have been scattered many times. In contrast, gravitational waves interact weakly with matter and arrive at the Earth in pristine condition. This means that gravitational waves can be used to probe regions of space that are opaque to electromagnetic waves. Unfortunately, this weak interaction with matter also makes the detection of gravitational waves an extremely hard task. c) Standard astronomy is based on deep imaging of small fields of view, while gravitational-wave detectors cover virtually the entire sky. d) The wavelength of the electromagnetic radiation is smaller than the size of the emitter, while the wavelength of a gravitational wave is usually larger than the size of the source. This means that we cannot use gravitational-wave data to create an image of the source. In fact, gravitational-wave observations are more like audio than visual.

## 2 Gravitational wave primer

The aim of the first part of our contribution is to provide a condensed text-book level introduction to gravitational waves. Although in no sense complete this description should prepare the reader for the discussion of high-frequency sources which follows.

The first aspect of gravitational waves that we need to appreciate is their *tidal* nature. This is important because it implies that we need to monitor, with extreme precision, the relative motion of test masses or the periodic (tidal) deformations of extended bodies. A gravitational wave, propagating in a flat spacetime, generates periodic distortions, which can be described in terms of the Riemann tensor which measures the curvature of the spacetime. In linearized theory the Riemann tensor takes the following gauge-independent form:

$$R_{\kappa\lambda\mu\nu} = \frac{1}{2}\left(\partial_{\nu\kappa}h_{\lambda\mu} + \partial_{\lambda\mu}h_{\kappa\nu} - \partial_{\kappa\mu}h_{\lambda\nu} - \partial_{\lambda\nu}h_{\kappa\mu}\right), \qquad (1)$$

which is considerably simplified by choosing the so called transverse and traceless gauge or *TT gauge*:

$$R^{\mathrm{TT}}_{j0k0} = -\frac{1}{2}\frac{\partial^2}{\partial t^2}h^{\mathrm{TT}}_{jk} \approx \frac{\partial^2 \Phi}{\partial x^j \partial x^k}, \qquad j,k = 1,2,3. \qquad (2)$$

where $h^{\mathrm{TT}}_{jk}$ is the gravitational wave field in the TT-gauge and $\Phi$ describes the gravitational potential in Newtonian theory. The Riemann tensor as is a pure geometrical object, but in general relativity has a simple physical interpretation: it is the tidal force field and describes the relative acceleration between two particles in free fall. If we assume two particles moving freely along geodesics of a curved spacetime with coordinates $x^\mu(\tau)$ and $x^\mu(\tau) + \delta x^\mu(\tau)$ (for a given value of the proper time $\tau$, $\delta^\mu(\tau)$

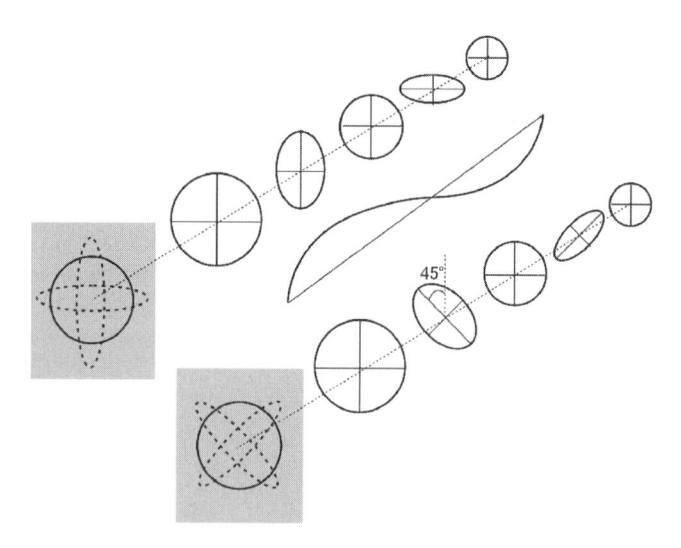

**Figure 1:** The effects of a gravitational wave travelling perpendicular the plane of a circular ring of particles, is sketched as a series of snapshots. The deformations due the two polarizations $h_+$ and $h_\times$ are shown.

is the displacement vector connecting the two events) it can be shown that, in the case of slowly moving particles,

$$\frac{d^2 \delta x^k}{dt^2} \approx -R^k{}_{0j0}{}^{TT} \delta x^j. \tag{3}$$

This is a simplified form of the equation of *geodesic deviation*. Hence, the tidal force acting on a particle is:

$$f^k \approx -m R^k_{0j0} \delta x^j, \tag{4}$$

where $m$ is the mass of the particle. Equation (4) corresponds to the standard Newtonian relation for the tidal force acting on a particle in a field $\Phi$. Then equation (4) integrates to

$$\delta x_j = \frac{1}{2} h^{TT}_{jk} x^k_0 \quad \text{or} \quad h \approx \frac{\Delta L}{L} \tag{5}$$

where $h$ is the dimensionless gravitational-wave strain.

Let us now assume that the waves propagate in the $z$-direction, i.e. that we have $h_{jk} = h_{jk}(t-z)$. Then one can show that we have only two independent components;

$$h_+ = h^{TT}_{xx} = -h^{TT}_{yy} \quad h_\times = h^{TT}_{xy} = h^{TT}_{yx} \tag{6}$$

What effect does $h_+$ have on matter? Consider a particle initially located at $(x_0, y_0)$ and let $h_\times = 0$ to find that

$$\delta x = \frac{1}{2} h_+ x_0 \quad \text{and} \quad \delta y = -\frac{1}{2} h_+ y_0 \tag{7}$$

*Gravitational Wave Astronomy*

That is, if $h_+$ varies periodically then an object will first experience a stretch in the $x$-direction accompanied by a squeeze in the $y$-direction. One half-cycle later, the squeeze is in the $x$-direction and the stretch in the $y$-direction. It is straightforward to show that the effect of $h_\times$ is the same, but rotated by 45 degrees. This is illustrated in Fig. 1. A general wave will be a linear combination of the two polarisations.

Up to this point we have shown the effect of propagating spacetime deformations (we called them gravitational waves) on two nearby particles. But we have not yet done the connection to Einstein's theory. Thus we will describe with a tensor $h_{\mu\nu}$ the variations of a flat spacetime from flatness, that is we will describe the spacetime with the metric $g_{\alpha\beta} = \eta_{\alpha\beta} + h_{\alpha\beta}$. Then after some algebra it can be shown that Einstein's equations will be reduced to the following form:

$$\left(-\frac{\partial^2}{\partial t^2} + \nabla^2\right)\tilde{h}^{\mu\nu} \equiv \partial_\lambda \partial^\lambda \tilde{h}^{\mu\nu} = 0 \quad \text{with} \quad \tilde{h}_{\mu\nu} \equiv h_{\mu\nu} - \frac{1}{2}\eta_{\mu\nu} h^\alpha_\alpha \quad (8)$$

where we have used a specific gauge choice $\partial_\mu \tilde{h}^{\mu\nu} = 0$, known as *Hilbert's gauge condition* (equivalent to the Lorentz gauge condition of electromagnetism), has been assumed.

The simplest solution to the wave equation (8) is a plane wave solution of the form

$$\tilde{h}^{\mu\nu} = A^{\mu\nu} e^{ik_\alpha x^\alpha}, \quad (9)$$

where $A^{\mu\nu}$ is a constant symmetric tensor, the *polarization tensor*, in which information about the amplitude and the polarization of the waves is encoded, while $k_\alpha$ is a constant vector, the *wave vector*, that determines the propagation direction of the wave and its frequency. In physical applications we will use only the real part of the above wave solution.

It is customary to write the gravitational wave solution in the TT gauge as $h^{TT}_{\mu\nu}$. That $A_{\mu\nu}$ has only two independent components means that a gravitational wave is completely described by two dimensionless amplitudes, $h_+$ and $h_\times$, say. If, for example, we assume a wave propagating along the z-direction, then the amplitude $A^{\mu\nu}$ can be written as

$$A^{\mu\nu} = h_+ \epsilon^{\mu\nu}_+ + h_\times \epsilon^{\mu\nu}_\times \quad (10)$$

where $\epsilon^{\mu\nu}_+$ and $\epsilon^{\mu\nu}_\times$ are the so-called *unit polarization tensors* defined by

$$\epsilon^{\mu\nu}_+ \equiv \begin{pmatrix} 0 & 0 & 0 & 0 \\ 0 & 1 & 0 & 0 \\ 0 & 0 & -1 & 0 \\ 0 & 0 & 0 & 0 \end{pmatrix} \qquad \epsilon^{\mu\nu}_\times \equiv \begin{pmatrix} 0 & 0 & 0 & 0 \\ 0 & 0 & 1 & 0 \\ 0 & 1 & 0 & 0 \\ 0 & 0 & 0 & 0 \end{pmatrix}. \quad (11)$$

Finally, we should point out that in the TT-gauge there is no difference between $h_{\mu\nu}$ (the perturbation of the metric) and $\tilde{h}_{\mu\nu}$ (the gravitational field).

## 2.1 Gravitational wave properties

Gravitational waves, once they are generated, propagate almost unimpeded. Indeed, it has been proven that they are even harder to stop than neutrinos! The only significant change they suffer as they propagate is the decrease in amplitude while they travel away from their source, and the *redshift* they feel (cosmological, gravitational or Doppler), as is the case for electromagnetic waves.

There are other effects that marginally influence the gravitational waveforms, for instance, *absorption* by interstellar or intergalactic matter intervening between the observer and the source, which is extremely weak (actually, the extremely weak coupling of gravitational waves with matter is the main reason that gravitational waves have not been observed). *Scattering* and *dispersion* of gravitational waves are also practically unimportant, although they may have been important during the early phases of the universe (this is also true for the absorption). Gravitational waves can be *focused* by strong gravitational fields and also can be *diffracted*, exactly as it happens with the electromagnetic waves.

## 2.2 Energy flux carried by gravitational waves

Gravitational waves carry energy and cause a deformation of spacetime. The stress-energy carried by gravitational waves cannot be localized within a wavelength. Instead, one can say that a certain amount of stress-energy is contained in a region of the space which extends over several wavelengths. It can be proven that in the TT gauge of linearized theory the stress-energy tensor of a gravitational wave (in analogy with the stress-energy tensor of a perfect fluid that we have defined earlier) is given by

$$t_{\mu\nu}^{GW} = \frac{1}{32\pi} \langle \partial_\mu h_{ij}^{TT} \cdot \partial_\nu h_{ij}^{TT} \rangle. \tag{12}$$

where the angular brackets are used to indicate averaging over several wavelengths. For the special case of a plane wave propagating in the $z$ direction, which we considered earlier, the stress-energy tensor has only three non-zero components, which take the simple form

$$t_{00}^{GW} = \frac{t_{zz}^{GW}}{c^2} = -\frac{t_{0z}^{GW}}{c} = \frac{1}{32\pi} \frac{c^2}{G} \omega^2 \left( h_+^2 + h_\times^2 \right), \tag{13}$$

where $t_{00}^{GW}$ is the energy density, $t_{zz}^{GW}$ is the momentum flux and $t_{0z}^{GW}$ the energy flow along the $z$ direction per unit area and unit time (for practical reasons we have restored the normal units). The energy flux has all the properties one would anticipate by analogy with electromagnetic waves: (a) it is conserved (the amplitude dies out as $1/r$, the flux as $1/r^2$), (b) it can be absorbed by detectors, and (c) it can generate curvature like any other energy source in Einstein's formulation of relativity.

The definition of the energy flux by equation (13) provides a useful formula

$$F = 3 \left( \frac{f}{1\,\text{kHz}} \right)^2 \left( \frac{h}{10^{-22}} \right)^2 \frac{\text{ergs}}{\text{cm}^2\text{sec}}, \tag{14}$$

which can be used to estimate the flux on Earth, given the amplitude of the waves (on Earth) and the frequency of the waves.

## 2.3 Generation of gravitational waves

As early as 1918, Einstein derived the quadrupole formula for gravitational radiation. This formula states that the wave amplitude $h_{ij}$ is proportional to the second time derivative of the quadrupole moment of the source:

$$h_{ij} = \frac{2}{r}\frac{G}{c^4}\ddot{Q}_{ij}^{TT}\left(t - \frac{r}{c}\right) \qquad (15)$$

where

$$Q_{ij}^{TT}(x) = \int \rho\left(x^i x^j - \frac{1}{3}\delta^{ij}r^2\right)d^3x \qquad (16)$$

is the quadrupole moment in the TT gauge, evaluated at the retarded time $t - r/c$ and $\rho$ is the matter density in a volume element $d^3x$ at the position $x^i$. This result is quite accurate for all sources, as long as the reduced wavelength $\tilde{\lambda} = \lambda/2\pi$ is much longer than the source size $R$. It should be pointed out that the above result can be derived via a quite cumbersome calculation in which we solve the wave equation (8) with a source term $T_{\mu\nu}$ on the right-hand side. In the course of such a derivation, a number of assumptions must be used. In particular, the observer must be located at a distance $r \gg \tilde{\lambda}$, far greater than the reduced wavelength, in the so called "radiation zone" and $T_{\mu\nu}$ must not change very quickly.

Using the formulae (12) and (13) for the energy carried by gravitational waves, one can derive the luminosity in gravitational waves as a function of the third-order time derivative of the quadrupole moment tensor. This is the quadrupole formula

$$L_{GW} = -\frac{dE}{dt} = \frac{1}{5}\frac{G}{c^5}\langle\frac{\partial^3 Q_{ij}}{\partial t^3}\frac{\partial^3 Q_{ij}}{\partial t^3}\rangle. \qquad (17)$$

Based on this formula, we derive some additional formulas, which provide order of magnitude estimates for the amplitude of the gravitational waves and the corresponding power output of a source. First, the quadrupole moment of a system is approximately equal to the mass $M$ of the part of the system that moves, times the square of the size $R$ of the system. This means that the third-order time derivative of the quadrupole moment is

$$\frac{\partial^3 Q_{ij}}{\partial t^3} \sim \frac{MR^2}{T^3} \sim \frac{Mv^2}{T} \sim \frac{E_{\rm ns}}{T}, \qquad (18)$$

where $v$ is the mean velocity of the moving parts, $E_{\rm ns}$ is the kinetic energy of the component of the source's internal motion which is non spherical, and $T$ is the time scale for a mass to move from one side of the system to the other. The time scale (or period) is actually proportional to the inverse of the square root of the mean density of the system

$$T \sim \sqrt{R^3/GM}. \qquad (19)$$

This relation provides a rough estimate of the characteristic frequency of the system $f = 2\pi/T$. For example, for a non-radially oscillating neutron star with a mass of roughly 1.4M$_\odot$ and a radius of 12 km, the frequency of oscillation which is directly related to the frequency of the emitted gravitational waves, will be roughly 2 kHz.

Similarly, for an oscillating black hole of the same mass we get a characteristic frequency of 10 kHz.

Then, the luminosity of gravitational waves of a given source is approximately

$$L_{GW} \approx \frac{G^4}{c^5}\left(\frac{M}{R}\right)^5 \sim \frac{G}{c^5}\left(\frac{M}{R}\right)^2 v^6 \sim \frac{c^5}{G}\left(\frac{R_{Sch}}{R}\right)^2 \left(\frac{v}{c}\right)^6 \qquad (20)$$

and

$$h \approx \frac{1}{c^2}\left(\frac{GM}{r}\right)\left(\frac{R_{Sch}}{R}\right) \approx \frac{2}{c^2}\left(\frac{GM}{r}\right)\left(\frac{v}{c}\right)^2 \qquad (21)$$

where $R_{Sch} = 2GM/c^2$ is the Schwarzschild radius of the source. It is obvious that the maximum values of the amplitude and the luminosity of gravitational waves can be achieved if the source's dimensions are of the order of its Schwarzschild radius and the typical velocities of the components of the system are of the order of the speed of light. This explains why we expect the best gravitational wave sources to be highly relativistic compact objects. The above formula sets also an upper limit on the power emitted by a source, which for $R \sim R_{Sch}$ and $v \sim c$ is

$$L_{GW} \sim c^5/G = 3.6 \times 10^{59} \text{ergs/sec.} \qquad (22)$$

This is an immense power, often called the *luminosity of the universe*.

Using the above order-of-magnitude estimates, we can get a rough estimate of the amplitude of gravitational waves at a distance $r$ from the source:

$$h \sim \frac{G}{c^4}\frac{E_{ns}}{r} \sim \frac{G}{c^4}\frac{\varepsilon E_{kin}}{r} \qquad (23)$$

where $\varepsilon E_{kin}$ (with $0 \leq \varepsilon \leq 1$), is the fraction of kinetic energy of the source that is able to produce gravitational waves. The factor $\varepsilon$ is a measure of the asymmetry of the source and implies that only a time varying quadrupole moment will emit gravitational waves. For example, even if a huge amount of kinetic energy is involved in a given explosion and/or implosion, if the event takes place in a spherically symmetric manner, there will be no gravitational radiation.

Another formula for the amplitude of gravitational waves relation can be derived from the flux formula (14). If, for example, we consider an event (perhaps a supernovae explosion) at the Virgo cluster during which the energy equivalent of $10^{-4} M_\odot$ is released in gravitational waves at a frequency of 1 kHz, and with signal duration of the order of 1 msec, the amplitude of the gravitational waves on Earth will be

$$h \approx 10^{-22} \left(\frac{E_{GW}}{10^{-4} M_\odot c^2}\right)^{1/2} \left(\frac{f}{1\,\text{kHz}}\right)^{-1} \left(\frac{\tau}{1\text{msec}}\right)^{-1/2} \left(\frac{r}{15\,\text{Mpc}}\right)^{-1}. \qquad (24)$$

For a detector with arm length of 4 km we are looking for changes in the arm length of the order of

$$\Delta \ell = h \cdot \ell = 10^{-22} \cdot 4\,\text{km} = 4 \times 10^{-17} \text{cm}\,!$$

This small number explains why all detection efforts till today were not successful.

## Gravitational Wave Astronomy

If the signal analysis is based on matched filtering, the *effective amplitude* improves roughly as the square root of the number of observed cycles $n$. Using $n \approx f\tau$ we get

$$h_c \approx 10^{-22} \left(\frac{E_{GW}}{10^{-3} M_\odot c^2}\right)^{1/2} \left(\frac{f}{1 \text{ kHz}}\right)^{-1/2} \left(\frac{r}{15 \text{ Mpc}}\right)^{-1} \quad (25)$$

We see that the "detector sensitivity" essentially depends only on the radiated energy, the characteristic frequency and the distance to the source. That is, in order to obtain a rough estimate of the relevance of a given gravitational-wave source at a given distance we only need to estimate the frequency and the radiated energy. Alternatively, if we know the energy released can work out the distance at which these sources can be detected.

## 2.4 Gravitational wave detection

One often classifies gravitational-wave sources by the nature of the waves. This is convenient because the different classes require different approaches to the data-analysis problem;

- *Chirps.* As a binary system radiates gravitational waves and loses energy the two constituents spiral closer together. As the separation decreases the gravitational-wave amplitude increases, leading to a characteristic "chirp" signal.

- *Bursts.* Many scenarios lead to burst-like gravitational waves. A typical example would be black-hole oscillations excited during binary merger.

- *Periodic.* Systems where the gravitational-wave backreaction leads to a slow evolution (compared to the observation time) may radiate persistent waves with a virtually constant frequency. This would be the gravitational-wave analogue of the radio pulsars.

- *Stochastic.* A stochastic (non-thermal) background of gravitational waves is expected to have been generated following the Big Bang. One may also have to deal with stochastic gravitational-wave signals when the sources are too abundant for us to distinguish them as individuals.

Given that the weak signals are going to be buried in detector noise, we need to obtain as accurate theoretical models as possible. The rough order of magnitude estimates we just derived will certainly not be sufficient, even though they provide an indication as to whether it is worth spending the time and effort required to build a detailed model. Such source models are typically obtained using either

- approximate perturbation techniques, eg. expansions in small perturbations away from a known solution to the Einstein equations, the archetypal case being black-hole and neutron star oscillations.

- post-Newtonian approximations, essentially an expansion in the ratio between a characteristic velocity of the system and the speed of light, most often used to model the inspiral phase of a compact binary system.

- numerical relativity, where the Einstein equations are formulated as an initial-value problem and solved on the computer. This is the only way to make progress in situations where the full nonlinearities of the theory must be included, eg. in the merger of black holes and neutron stars or a supernova core collapse.

The first attempt to detect gravitational waves was undertaken by the pioneer Joseph Weber during the early 1960s. He developed the first resonant mass (bar) detector and inspired many other physicists to build new detectors and to explore from a theoretical viewpoint possible cosmic sources of gravitational radiation. When a gravitational wave hits such a device, it causes the bar to vibrate. By monitoring this vibration, we can reconstruct the true waveform. The next step, was to monitor the change of the distance between two fixed points by a passing-by gravitational wave. This can be done by using laser interferometry. The use of interferometry is probably the most decisive step in our attempt to detect gravitational wave signals. Although the basic principle of such detectors is very simple, the sensitivity of detectors is limited by various sources of noise. The internal noise of the detectors can be Gaussian or non-Gaussian. The non-Gaussian noise may occur several times per day such as strain releases in the suspension systems which isolate the detector from any environmental mechanical source of noise, and the only way to remove this type of noise is via comparisons of the data streams from various detectors. The so-called Gaussian noise obeys the probability distribution of Gaussian statistics and can be characterized by a *spectral density* $S_n(f)$. The observed signal at the output of a detector consists of the true gravitational wave strain $h$ and Gaussian noise. The optimal method to detect a gravitational wave signal leads to the following signal-to-noise ratio:

$$\left(\frac{S}{N}\right)^2_{\text{opt}} = 2 \int_0^\infty \frac{|\tilde{h}(f)|^2}{S_n(f)} df, \qquad (26)$$

where $\tilde{h}(f)$ is the Fourier transform of the signal waveform. It is clear from this expression that the sensitivity of gravitational wave detectors is limited by noise.

In reality, the efficiency of a resonant bar detector depends on several other parameters. Here, we will discuss only the more fundamental ones. Assuming perfect isolation of the resonant bar detector from any external source of noise (acoustical, seismic, electromagnetic), the thermal noise is the only factor limiting our ability to detect gravitational waves. Thus, in order to detect a signal, the energy deposited by the gravitational wave every $\tau$ seconds should be larger than the energy $kT$ due to thermal fluctuations. This leads to a formula for the minimum detectable energy flux of gravitational waves, which, following equation (13), leads into a minimum detectable strain amplitude

$$h_{\min} \leq \frac{1}{\omega_0 L Q} \sqrt{\frac{15kT}{M}} \qquad (27)$$

where $L$ and $M$ are the size and the mass of the resonant bar correspondingly and $Q$ is the quality factor of the material. During the last 20 years, a number of resonant bar detectors have been in nearly continuous operation in several places around the world. They have achieved sensitivities of a few times $10^{-21}$, but still there has been no clear evidence of gravitational wave detection.

A laser interferometer is an alternative gravitational wave detector that offers the possibility of very high sensitivities over a broad frequency band. Originally, the idea was to construct a new type of resonant detector with much larger dimensions. Gravitational waves that are propagating perpendicular to the plane of the interferometer will increase the length of one arm of the interferometer, and at the same time will shorten the other arm, and vice versa. This technique of monitoring the waves is based on Michelson interferometry. L-shaped interferometers are particularly suited to the detection of gravitational waves due to their quadrupolar nature. For extensive reviews refer to [4–7].

The US project named LIGO [8](Laser Interferometer Gravitational Observatory) consists of two detectors with arm length of 4 Km, one in Hanford, Washington, and one in Livingston, Louisiana. The detector in Hanford includes, in the same vacuum system, a second detector with arm length of 2 km. The detectors are already in operation and they achieved the designed sensitivity. The Italian/French EGO (VIRGO) detector [9] of arm-length 3 km at Cascina near Pisa is designed to have better sensitivity at lower frequencies. GEO600 is the German/British detector build in Hannover [10]. The TAMA300[11] detector in Tokyo has arm length of 300 m and it was the first major interferometric detector in operation. There are already plans for improving the sensitivities of all the above detectors and the construction of new interferometers in the near future.

Up to now LIGO has completed four science runs, S1 from August–September 2002, S2 from February–April 2003, S3 from October 2003–January 2004, and S4 February–March 2005. These short science runs where interrupted with improvements which led LIGO to operate now with its designed sensitivity. The fifth science run, S5, (November 2005–September 2007) surveyed a considerable larger volume of the Universe and set upper limits to a number of gravitational wave sources [13–17]. The "enchanced" LIGO interferometer is expected to commence an S6 science run in 2009 and will survey a volume of space eight times as great as the current LIGO. In 2011, the LIGO interferometers will be shut down for decommisioning in order to install advanced interferometers. With these advanced interferometers LIGO are expected to operate with ten times the current sensitivity which means a factor of 1000 increase in the volume of the Universe surveyed by 2014.

# 3 Sources of gravitational waves

## 3.1 Radiation from binary systems

Among the most interesting sources of gravitational waves are binaries. The inspiralling of such systems, consisting of black holes or neutron stars, is, as we will discuss later, the most promising source for the gravitational wave detectors. Binary systems are also the sources of gravitational waves whose dynamics we understand the best. They emit copious amounts of gravitational radiation, and for a given system we know quite accurately the amplitude and frequency of the gravitational waves in terms of the masses of the two bodies and their separation.

The gravitational-wave signal from inspiraling binaries is approximatelly sinusoidal, with a frequency which is twice the orbital frequency of the binary. According to equation (17) the gravitational radiation luminosity of the system is

$$L^{\text{GW}} = \frac{32}{5} \frac{G}{c^5} \mu^2 a^4 \Omega^6 = \frac{32}{5} \frac{G^4}{c^5} \frac{M^3 \mu^2}{a^5}, \qquad (28)$$

where $\Omega$ is the orbital angular velocity, $a$ is the distance between the two bodies, $\mu = M_1 M_2/M$ is the reduced mass of the system and $M = M_1 + M_2$ its total mass. In order to obtain the last part of the relation, we have used Kepler's third law, $\Omega^2 = GM/a^3$. As the gravitating system loses energy by emitting radiation, the distance between the two bodies shrinks and the orbital frequency increases accordingly ($\dot{T}/T = 1.5\dot{a}/a$). Finally, the amplitude of the gravitational waves is

$$h = 5 \times 10^{-22} \left(\frac{M}{2.8 M_\odot}\right)^{2/3} \left(\frac{\mu}{0.7 M_\odot}\right) \left(\frac{f}{100\,\text{Hz}}\right)^{2/3} \left(\frac{15\,\text{Mpc}}{r}\right). \qquad (29)$$

In all these formulae we have assumed that the orbits are circular.

As the binary system evolves the orbit shrinks and the frequency increases in the characteristic chirp. Eventually, depending on the masses of the binaries, the frequency of the emitted gravitational waves will enter the bandwidth of the detector at the low-frequency end and will evolve quite fast towards higher frequencies. A system consisting of two neutron stars will be detectable by LIGO when the frequency of the gravitational waves is $\sim$10 Hz until the final coalescence around 1 kHz. This process will last for about 15 min and the total number of observed cycles will be of the order of $10^4$, which leads to an enhancement of the detectability by a factor 100 (remember $h_c \sim \sqrt{n}h$). Binary neutron star systems and binary black hole systems with masses of the order of 50 $M_\odot$ are the primary sources for LIGO. Given the anticipated sensitivity of LIGO, binary black hole systems are the most promising sources and could be detected as far as 200 Mpc away. For the present estimated sensitivity of LIGO the event rate is probably a few per year, but future improvements of detector sensitivity (the LIGO II phase) could lead to the detection of at least one event per month. Supermassive black hole systems of a few million solar masses are the primary source for LISA. These binary systems are rare, but due to the huge amount of energy released, they should be detectable from as far away as the boundaries of the observable universe. Finally, the recent discovery of the highly relativistic binary pulsar J0737-3039 [18] enchanced considerably the expected coalescence event rate of NS-NS binaries [19]. The event rate for initial LIGO is in the best case 0.2 per year while advanced LIGO might be able to detect 20–1000 events per year.

Depending on the high-density EOS and their initial masses, the outcome of the merger of two neutron stars may not always be a black hole, but a hypermassive, differentially rotating compact star (even if it is only temporarily supported against

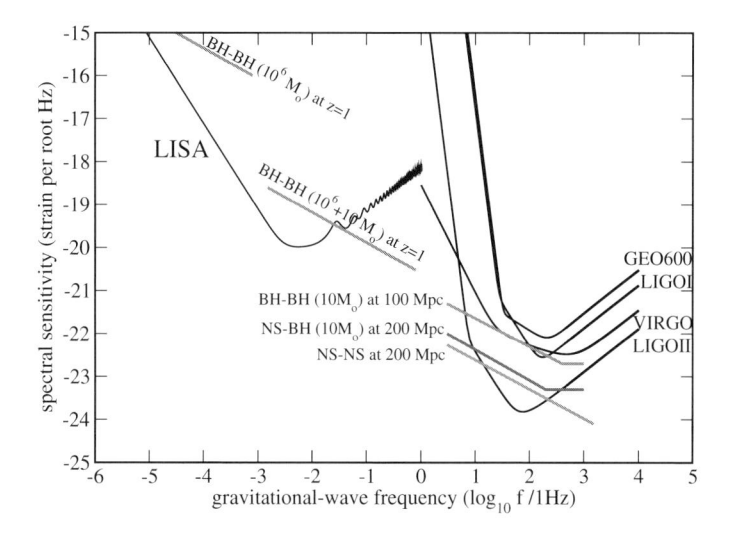

**Figure 2:** Estimated signal strengths for various inspiralling binaries relevant for ground- and space-based detectors.

collapse by differential rotation). A recent detailed simulation [20] in full GR has shown that the hypermassive object created in a binary NS merger is nonaxisymmetric. The nonaxisymmetry lasts for a large number of rotational periods, leading to the emission of gravitational waves with a frequency of 3 kHz and an effective amplitude of $\sim 6$–$7 \times 10^{-21}$ at a large distance of 50 Mpc. Such large effective amplitude may be detectable even by LIGO II at this high frequency.

The tidal disruption of a NS by a BH [21] or the merging of two NSs [22] may give valuable information for the radius and the EoS if we can recover the signal at frequencies higher than 1 kHz.

## 3.2 Gravitational collapse

One of the most spectacular astrophysical events is the core collapse of massive stars, leading to the formation of a neutron star (NS) or a black hole (BH). The outcome of core collapse depends sensitively on several factors: mass, angular momentum and metallicity of progenitor, existence of a binary companion, high-density equation of state, neutrino emission, magnetic fields, etc. Partial understanding of each of the above factors is emerging, but a complete and consistent theory for core collapse is still years away.

Roughly speaking, isolated stars more massive than $\sim 8$–$10 M_\odot$ end in core collapse and $\sim 90\%$ of them are stars with masses $\sim 8$–$20 M_\odot$. After core bounce, most of the material is ejected and if the progenitor star has a mass $M \lesssim 20 M_\odot$ a neutron star is left behind. On the other hand, if $M \gtrsim 20 M_\odot$ fall-back accretion increases the mass of the formed proton-neutron star (PNS), pushing it above the maximum mass limit, which results in the formation of a black hole. Furthermore, if the progenitor star has a mass of roughly $M \gtrsim 45 M_\odot$, no supernova explosion is launched and the star collapses directly to a BH [23].

The above picture is, of course, greatly simplified. In reality, the metallicity of the progenitor, the angular momentum of the pre-collapse core and the presence of a binary companion will decisively influence the outcome of core collapse [24]. Rotation influences the collapse by changing dramatically the properties of the convective region above the proto-neutron star core. Centrifugal forces slow down infalling material in the equatorial region compared to materiall falling in along the polar axis, yielding a weaker bounce. This asymmetry between equator and poles also strongly influences the neutrino emmission and the revival of the stalled shock by neutrinos [25, 26].

The supernova event rate is 1–2 per century per galaxy [27] and about 5–40% of them produce BHs in delayed collapse (through fall-back accretion), or direct collapse [28].

Of considerable importance is the *initial rotation rate* of proto-neutron stars, since (as will be detailed in the next sections) most mechanisms for emission of detectable gravitational waves from compact objects require very rapid rotation at birth (rotational periods of the order of a few milliseconds or less). Since most massive stars have non-negligible rotation rates (some even rotate near their break-up limit), simple conservation of angular momentum would suggest a proto-neutron star to be strongly differentially rotating with very high rotation rates and this picture is supported by numerical simulations of rotating core collapse [29, 30].

Other ways to form a rapidly rotating proto-neutron star would be through *fall-back accretion*[33], through the *accretion-induced collapse of a white dwarf* [34–37] or through the merger of binary white dwarfs in globular clusters [38]. It is also relevant to take into account current gamma-ray-burst models. The *collapsar* [39] model requires high rotation rates of a proto-black hole [32]. In addition, a possible formation scenario for magnetars involves a rapidly rotating protoneutron star formed through the collape of a very massive progenitor and some observational evidence is already emerging [40].

Gravitational waves from core collapse have a rich spectrum, reflecting the various stages of this event. The initial signal is emitted due to the *changing axisymmetric quadrupole moment* during collapse. In the case of neutron star formation, the quadrupole moment typically becomes larger, as the core spins up during contraction. In contrast, when a rapidly rotating neutron star collapses to form a Kerr black hole, the axisymmetric quadrupole moment first increases but is finally reduced by a large factor when the black hole is formed.

A second part of the gravitational wave signal is produced when gravitational collapse is halted by the stiffening of the equation of state above nuclear densities and the core bounces, driving an outwards moving shock. The dense fluid undergoes motions with relativistic speeds ($v/c \sim 0.2$–$0.4$) and a rapidly rotating proto-neutron star thus oscillates in several of its axisymmetric *normal modes of oscillation*. This quasi-periodic part of the signal could last for hundreds of oscillation periods, before being effectively damped. If, instead, a black hole is directly formed, then black hole quasinormal modes are excited, lasting for only a few oscillation periods. A combination of neutron star and black hole oscillations will appear if the proton-neutron star is not stable but collapses to a black hole.

In a rotating proto-neutron star, nonaxisymmetric processes can yield additional types of gravitational wave signals. Such processes are *dynamical instabilities, secular gravitational-wave driven instabilities* or *convection* inside the proto-neutron star and in its surrounding hot envelope. *Anisotropic neutrino emission* is accompanied by a gravitational wave signal. *Nonaxisymmetries* could already be present in the pre-collapse core and become amplified during collapse [41]. Furthermore, if there is persistent fall-back accretion onto a proto-neutron star or black hole, these can be brought into *ringing*.

Below, we discuss in more detail those processes which result in high frequency gravitational radiation.

### 3.2.1 Neutron star formation

Core collapse as a potential source of GWs has been studied for more than three decades (some of the most recent simulations can be found in [30, 36, 42–44, 49–53]). The main differences between the various studies are the progenitor models (slowly or rapidly rotating), equation of state (polytropic or realistic), gravity (Newtonian or relativistic) and neutrino emission (simple, sophisticated or no treatment). In general, the gravitational wave signal from neutron star formation is divided into a core bounce signal, a signal due to convective motions and a signal due to anisotropic neutrino emission.

The core bounce signal is produced due to rotational flattening and excitation of normal modes of oscillations, the main contributions coming from the axisymmetric quadrupole ($l = 2$) and quasi-radial ($l = 0$) modes (the latter radiating through its rotationally acquired $l = 2$ piece). If detected, such signals will be a unique probe for the high-density EOS of neutron stars [54, 55]. The strength of this signal is sensitive to the available angular momentum in the progenitor core. If the progenitor core is rapidly rotating, then core bounce signals from Galactic supernovae ($d \sim 10$ kpc) are detectable even with the initial LIGO/Virgo sensitivity at frequencies $\lesssim 1$ kHz. In the best-case scenario, advanced LIGO could detect signals from distances of 1 Mpc, but not from the Virgo cluster ($\sim 15$ Mpc), where the event rate would be high. The typical GW amplitude from 2D numerical simulations [30, 44–48] for an observer located in the equatorial plane of the source is [51]

$$h \approx 9 \times 10^{-21} \varepsilon \left( \frac{10 \, \text{kpc}}{d} \right) \quad (30)$$

where $\varepsilon \sim 1$ is the normalized GW amplitude. For such rapidly rotating initial models, the total energy radiated in GWs during the collapse is $\lesssim 10^{-6}$–$10^{-8} M_\odot c^2$. If, on the other hand, progenitor cores are slowly rotating (due to e.g. magnetic torques [31]), then the signal strength is significantly reduced, but, in the best case, is still within reach of advanced LIGO for galactic sources.

Normal mode oscillations, if excited in an equilibrium star at a small to moderate amplitude, would last for hundreds to thousands of oscillation periods, being damped only slowly by gravitational wave emission or viscosity. However, the proto-neutron star immediately after core bounce has a very different structure than a cold equilibrium star. It has a high internal temperature and is surrounded by an extended, hot envelope. Nonlinear oscillations excited in the core after bounce can penetrate into the hot envelope. Through this damping mechanism, the normal mode oscillations are damped on a much shorter timescale (on the order of ten oscillation periods), which is typically seen in the core collapse simulations mentioned above.

*Convection signal.* The post-shock region surrounding a proto-neutron star is convectively unstable to both low-mode and high-mode convection. Neutrino emission also drives convection in this region. The most realistic 2D simulations of core collapse to date [50] have shown that the gravitational wave signal from convection significantly exceeds the core bounce signal for slowly rotating progenitors, being detectable with advanced LIGO for galactic sources, and is detectable even for non-rotating collapse. For slowly rotating collapse, there is a detectable part of the signal in the high-frequency range of 700 Hz–1 kHz, originating from convective motions that dominate around 200 ms after core bounce. Thus, if both core a bounce signal and a convection signal would be detected in the same frequency range, these would be well separated in time.

*Neutrino signal.* In many simulations the gravitational wave signature of anisotropic neutrino emission has also been considered [56–58]. This type of signal can be detectable by advanced LIGO for galactic sources, but the main contribution is at low frequencies for a slowly rotating progenitor [50]. For rapidly rotating progenitors, stronger contributions at high frequencies could be present, but would probably be burried within the high-frequency convection signal.

Numerical simulations of neutron star formation have gone a long way, but a fully consistent 3D simulation including relativistic gravity, neutrino emission and magnetic fields is still missing. The combined treatment of these effects might not change the above estimations by orders of magnitude but it will provide more conclusive answers. There are also issues that need to be understood such as pulsar kicks (velocities exceeding 1000 km/s) which suggest that in a fraction of newly-born NSs (and probably BHs) the formation process may be strongly asymmetric [59]. Better treatment of the microphysics and construction of accurate progenitor models for the angular momentum distributions are needed. All these issues are under investigation by many groups.

### 3.2.2 Black hole formation

The gravitational-wave emission from the formation of a Kerr BH is a sum of two signals: the *collapse signal* and the *BH ringing*. The collapse signal is produced due to the changing multipole moments of the spacetime during the transition from a rotating iron core or proto-neutron star to a Kerr BH. A uniformly rotating neutron star has an axisymmetric quadrupole moment given by [60]

$$Q = -a\frac{J^2}{M} \tag{31}$$

where $a$ depends on the equation of state and is in the range of 2–8 for $1.4\,M_\odot$ models. This is several times larger in magnitude than the corresponding qudrupole moment of a Kerr black hole ($a = 1$). Thus, the *reduction* of the axisymmetric quadrupole moment is the main source of the collapse signal. Once the BH is formed, it continues to oscillate in its axisymmetric $l = 2$ quasinormal mode (QNM), until all oscillation energy is radiated away and the stationary Kerr limit is approached.

The numerical study of rotating collapse to BHs was pioneered by Nakamura [61] but first waveforms and gravitational-wave estimates were obtained by Stark and Piran [62]. These simulations we performed in 2D, using approximate initial data (essentially a spherical star to which angular momentum was artificially added). A new 3D computation of the gravitational wave emission from the collapse of unstable uniformly rotating relativistic polytropes to Kerr BHs [63] finds that the energy emitted is

$$\Delta E \sim 1.5 \times 10^{-6}(M/M_\odot), \qquad (32)$$

significantly less than the result of Stark and Piran. Still, the collapse of an unstable $2M_\odot$ rapidly rotating neutron star leads to a characteristic gravitational-wave amplitude $h_c \sim 3 \times 10^{-21}$, at a frequency of $\sim 5.5$ kHz, for an event at 10 kpc. Emission is mainly through the "+" polarization, with the "×" polarization being an order of magnitude weaker.

Whether a BH forms promptly after collapse or a delayed collapse takes place depends sensitively on a number of factors, such as the progenitor mass and angular momentum and the high-density EOS. The most detailed investigation of the influence of these factors on the outcome of collapse has been presented recently in [53], where it was found that shock formation increases the threshold for black hole formation by $\sim 20$–$40\%$, while rotation results in an increase of at most 25%.

### 3.2.3 Black hole ringing through fall-back

A black hole can form after core collapse, if fall-back accretion increases the mass of the proto-neutron star above the maximum mass allowed by axisymmetric stability. Material falling back after the black hole is formed excites the black hole quasi-normal modes of oscillation. If, on the other hand, the black hole is formed directly through core collapse (without a core bounce taking place) then most of the material of the progenitor star is accreted at very high rates ($\sim 1$–$2M_\odot$/s) into the hole. In such *hyper-accretion* the black hole's quasi-normal modes (QNM) can be excited for as long as the process lasts and until the black hole becomes stationary. Typical frequencies of the emitted GWs are in the range 1–3 kHz for $\sim 3$–$10\,M_\odot$ BHs.

The frequency and the damping time of the oscillations for the $l = m = 2$ mode can be estimated via the relations [64]

$$\sigma \approx 3.2\,\text{kHz}\,M_{10}^{-1}\left[1 - 0.63(1 - a/M)^{3/10}\right] \qquad (33)$$

$$Q = \pi\sigma\tau \approx 2\,(1-a)^{-9/20} \qquad (34)$$

These relations together with similar ones either for the 2nd QNM or the $l = 2$, $m = 0$ mode can uniquely determine the mass $M$ and angular momentum parameter $a$ of the BH if the frequency and the damping time of the signal have been accurately extracted [65–67]. The amplitude of the ring-down waves depends on the BH's initial distortion, i.e. on the nonaxisymmetry of the blobs or shells of matter falling into the BH. If matter of mass $\mu$ falls into a BH of mass $M$, then the gravitational wave energy is roughly

$$\Delta E \gtrsim \epsilon \mu c^2 (\mu/M) \tag{35}$$

where $\epsilon$ is related to the degree of asymmetry and could be $\epsilon \gtrsim 0.01$ [68]. This leads to an effective GW amplitude

$$h_{\text{eff}} \approx 2 \times 10^{-21} \left(\frac{\epsilon}{0.01}\right) \left(\frac{10\,\text{Mpc}}{d}\right) \left(\frac{\mu}{M_\odot}\right) \tag{36}$$

## 3.3 Rotational instabilities

If proto-neutron stars rotate rapidly, nonaxisymmetric *dynamical instabilities* can develop. These arise from non-axisymmetric perturbations having angular dependence $e^{im\phi}$ and are of two different types: the *classical bar-mode* instability and the more recently discovered *low-T/|W| bar-mode* and *one-armed spiral* instabilities, which appear to be associated to the presence of corotation points. Another class of nonaxisymmetric instabilities are *secular instabilities*, driven by dissipative effects, such as fluid viscosity or gravitational radiation.

### 3.3.1 Dynamical instabilities

The classical $m = 2$ bar-mode instability is excited in Newtonian stars when the ratio $\beta = T/|W|$ of the rotational kinetic energy $T$ to the gravitational binding energy $|W|$ is larger than $\beta_{\text{dyn}} = 0.27$. The instability grows on a dynamical time scale (the time that a sound wave needs to travel across the star) which is about one rotational period and may last from 1 to 100 rotations depending on the degree of differential rotation in the PNS.

The bar-mode instability can be excited in a hot PNS, a few milliseconds after core bounce, or, alternatively, it could also be excited a few tenths of seconds later, when the PNS cools due to neutrino emission and contracts further, with $\beta$ becoming larger than the threshold $\beta_{dyn}$ ($\beta$ increases roughly as $\sim 1/R$ during contraction). The amplitude of the emitted gravitational waves can be estimated as $h \sim MR^2\Omega^2/d$, where $M$ is the mass of the body, $R$ its size, $\Omega$ the rotation rate and $d$ the distance of the source. This leads to an estimation of the GW amplitude

$$h \approx 9 \times 10^{-23} \left(\frac{\epsilon}{0.2}\right) \left(\frac{f}{3\,\text{kHz}}\right)^2 \left(\frac{15\,\text{Mpc}}{d}\right) M_{1.4} R_{10}^2. \tag{37}$$

where $\epsilon$ measures the ellipticity of the bar, $M$ is measured in units of $1.4\,M_\odot$ and $R$ is measured in units of 10 km. Notice that, in uniformly rotation Maclaurin spheroids, the GW frequency $f$ is twice the rotational frequency $\Omega$. Such a signal is detectable only from sources in our galaxy or the nearby ones (our Local Group). If the sensitivity of the detectors is improved in the kHz region, signals from the Virgo cluster could be detectable. If the bar persists for many ($\sim$ 10–100) rotation periods, then even signals from distances considerably larger than the Virgo cluster will be detectable. Due to the requirement of rapid rotation, the event rate of the classical dynamical instability is considerably lower than the SN event rate.

The above estimates rely on Newtonian calculations; GR enhances the onset of the instability, $\beta_{\rm dyn} \sim 0.24$ [69, 70] and somewhat lower than that for large compactness (large $M/R$). Fully relativistic dynamical simulations of this instability have been obtained, including detailed waveforms of the associated gravitational wave emission. A detailed investigation of the required initial conditions of the progenitor core, which can lead to the onset of the dynamical bar-mode instability in the formed PNS, was presented in [51]. The amplitude of gravitational waves was due to the bar-mode instability was found to be larger by an order of magnitude, compared to the axisymmetric core collapse signal.

**Low-$T/|W|$ instabilities.** The *bar-mode* instability may be excited for significantly smaller $\beta$, if centrifugal forces produce a peak in the density off the source's rotational center [71]. Rotating stars with a high degree of differential rotation are also dynamically unstable for significantly lower $\beta_{\rm dyn} \gtrsim 0.01$ [72, 73]. According to this scenario the unstable neutron star settles down to a non-axisymmetric quasi-stationary state which is a strong emitter of quasi-periodic gravitational waves

$$h_{\rm eff} \approx 3 \times 10^{-22} \left(\frac{R_{\rm eq}}{30\,{\rm km}}\right) \left(\frac{f}{800\,{\rm Hz}}\right)^{1/2} \left(\frac{100\,{\rm Mpc}}{d}\right) M_{1.4}^{1/2}. \quad (38)$$

The bar-mode instability of differentially rotating neutron stars is an excellent source of gravitational waves, provided the high degree of differential rotation that is required can be realized. One should also consider the effects of viscosity and magnetic fields. If magnetic fields enforce uniform rotation on a short timescale, this could have strong consequences regarding the appearance and duration of the dynamical nonaxisymmetric instabilities.

An $m = 1$ *one-armed spiral* instability has also been shown to become unstable in proto-neutron stars, provided that the differential rotation is sufficiently strong [71, 74]. Although it is dominated by a "dipole" mode, the instability has a spiral character, conserving the center of mass. The onset of the instability appears to be linked to the presence of corotation points [75] (a similar link to corotation points has been proposed for the low-$T/|W|$ bar mode instability [76, 77]) and requires a very high degree of differential rotation (with matter on the axis rotating at least 10 times faster than matter on the equator). The $m = 1$ spiral instability was recently observed in simulations of rotating core collapse, which started with the core of an evolved $20M_\odot$ progenitor star to which differential rotation was added [78]. Growing from noise level ($\sim 10^{-6}$) on a timescale of 5ms, the $m = 1$ mode reached its maximum amplitude after $\sim$ 100ms. Gravitational waves were emitted through the excitation of an $m = 2$ nonlinear harmonic at a frequency of $\sim$800 Hz with an amplitude comparable to the core-bounce axisymmetric signal.

### 3.3.2 Secular gravitational-wave-driven instabilities

In a nonrotating star, the forward and backward moving modes of same $(l, |m|)$ (corresponding to $(l, +m)$ and $(l, -m)$) have eigenfrequencies $\pm|\sigma|$. Rotation splits this degeneracy by an amount $\delta\sigma \sim m\Omega$ and both the prograde and retrograde modes are dragged forward by the stellar rotation. If the star spins sufficiently rapidly, a mode which is retrograde (in the frame rotating with the star) will appear as prograde in the inertial frame (a nonrotating observer at infinity). Thus, an inertial observer sees GWs with positive angular momentum emitted by the retrograde mode, but since the perturbed fluid rotates slower than it would in the absence of the perturbation, the angular momentum of the mode in the rotating frame is negative. The emission of GWs consequently makes the angular momentum of the mode increasingly negative, leading to the instability. A mode is unstable when $\sigma(\sigma - m\Omega) < 0$. This class of *frame-dragging instabilities* is usually referred to as Chandrasekhar–Friedman–Schutz [79, 80] (CFS) instabilities.

*f*-**mode instability.** In the Newtonian limit, the $l = m = 2$ $f$-mode (which has the shortest growth time of all polar fluid modes) becomes unstable when $T/|W| > 0.14$, which is near or even above the mass-shedding limit for typical polytropic EOSs used to model uniformly rotating neutron stars. Dissipative effects (e.g. shear and bulk viscosity or mutual friction in superfluids) [81–84] leave only a small instability window near mass-shedding, at temperatures of $\sim 10^9$ K. However, relativistic effects strengthen the instability considerably, lowering the required $\beta$ to $\approx 0.06$–$0.08$ [85, 86] for most realistic EOSs and masses of $\sim 1.4 M_\odot$ (for higher masses, such as hypermassive stars created in a binary NS merger, the required rotation rates are even lower).

Since PNSs rotate differentially, the above limits derived under the assumption of uniform rotation are too strict. Unless uniform rotation is enforced on a short timescale, due to e.g. magnetic braking [87], the $f$-mode instability will develop in a differentially rotating background, in which the required $T/|W|$ is only somewhat larger than the corresponding value for uniform rotation [88], but the mass-shedding limit is dramatically relaxed. Thus, in a differentially rotating PNS, the $f$-mode instability window is huge, compared to the case of uniform rotation and the instability can develop provided there is sufficient $T/|W|$ to begin with.

The $f$-mode instability is an excellent source of GWs. Simulations of its nonlinear development in the ellipsoidal approximation [89] have shown that the mode can grow to a large nonlinear amplitude, modifying the background star from an axisymmetric shape to a differentially rotating ellipsoid. In this modified background the $f$-mode amplitude saturates and the ellipsoid becomes a strong emitter of gravitational waves, radiating away angular momentum until the star is slowed-down towards a stationary state. In the case of uniform density ellipsoids, this stationary state is the Dedekind ellipsoid, i.e. a nonaxisymmetric ellipsoid with internal flows but with a stationary (nonradiating) shape in the inertial frame. In the ellipsoidal approximation, the nonaxisymmetric pattern radiates gravitational waves sweeping through the LIGO II sensitivity window (from 1 kHz down to about 100 Hz) which could become detectable out to a distance of more than 100 Mpc.

Two recent hydrodynamical simulations [90, 91] (in the Newtonian limit and using a post-Newtonian radiation-reaction potential) essentially confirm this picture. In [90] a differentially rotating, $N = 1$ polytropic model with a large $T/|W| \sim 0.2$–$0.26$ is chosen as the initial equilibrium state. The main difference of this simulation compared to the ellipsoidal approximation comes from the choice of EOS. For $N = 1$ Newtonian polytropes it is argued that the secular evolution cannot lead to a stationary Dedekind-like state. Instead, the $f$-mode instability will continue to be active until all nonaxisymmetries are radiated away and an axisymmetric shape is reached. This conclusion should be checked when relativistic effects are taken into account, since, contrary to the Newtonian case, relativistic $N = 1$ uniformly rotating polytropes *are* unstable to the $l = m = 2$ $f$-mode [85] – however it is not still possible up to date, to construct relativistic analogs of Dedekind ellipsoids.

In the other recent simulation [91], the initial state was chosen to be a uniformly rotating, $N = 0.5$ polytropic model with $T/|W| \sim 0.18$. Again, the main conclusions reached in [89] are confirmed, however, the assumption of uniform initial rotation limits the available angular momentum that can be radiated away, leading to a detectable signal only out to about $\sim 40$ Mpc. The star appears to be driven towards a Dedekind-like state, but after about 10 dynamical periods, the shape is disrupted by growing short-wavelength motions, which are suggested to arise because of a shearing type instability, such as the elliptic flow instability [92].

**$r$-mode instability.** Rotation does not only shift the spectra of polar modes; it also lifts the degeneracy of axial modes, give rise to a new family of *inertial* modes, of which the $l = m = 2$ $r$-mode is a special member. The restoring force, for these oscillations is the Coriolis force. Inertial modes are primarily velocity perturbations. The frequency of the $r$-mode in the rotating frame of reference is $\sigma = 2\Omega/3$. According to the criterion for the onset of the CFS instability, the $r$-mode is unstable for any rotation rate of the star [93, 94]. For temperatures between $10^7$–$10^9$ K and rotation rates larger than 5–10% of the Kepler limit, the growth time of the unstable mode is smaller than the damping times of the bulk and shear viscosity [95, 96]. The existence of a solid crust or of hyperons in the core [97] and magnetic fields [98, 99], can also significantly affect the onset of the instability (for extended reviews see [100, 101]). The suppression of the $r$-mode instability by the presence of hyperons in the core is not expected to operate efficiently in rapidly rotating stars, since the central density is probably too low to allow for hyperon formation. Moreover, a recent calculation [102] finds the contribution of hyperons to the bulk viscosity to be two orders of magnitude smaller than previously estimated. If accreting neutron stars in Low Mass X-Ray Binaries (LMXB, considered to be the progenitors of millisecond pulsars) are shown to reach high masses of $\sim 1.8 M_\odot$, then the EOS could be too stiff to allow for hyperons in the core (for recent observations that support a high mass for some millisecond pulsars see [103]).

The unstable $r$-mode grows exponentially until it saturates due to nonlinear effects at some maximum amplitude $\alpha_{max}$. The first computation of nonlinear mode couplings using second-order perturbation theory suggested that the $r$-mode is limited to very small amplitudes (of order $10^{-3}$–$10^{-4}$) due to transfer of energy to a large number of other inertial modes, in the form of a cascade, leading to an equilibrium distribution of mode amplitudes [104]. The small saturation values for the amplitude are supported by recent nonlinear estimations [105, 106] based on the drift, induced by the r-modes, causing differential rotation. On the other hand, hydrodynamical simulations of limited resolution showed that an initially large-amplitude r-mode does not decay appreciably over several dynamical timescales [108], but on a somewhat longer timescale a catastrophic decay was observed [109] indicating a transfer of energy to other modes, due to nonlinear mode couplings and suggesting that a hydrodynamical instability may be operating. A specific resonant 3-mode coupling was identified [110] as the cause of the instability and a perturbative analysis of the decay rate suggests a maximum saturation amplitude $\alpha_{max} < 10^{-2}$. A new computation using second-order perturbation theory finds that the catastrophic decay seen in the hydrodynamical simulations [109, 110] can indeed be explained by a parametric instability operating in 3-mode couplings between the r-mode and two other inertial modes [107, 111–113]. Whether the maximum saturation amplitude is set by a network of 3-mode couplings or a cascade is reached, is, however, still unclear.

A neutron star spinning down due to the r-mode instability will emit gravitational waves of amplitude

$$h(t) \approx 10^{-21} \alpha \left(\frac{\Omega}{1\,\text{kHz}}\right) \left(\frac{100\,\text{kpc}}{d}\right) \tag{39}$$

Since $\alpha$ is small, even with LIGO II the signal is undetectable at large distances (VIRGO cluster) where the SN event rate is appreciable, but could be detectable after long-time integration from a galactic event. However, if the compact object is a strange star, then the instability may not reach high amplitudes ($\alpha \sim 10^{-3} - 10^{-4}$) but it will persist for a few hundred years (due to the different temperature dependence of viscosity in strange quark matter) and in this case there might be up to ten unstable stars in our galaxy at any time [114]. Integrating data for a few weeks could lead to an effective amplitude $h_{\text{eff}} \sim 10^{-21}$ for galactic signals at frequencies $\sim$700–1000 Hz. The frequency of the signal changes only slightly on a timescale of a few months, so that the radiation is practically monochromatic.

*Other unstable modes.* The CFS instability can also operate for core g-mode oscillations [115] but also for w-mode oscillations, which are basically spacetime modes [116]. In addition, the CFS instability can operate through other dissipative effects. Instead of the gravitational radiation, any radiative mechanism (such as electromagnetic radiation) can in principle lead to an instability.

## 3.4 Accreting neutron stars in LMXBs

Spinning neutron stars with even tiny deformations are interesting sources of gravitational waves. The deformations might results from various factors but it seems that the most interesting cases are the ones in which the deformations are caused by accreting material. A class of objects called Low-Mass X-Ray Binaries (LMXB) consist of a fast rotating neutron star (spin $\approx$270–650 Hz) torqued by accreting material from a companion star which has filled up its Roche lobe. The material adds both mass and angular momentum to the star, which, on timescales of the order of tenths of Megayears could, in principle, spin up the neutron star to its break up limit. One viable scenario [118] suggests that the accreted material (mainly hydrogen and helium) after an initial phase of thermonuclear burning undergoes a non-uniform crystallization, forming a crust at densities $\sim 10^8$–$10^9 \mathrm{g/cm^3}$. The quadrupole moment of the deformed crust is the source of the emitted gravitational radiation which slows-down the star, or halts the spin-up by accretion.

An alternative scenario has been proposed by Wagoner [119] as a follow up of an earlier idea by Papaloizou-Pringle [120]. The suggestion was that the spin-up due to accretion might excite the $f$-mode instability, before the rotation reaches the breakup spin. The emission of gravitational waves will torque down the star's spin at the same rate as the accretion will torque it up, however, it is questionable whether the $f$-mode instability will ever be excited for old, accreting neutron stars. Following the discovery that the $r$-modes are unstable at any rotation rate, this scenario has been revived independently by Bildsten [118] and Andersson, Kokkotas and Stergioulas [121]. The amplitude of the emitted gravitational waves from such a process is quite small, even for high accretion rates, but the sources are persistent and in our galactic neighborhood the expected amplitude is

$$h \approx 10^{-27} \left( \frac{1.6 \, \mathrm{ms}}{P} \right)^5 \frac{1.5 \, \mathrm{kpc}}{D} . \qquad (40)$$

This signal is within reach of advanced LIGO with signal recycling tuned at the appropriate frequency and integrating for a few months [1]. This picture is in practice more complicated, since the growth rate of the $r$-modes (and consequently the rate of gravitational wave emission) is a function of the core temperature of the star. This leads to a thermal runaway due to the heat released as viscous damping mechanisms counteract the r-mode growth [122]. Thus, the system executes a limit cycle, spinning up for several million years and spinning down in a much shorter period. The duration of the unstable part of the cycle depends critically on the saturation amplitude $\alpha_{max}$ of the $r$-modes [123, 124]. Since current computations [104, 106] suggest an $\alpha_{max} \sim 10^{-3}$–$10^{-4}$, this leads to a quite long duration for the unstable part of the cycle of the order of $\sim 1$ Myear.

The instability window depends critically on the effect of the shear and bulk viscosity and various alternative scenarios might be considered. The existence of hyperons in the core of neutron stars induces much stronger bulk viscosity which suggests a much narrower instability window for the $r$-modes and the bulk viscosity prevails over the instability even in temperatures as low as $10^8$ K [97]. A similar picture can be drawn if the star is composed of "deconfined" $u$, $d$ and $s$ quarks – a strange star [125]. In this case, there is a possibility that the strange stars in LMXBs evolve into a quasi-steady state with nearly constant rotation rate, temperature and mode amplitude [114] emitting gravitational waves for as long as the accretion lasts. This result has also been found later for stars with hyperon cores [126, 127]. It is interesting that the stalling of the spin up in millisecond pulsars (MSPs) due to $r$-modes is in good agreement with the minimum observed period and the clustering of the frequencies of MSPs [123].

## 3.5 Gravitational-wave asteroseismology

If various types of oscillation modes are excited during the formation of a compact star and become detectable by gravitational wave emission, one could try to identify observed frequencies with frequencies obtained by mode-calculations for a wide parameter range of masses, angular momenta and EOSs [55, 128–132]. Thus, *gravitational wave asteroseismology* could enable us to estimate the mass, radius and rotation rate of compact stars, leading to the determination of the "best-candidate" high-density EoS, which is still very uncertain. For this to happen, accurate frequencies for different mode-sequences of rapidly rotating compact objects have to be computed.

For slowly rotating stars, the frequencies of $f-$, $p-$ and $w-$ modes are still unaffected by rotation, and one can construct approximate formulae in order to relate observed frequencies and damping times of the various stellar modes to stellar parameters. For example, for the fundamental oscillation ($l = 2$) mode ($f$-mode) of non-rotating stars one obtains [55]

$$\sigma(\text{kHz}) \approx 0.8 + 1.6 M_{1.4}^{1/2} R_{10}^{-3/2} + \delta_1 m \bar{\Omega} \qquad (41)$$

$$\tau^{-1}(\text{secs}^{-1}) \approx M_{1.4}^3 R_{10}^{-4} \left(22.9 - 14.7 M_{1.4} R_{10}^{-1}\right) + \delta_2 m \bar{\Omega} \qquad (42)$$

where $\bar{\Omega}$ is the normalized rotation frequency of the star, and $\delta_1$ and $\delta_2$ are constants estimated by sampling data from various EOSs. The typical frequencies of NS oscillation modes are larger than 1 kHz. Since each type of mode is sensitive to the physical conditions where the amplitude of the mode is largest, the more oscillations modes can be identified through gravitation waves, the better we will understand the detailed internal structure of compact objects, such as the existence of a possible superfluid state of matter [133].

If, on the other hand, some compact stars are born rapidly rotating with moderate differential rotation, then their central densities will be much smaller than the central density of a nonrotating star or same baryonic mass. Correspondingly, the typical axisymmetric oscillation frequencies will be smaller than 1 kHz, which is more favorable for the sensitivity window of current interferometric detectors [134]. Indeed, axisymmetric simulations of rotating core-collapse have shown that if a rapidly rotating NS is created, then the dominant frequency of the core-bounce signal (originating from the fundamental $l = 2$ mode or the $l = 2$ piece of the fundamental quasi-radial mode) is in the range 600 Hz–1 kHz [30].

If different type of signals are observed after core collapse, such as both an axisymmetric core-bounce signal and a nonaxisymmetric one-armed instability signal, with a time separation of the order of 100ms, this would yield invaluable information about the angular momentum distribution in the proto-neutron stars.

## Acknowledgements

I am grateful to N. Stergioulas and A. Colaiuda for suggestions which improved the original manuscript. This work was supported by the German Foundation (DFG) via SFB/TR7.

# References

[1] Cutler C, Thorne K S 2002: *in proceedings of GR16 (Durban South Africa, 2001)*, gr-qc/0204090
[2] Schnabel R, Harms J, Strain K A, Danzmann K 2004, *Class. Quantum Grav.* **21**, S1155
[3] Bonaldi M, Cerdonio M, Conti L, Prodi G A, Taffarello L, Zendri J P : 2004 *Class. Quantum Grav.* **21**, S1155
[4] Saulson P R, 1994 *Fundamentals of Interferometric Gravitational Wave Detectors*, World Scientific
[5] Blair D. G., 1991, *The Detection of Gravitational Waves*, Cambridge University Press
[6] Hough J and Rowan S 2000 *Gravitational Wave Detection by Interferometry (Ground and Space)*, *Living Rev. Relativity* **3** 1.
[7] Maggiore M 2008 *Gravitational Waves*, Oxford
[8] LIGO http://www.ligo.caltech.edu
[9] VIRGO http://www.pi.infn.it/virgo/virgoHome.html
[10] GEO http://www.geo600.uni-hannover.de
[11] TAMA http://tamago.mtk.nao.ac.jp
[12] LISA http://lisa.jpl.nasa.gov
[13] By LIGO Collaboration (B. Abbott et al.) *Phys. Rev. Lett.* **95**: 221101,2005.
[14] By LIGO Collaboration (B. Abbott et al.) *Phys. Rev. D* **72**: 082001,2007.
[15] By LIGO Collaboration (B. Abbott et al.) *Phys. Rev. D* **73**: 062001,2007.
[16] By LIGO Collaboration (B. Abbott et al.) *Astrophys.J.* **659**:918-930,2007.
[17] By LIGO Collaboration (B. Abbott et al.) *Phys. Rev. D* **76**: 042001,2007.
[18] Burgay M, D'Amico N, Possenti A, Manchester R N, Lyne A G, Joshi B C, McLaughlin M A, Kramer M, Sarkissian J M, Camilo F, Kalogera V, Kim C, Lorimer D R: 2003, Nature, **426**, 531
[19] Kalogera V, Kim C, Lorimer D R, Burgay M, D'Amico N, Possenti A, Manchester R N, Lyne A G, Joshi B C, McLaughlin M A, Kramer M, Sarkissian J M, Camilo F : 2004 *Astrophys.J.* **601**, L179
[20] Shibata M 2005 *Phys. Rev. Lett.* **94** 201101
[21] Vallisneri M 2000 *Phys. Rev. Lett.* 84 3519
[22] Faber J A, Glandcement P, Rasio F A and Taniguchi K 2002 *Phys. Rev. Lett.* **89** 231102
[23] Fryer C L 1999 *Astrophys. J.* **522** 413
[24] Fryer C L 2003 *Class. Quantum Gravity* **20** S73
[25] Fryer C L and Warren M S 2004 *Astrophys. J.* **601** 391
[26] Burrows A, Walder R, Ott C D and Livne E 2005 *Nucl. Phys. A* **752** 570
[27] Cappellaro E, Turatto M, Tsvetov D Yu, Bartunov O S, Pollas C, Evans R and Hamuy M 1999 *A&A* **351** 459
[28] Fryer C L and Kalogera V 2001 *Astrophys.J* **554**, 548
[29] Fryer C L and Heger A 2000 *Astrophys. J.* **541** 1033
[30] Dimmelmeier H, Font J A and Müller E 2002 *A&A* **393** 523
[31] Spruit H C 2002 *Astron. Astrophys.* **381** 923
[32] Petrovic J, Langer N, Yoon S C and Heger A 2005 astro-ph/0504175
[33] Watts A L and Andersson N 2002 *MNRAS* **333** 943
[34] Hillebrandt W, Wolff R G and Nomoto K 1984 *Astrophys. J.* **133** 175
[35] Liu Y T and Lindblom L 2001 *MNRAS* **324** 1063
[36] Fryer C L, Holz D E, Hughes S A 2002 *Astrophys. J.* **565** 430
[37] Yoon S C and Langer N 2005 astro-ph/0502133
[38] Middleditch J 2004 *Astrophys. J.* **601** L167
[39] Woosley S E 1993 *Bulletin of the American Astronimical Society* **25** 894

[40] Gaensler B M et al. 2005 *Nature* **434** 1104
[41] Fryer C L, Holz D E,Hughes S A 2004 *Astrophys. J.* **609** 288
[42] Zwerger T, Müller E 1997 *A & A* **320** 209
[43] Rampp M, Müller E and Ruffert M 1998 *A& A* **332** 969
[44] Ott C D, Burrows A, Livne E, Walder R, 2004 *Astrophys. J. Lett.* **625**, 119
[45] Burrows A, Dessart L, Livne E, Ott C D, Murphy J 2007, *Astrophys. J.* **664**, 416
[46] Ott, C. D.; Dimmelmeier, H.; Marek, A.; Janka, H.-T.; Hawke, I.; Zink, B.; Schnetter, E. 2007, *Phys. Rev. Lett.* **98** 261101
[47] Dimmelmeier, H.; Ott, C. D.; Janka, H.-T.; Marek, A.; Müller, E. 2007, *Phys. Rev. Lett.* **98** 261101
[48] Ott C D, Dimmelmeier H, Marek A, Janka, H.-T, Zink B, Hawke I, Schnetter E 2007, *CQG* **24** 139
[49] Kotake K, Yamada S and Sato K, 2003 *Phys. Rev. D* **68** 044023
[50] Müller E, Rampp M, Buras R, Janka H-Th and Shoemaker D H 2004 *Astrophys. J.* **603** 221
[51] Shibata M and Sekiguchi Y 2004 *Phys. Rev. D* **69** 084024
[52] Shibata M and Sekiguchi Y 2005 *Phys. Rev. D* **71** 044017
[53] Sekiguchi Y and Shibata M 2005 *Phys. Rev. D* **71** 084013
[54] Andersson N, Kokkotas K D 1996 *Phys. Rev. Lett.* **77** 4134
[55] Andersson N, Kokkotas K D 1998 *M.N.R.A.S.* **299** 1059
[56] Epstein R 1978 *Astrophys. J.* **223** 1037
[57] Burrows A and Hayes J 1996 *Phys. Rev. Lett.* **76** 352
[58] Müller E and Janka H T 1997 *Astron. Astrophys.* **317** 140
[59] Hoeflich P, Khokhlov A, Wang L, Wheeler J C and Baade D 2002, IAU Symposium 212 on Massive Stars, D. Reidel Conf. Series, ed. E. van den Hucht, astro-ph/0207272
[60] Laarakkers W G and Poisson E 1999 *Astrophys. J.* **512** 282
[61] Nakamura T 1981 *Prog. Theor. Phys.* **65** 1876
[62] Stark R F and Piran T 1985 *Phys. Rev. Lett.* **55** 891
[63] Baiotti L, Hawke I, Rezzolla L and Schnetter E 2005 *Phys. Rev. Lett.* **94** 131101
[64] Echeverria E 1980 *Phys. Rev. D* **40** 3194
[65] Finn L S 1992 *Phys. Rev. D* **8** 3308
[66] Nakano H, Takahashi H, Tagoshi H and Sasaki M 2003 *Phys. Rev. D* **68** 102003
[67] Dryer O, Kelly B, Krishnan B, Finn L S, Garrison D, Lopez-Aleman R 2004 *Class. Quantum Grav.* **21** 787
[68] Davis M, Ruffini R, Press W H and Price R H 1971 **27** 1466
[69] Shibata M, Baumgarte T W and Shapiro S L 2000 *Astrophys. J.* **542** 453
[70] Saijo M, Shibata M, Baumgarte T W and Shapiro S L 2001 *Astrophys. J.* **548** 919
[71] Centrella J M, New K C B., Lowe L L , Brown J D 2001 *Astrophys. J. Lett.* **550** 193
[72] Shibata M, Karino S, Eriguchi Y 2002 *MNRAS* **334** L27
[73] Shibata M, Karino S, Eriguchi Y 2003 *MNRAS* **343** 619
[74] Saijo M, Baumgarte T W and Shapiro S L, 2002 *Astrophys. J* **595** 352
[75] Saijo M and Yoshida Y 2006 *MNRAS* **368** 1429
[76] Watts A L, Andersson N and Williams R L 2004 *MNRAS* **350** 927
[77] Passamonti A, Stavridis A, Kokkotas KD 2008 *Phys. Rev. D* in press, gr-qc/0706.0991
[78] Ott C D, Ou S, Tohline J E, Burrows A 2006 *Astroph. J. Lett.* **625**, 119
[79] Chandrasekhar S 1970 *Phys. Rev. Lett.* **24** 611
[80] Friedman J L and Schutz B F 1978 *Astrophys. J.* **222** 281
[81] Cutler C and Lindblom L 1987 *Astrophys. J.* **314** 234
[82] Lindblom L and Detweiler S 1979 *Astrophys. J.* **232** L101
[83] Ipser J R and Lindblom L 1991 *Astrophys. J.* **373** 213

[84] Lindblom L and Mendell G 1995 *Astrophys. J.* **444** 804
[85] Stergioulas N, Friedman J L, 1998 *Astrophys. J.* **492** 301
[86] Morsink S, Stergioulas N and Blattning S 1999 *Astrophys. J.* **510** 854
[87] Liu Y T and Shapiro S L 2004 *Phys. Rev. D* **69** 044009
[88] Yoshida S, Rezzolla L, Karino S and Eriguchi Y, 2002 *Astrophys. J. Lett.* **568** 41
[89] Lai D and Shapiro S L 1995 *Astrophys. J.* **442** 259
[90] Shibata M and Karino S 2004 *Phys. Rev. D* **70** 084022
[91] Ou S, Tohline J E and Lindblom L 2004 *Astrophys. J.* **617** 490
[92] Lifschitz A and Lebovitz N 1993 *Astrophys. J.* **408** 603
[93] Andersson N 1998 *Astrophys. J.* **502** 708
[94] Friedman J L and Morsink S 1998 *Astrophys. J.* **502** 714
[95] Linblom L, Owen B J and Morsink S M 1998 *Phys. Rev. Lett.* **80** 4843
[96] Andersson N, Kokkotas K D and Schutz B F 1999 *Astrophys. J.* **510** 846
[97] Lindblom L, Owen B, 2002 *Phys. Rev. D* **65** 063006
[98] Rezzolla L, Lamb, F K, Markovic D and Shapiro S L 2001 *Phys. Rev. D* **64** 104013
[99] Rezzolla L, Lamb, F K, Markovic D and Shapiro S L 2001 *Phys. Rev. D* **64** 104014
[100] Andersson N and Kokkotas K D 2001 *Int. J. Modern Phys.* **D10** 381
[101] Andersson N 2003 *Class. Quantum Grav.* **20** R105
[102] van Dalen E N E and Dieperink A E L 2003 *Phys. Rev. C* **69** 025802
[103] Nice, D. J., Splaver, E. M., Stairs, I. H., 2003, astro-ph/0311296
[104] Schenk A K, Arras P, Flanagan E E, Teukolsky S A, Wasserman I 2002 *Phys. Rev. D* **65** 024001
[105] Sa P M 2004 *Phys. Rev. D* **69** 084001
[106] Sa P M and Tome B 2005 *Phys. Rev. D* **71** 044007
[107] Bondarescu R, Teukolsky S A and Ira Wasserman I 2007 *Phys. Rev. D.* **76** 064019
[108] Stergioulas N and Font J A 2001 *Phys. Rev. Lett.* **86** 1148
[109] Gressman P, Lin L M, Suen W M, Stergioulas N and Friedman J L 2002 *Phys. Rev. D.* **66** 041303(R)
[110] Lin L M and Suen W M 2004 gr-qc/0409037
[111] Brink J, Teukolsky S A and Wasserman I 2004 *Phys. Rev. D.* **70** 121501
[112] Brink J, Teukolsky S A and Wasserman I 2004 *Phys. Rev. D.* **70** 124017
[113] Brink J, Teukolsky S A and Wasserman I 2005 *Phys. Rev. D.* **71** 064029
[114] Andersson N, Jones D I and Kokkotas K D 2002 *MNRAS* **337** 1224
[115] Lai D 1999 *MNRAS* **307** 1001
[116] Kokkotas K D, Ruoff J and Andersson N 2004 *Phys. Rev. D* **70** 043003
[117] Roberts P H and Stewartson K 1963 *Astrophys. J.* **137** 777
[118] Bildsten L. 1998 *Astrophys. J.* **501** L89
[119] Wagoner R.V., 1984 *Astrophys. J.* **278** 345
[120] Papaloizou J., Pringle J.E., 1978 *MNRAS* **184**, 501
[121] Andersson N, Kokkotas K D and Stergioulas N 1999 *Astrophys. J.* **307** 314
[122] Levin Y., 1999 *Astrophys. J.* **517** 328
[123] Andersson N, Jones D I, Kokkotas K D and Stergioulas N 2000 *Astrophys. J. Lett.* **534** 75
[124] Heyl J 2002 *Astrophys. J. Lett.* **574** L57
[125] Madsen J., 1998 *Phys. Rev. Lett.* **81**, 3311
[126] Wagoner R V 2002 *Astrophys. J. Lett.* **578** L63
[127] Reisenegger A., Bonacic A., 2003 *Phys. Rev. Lett.* **91** 201103
[128] Benhar O, Berti E and Ferrari V 1999 *MNRAS* **310** 797
[129] Kokkotas K D, Apostolatos T, Andersson N 2001 *M.N.R.A.S.* **320** 307
[130] Sotani H, Kohri K and Harada T 2004 *Phys. Rev. D* **69** 084008

[131] Sotani H and Kokkotas K D 2004 *Phys. Rev. D* **70** 084026
[132] Benhar O, Ferrari V and Gualtieri L 2004 *Phys. Rev. D* **70** 124015
[133] Andersson N and Comer G L 2001 *Phys. Rev. Lett.* **87**, 241101
[134] Stergioulas N, Apostolatos T A and Font J A 2004 *MNRAS* **352**, 1089

# High-(Energy)-Lights
# The Very High Energy Gamma-Ray Sky

Dieter Horns

Institute for Experimental Physics, University of Hamburg
Luruper Chaussee 149, 22761 Hamburg, Germny
dieter.horns@desy.de

**Abstract**

*The high-lights of ground-based very-high-energy (VHE, $E > 100$ GeV) gamma-ray astronomy are reviewed. The summary covers both Galactic and extra-galactic sources. Implications for our understanding of the non-thermal Universe are discussed. Identified VHE sources include various types of supernova remnants (shell-type, mixed morphology, composite) including pulsar wind nebulae, and X-ray binary systems. A diverse population of VHE-emitting Galactic sources include regions of active star formation (young stellar associations), and massive molecular clouds. Different types of active galactic nuclei have been found to emit VHE gamma-rays: besides predominantly Blazar-type objects, a radio-galaxy and a flat-spectrum radio-quasar have been discovered. Finally, many (presumably Galactic) sources have no convincing counterpart and remain at this point unidentified. A total of at least 70 sources are currently known. The next generation of ground based gamma-ray instruments aims to cover the entire accessible energy range from as low as $\approx 10$ GeV up to $10^5$ GeV and to improve the sensitivity by an order of magnitude in comparison with current instruments.*

## 1 Introduction

The highest energy photons known are produced in astrophysical processes involving even more energetic particles presumably accelerated through stochastic acceleration mechanisms as suggested initially by Fermi [108]. Therefore, observations of VHE photons provide a direct view of the astrophysical accelerators of charged particles and allow to identify the individual sources of cosmic rays: VHE photons open our view to the "accelerator sky". The origin of the Galactic population of charged cosmic rays has remained since its discovery by V. Hess in 1912 until today a long-standing question of astro- and particle physics. A widely favored model on the origin of cosmic rays assumes that diffusive shock acceleration takes place in the expanding blast waves of supernova remnants converting 10–20 % of the kinetic energy into cosmic rays. Under this assumption, Galactic supernova remnants provide sufficient power ($\approx 10^{41}$ ergs s$^{-1}$) in order to balance the escape losses of Galactic cosmic rays as well as produce a power-law type distribution of particle energy that closely resembles the cosmic ray spectrum measured locally [see e.g. 120, 127].

Observations of VHE-gamma-rays, mainly by imaging air Cherenkov telescopes in the last decade, have surpassed the anticipated detection of a few supernova remnants and have established a rich and diverse collection of VHE sources. Specifically, the current generation of imaging air Cherenkov telescopes (CANGAROO III, H.E.S.S., MAGIC, and VERITAS) have fulfilled and by far exceeded the expectations that were based upon the pioneering previous generation of experiments: the results obtained in the last years have shown that VHE emission is common to a variety of different source types – not only shell-type SNRs.

The current view of the source distribution in the VHE sky is shown in Figure 1 where all known sources are displayed (status of early fall 2007). Note, the sensitivity achieved varies greatly across the sky. The best sensitivity is reached in the inner Galactic disk where a dedicated survey with the air Cherenkov telescopes of the H.E.S.S. experiment has been performed (see Section 2.2).

The most remarkable feature of the gamma-ray sky is not evident in this picture: each source shown is an accelerator of particles up to and beyond TeV energies ("accelerator sky"). A similar conclusion can not be drawn when looking at source populations detected in other wavelength bands. This makes the VHE band a sensitive window to detect non-thermal particle accelerators. Future neutrino telescopes will almost certainly have the potential to detect some of the brighter VHE sources (and discover new sources that are optically thick for VHE gamma-rays). The detection of high energy neutrinos is experimentally a challenging task and requires tremendous efforts (see e.g. the Ice-CUBE neutrino telescope in the Antarctic ice). However, the observation of a neutrino source will ultimately demonstrate the presence of accelerated nuclei – largely model-independent. Different to the clear observational signature in the neutrino channel, VHE gamma-rays are sensitive to accelerated electrons (through inverse Compton scattering) as well as to accelerated nuclei (through neutral meson production and decay).

The scope of this article is limited to observational highlights obtained with ground based instruments and does not aim at presenting a complete review of the field of VHE astrophysics. The paper is structured in the following way: In section 2, the experimental techniques of ground-based instruments (both imaging and non-imaging) are described before continuing with the census of today's VHE sky in section 3. Section 4 provides a a short overview on VHE gamma-rays as probes of the interstellar medium including Lorenz-invariance violating effects related to structure of space-time at Planck-scales. Finally, in section 5 this review is concluded with some comments on the future of the field.

## 2 Observational techniques

Non-thermal gamma-ray sources produce typically power-law type energy spectra in which the flux drops with increasing energy. As a consequence of this, space based detection techniques are limited by the small detection rate of photons at high energies ($E < 10$–$100$ GeV). The detection of VHE ($E > 100$ GeV) photons requires

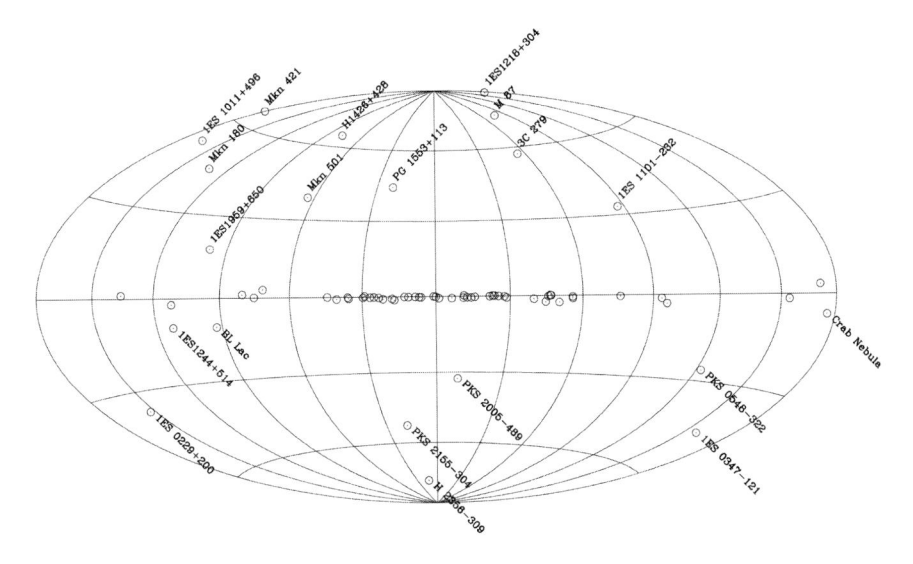

**Figure 1:** All VHE sources as known in September 2007 are shown as circles in a Hammer–Aitoff projection of a Galactic coordinate system. The Galactic center is in the center of the picture, Galactic longitude runs from right to left.

collection areas that are substantially larger than the $\mathcal{O}(\mathrm{m}^2)$ of satellite based detectors. One approach towards large collection areas are ground based observation techniques which however are limited to energies sufficiently high ($E > 5$ GeV) to produce a detectable air shower.

Ground based gamma-ray observations are always based upon air shower detection techniques. High energy particles (photons, electrons, and nuclei) initiate extensive air showers in the atmosphere which acts as a natural calorimeter: secondary particles are detected either when they reach the ground or by the Cherenkov light produced in the atmosphere. At ultra-high energies, it becomes feasible to detect fluorescence light as well as radio signals which are not discussed here (see e.g. the contribution by J. Hörandel at this conference).

Generally, the air shower detection technique relies on one of the following approaches:

- Shower-front sampling of particles
- Shower-front sampling of Cherenkov light
- Imaging of the air shower using Cherenkov light.

The benefits (marked '+') and draw-backs ('−') of the shower-front sampling of particles and the imaging air Cherenkov technique are mainly the following[1]:

- Imaging technique:
  - \+ sensitivity ($\approx 10^{-13}$ ergs/(cm$^2$ s)$^{-1}$ at TeV-energies, 50 hrs exposure)
  - \+ angular resolution (few arc minutes per event)
  - \+ spectroscopy ($\Delta E/E \approx 15\ \%$)
  - \+ low energy threshold
  - − field of view (a few degrees, $\approx$ msrad)
  - − duty cycle ($\approx 10\ \%$)

- Shower front sampling of particles:
  - \+ high duty cycle ($\approx 95\ \%$)
  - \+ large field of view ($\approx 2$ srad)
  - − high energy threshold
  - − sensitivity ($\approx 10^{-12}$ ergs cm$^{-2}$ s$^{-1}$, 1 year exposure)
  - − angular resolution (0.3–0.7°)
  - − energy resolution ($\Delta E/E \approx 30\text{–}100\ \%$)

The obvious complementarity of the two techniques justifies the further development of both of them in the future: non-imaging techniques cover simultaneously a large field of view and can be operated 24 hours a day. On the other hand, air Cherenkov telescopes have a narrow field of view (a few milli steradians), reaching a supreme sensitivity, but can only be operated in clear, dark nights.

In addition to the indirect ground based techniques discussed until now, high energy gamma-rays can be detected above the atmosphere directly by pair-conversion detectors. With these detectors, events initiated by charged particles can be effectively suppressed by e.g. a scintillation veto shield. Obviously, this is not possible for ground based indirect detection techniques where a separation of gamma-ray and cosmic-ray events has to be done using the information of the air shower itself (gamma-hadron separation).

With the launch of the upcoming GLAST mission [74], ground-based gamma-ray astronomy will benefit from the simultaneous operation of essentially a sensitive all-sky monitor operating at lower energies. The sensitivity of GLAST is well matched by the sensitivity of ground-based experiments so that the two instruments will be able to provide detection or constraints on the spectral shape across 6–7 orders of magnitude in energy.

---

[1] Shower-front sampling of Cherenkov light is discussed separately below.

## 2.1 Shower-front sampling techniques

**Particle arrays.** The experimental approach for the measurement of cosmic rays as well as the search for gamma-ray sources had been largely dominated by air shower arrays until the beginning of the 1990's. These arrays of particle detectors measure the arrival time of the shower front and the lateral particle density distribution. The total number of particles in the air shower is used to reconstruct the total energy of the primary particle. The relative timing of the individual particle detectors in the array is used to determine the direction of the air shower (typical angular resolutions of $1°$ can be reached) while both the timing and particle density measured on the ground are in principle useful to discern electromagnetic from hadronic air-showers (gamma-hadron separation). However, the gamma-hadron separation is naturally limited by the filling factor[2] of the detector array which typically amounts to less than 1%. An important development in the field was the innovation introduced by the MILAGRO collaboration [see e.g. 72]: they use a pond of water which is instrumented with photo-multiplier tubes (PMTs). The PMTs detect Cherenkov light from the secondary particles entering the pond's water. This way, the filling factor is increased to almost 100% and the entire particle distribution is sampled by the detector. Using the *clumpiness* of the particle distribution as an indicator for a hadronic air shower[3], it has been possible to reach sufficient sensitivities to detect VHE gamma-ray sources.

Future projects along this direction aim at installing such a detector at high altitude, where the energy threshold can be decreased to reach values below 1 TeV. Parallel to the MILAGRO group, the ARGO collaboration has developed a new type of air shower detector using resistive plate chambers to increase the filling factor. Their installation is already located at high altitude and is starting operation [51]. For a review of ground based non-imaging detectors, see Lorenz [155].

**Shower-front sampling with Cherenkov light.** The experimental technique used by non-imaging air Cherenkov detectors marks the transition from non-imaging to imaging observations. These detectors sample the arrival time and density of air Cherenkov photons using either open PMTs like e.g. the THEMISTOCLE experiment [110], AIROBICC array [139] or large reflecting surfaces as e.g. heliostats provided by solar power plants. The latter approach has been pursued by a number of groups including the C.A.C.T.U.S. [159], CELESTE [165] as well as STACEE [95]. While arrays of open PMTs retain the large field of view of classical air shower sampling arrays, the heliostat arrays have a very small field of view but a substantially smaller energy threshold reaching below 100 GeV as achieved e.g. by the CELESTE experiment. The potential for discovering faint sources is limited by the poor gamma-hadron rejection power of these shower sampling instruments.

---

[2] the fraction of the total surface covered with active detector surface

[3] for electromagnetic air showers, the lateral distribution is smoother than for a hadronic shower which contains muons and sub-showers

## 2.2 Imaging air Cherenkov telescopes

The field of ground-based gamma astronomy has been largely driven by the remarkable results obtained with the imaging air Cherenkov telescopes (IACTs). Extensive air showers emit in the forward-direction a beam of atmospheric Cherenkov light with an opening angle of $\approx 1°$. This beam illuminates almost homogeneously a circular region on the ground with a diameter of 200–300 m (depending on the altitude and inclination of the shower axis). An optical telescope pointing parallel to the shower axis and located within the illuminated footprint of the shower can make an image of the air shower against the background light of the night sky, provided the camera is sufficiently fast to integrate the short Cherenkov flash of only a few nanoseconds. The image provides information on the original particle's energy, direction, and on its nature (nucleus or photon). The reconstruction of these parameters is improved considerably when more than one telescope is used in a stereoscopic set-up where the telescopes are separated by 50–150 m in order to provide a baseline for triangulating the atmospheric air shower. The stereoscopic technique has become the nominal standard for all current and future installations.

The performance of today's telescopes is essentially characterized by the sensitivity to detect VHE sources with an energy flux down to $10^{-13}$ ergs (cm$^2$ s)$^{-1}$ in 50 hrs of observation time; this corresponds to a minimum detectable luminosity of $L_{\min} \approx 10^{31}$ ergs s$^{-1}$ $(d/1 \text{ kpc})^2$ at a distance of 1 kpc or more suitable for extragalactic objects $L_{\min} \approx 10^{41}$ ergs s$^{-1}$ $(d/100 \text{ Mpc})^2$ at a distance of 100 Mpc.

The angular resolution of each reconstructed primary $\gamma$-ray is typically better than 6 arc min. The relative energy resolution is comparably good and reaches values of $\Delta E/E \approx 10\text{–}20\,\%$. This is sufficient to detect and characterize spectral features like curvature or even lines from e.g. self-annihilation of Dark matter particles. For a list of currently operating IACTs see e.g. Hinton [128].

## 3 A census of today's VHE gamma-ray sky

The currently known list of VHE sources encompasses 70 sources (see Fig. 1). The number of sources is modest in comparison with catalogues assembled at lower energies. However, it is remarkable, how many different types of objects are actually emitting VHE gamma-rays and are therefore accelerators of multi-TeV particles.

Before discussing the different source types that have been detected, it is enlightening to list promising source types which have not been detected in the VHE band so far (the possible reason(s) for non-detection listed in parenthesis):

- **Star burst galaxies (too faint):** The high star forming rate should lead ultimately to an enhanced rate of supernova explosions. The upper limits on VHE emission from e.g. NGC 253 [28] are still higher than the expected flux [100] but deeper observations (ca. 50 hrs) will eventually reach the required sensitivity to either detect a signal or put meaningful constraints on the models[4].

---

[4] The CANGAROO collaboration published a claim for a detection from the starburst galaxy NGC 253 which was, however, later retracted [136].

- **Gamma-ray bursts (too short, too far):** These extremely powerful explosions are transient events requiring fast turn-around times and/or wide-field-of-view instruments in order to capture these objects during their outbursts. No convincing GRB detection has been claimed so far.[5] The ultra-fast slewing MAGIC telescope has succeeded in observing (but not detecting) a GRB while the outburst was still ongoing [55]. Non-imaging all-sky survey instruments like MILAGRO are probably more likely to detect these transient events including also the distinct class of short GRBs [4]. The visibility of distant GRB events is limited due to absorption of the energetic photons emitted at cosmological distances in pair-production processes with the optical-to-infra-red background light. An energy threshold well below 100 GeV is crucial to be able to observe GRBs at red-shifts $z > 0.1$.

- **Clusters of galaxies (too extended, too faint):** So far, no group or cluster of galaxies has been detected in the VHE band. Clusters of galaxies confine effectively cosmic rays and are expected to produce VHE-emission via inelastic scattering of nuclei with the intra-cluster gas [168, 200]. In addition to cosmic rays injected by normal galaxies and accelerated in large-scale shocks, AGN activity could also lead to an additional contribution to the intra-cluster cosmic ray density [131]. Current upper limits do not constrain severely the non-thermal population of cosmic rays in the clusters [99].

- **Pulsars (early cut-off in energy spectrum):** Fast rotating neutron stars are expected to give rise to acceleration within the magnetosphere [for a review, see e.g. 140]. This leads to pulsed emission (mainly from curvature radiation) that has been detected at least from six isolated pulsars up to GeV energies with the EGRET spark chamber detector on board the Compton Gamma-Ray Observatory [192]. However, the energy spectra of pulsars are also expected to show a sudden cut-off at a few GeV which makes them invisible to current ground-based telescopes with an energy threshold around 100 GeV. Searches for pulsed emission have so-far not been successful in finding VHE emission from pulsars – consistent with the expectations [43, 59]. Additionally, pulsed emission has been suggested to be produced via bulk Compton-scattering processes in the un-shocked pulsar wind-zone [87]. Non-detection of pulsed emission from the Crab pulsar has been used to constrain the formation region of the ultra-relativistic wind [11].

The breakdown of currently 70 known VHE sources includes 50 Galactic objects: 4 shell-type SNRs, 2 mixed-morphology SNRs, 2 composite SNRs, 20 pulsar-wind nebulae, 2 stellar associations, 4 X-ray binary systems, and roughly 18 sources without clear association to known objects. The remaining 20 objects are of extra-galactic origin: 18 Blazars, one Fanaroff-Riley Type I (M 87), and one flat-spectrum radio quasar (3C279).

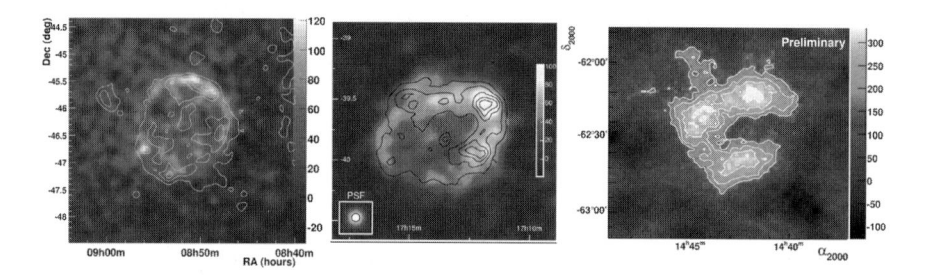

**Figure 2:** From left to right: RX J0852.0-4622 ("Vela Jr"), RX J1713.7-3946, RCW 86. The color-coded images are the excess maps as obtained with the H.E.S.S. telescopes (see text for references). Overlaid in white contours (RX J0852.0-4622) is the smoothed ROSAT image obtained above 1.3 keV. The black contours overlaid on the VHE-image of RX J1713.7-3946 follow the ASCA image (1–3 keV). The white contours overlaid on RCW 86 are the significance contours (starting at 3 standard deviations, increasing by 1).

## 3.1 Shell-type supernova remnants

Shell-type supernova remnants are commonly considered to be the best candidates to accelerate Galactic cosmic rays. During the supernova explosion (either a thermo-nuclear explosion or a core-collapse event), the stellar atmosphere is ejected with an initial velocity of up to 3000 km s$^{-1}$ carrying $10^{51}$ erg in kinetic energy and driving a shock front in the interstellar medium. While slowing down during the Sedov phase[6], the expanding shock front heats up the ambient medium giving rise to thermal X-ray emission. The shock is expected to dissipate a good fraction (10–30 %) of its energy during the Sedov phase in the form of accelerated particles (electrons and nuclei). In this case, the total power injected by roughly 1 supernova explosion every 30 years is sufficient to balance the escape losses of Galactic cosmic rays. Radio- as well as X-ray synchrotron emission is detected and attributed to electrons accelerated by the forward shock which in projection produces the characteristic shell-like morphology. For a few objects, the non-thermal X-ray component dominates entirely the observed X-ray spectrum (e.g. SN 1006, RX J1713.7-3946).

However, radio and X-ray observations are sensitive mainly to electrons and only indirectly to the presence of energetic nuclei: X-ray observations with sub-arcsecond spatial resolution have revealed indirectly, that the forward shock in many SNRs is very likely modified by cosmic ray streaming instabilities leading to enhanced magnetic field compression [156]. This in turn leads to observable effects: electrons produce X-ray synchrotron emission in small regions with increased magnetic field upstream of the shock that appear as narrow X-ray filaments ($d < 1$ pc) that have been first detected using the superior angular resolution of the Chandra X-ray telescope [79]. The inferred magnetic field strength in the X-ray filaments of e.g. SN1006 is of the order of 100 μG [83] which is much larger than in the case of a linear shock

---

[5] Some evidence for GRB detections have been published in the past [see e.g. 71, 164].

[6] once the SNR has swept up a comparable amount of matter from the interstellar medium as its ejecta mass, the free expansion phase ends and the Sedov phase of SNR evolution starts

thus indirectly revealing efficient acceleration of nuclei. Recent observations of fast variability of X-ray emission in filamentary structures seems to provide additional evidence for strong magnetic fields in RX J1713.7-3946 [195].

VHE gamma-ray observations are sensitive to both, electrons as well as nuclei. Sufficiently energetic electrons can radiate VHE gamma-rays through inverse Compton-up-scattering of soft seed photons (leptonic origin of VHE-emission) while nuclei produce neutral mesons decaying into gamma-rays in inelastic scattering events with the ambient gas (hadronic origin of VHE-emission).

In the hadronic scenario, the energy loss-time for accelerated protons ($\tau_{pp\to\pi^0} \approx 4.5 \times 10^{15}\, n^{-1}$ s) varies only slowly with energy and depends mainly on the ambient medium density ($n_H = n$ cm$^{-3}$) which includes cold molecular and hot ionized gas as well as the ejecta of the progenitor star. The actual density of gas near the shock can be inferred e.g. by modeling thermal X-ray emission from heated and partially ionized gas. However, for SNR without thermal X-rays, the density can only be constrained rather loosely and uncertainties close to an order of magnitude can remain. The typical integrated luminosity in VHE gamma-rays (1–10 TeV) is from a population of energetic protons with total energy $W_p$ $L_\gamma \approx W_p\, \tau^{-1} = 3.7 \cdot 10^{33} \cdot \eta/0.1 \cdot (E_{\text{SN}}/10^{51}\, \text{ergs}) n$ ergs s$^{-1}$ for an efficiency of shock acceleration $\eta = 0.1$, ie. the fraction of the kinetic energy of the blast wave converted into charged particles accelerated [7]. A similar result on the SNR luminosity has been obtained by Drury et al. [102].

Currently four VHE-emitting shell-type SNRs are detected and upper limits have been derived for a few more objects (see Table 1 for a list). Except for Cassiopeia A, the spatial extension of the SNR has been resolved (see Fig. 2). Generally, the observed luminosities of the shell-type SNRs are consistent with the expected luminosity in the hadronic scenario. Therefore, the VHE observations are consistent with the indirect evidence from X-ray observations for efficient shock acceleration of nuclei. A leptonic origin can not be excluded but is disfavored as it would lead to a comparably small efficiency for acceleration of nuclei which in turn would not produce non-linear modifications as are observed in X-rays.

A second crucial test of the SNR origin of cosmic rays is the spectral shape and maximum energy of particles accelerated[8]. The observed VHE energy spectra from RX J1713.7-3946 [31] and RX J0852.0-4622 [42] indicate that the accelerated proton spectrum follows an $E_p^{-2}$ type power-law and cuts off at energies of roughly 100–200 TeV which is about one order of magnitude smaller than the energy of the "knee" in the cosmic ray all-particle spectrum at a few $10^{15}$ eV. This discrepancy can be accommodated if these two SNRs are already too old and the shock has decelerated such that the maximum energy has dropped to the currently observed value [170].

In this context, the young SNR Cassiopeia A[9] and its VHE properties are crucial to our understanding of the SNR contribution to cosmic rays. Cassiopeia A has been recently confirmed by the MAGIC collaboration to be a VHE emitting source [63] after the initial discovery by the HEGRA collaboration [6]. The spectra of the

---

[7] The accelerated spectrum is assumed to follow a power-law with $dn/dE \propto E^{-2}$
[8] which can not be inferred from observations in other wavelengths
[9] and also Vela Jr if the age is in the range of 200–300 yrs

**Table 1:** List of supernova-remnants (shell-type and mixed-morphology) observed in VHE gamma-rays.

| Name | distance [kpc] | $L_\gamma$ [$10^{33}$ ergs/s] | age [kyrs] | Type |
|---|---|---|---|---|
| Cas A | 3.4[a] | 1.4[b] | 0.33[c] | Shell, II |
| RX J1713.7-3946 | $\simeq 1$[d] | 5.7[E] | 1.6[e] | Shell, II/Ib[f] |
| Vela Jr | 0.2[g] | 0.3[G] | 0.7[g] | Shell, II |
|  | 0.33[h] | 0.6[G] | 0.66[h] |  |
|  | $1-2$[i] | 6-26[G] | 4-19[i] |  |
| RCW 86 | 2.8[j] | 5.5[k] | 1.8 (SN185?)[j] | Shell, ? |
| W28 | 1.9[l] | 0.5[m] | 33[m] | MM |
| IC443 | 1.5[n] | 0.4[o] | 20[p] | MM |
| Without detection (historical SNe) | | | | |
| Tycho | 2.2[q] | $< 0.5$[T] | 0.436 | Shell, Ia |
|  | 4.5[r] | $< 2$[T] | SN1572 |  |
| SN1006 | 2.2[t] | $< 1.7$[u] | 1 | Shell, Ia |

[a] Reed et al. [177], [b] Aharonian et al. [6][63], [c] Thorstensen et al. [193], [d] Koyama et al. [148], [E] Aharonian et al. [31], [e] Koyama et al. [148], [f] Cassam-Chenaï et al. [90], [G] Aharonian et al. [42], [g] Aschenbach et al. [70], [h] Bamba et al. [78], [i] Slane et al. [189], [j] Vink et al. [198], [k] Hoppe et al. [132], [l] Velázquez et al. [197], [m] Aharonian et al. [50], [n] Claussen et al. [92], [o] Albert et al. [60], [p] Lee et al. [153], [q] Albinson et al. [68], [r] Schwarz et al. [187], [s] Aharonian et al. [46], [T] Aharonian et al. [46], [t] Winkler et al. [204], [u] Aharonian et al. [20]

two measurements are in agreement but require a rather soft energy spectrum with a photon index close to 2.5. The total energy in protons $W_p \approx 2 \times 10^{49}$ ergs is only a small fraction of the kinetic energy of the blast wave and only a factor of 4–8 larger than the inferred energy in accelerated electrons [73]. Given the young age of the source which is at the beginning of the Sedov phase, the small value of $W_p$ may still be re-reconciled with Cassiopeia A to be a typical SNR. It will be quite interesting to measure with better accuracy the energy spectrum to higher energies in order to probe the maximum energy of accelerated protons which should be close to PeV energies. Furthermore, as indicated in Table 1, future detections of other historical SNRs like Tycho or SN 1006 will provide important clues on cosmic ray acceleration in these objects.

## 3.2 Mixed morphology supernova remnants

Besides young ($t \approx O(1000)$ yrs) shell-type supernova-remnants, VHE-gamma-ray emission has also been detected from the direction of mixed-morphology (MM) type supernova-remnants (SNRs) which are considered evolved systems. MM systems are characterized by a shell-like emission observed *only* in the radio-band with thermal, centrally-peaked X-ray emission predominantly from the interior of the SNR [see e.g 181]. The mechanism responsible for heating the gas is not well-understood. It has been suggested that mixed morphology SNR have entered a post-Sedov stage

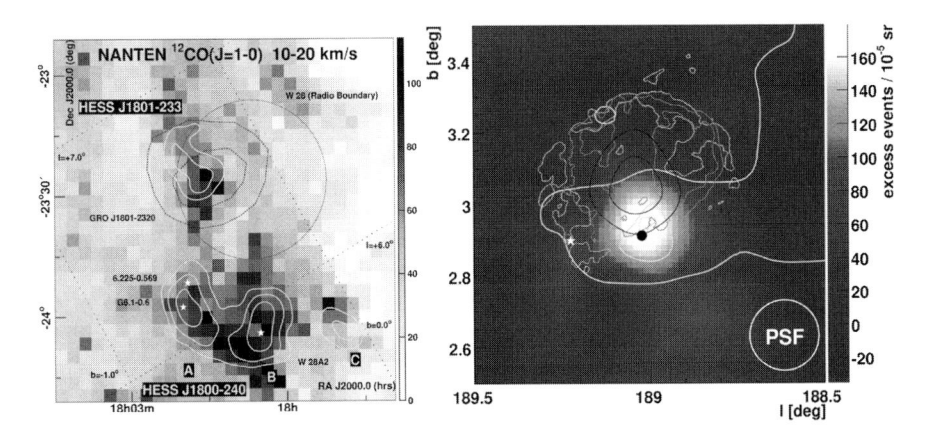

**Figure 3:** Left panel (a): The integrated CO emission in the velocity range 10–20 km/s (color scale) with significance contours from H.E.S.S. observations of W28 overlaid (green). The black contours follow the 95 % and 99 % confidence level regions for the unidentified EGRET source GRO J1601-2320 [50]. Right panel (b): The color scale indicates the smoothed excess map obtained from MAGIC observations of IC 443. The molecular gas density (CO) is traced with the cyan colored contours, X-ray contours (purple) from ROSAT observations, gamma-ray contours (EGRET) in black. See captions of Fig. 1 in Albert et al. [60] for references.

of their evolution which is characterized by an equal electron and ion temperature [142]. The fact that roughly 20 % of the known Galactic SNR are of MM supports this scenario.

The observation of VHE gamma-rays from MM SNR like W28 [50] and IC443 [60] is – independent of the underlying radiation mechanism – direct evidence for particle acceleration to multi-TeV energies in these objects (see Fig. 3). However, one would expect that the evolved and slow shock present in these objects is not strong enough to accelerate particles at the current stage of evolution. Furthermore, the observed Gamma-ray emission seems to coincide well with regions of dense molecular gas. In the north-eastern region of W28, some of the clouds must have already been interacting with the expanding blast wave. The interaction can be traced by maser emission in the shocked gas (see Fig. 3a). The observation of VHE-emission from IC 443, another MM type SNR, supports a similar scenario of an evolved SNR interacting with dense and cold molecular gas (see Fig. 3b). In this scenario, electrons can not produce the observed VHE-emission because of the rapid cooling through inverse Compton and synchrotron emission, as well as Bremsstrahlung in the dense gas. Typical cooling times of electrons at multi-TeV energies would be of the order of a few hundred years and by far too short to allow for a sizeable population of electrons to be present in an evolved SNR without ongoing acceleration. An alternative and much more favorable interpretation for MM SNRs requires that accelerated nuclei have been partially confined to the SNR and its environment and produce efficiently gamma-rays through $\pi^0$-decay in the dense target material of the molecular clouds. Gabici & Aharonian [113] have investigated the effect of cosmic rays that have left the accelerator and produce an observable VHE emission in a nearby cloud of molecular gas.

## 3.3 Pulsar wind nebulae and composite supernova remnants

While shell type SNR are considered to be the likely source of the Galactic cosmic rays, the blind survey of the Galactic plane [15, 38] performed with the H.E.S.S. telescopes has revealed a comparably small number of shell-type SNR while VHE emission from pulsar wind nebula (PWN) systems has been observed more frequently [115]. In total, 20 VHE sources (including candidate associations) have been discovered which are very likely powered by isolated pulsars.

The population of "TeV-Plerions" [134] includes young objects like Kes-75/ PSR J1846-0258 (spin-down age $\tau = 723$ yrs) [1], MSH 15-52 [16] driven by PSR J1513-5908 ($\tau = 1.55$ kyrs) as well as evolved systems like the PWN driven by PSR B1823-13 ($\tau = 21.4$ kyrs) associated with HESS J1825-137 [47]. Apparently, PWN systems are active VHE sources for a few 10 000 years thus exceeding the time that a SNR evolves through the Sedov-stage of shell type SNR during which presumably VHE emission is most likely produced. The longer time during which PWNe are active particle accelerators explains the larger number of PWN systems active at any given time in the Galaxy in comparison to shell-type SNRs.

However, the contribution of PWN to the Galactic cosmic rays remains an open question as long as the nature of the particles accelerated in these systems is not clarified. Currently, for most of the VHE emitting PWNe, a leptonic origin of the VHE emission can qualitatively explain the multiwavelength morphology and energy spectra. As an example, HESS J1825-137 has been studied intensively both in X-rays as well as in VHE-gamma-rays where the source extension appears to be larger by a factor of $\approx 6$ than in the X-ray band. This larger size can be explained by a greater loss time of electrons radiating VHE gamma-rays via inverse Compton scattering than the loss time of electrons radiating synchrotron X-rays [35]. The energetic electrons injected by the pulsar PSR B1823-13 loose rapidly energy in the high magnetic field environment close to the pulsar where the X-ray nebula is observed [114]. While the particles leave the high magnetic field environment, they presumably radiate only at wave-lengths longer than the X-ray band in the synchrotron channel while inverse Compton scattering off the cosmic microwave background is the dominant energy loss process leading to observable, extended VHE emission. In this model, the slowly decaying spin-down power of the pulsar is accumulated in an expanding bubble containing the TeV-emitting electrons. This scenario is supported by softening of the observed VHE spectra with increasing distance to the compact X-ray PWN system [35].

The observed morphology of the PWN is often asymmetric and not centered on the pulsar. While in some cases, the velocity of the pulsar can accommodate for the offset, other objects require an alternative explanation (e.g. Vela X): here, the asymmetric reverse shock of the SNR shell interacts with the relativistic PWN gas "crushing" and pushing it off-centered [86, 109].

A contribution to the observed VHE gamma-ray flux from nuclei accelerated by the pulsar is not excluded but seems to be less important than the contribution of electrons for most systems.

The Vela X PWN system may be an exception. Here, the spin down power of the pulsar is by far larger than the acceleration rate of electrons required to drive the

PWN system. In this sense, there is missing energy that can be naturally explained by a nucleonic component in the wind that drives the acceleration of electrons through resonant wave absorption [69] and produces the bulk of the observed VHE emission [36]. The VHE spectrum can be explained quite well in terms of over-all energetics (thus solving the problem of "missing" energy) as well as its shape [134]. A crucial test will be the detection of high energy neutrinos from this object that may be the brightest steady neutrino source in the sky [137, 138].

Finally, composite systems where a plerionic, non-thermal X-ray component is observed to be embedded in a radio-emitting shell like G0.9+0.1 have been found to emit VHE gamma-rays [21]. In addition to known composite SNRs, VHE observations have helped to discover and identify new systems: X-ray observations of HESS J1813-178 [15] carried out with XMM-Newton have revealed a previously unknown extended X-ray source inside the radio-shell of SNR G12.82-0.02 [112] putting this object in the category of composite SNR systems. Most likely, the VHE-emission is produced in the central PWN systems while the radio-shell is VHE-dim. However, it should be noted, that the spatial extension of these two systems is barely resolved at VHE energies.

## 3.4 X-ray binary systems

Before the imaging air Cherenkov technique was widely established, several groups had claimed a number of X-ray binary sources to be periodic (e.g Her X-1, Cyg X-3). Cyg X-3 had been claimed to be a transient source of energetic particles up to PeV energies [183], for a review of early claims of detections see Weekes [203]. With the notable exception of a possible burst observed from the direction of GRS 1915+105 [49], none of the XRB systems had been confirmed until recently to be VHE gamma-ray emitters. Finally, the lack of detections from XRBs has been overcome with the detection of 4 XRB systems listed in Table 2. All of these objects show variability in the VHE band either linked to the orbital motion as for LS I +61 303 [58, 157], LS 5039 [30], and PSR B1259-63 [17], or transient activity as for Cyg X-1 [65].

Given that these objects are currently the only variable Galactic sources of VHE gamma-rays, the interpretation of their properties is quite different from the other source types. The orbital parameters as well as the physical conditions in these systems are reasonably well-known and allow for a detailed modelling of the processes leading to acceleration and VHE emission. In this sense, one can consider the systems as "laboratories".

The pulsar PSR B1259-63 ($P = 48$ ms) is in a highly eccentric orbit ($P_{\rm orbit} = 3.4$ yrs) around the Be-type companion star SS2883. VHE emission is predominantly produced around the periastron where the effects of adiabatic and radiative cooling due to inverse Compton as well as synchrotron emission are strongest [17]. The IC component had been predicted successfully prior to the detection [145], albeit the observed temporal behavior of the source as it passed through the periastron in 2004 was quite different from the expectation [17]. Refined models including a more detailed treatment of inverse Compton scattering in energetic and anisotropic radiation fields as well as adiabatic energy losses can reproduce the observed light curve and provide some crucial predictions to be probed with new observations [144]. Alter-

**Table 2:** X-ray binary systems detected with imaging air Cherenkov telescopes.

| Identifier | Variability | D [kpc] | $L^a$ [$10^{33}$ ergs/s] | System[b] |
|---|---|---|---|---|
| PSR B1259-63 | $P = 3.4$ a | 1.5 | 1.6 | PSR(47 ms)/B2e |
| LS 5039 | $P = 3.9$ d | 2.5 | 8.7 | BH or NS/O6.5V |
| LS I +61 303 | $P = 26.5$ d | 2 | 2.5 | NS/B0Ve |
| Cyg X-1 | $T_{\text{var}} \approx 80$ min | 2.2 | 1.6 | BH/O9.7 Iab |

[a]Integrated from 1 to 10 TeV
[b]BH: Black hole candidate, NS: Neutron star, PSR: Pulsar

native models invoking the presence of a nucleonic component that leads to VHE emission through the interaction with the disk-outflow of the Be-star have been proposed as well [141, 163].

Unfortunately, the periastron passage in 2007 took place during the time when PSR B1259+63 culminated during the day time. A crucial observation will be only possible during the periastron passage in 2011 when the system will be visible for ground based VHE instruments as well as for GLAST.

The two systems LS 5039 as well as LS I +61 303 have shorter orbital periods of 3.9 and 26.5 days, respectively. Initially, both objects have been considered to be micro-quasars where a steady radio-jet has been observed. There is an ongoing debate whether this picture may have to be modified. Radio observations of LS I+61 303 indicate that the orientation of the "jet" changes during the orbit [97] which is more in line with the cometary tail predicted by Dubus [104] as a consequence of the interaction of a pulsar wind with the stellar outflow. For LS 5039, the current high resolution radio observations also indicate a change in the orientation of the jet-like structure [182]. However, since the observations do not cover a full orbital period, the results are not conclusive yet.

Given the uncertainty of the nature of the compact object, the VHE emission from LS 5039 has been interpreted in the context of a micro-quasar model [88] as well as in an interacting PWN model [104].

A crucial consideration is the absorption of VHE photons due to pair-production in the photon field of the stellar companion. If the line-of-sight towards the VHE emission region passes within roughly $10^{14}$ cm of the companion star of LS 5039, absorption will produce visible effects [103] varying with the phase of the orbit. The absorption effect is energy dependent and should lead to a pronounced variability at energies between 200–400 GeV. However, the observations indicate that the flux at 200–400 GeV remains almost unchanged along the orbit while the modulation is most pronounced at TeV-energies (see Fig. 4).

Secondary particle production in cascades can in principle reproduce the observed variability [81] as well as models where the emission takes place at a larger distance to the stellar companion [143].

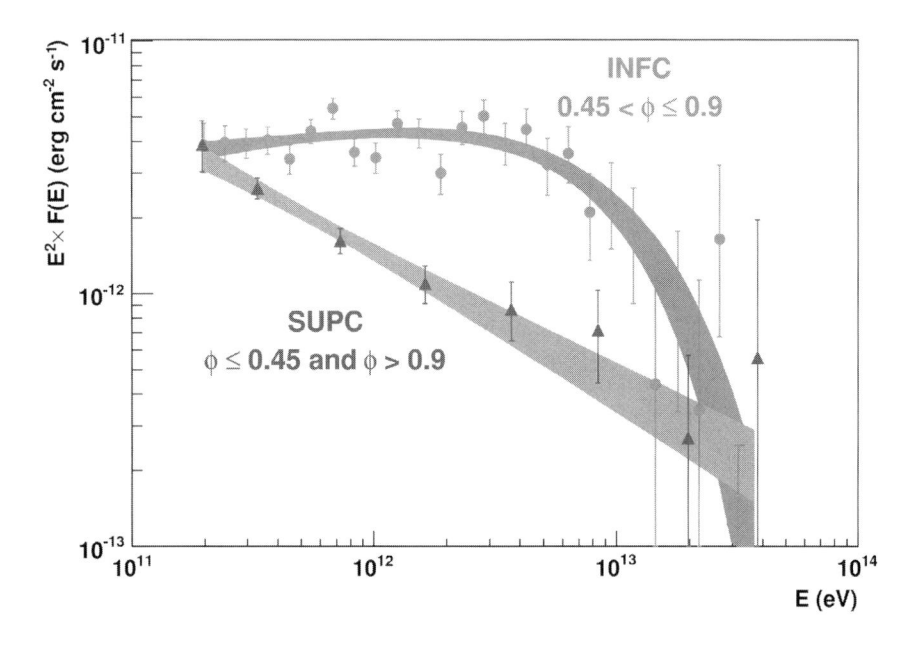

**Figure 4:** The VHE energy spectrum of LS 5039 for two distinct orbital phases. The *INFC* corresponds to the inferior conjunction, while *SUPC* is the phase interval at the superior conjunction with the periastron passage [30].

## 3.5 Star-forming regions

So far, the objects under scrutiny are related to the final stages of stellar evolution and their remnants. However, gamma-rays are also produced in molecular clouds which are the cradles of star formation and possibly during the final mega-year of stellar evolution where massive stars ($> 15\ M_\odot$) drive powerful winds. The wind-phase of stellar evolution is characterized by mass loss rates of up to $10^{-5}\ M_\odot$/year and terminal velocities of a few 1000 km/s. These early-type stars are usually born in open stellar associations in the spiral arms of the Galactic disk. The massive O stars in these associations often evolve into the Wolf-Rayet phase where strong stellar winds produce a hot tenuous cavity in the interstellar medium. In such systems up to $10^{39}$ ergs/s are dissipated in the form of kinetic energy from stellar winds, even before member stars evolve into supernovae. Provided that shocks will form either in the termination region of cumulative winds [101] or in wind-wind interaction regions [96, 178], some of this power can be converted into cosmic-ray acceleration.

The detection of VHE-emission from Westerlund-2 is a crucial step towards establishing the rôle of stellar driven cosmic ray acceleration (see Fig. 5 from Aharonian et al. [40]). Westerlund-2 is a young star forming region with roughly 20 O-type stars and the Wolf-Rayet systems WR 20a&b [176]. The total power released in the stellar winds is estimated to be $5 \times 10^{37}$ ergs/s. The observed VHE emission can be easily produced if roughly 1 % of the kinetic energy is converted into the acceleration of electrons [158, 180].

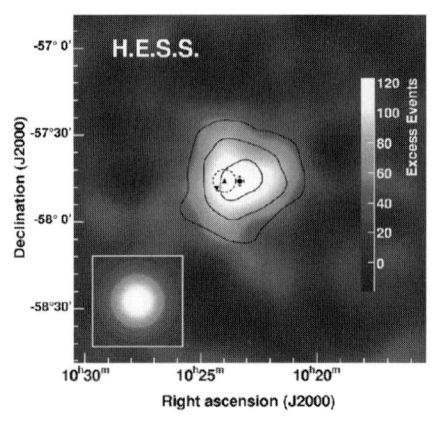

**Figure 5:** Westerlund 2 [40]: The color scale indicates the smoothed VHE excess map (see the bottom left inlaid figure for the point-spread function), the contours are the significance levels starting at 5 $\sigma$ and increasing by 2. The filled circle and error bars mark the position and uncertainty of the VHE source (1 $\sigma$). The dashed circle marks the position and extension of Westerlund 2, the upright triangle is at the position of the Wolf-Rayet binary system WR 20a and the downward triangle is at the position of WR 20.

Prior to the discovery of VHE emission from Westerlund-2, the first unidentified VHE source TeV J2032+4130 had been associated with the remarkably powerful OB association in the Cygnus arm. The HEGRA discovery of TeV J2032+4130 [7, 12] has been recently confirmed by MAGIC [66] as well as by the MILAGRO and VERITAS groups [2, 3, 146, 150]. Remarkably, observations with the MILAGRO detector show an extended ($\approx 3°$) source of $>$ 10 TeV photons centered on TeV J2032+4130 [2]. This source has received considerable multi-wavelength coverage including deep radio observations which show indications for extended radio emission coinciding with the VHE source [166]. Recent X-ray observations with XMM-Newton have revealed faint non-thermal extended X-ray emission co-located with the VHE-source [135]. Future observations of this region with the VERITAS telescope array will be very helpful to disentangle possible multiple-sources and to study the source size at different energies.

Besides Westerlund-2 and Cyg OB2, the MILAGRO source MGRO J2019+37 [2] has been associated with the young stellar association Berk-87 including the Wolf-Rayet object WR 142 [80].

A systematic search for VHE emission with the HEGRA telescopes from young open clusters has so far produced only upper limits [13].

## 3.6 Unidentified Galactic sources

The sources discussed in the previous section have been identified or are associated with known objects. However, a number of objects remain to be identified. A total of 10–20 objects have no clear or even multiple counterparts. Among these objects, the Galactic center source and the point-like source in the Monoceros region are highlighted in the following. For a recent summary of unidentified sources, see e.g. Aharonian et al. [26], Funk [111].

**Galactic center region.** The central region of our Galaxy contains a super-massive black hole with a mass of $3 \times 10^6$ $M_\odot$ [119, 185]. Given the faintness of this object, the accretion rate must be $< 10^{-8}$ of the Eddington-rate. The mm, IR, and X-ray sources associated with Sgr A* have been observed to show variability including flares and outbursts with a time-scale of hours [117]. The shortness of the flares constrains the emission region to be compact with a spatial extension $r < t_{\text{var}} \cdot c \approx 10^{14}$ cm $(t_{var}/hr) \approx 100 r_G (M/3 \cdot 10^6 M_\odot)^{-1} t_{var}/\text{hr}$. Observations with the CANGAROO III telescopes as well as at large zenith angles with the Whipple 10m telescope revealed a VHE gamma-ray source co-located with the Galactic center [147, 194]. The positional accuracy as well as spectral measurements have been considerably improved with the observations of the H.E.S.S.-telescopes [14]. The MAGIC telescope has confirmed the earlier findings in independent observations [56].

The detection of a point-like source with a hard power-law spectrum was shortly after the discovery discussed in the context of emission scenarios near to the supermassive black hole (at distances larger than $\approx 10\ r_G$ in order to avoid internal absorption) from e.g. ultra-high energy protons emitting synchrotron and curvature radiation in the high magnetic fields that are believed to be threaded in the accretion disk [45] or $\pi^0$-decay from energetic protons [76, 154]. At a larger distance from Sgr A*, acceleration in SNRs [107] including Sgr A East [94] as well as acceleration in stellar winds have been proposed [173].

The initial data were consistent with a Dark Matter annihilation scenario albeit requiring uncomfortably large masses beyond 20 TeV for the annihilating particles [133].

With more data, the H.E.S.S. experiment has constrained the energy spectrum to extend beyond 10 TeV and to deviate from a WIMP-annihilation scenario [37][10]. With the increased exposure, faint spatially extended VHE emission along the Galactic ridge was detected [34]. The positional accuracy of the point source has been improved to the level of the systematic pointing uncertainty of 6 arc seconds [196] and excludes Sgr A East. The error box for the VHE source still encompasses at least three possible candidates for VHE emission (see Fig. 6a): Sgr A*, a low-mass X-ray binary system (LMXRB) [162], the stellar system IRS 13 [75], and the PWN G359.95-0.04 [202]. The PWN system has been argued to be a possible candidate to explain the observed VHE emission [130].

An association with a variable source like Sgr A* would be supported if simultaneous multi-wavelength (MWL) observations of Sgr A* during an outburst would show a correlation between e.g. X-ray and gamma-ray flux. During a dedicated MWL observation of Sgr A* in summer 2005 with H.E.S.S. and the Chandra X-ray telescope, a typical X-ray flare was detected with the Chandra instrument [129]. The simultaneously taken H.E.S.S.-data show, however, no variability even though the observed rate of VHE photons is sufficiently large to detect a flare of similar strength as seen in the Chandra data (see Fig. 6b). The absence of correlation would disfavor at least a common origin of the X-ray and VHE emission from Sgr A*.

An association of the VHE source with the LMXRB could in principle be demonstrated by correlated activity in different wavelength-bands. However, so far no such observations are available.

---

[10] see, however, Bringmann et al. [89] for a discussion on the effects of internal Bremsstrahlung photons which are important for heavy WIMP annihilation in order to reduce the helicity suppression for the annihilation channel

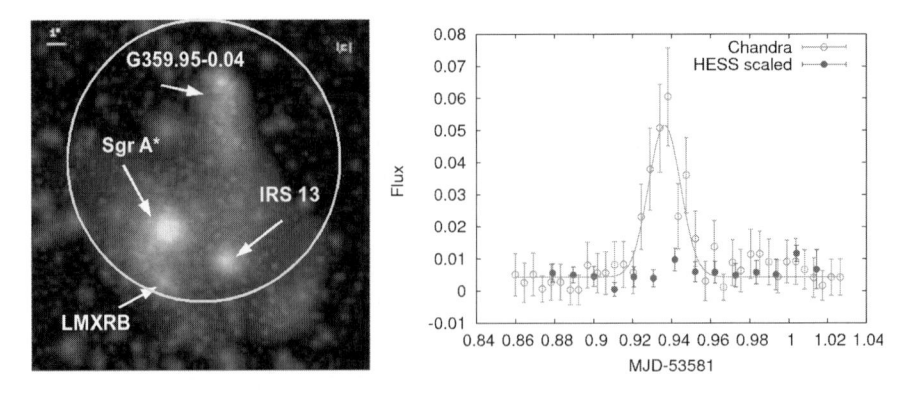

**Figure 6: Left panel:** Chandra X-ray and SINFONI near-IR composite from Wang et al. [202] with additional annotations. The circular region corresponds to the combined systematic and statistical uncertainty for the HESS-source associated with the Galactic center. **Right panel:** X-ray and VHE-gamma-ray light curve from simultaneous observations of Sgr A*. The H.E.S.S. light curve has been scaled to match the Chandra light curve in units of counts per second. Note the comparably small error bars of the H.E.S.S. observations – a flare of similar magnitude as observed in the X-ray band would have been clearly detected.

At this point it is not possible to discern between the different source candidates nor exclude a dark matter origin.

**Monoceros source.** The newly discovered population of unidentified VHE sources in the Galactic plane form a quite homogeneous sample of objects with respect to their spatial extension and energy spectra. It should be emphasized, however, that the similarity of the VHE characteristics does not necessarily imply that the underlying sources are similar. Certainly, observational biases are introduced by e.g. the limited sensitivity for detecting sources with low surface brightness.

Among the unidentified sources, there is – besides the Galactic center – only one more unresolved object: HESS J0632+058 which is close to the rim of the Monoceros Loop SNR [48]. The radial extension of this object is constrained to be smaller than $2'$ (95% c.l.). No strong indications for variability are observed. The energy spectrum of the source is well-described by a power-law ($dN/dE \propto E^{-2.5\pm0.3}$).

For an unresolved source ($r < 2'$), the search for a counterpart is simpler than for extended objects. In Fig. 7, the position of HESS J0632+058 is combined with multi-wavelength observations. Only a few candidate objects coincide spatially with the VHE source including an unidentified EGRET source 3EG J0634+0521 [123], an unidentified X-ray source 1RXS J063258.3+054857 [199], and a B0pe type star MWC 148 which may be a binary system similar to PSR B1259-63. Alternatively, it could be a new type of VHE source associated with an isolated stellar system.

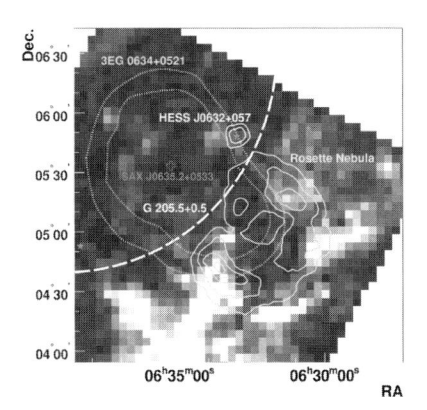

**Figure 7:** Monoceros source [48]: The grey-scale illustrates the velocity-integrated (0–30 km s$^{-1}$) $^{12}$CO emission as measured with the NANTEN telescope [161]. The yellow-contours trace the 4 and 6 $\sigma$ level of detection with the H.E.S.S. telescopes. The cyan contours indicate the 8.35 GHz radio measurements from Langston et al. [151] while the green contours mark the 95 % and 99 % confidence regions for the position of the EGRET source 3EG 0634+0521. SAX J0635.2+0533 is a pulsar binary system. The positions of Be-stars are indicated with pink stars.

## 3.7 Extra-galactic sources

The blind survey of the Galactic plane has revealed an interesting and previously unknown population of VHE-sources. Unfortunately, a survey with similar sensitivity of the extra-galactic sky has to be postponed until wide-field-of-view Cherenkov telescopes or the next generation of Water-Cherenkov-experiments like HAWC will come online (see also next section). Currently, the observations have focussed on Blazars ("BL Lac type Quasars").

Historically, the discovery of VHE emission from the Blazars Mkn 421 [171] and Mkn 501 [174] have opened the field of TeV-Blazar observations. These objects are in general characterized by a featureless optical spectrum, non-thermal X-ray emission and violent variability in the optical as well as in the X-rays. The broad-band spectral energy distribution is dominated by two broad maxima – one in the optical-to-X-rays and one in the gamma-ray-to-VHE-gamma-ray domain. The low-energy maximum is usually assumed to be produced by synchrotron emitting electrons while the high energy bump would be due to inverse-Compton scattering.

TeV-Blazars can be considered "clean" systems where the main source of seed photons are actually the synchrotron photons emitted by the electrons themselves, commonly called Synchrotron-Self-Compton (SSC) mechanism. Within the unified scheme, Blazars are believed to be Fanaroff-Riley Type I galaxies with a jet axis pointing close to the line-of-sight.

In a simple one-zone-SSC model, the VHE luminosity is closely linked to the synchrotron X-ray luminosity which in turn is related to the particle and seed photon density. A simple but very successful selection scheme of possible VHE emitting candidates is based on the X-ray or radio flux of Blazars [93]. Most of the Blazars that have been predicted to be detectable following this scheme have been actually discovered by the H.E.S.S. and MAGIC collaboration.

With the growing number of known Blazars at different red-shifts (see Table 3 for a complete list of extra-galactic VHE-gamma-ray sources), a number of issues and questions can be addressed:

- Is a simple one-zone SSC model sufficient to explain all (multi-wavelength) data including correlated variability patterns or is a more complicated model required (stratified jets, multiple zones model)?

- Where does the acceleration take place?

- What is the underlying acceleration mechanism in Blazars and what is the composition of the jet? This is a fundamental question related to the jet physics and the energetics of AGN.

- What is the duty cycle of Blazars?

- How strong is the beaming? In the relativistic outflow, a co-moving isotropic emission is strongly beamed in the forward direction. With increasing bulk Lorentz factor $\Gamma$, the opening angle gets narrower $\propto \Gamma^{-1}$. The higher the beaming, the smaller the likelihood that we can observe the emission. In consequence, a high beaming factor can lead to dramatically increased number of Blazars that are pointing *not* in our direction.

- Is there internal absorption?

- How strong is the absorption on the extra-galactic background light (EBL)?

In the following, I will highlight observations from Blazars as well as from non-Blazar VHE sources including M87 and the flat-spectrum-radio quasar 3C279.

## 3.8 Blazars

The most spectacular feature of VHE emission from Blazars is the unprecedented short time-scale variability. The most impressive examples of short flares have been observed in 2006 from the nearby object PKS 2155-305 (red-shift $z = 0.116$). This object was the first Blazar discovered in the southern hemisphere by the Durham group using their Mark VI telescope in Australia [91]. The H.E.S.S. telescopes observed their "first gamma-light" from this source already in 2002 when the first of the four H.E.S.S. telescopes went online [172]. Since PKS 2155-305 is a remarkably bright object, it can be easily detected in quiescent state with the system of four H.E.S.S. telescopes. Detailed measurements of the energy spectrum show even for widely different flux states marginal deviations from a soft power-law with photon index $\approx 3.3$ [19]. PKS 2155-305 has been the target of a number of simultaneous multi-wavelength-campaigns together with the RXTE X-ray satellite as well as in conjunction with the Chandra, XMM-Newton and the SWIFT X-ray telescopes [29]. During summer 2006, the source went into an unprecedented high state reaching flux levels more than a factor of 100 higher than the quiescent flux [39]. During the peak flares, the rate of detected VHE gamma-rays reached values exceeding 1 Hz which marks the highest rate ever detected at VHE energies. During the night with the highest flux, a sequence of flares can be discerned with rise and decay timescales of minutes (see Fig. 8). This immediately constrains the emission region to be very compact and to be moving with a large bulk Lorentz factor to overcome pair-opacity and to relax the size constraint [82].

*The Very High Energy Gamma-Ray Sky*  187

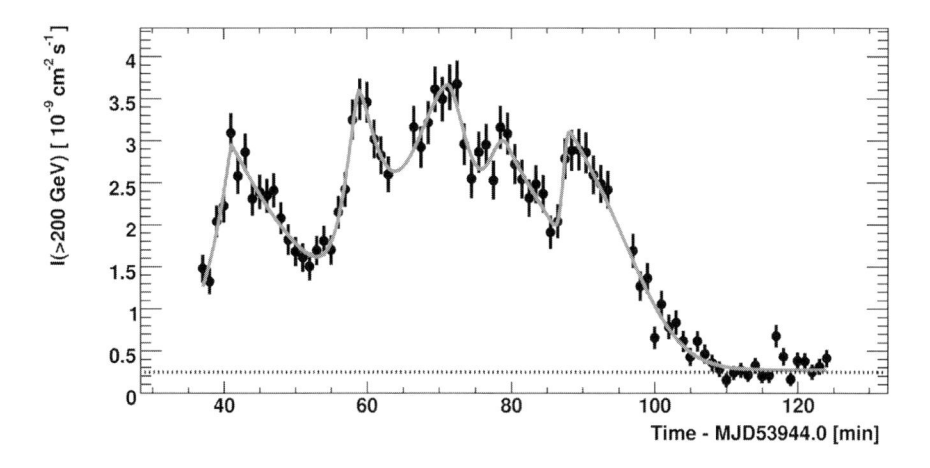

**Figure 8:** The light curve from PKS 2155-304 in the night July-28 2006 as measured with H.E.S.S. above 200 GeV. The data are binned in 1-minute intervals. For comparison, the horizontal dotted line represents I($>$200 GeV) observed from the Crab Nebula. The curve is a fit of a superposition of five individual bursts and a constant flux. See Aharonian et al. [39] for additional details.

## 3.9 Other active galactic nuclei

Besides the Blazars, only two non-Blazar extra-galactic objects have been discovered so far: M87 and 3C279 [11]

**M87** Besides the collective study of a Blazar sample [201], it is interesting to closely examine nearby FR-I radio galaxies as these objects are considered to be mis-aligned Blazars. The closest FR-I objects are M87 and Cen-A. VHE gamma-ray emission from M87 had been initially reported from the HEGRA group [10] and later verified by the Whipple collaboration [152], as well as by the H.E.S.S. group [27] and most recently with the VERITAS telescopes [5].

While the initial sensitivity was only sufficient to detect a signal and coarsely constrain the energy spectrum, following observations mainly with the H.E.S.S. telescopes have lead to a number of important discoveries: the energy spectrum of M87 extends to energies beyond 10 TeV and even more spectacular, the observed emission is highly variable on time-scales as short as days [27]. The hard spectrum is a surprise and disfavors the models suggested by Georganopoulos et al. [118] and Reimer et al. [179] where the predicted gamma-ray spectrum is soft as a consequence of the assumption of a small Doppler-factor derived from radio measurements of the jet dynamics. It should be noted, however, that also other nearby Blazars that have been observed with high resolution radio interferometers do not show strong superluminal motion indicative for large Doppler-factors. It seems that there is growing

---

[11] A signal from Centaurus A at energies above 300 GeV had been claimed on the basis of non-imaging observations of air Cherenkov light at the level of 4.5 $\sigma$ [122]. This claim awaits confirmation.

**Table 3:** List of extragalactic VHE-sources sorted by red-shift. Whenever available, different source activity states are characterized.

| Source | Type | $z$ | $d_L{}^a$ | $\log_{10}(L_\gamma)^b$ | $\Gamma_{measured}$ | Ref. |
|---|---|---|---|---|---|---|
| M87 | FR I | 0.0044 | 23 | 41.1 | $2.6 \pm 0.4$ | c |
|  |  |  |  | 41.6 | $2.2 \pm 0.2$ | c |
| Mkn 421 | HBL | 0.030 | 130 | 44.8 | $3.0 \pm 0.2$ | d |
|  |  |  |  | 45.3 | $2.06 \pm 0.03$ | d |
| Mkn 501 | HBL | 0.034 | 142 | 44.3 | $2.45 \pm 0.07$ | e |
|  |  |  |  | 45.2 | $2.09 \pm 0.03$ | e |
| 1ES 2344+514 | HBL | 0.044 | 183 | 43.9 | $2.9 \pm 0.2$ | f |
|  |  |  |  | 45.2 | $2.5 \pm 0.2$ | g |
| Mkn 180 | HBL | 0.045 | 194 | 44.0 | $3.3 \pm 0.7$ | h |
| 1ES 1959+650 | HBL | 0.047 | 198 | 44.2 | $2.7 \pm 0.1$ | i |
|  |  |  |  | 44.9 | $1.8 \pm 0.2$ | j |
| BL Lacertæ | LBL | 0.069 | 293 | 44.0 | $3.6 \pm 0.5$ | k |
| PKS 0548-322 | HBL | 0.069 | 300 | 43.6 | $2.8 \pm 0.3$ | l |
| PKS 2005-489 | HBL | 0.071 | 306 | 44.2 | $4.0 \pm 0.4$ | m |
| RGB J0152+017 | HBL | 0.080 | 345 | 44.0 | $3.0 \pm 0.4$ | n |
| PKS 2155-304 | HBL | 0.116 | 515 | 44.1 | $3.4 \pm 0.1$ | o |
|  |  |  |  | 46.8 | $2.7 \pm 0.1$ | p |
|  |  |  |  | Break at 0.43 TeV | $3.5 \pm 0.1$ | p |
| H 1426+428 | HBL | 0.129 | 585 | 45.9 | $3.7 \pm 0.4$ | q |
| 1ES 0229+200 | HBL | 0.139 | 633 | 44.3 | $2.5 \pm 0.2$ | r |
| H 2356-309 | HBL | 0.165 | 758 | 44.5 | $3.1 \pm 0.2$ | s |
| 1ES 1218+304 | HBL | 0.182 | 855 | 45.2 | $3.0 \pm 0.4$ | t |
| 1ES 0347-121 | HBL | 0.185 | 864 | 44.8 | $3.1 \pm 0.3$ | u |
| 1ES 1101-232 | HBL | 0.186 | 877 | 44.8 | $2.9 \pm 0.2$ | v |
| 1ES 1011+496 | HBL | 0.212 | 1008 | 45.5 | $4.0 \pm 0.5$ | w |
| PG 1553+113 | HBL | > 0.3 | > 1498 | > 46.1 | $4.5 \pm 0.3$ | x |
|  |  | < 0.74 | < 4437 | < 47.1 |  | x |
| 3C279 | FSRQ | 0.536 | 2998 | 46.0 | – | y |

$^a$Assuming $\Omega_m = 0.27$, $\Omega_\Lambda = 0.73$, $H_0 = 73\,\mathrm{km\,s^{-1} Mpc^{-1}}$
$^b$Isotropic luminosity $L_\gamma$ in ergs s$^{-1}$ integrated between 0.1 and 1 TeV
$^c$Aharonian et al. [27], $^d$Aharonian et al. [8], $^e$Albert et al. [64], $^f$Albert et al. [62], $^g$Schroedter et al. [186], $^h$Albert et al. [54], $^i$Albert et al. [57], $^j$Aharonian et al. [9], $^k$Albert et al. [61], $^l$The H. E. S. S. Collaboration [191], $^m$Aharonian et al. [18], $^n$Aharonian et al. [24], $^o$Aharonian et al. [19], $^p$Aharonian et al. [39], $^q$Djannati-Ataï et al. [98], $^r$Aharonian et al. [23], $^s$Aharonian et al. [33], $^t$Albert et al. [53], $^u$Aharonian et al. [22], $^v$Aharonian et al. [41], $^w$Albert et al. [52], $^x$Aharonian et al. [25], $^y$Teshima et al. [190]

*The Very High Energy Gamma-Ray Sky* 189

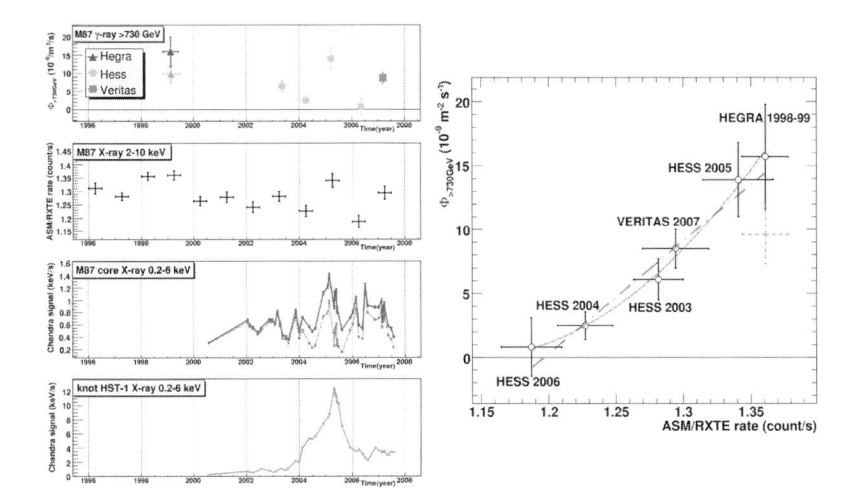

**Figure 9:** From Acciari et al. [5]: The left panel combines the VHE and X-ray light curves from observations of M87 with ground based Cherenkov telescopes as well as with satellite instruments on-board the RXTE satellite (ASM) and the Chandra observatory. The right panel correlates the yearly VHE fluxes with the appropriately averaged ASM fluxes. The dashed curve is a linear correlation and the solid curve a squared function.

evidence for a discrepancy between on high Doppler factors implied by modelling the VHE emission of Blazars with single zone SSC model and on the other side the lack of evidence for strong super-luminal motion in these objects from radio observations [169]. The fast variability observed from M87 leads to interesting implications. First of all, the variability excludes models where the VHE gamma-rays are produced in processes like Cosmic ray interactions in the host galaxy [167] or even Dark Matter annihilation [77]. These processes should not lead to temporal variation of the observed emission. Given that the variability time scale can not be smaller than the light crossing time, the constraint on the size of the emission region is severe and excludes regions in the outer radio jet as well. A possible candidate for the emission could be the notoriously variable HST-1 knot located less than 1 arc sec off the core. The most recent long-term light-curve obtained with the all-sky-monitor on-board the RXTE satellite as well as a dedicated Chandra monitoring seems to support a correlation of the HST-1 activity in X-rays and the observed VHE emission (see Fig. 9).

**3C279** This object belongs to the so-called flat-spectrum radio quasars (FSRQs). These objects are long-known to be emitters of GeV gamma-rays that have been detected with EGRET [124, 125]. There are two aspects which single out this source: (a) 3C279 has a comparably large red-shift of $z = 0.536$ and (b) it is a new type of AGN which shows acceleration to TeV energies.

The EGRET Blazars are mostly members of the FSRQ-class and are characterized by a spectral energy distribution with a peak in the EGRET energy range, dominating the total luminosity. The observed emission is commonly attributed to external Comptonization of photons from the broad-line region and possibly from the accretion disk. Given that FSRQs have intrinsically a higher gamma-ray luminosity than Blazars, they can be in principle observed to higher red-shifts opening up the possibility to finally detect absorption features in the VHE gamma-ray spectra (see also next section) of extra-galactic objects (see also next section). The detection of VHE emission from 3C279 with the MAGIC telescope [190] indicates variability which seems to be correlated with changes in the optical flux. Further observations of this source and a measurement of the VHE energy spectrum could provide important clues on the level of extra-galactic background light (see next section). Even in the case of a low level of EBL, only $10^{-3}$ of the emitted flux will be observable at 1 TeV [175].

## 4 "Secondary physics": Propagation of gamma-rays

### 4.1 EBL absorption

VHE emission from AGN appears to be more common than what could have been hoped for and the temporal variability observed at the highest energy is providing us an important and unique in-sight into the "engine-room" of AGN. The understanding of processes leading to VHE-emission hinges, however, on a good understanding of pair-absorption processes of the type $\gamma_{\text{VHE}}\gamma_{\text{EBL}} \rightarrow e^+e^-$. This process has been known for a long time to be relevant for the propagation of TeV-photons over cosmological distances [121].

With the growing number of Blazars observed at various red-shifts it has become feasible to use VHE-photons as *probes* to constrain or even measure the background photon density in extra-galactic space. Until now, it has, however, not been possible to clearly establish absorption in the measured Blazar spectra because the intrinsic emission spectrum is not very well known and may suffer additional internal absorption [44]. When invoking a constraint on the shape of the intrinsic spectrum, it is, however, possible to derive (model-dependent) upper limits on the EBL density [32, 105, 160]. The situation will be greatly improved when the first broad-band spectra of Blazars will be obtained that should show a transition from the optically thin to the optically thick part. The detection of such an absorption feature will in turn lead to a detection of the EBL. In combination with increasing statistics and coverage over varying red-shift, it will be feasible to investigate the evolution of the EBL and finally, derive in a completely independent way constraints for cosmological parameters [84, 85].

### 4.2 Lorentz invariance violation

VHE photons are useful probes for Lorentz invariance violating (LIV) processes. A number of theoretical approaches towards a quantum theory of Gravity predict a structure of space-time at the Planck scale. This will result in a modified dispersion relation for photons of the form $c^2 p^2 = E^2 (1 + \xi(E/M_{\text{QG}}) + \mathcal{O}(E/M_{\text{QG}})^2)$ see e.g. Sarkar [184] for a review. This dispersion relation affects the time-of-flight of photons of different energies and the kinematics of interaction, suppressing e.g. the pair-production process of photons.

Currently, fast flares from Mkn 421 [116], Mkn 501 [64], and PKS 2155-304 [39] have been observed. The latter is potentially the most constraining observation as the red-shift is larger and the variability faster as well as more photons have been collected than for the other sources. Current limits on $M_{QG} > 3 \times 10^{17}$ GeV [67] are already an order of magnitude better than limits obtained from GRBs [106]. It is important to point out that if a dispersion of arrival times from flares is detected, source intrinsic spectral variations can be excluded once consistent time lags are detected from a sample of flares from different sources at different red-shifts. Ultimately, the inferred LIV characteristics should be consistent with a suppression of EBL absorption processes.

# 5 Concluding remarks on the future perspectives of the field

The observational field of VHE gamma-ray astrophysics has been successfully driven in the last years by ground based imaging air Cherenkov telescopes. The "high energy frontier" of Astrophysics will be expanded by the upcoming launch of the GLAST satellite. The all-sky sensitivity of the GLAST Large-Area-Telescope (LAT) will most likely lead to many interesting new discoveries and surprises. The inter-play of the GLAST LAT detections from space and follow-up observations from ground will provide mutual benefits for the communities as well as certainly answer many questions as well as stimulate new ones.

While space-based observations of gamma-rays will be for a long time limited to the results obtained with GLAST (until maybe pair conversion telescopes will be deployed on the surface of the moon), ground based observatories will continue to be developed and constructed during the next decade(s). To my knowledge, besides GLAST, no further space-based gamma-ray telescope, operating in the GeV energy range, is planned. Consequently, ground based gamma-ray detection techniques will be the only available experimental approach to observe high energy gamma-rays after the termination of the GLAST mission.

The extensions of H.E.S.S. and MAGIC into phase II until 2009 will lower the energy threshold and improve existing sensitivity moderately. The next generation of ground based installations will become fully operational at the end of the next decade as envisaged in the AGIS[12][149] and CTA[13] [126] projects. These installations aim at an improvement in sensitivity by a factor of 10 and a widened reach in energy. At that time, the new ground based instruments will explore new energy windows above $\approx 10$ GeV as well as above 10 TeV.

However, there still remains a "blind spot": the transient sky above 10 GeV will neither be explored very well with GLAST (limited photon rate) nor with ground

---

[12]Advanced gamma ray imaging system
[13]Cherenkov telescope array

based instruments of the current generation (limited field of view and energy threshold). New installations like HAWC [188] will improve the situation in the energy range above a few 100 GeV. The lesson from the detection of fast transient events as seen from Cyg X-1 and PKS 2155-304 however is that we are very likely missing the fast variability of XRBs and AGN. Proper coverage of these objects requires instruments with a large field of view and an energy threshold well below 100 GeV.

# References

[1] A. Djannati-Atai et al. 2007, ArXiv e-prints, 0710.2247
[2] Abdo, A. A., Allen, B., Berley, D., et al. 2007, ApJ, 658, L33
[3] —. 2007, ApJ, 664, L91
[4] Abdo, A. A., Allen, B. T., Berley, D., et al. 2007, ApJ, 666, 361
[5] Acciari, V. A., Beilicke, M., Blaylock, G., et al. 2008, ArXiv e-prints, 0802.1951
[6] Aharonian, F., Akhperjanian, A., Barrio, J., et al. 2001, A&A, 370, 112
[7] Aharonian, F., Akhperjanian, A., Beilicke, M., et al. 2002, A&A, 393, L37
[8] —. 2002, A&A, 393, 89
[9] —. 2003, A&A, 406, L9
[10] —. 2003, A&A, 403, L1
[11] —. 2004, ApJ, 614, 897
[12] —. 2005, A&A, 431, 197
[13] —. 2006, A&A, 454, 775
[14] Aharonian, F., Akhperjanian, A. G., Aye, K.-M., et al. 2004, A&A, 425, L13
[15] —. 2005, Science, 307, 1938
[16] —. 2005, A&A, 435, L17
[17] —. 2005, A&A, 442, 1
[18] —. 2005, A&A, 436, L17
[19] —. 2005, A&A, 430, 865
[20] —. 2005, A&A, 437, 135
[21] —. 2005, A&A, 432, L25
[22] Aharonian, F., Akhperjanian, A. G., Barres de Almeida, U., et al. 2007, A&A, 473, L25
[23] —. 2007, A&A, 475, L9
[24] —. 2008, ArXiv e-prints, 0802.4021
[25] —. 2008, A&A, 477, 481
[26] —. 2008, A&A, 477, 353
[27] Aharonian, F., Akhperjanian, A. G., Bazer-Bachi, A. R., & et al. 2006, Science, 314, 1424
[28] Aharonian, F., Akhperjanian, A. G., Bazer-Bachi, A. R., et al. 2005, A&A, 442, 177
[29] —. 2005, A&A, 442, 895
[30] —. 2006, A&A, 460, 743
[31] —. 2006, A&A, 449, 223
[32] —. 2006, Nature, 440, 1018
[33] —. 2006, A&A, 455, 461
[34] —. 2006, Nature, 439, 695
[35] —. 2006, A&A, 460, 365
[36] —. 2006, A&A, 448, L43
[37] —. 2006, Physical Review Letters, 97, 221102
[38] —. 2006, ApJ, 636, 777

[39] —. 2007, ApJ, 664, L71
[40] —. 2007, A&A, 467, 1075
[41] —. 2007, A&A, 470, 475
[42] —. 2007, ApJ, 661, 236
[43] —. 2007, A&A, 466, 543
[44] Aharonian, F., Khangulyan, D., & Costamante, L. 2008, ArXiv e-prints, 0801.3198
[45] Aharonian, F. & Neronov, A. 2005, ApJ, 619, 306
[46] Aharonian, F. A., Akhperjanian, A. G., Barrio, J. A., et al. 2001, A&A, 373, 292
[47] Aharonian, F. A., Akhperjanian, A. G., Bazer-Bachi, A. R., et al. 2005, A&A, 442, L25
[48] —. 2007, A&A, 469, L1
[49] Aharonian, F. A. & Heinzelmann, G. 1998, Nuclear Physics B Proceedings Supplements, 60, 193
[50] Aharonian, F. A. et al. 2008, ArXiv e-prints, 0801.3555
[51] Aielli, G. & The Argo-YBJ Collaboration. 2007, Nuclear Physics B Proceedings Supplements, 166, 96
[52] Albert, J., Aliu, E., Anderhub, H., Antoranz, P., et al. 2007, ApJ, 667, L21
[53] Albert, J., Aliu, E., Anderhub, H., et al. 2006, ApJ, 642, L119
[54] —. 2006, ApJ, 648, L105
[55] —. 2006, ApJ, 641, L9
[56] —. 2006, ApJ, 638, L101
[57] —. 2006, ApJ, 639, 761
[58] —. 2006, Science, 312, 1771
[59] —. 2007, ApJ, 669, 1143
[60] —. 2007, ApJ, 664, L87
[61] —. 2007, ApJ, 666, L17
[62] —. 2007, ApJ, 662, 892
[63] —. 2007, A&A, 474, 937
[64] —. 2007, ApJ, 669, 862
[65] —. 2007, ApJ, 665, L51
[66] Albert, J. & for the MAGIC Collaboration. 2008, ArXiv e-prints, 0801.2391
[67] Albert, J., for the MAGIC Collaboration, Ellis, J., Mavromatos, N. E., Nanopoulos, D. V., Sakharov, A. S., & Sarkisyan, E. K. G. 2007, ArXiv e-prints, 0708.2889
[68] Albinson, J. S., Tuffs, R. J., Swinbank, E., & Gull, S. F. 1986, MNRAS, 219, 427
[69] Arons, J. & Tavani, M. 1994, ApJS, 90, 797
[70] Aschenbach, B., Iyudin, A. F., & Schönfelder, V. 1999, A&A, 350, 997
[71] Atkins, R., Benbow, W., Berley, D., et al. 2000, ApJ, 533, L119
[72] —. 2004, ApJ, 608, 680
[73] Atoyan, A. M., Aharonian, F. A., Tuffs, R. J., & Völk, H. J. 2000, A&A, 355, 211
[74] Atwood, W. B., Bagagli, R., Baldini, L., et al. 2007, Astroparticle Physics, 28, 422
[75] Baganoff, F. K., Maeda, Y., Morris, M., et al. 2003, ApJ, 591, 891
[76] Ballantyne, D. R., Melia, F., Liu, S., & Crocker, R. M. 2007, ApJ, 657, L13
[77] Baltz, E. A., Briot, C., Salati, P., Taillet, R., & Silk, J. 2000, Phys. Rev. D, 61, 023514
[78] Bamba, A., Yamazaki, R., & Hiraga, J. S. 2005, ApJ, 632, 294
[79] Bamba, A., Yamazaki, R., Ueno, M., & Koyama, K. 2003, ApJ, 589, 827
[80] Bednarek, W. 2007, MNRAS, 382, 367
[81] —. 2007, A&A, 464, 259
[82] Begelman, M. C., Fabian, A. C., & Rees, M. J. 2008, MNRAS, 384, L19
[83] Berezhko, E. G., Ksenofontov, L. T., & Völk, H. J. 2003, A&A, 412, L11
[84] Blanch, O. & Martinez, M. 2005, Astroparticle Physics, 23, 598
[85] —. 2005, Astroparticle Physics, 23, 608

[86] Blondin, J. M., Chevalier, R. A., & Frierson, D. M. 2001, ApJ, 563, 806
[87] Bogovalov, S. V. & Aharonian, F. A. 2000, MNRAS, 313, 504
[88] Bosch-Ramon, V., Romero, G. E., & Paredes, J. M. 2006, A&A, 447, 263
[89] Bringmann, T., Bergström, L., & Edsjö, J. 2008, Journal of High Energy Physics, 1, 49
[90] Cassam-Chenaï, G., Decourchelle, A., Ballet, J., et al. 2004, A&A, 427, 199
[91] Chadwick, P. M., Lyons, K., McComb, T. J. L., Orford, K. J., Osborne, J. L., Rayner, S. M., Shaw, S. E., Turver, K. E., & Wieczorek, G. J. 1999, ApJ, 513, 161
[92] Claussen, M. J., Frail, D. A., Goss, W. M., & Gaume, R. A. 1997, ApJ, 489, 143
[93] Costamante, L. & Ghisellini, G. 2002, A&A, 384, 56
[94] Crocker, R. M., Fatuzzo, M., Jokipii, J. R., Melia, F., & Volkas, R. R. 2005, ApJ, 622, 892
[95] D. M. Gingrich, Boone, L. M., Bramel, D., et al. 2005, ArXiv Astrophysics e-prints
[96] De Becker, M. 2007, A&A Rev., 14, 171
[97] Dhawan, V., Mioduszewski, A., & Rupen, M. 2006, in VI Microquasar Workshop: Microquasars and Beyond
[98] Djannati-Ataï, A., Khelifi, B., Vorobiov, S., et al. 2002, A&A, 391, L25
[99] Domainko, W., Benbow, W., Hinton, J. A., & others for the H. E. S. S. Collaboration. 2007, ArXiv e-prints, 708.1384
[100] Domingo-Santamaría, E. & Torres, D. F. 2005, A&A, 444, 403
[101] —. 2006, A&A, 448, 613
[102] Drury, L. O., Aharonian, F. A., & Voelk, H. J. 1994, A&A, 287, 959
[103] Dubus, G. 2006, A&A, 451, 9
[104] —. 2006, A&A, 456, 801
[105] Dwek, E. & Krennrich, F. 2005, ApJ, 618, 657
[106] Ellis, J., Mavromatos, N. E., Nanopoulos, D. V., & Sakharov, A. S. 2003, A&A, 402, 409
[107] Erlykin, A. D. & Wolfendale, A. W. 2007, Journal of Physics G Nuclear Physics, 34, 1813
[108] Fermi, E. 1949, Physical Review, 75, 1169
[109] Ferreira, S. E. S. & de Jager, O. C. 2008, A&A, 478, 17
[110] Fontaine, G., Espigat, P., Ghesquiere, C., et al. 1990, Nuclear Physics B Proceedings Supplements, 14, 79
[111] Funk, S. 2007, Ap&SS, 309, 11
[112] Funk, S., Hinton, J. A., Moriguchi, Y., et al. 2007, A&A, 470, 249
[113] Gabici, S. & Aharonian, F. A. 2007, ApJ, 665, L131
[114] Gaensler, B. M., Schulz, N. S., Kaspi, V. M., Pivovaroff, M. J., & Becker, W. E. 2003, ApJ, 588, 441
[115] Gaensler, B. M. & Slane, P. O. 2006, ARA&A, 44, 17
[116] Gaidos, J. A., Akerlof, C. W., Biller, S. D., et al. 1996, Nature, 383, 319
[117] Genzel, R., Schödel, R., Ott, T., Eckart, A., Alexander, T., Lacombe, F., Rouan, D., & Aschenbach, B. 2003, Nature, 425, 934
[118] Georganopoulos, M., Perlman, E. S., & Kazanas, D. 2005, ApJ, 634, L33
[119] Ghez, A. M., Salim, S., Hornstein, S. D., Tanner, A., Lu, J. R., Morris, M., Becklin, E. E., & Duchêne, G. 2005, ApJ, 620, 744
[120] Ginzburg, V. L. & Syrovatskii, S. I. 1964, The Origin of Cosmic Rays (The Origin of Cosmic Rays, New York: Macmillan, 1964)
[121] Gould, R. J. & Schréder, G. 1966, Physical Review Letters, 16, 252
[122] Grindlay, J. E., Helmken, H. F., Brown, R. H., Davis, J., & Allen, L. R. 1975, ApJ, 197, L9
[123] Hartman, R. C., Bertsch, D. L., Bloom, S. D., et al. 1999, ApJS, 123, 79

[124] Hartman, R. C., Bertsch, D. L., Fichtel, C. E., et al. 1992, ApJ, 385, L1
[125] Hartman, R. C., Böttcher, M., Aldering, G., et al. 2001, ApJ, 553, 683
[126] Hermann, G. 2007, Astronomische Nachrichten, 328, 600
[127] Hillas, A. M. 2006, ArXiv astro-ph/0607109
[128] Hinton, J. 2007, arXiv e-prints, 0712.3352
[129] Hinton, J., Vivier, M., Bühler, R., Pühlhofer, G., & Wagner, S. 2007, ArXiv e-prints, 0710.1537
[130] Hinton, J. A. & Aharonian, F. A. 2007, ApJ, 657, 302
[131] Hinton, J. A., Domainko, W., & Pope, E. C. D. 2007, MNRAS, 382, 466
[132] Hoppe, S., Lemoine-Goumard, M., & for the H. E. S. S. Collaboration. 2007, ArXiv e-prints, 0709.4103
[133] Horns, D. 2005, Physics Letters B, 607, 225
[134] Horns, D., Aharonian, F., Santangelo, A., Hoffmann, A. I. D., & Masterson, C. 2006, A&A, 451, L51
[135] Horns, D., Hoffmann, A. I. D., Santangelo, A., Aharonian, F. A., & Rowell, G. P. 2007, A&A, 469, L17
[136] Itoh, C., Enomoto, R., Yanagita, S., et al. 2007, A&A, 462, 67
[137] Kappes, A., Hinton, J., Stegmann, C., & Aharonian, F. A. 2007, ApJ, 661, 1348
[138] —. 2007, ApJ, 656, 870
[139] Karle, A., Merck, M., Plaga, R., et al. 1995, Astroparticle Physics, 3, 321
[140] Kaspi, V. M., Roberts, M. S. E., & Harding, A. K. 2006, Compact stellar X-ray sources, 279
[141] Kawachi, A., Naito, T., Patterson, J. R., et al. 2004, ApJ, 607, 949
[142] Kawasaki, M., Ozaki, M., Nagase, F., Inoue, H., & Petre, R. 2005, ApJ, 631, 935
[143] Khangulyan, D., Aharonian, F., & Bosch-Ramon, V. 2008, MNRAS, 383, 467
[144] Khangulyan, D., Hnatic, S., Aharonian, F., & Bogovalov, S. 2007, MNRAS, 380, 320
[145] Kirk, J. G., Ball, L., & Skjaeraasen, O. 1999, Astroparticle Physics, 10, 31
[146] Konopelko, A., Atkins, R. W., Blaylock, G., et al. 2007, ApJ, 658, 1062
[147] Kosack, K., Badran, H. M., Bond, I. H., et al. 2004, ApJ, 608, L97
[148] Koyama, K., Kinugasa, K., Matsuzaki, K., Nishiuchi, M., Sugizaki, M., Torii, K., Yamauchi, S., & Aschenbach, B. 1997, PASJ, 49, L7
[149] Krawczynski, H., Buckley, J., Byrum, K., et al. 2007, ArXiv e-prints, 0709.0704
[150] Lang, M. J., Carter-Lewis, D. A., Fegan, D. J., et al. 2004, A&A, 423, 415
[151] Langston, G., Minter, A., D'Addario, L., Eberhardt, K., Koski, K., & Zuber, J. 2000, AJ, 119, 2801
[152] Le Bohec, S., Badran, H. M., Bond, I. H., et al. 2004, ApJ, 610, 156
[153] Lee, J.-J., Koo, B.-C., Yun, M. S., Stanimirović, S., Heiles, C., & Heyer, M. 2008, AJ, 135, 796
[154] Liu, S., Melia, F., Petrosian, V., & Fatuzzo, M. 2006, ApJ, 647, 1099
[155] Lorenz, E. 2007, Journal of Physics Conference Series, 60, 1
[156] Lucek, S. G. & Bell, A. R. 2000, MNRAS, 314, 65
[157] Maier, G. 2007, ArXiv e-prints, 0709.3661
[158] Manolakou, K., Horns, D., & Kirk, J. G. 2007, A&A, 474, 689
[159] Marleau, P., Alfonso, P., Chertok, M., et al. 2005, in Cherenkov 2005
[160] Mazin, D. & Raue, M. 2007, A&A, 471, 439
[161] Mizuno, A. & Fukui, Y. 2004, in Astronomical Society of the Pacific Conference Series, Vol. 317, Milky Way Surveys: The Structure and Evolution of our Galaxy, ed. D. Clemens, R. Shah, & T. Brainerd, 59–+
[162] Muno, M. P., Lu, J. R., Baganoff, F. K., Brandt, W. N., et al. 2005, ApJ, 633, 228
[163] Neronov, A. & Chernyakova, M. 2007, Ap&SS, 309, 253

[164] Padilla, L., Funk, B., Krawczynski, H., et al. 1998, A&A, 337, 43
[165] Paré, E., Balauge, B., Bazer-Bachi, R., et al. 2002, Nuclear Instruments and Methods in Physics Research A, 490, 71
[166] Paredes, J. M., Martí, J., Ishwara Chandra, C. H., & Bosch-Ramon, V. 2007, ApJ, 654, L135
[167] Pfrommer, C. & Enßlin, T. A. 2003, A&A, 407, L73
[168] —. 2004, A&A, 413, 17
[169] Piner, B. G., Pant, N., & Edwards, P. G. 2008, ArXiv e-prints, 801
[170] Ptuskin, V. S. & Zirakashvili, V. N. 2005, A&A, 429, 755
[171] Punch, M., Akerlof, C. W., Cawley, M. F., et al. 1992, Nature, 358, 477
[172] Punch, M. f. t. H. c. 2007, in $30^{th}$ ICRC, Merida Mexico, Vol. 1, Proccedings of the 30th ICRC
[173] Quataert, E. & Loeb, A. 2005, ApJ, 635, L45
[174] Quinn, J., Akerlof, C. W., Biller, S., et al. 1996, ApJ, 456, L83+
[175] Raue, M. & Mazin, D. 2008, ArXiv e-prints, 0802.0129
[176] Rauw, G., Manfroid, J., Gosset, E., Nazé, Y., Sana, H., De Becker, M., Foellmi, C., & Moffat, A. F. J. 2007, A&A, 463, 981
[177] Reed, J. E., Hester, J. J., Fabian, A. C., & Winkler, P. F. 1995, ApJ, 440, 706
[178] Reimer, A., Pohl, M., & Reimer, O. 2006, ApJ, 644, 1118
[179] Reimer, A., Protheroe, R. J., & Donea, A.-C. 2004, A&A, 419, 89
[180] Reimer, O., Aharonian, F., Hinton, J., Hofmann, W., Hoppe, S., Raue, M., & Reimer, A. 2007, ArXiv e-prints, 0710.3418
[181] Rho, J. & Petre, R. 1998, ApJ, 503, L167+
[182] Ribo, M., Paredes, J. M., Moldon, J., Marti, J., & Massi, M. 2008, ArXiv e-prints, 0801.2940
[183] Samorski, M. & Stamm, W. 1983, ApJ, 268, L17
[184] Sarkar, S. 2002, Modern Physics Letters A, 17, 1025
[185] Schödel, R., Ott, T., Genzel, R., et al. 2002, Nature, 419, 694
[186] Schroedter, M., Badran, H. M., Buckley, J. H., et al. 2005, ApJ, 634, 947
[187] Schwarz, U. J., Goss, W. M., Kalberla, P. M., & Benaglia, P. 1995, A&A, 299, 193
[188] Sinnis, G. & Hawc Collaboration. 2005, in American Institute of Physics Conference Series, Vol. 745, High Energy Gamma-Ray Astronomy, ed. F. A. Aharonian, H. J. Völk, & D. Horns, 234–245
[189] Slane, P., Hughes, J. P., Edgar, R. J., Plucinsky, P. P., Miyata, E., Tsunemi, H., & Aschenbach, B. 2001, ApJ, 548, 814
[190] Teshima, M., Prandini, E., Bock, R., et al. 2007, ArXiv e-prints, 0709.1475
[191] The H. E. S. S. Collaboration. 2007, ArXiv e-prints, 0710.4057
[192] Thompson, D. J., Bailes, M., Bertsch, D. L., et al. 1999, ApJ, 516, 297
[193] Thorstensen, J. R., Fesen, R. A., & van den Bergh, S. 2001, AJ, 122, 297
[194] Tsuchiya, K., Enomoto, R., Ksenofontov, L. T., et al. 2004, ApJ, 606, L115
[195] Uchiyama, Y., Aharonian, F. A., Tanaka, T., Takahashi, T., & Maeda, Y. 2007, Nature, 449, 576
[196] van Eldik, C., Bolz, O., Braun, I., Hermann, G., Hinton, J., & Hofmann, W. 2007, ArXiv e-prints, 0709.3729
[197] Velázquez, P. F., Dubner, G. M., Goss, W. M., & Green, A. J. 2002, AJ, 124, 2145
[198] Vink, J., Bleeker, J., van der Heyden, K., Bykov, A., Bamba, A., & Yamazaki, R. 2006, ApJ, 648, L33
[199] Voges, W., Aschenbach, B., Boller, T., et al. 2000, VizieR Online Data Catalog, 9029, 0
[200] Völk, H. J., Aharonian, F. A., & Breitschwerdt, D. 1996, Space Science Reviews, 75,

279
[201] Wagner, R. M. 2008, MNRAS, 146
[202] Wang, Q. D., Lu, F. J., & Gotthelf, E. V. 2006, MNRAS, 367, 937
[203] Weekes, T. C. 1992, Space Science Reviews, 59, 315
[204] Winkler, P. F., Gupta, G., & Long, K. S. 2003, ApJ, 585, 324

# Astronomy with Ultra High-Energy Particles

Jörg R. Hörandel

Department of Astrophysics, Radboud University Nijmegen
P.O. Box 9010, 6500 GL Nijmegen, The Netherlands
J.Horandel@astro.ru.nl, http://particle.astro.kun.nl

### Abstract

*Recent measurements of the properties of cosmic rays above $10^{17}$ eV are summarized and implications on our contemporary understanding of their origin are discussed. Cosmic rays with energies exceeding $10^{20}$ eV have been measured, they are the highest-energy particles in the Universe. Particles at highest energies are expected to be only marginally deflected by magnetic fields and they should point towards their sources on the sky. Recent results of the Pierre Auger Observatory have opened a new window to the Universe — astronomy with ultra high-energy particles.*

## 1 Introduction

The Earth is permanently exposed to a vast flux of high-energy particles from outer space. Most of these particles are fully ionized atomic nuclei with relativistic energies. The extraterrestrial origin of these particles has been demonstrated by V. Hess in 1912 [87] and he named the particles "Höhenstrahlung" (high-altitude radiation) or "Ultrastrahlung" (ultra radiation). In 1925 R. Millikan coined the term "Cosmic Rays". They have a threefold origin. Particles with energies below 100 MeV[1] originate from the Sun [100, 137]. Cosmic rays in narrower sense are particles with energies from the 100 MeV domain up to energies beyond $10^{20}$ eV. Up to several 10 GeV the flux of the particles observed is modulated on different time scales by the heliospheric magnetic fields [58, 86]. Particles with energies below $10^{17}$ to $10^{18}$ eV are usually considered to be of galactic origin [65, 66, 95, 97, 141]. The Larmor radius of a particle with energy $E_{15}$ (in units of $10^{15}$ eV) and charge $Z$ in a magnetic field $B_{\mu G}$ (in μG) is

$$r_L = 1.08 \frac{E_{15}}{Z B_{\mu G}} \text{ pc}, \quad (1)$$

yielding a value of $r_L = 360$ pc for a proton with an energy of $10^{18}$ eV in the galactic magnetic field ($B_{\mu G} \approx 3$). This radius is comparable to the thickness of the galactic disc and illustrates that particles (at least with small charge $Z$) at the highest energies

---

[1] In this review we use the particle physics energy units MeV= $10^6$ eV, GeV= $10^9$ eV, TeV= $10^{12}$ eV, PeV= $10^{15}$ eV, and EeV= $10^{18}$ eV; 1 eV= $1.6 \cdot 10^{-19}$ J.

*Reviews in Modern Astronomy 20.* Edited by S. Röser
Copyright © 2008 WILEY-VCH Verlag GmbH & Co. KGaA, Weinheim
ISBN: 978-3-527-40820-7

can not be magnetically bound to the Galaxy. Hence, they are considered to be of extragalactic origin [39, 105, 118].

In the present article, we focus on recent results concerning the origin of the extragalactic particles. Cosmic rays with energies exceeding $10^{20}$ eV are the highest-energy particles in the Universe. Particles at highest energies are only marginally deflected in the galactic magnetic fields, following (1) they have a Larmor radius $r_L > 36$ kpc, exceeding the diameter of the Milky Way. Thus, they should point back to their sources, enabling astronomical observations with charged particles.

Several questions arise, concerning the origin of highest-energy cosmic rays. Among them are:

– What are the energies of the particles? (Sect. 4)

– What are these particles? Are they protons, nuclei of heavy atoms like oxygen or iron, furthermore are they photons or neutrinos? (Sect. 5)

– Where do they come from? Can we learn something by studying their arrival directions? (Sect. 6)

– How do they propagate to us? Do they suffer any interactions? (Sect. 4)

In the following sections (4 to 6) recent experimental results are compiled and their implications to answer the questions raised above are discussed. Before, possible scenarios for the sources of the particles are summarized (Sect. 2.1) and mechanisms are discussed which are important during the propagation of the particles through the Universe (Sect. 2.2). The detections methods applied are sketched in Sect. 3.

## 2 Sources and Propagation

### 2.1 Sources

The energy density contained in the flux of extragalactic cosmic rays can be inferred from the measured differential energy spectrum $dN/dE$ [80]

$$\rho_E = \frac{4\pi}{c} \int \frac{E}{\beta} \frac{dN}{dE} dE, \qquad (2)$$

where $\beta c$ is the velocity of particles with energy $E$. To estimate the energy content of the extragalactic component, assumptions have to be made about the contribution of galactic cosmic rays at energies in the transition region ($10^{17} - 10^{18}$ eV). The extragalactic component needed according to the poly-gonato model [92] to sustain the observed all-particle flux at highest energies has an energy density of $\rho_E = 3.7 \cdot 10^{-7}$ eV/cm$^3$. The power required for a population of sources to generate this energy density over the Hubble time of $10^{10}$ years is $5.5 \cdot 10^{37}$ erg/(s Mpc$^3$). This leads to $\approx 2 \cdot 10^{44}$ erg/s per active galaxy or $\approx 2 \cdot 10^{52}$ erg per cosmological gamma ray burst [64]. The coincidence between these numbers and the observed

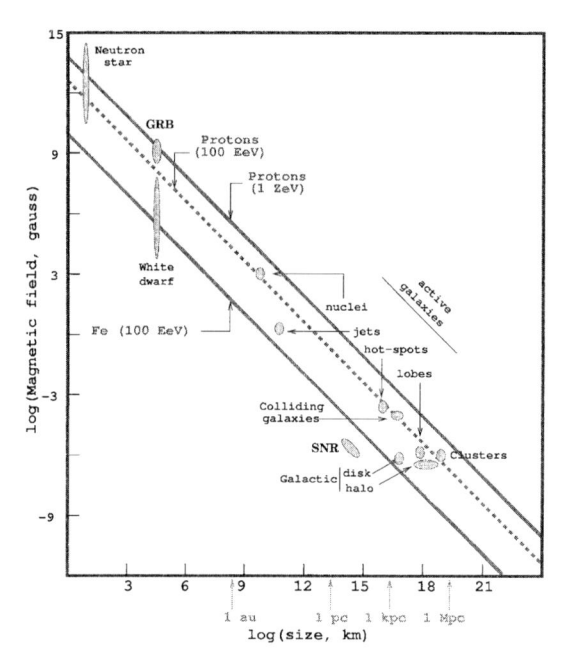

**Figure 1:** Size and magnetic field strength of possible sites of particle acceleration (Hillas diagram). Acceleration of cosmic rays up to a given energy requires conditions above the respective line [129].

output in electromagnetic energy of these sources explains why they are considered as promising candidates to accelerate highest-energy cosmic rays.

The characteristic size of an accelerating region can be estimated for models of gradual acceleration, where the particles make many irregular loops in a magnetic field while gaining energy [90]. The size $L$ of the essential part of the accelerating region containing the magnetic field must be grater than $2r_L$. A closer look reveals that a characteristic velocity $\beta c$ of scattering centers is of virtual importance [90], which yields the expression

$$B_{\mu G} L_{pc} > 2E_{15}/(Z\beta). \qquad (3)$$

It relates the characteristic size $L_{pc}$ (in pc) and magnetic fields $B_{\mu G}$ of objects being able to accelerate particles to energies $E_{15}$. Several possible acceleration sites are explored in Fig. 1, where the magnetic field strength is plotted as function of their typical sizes [129]. The lines according to (3) represent the conditions for protons and iron nuclei of different energies, as indicated. Objects capable to accelerate particles above a respective energy should be above the respective line. As can be inferred from the figure, most promising candidates to accelerate highest-energy cosmic rays are gamma ray bursts and active galactic nuclei (AGN) [70, 90]. These objects are typically in a distance of several tens of Mpc to the Earth. Interactions in the source itself or in the vicinity of the source of hadronic particles (protons, nuclei) yield neutral and charged pions, which subsequently decay into high-energy photons and neutrinos.

Alternatively, so called "top-down models" are discussed in the literature [40, 88, 124]. They have been motivated by events seen by the AGASA experiment above the threshold for the GZK effect [142]. It is proposed that ultra high-energy particles (instead of being accelerated, "bottom-up scenario") are the decay products of exotic, massive particles originating from high-energy processes in the early Universe. Such super-massive particles (with $m_X \gg 10^{11}$ GeV) decay e.g. via W and Z bosons into high-energy protons, photons, and neutrinos.

## 2.2 Propagation

On the way from their sources to Earth the particles propagate mostly outside galaxies in intergalactic space with very low particle densities. In this environment the most important interactions of cosmic rays occur with photons of the 2.7-K microwave background radiation, namely pair production and pion photoproduction [89].

On the last part of their way to Earth they propagate through the Galaxy. However, since particles at the highest energies travel almost along straight lines they accumulate a negligible amount of material during their short travel through regions with relatively high densities. Thus, interactions with the interstellar material can be neglected.

The Universe is filled with about 412 photons/cm$^3$ of the 2.7° K microwave background radiation. Shortly after the discovery of the microwave background it was proposed that ultra high-energy cosmic rays should interact with the photons, leading to a suppression of the observed flux at highest energies [78, 153]. This effect is called after its proposers the Greisen-Zatsepin-Kuz'min (GZK) effect. A nucleon of energies exceeding $E_{GZK} \approx 6 \cdot 10^{19}$ eV colliding head-on with a 2.7° K photon comprises a system of sufficient energy to produce pions by the photoproduction reaction

$$p + \gamma_{3K} \to \Delta^+ \to \begin{matrix} p + \pi^0 \\ n + \pi^+ \end{matrix} .  \qquad (4)$$

The energy loss of the nucleon is a significant fraction of the initial energy. The pion photoproduction cross section is quite large above threshold due to resonance production ($\Delta$ resonance), rising quickly to 500 mb for photon laboratory energies of about 0.3 GeV. Subsequent decays of the neutral and charged pions produced in (4) yield high-energy photons and neutrinos.

The center of mass energy of interactions of cosmic rays with energies exceeding $10^{18}$ eV colliding with microwave-background photons is sufficient to generate electron-positron pairs $p + \gamma_{3K} \to p + e^+ + e^-$. As a consequence, the cosmic-ray particles loose energy which leads presumably to a reduction of the flux or a dip in the spectrum between $10^{18}$ and $10^{19}$ eV [35, 38, 89].

The effect of both processes on the observed energy spectrum is frequently expressed by a modification factor $f(E) = I_p(E)/I_0(E)$, describing the ratio of the observed spectrum $I_p$ and the initial spectrum $I_0$ as function of energy. The modification factor according to recent calculations is shown in Fig. 2 (*left*) [23]. A twofold structure can be recognized. A depression (the dip) at energies between $10^{18}$ and $10^{19}$ eV and the GZK feature at energies exceeding $5 \cdot 10^{19}$ eV. The two cases (1 and 2) represent initial spectra with a spectral index of 2.0 and 2.7, respectively.

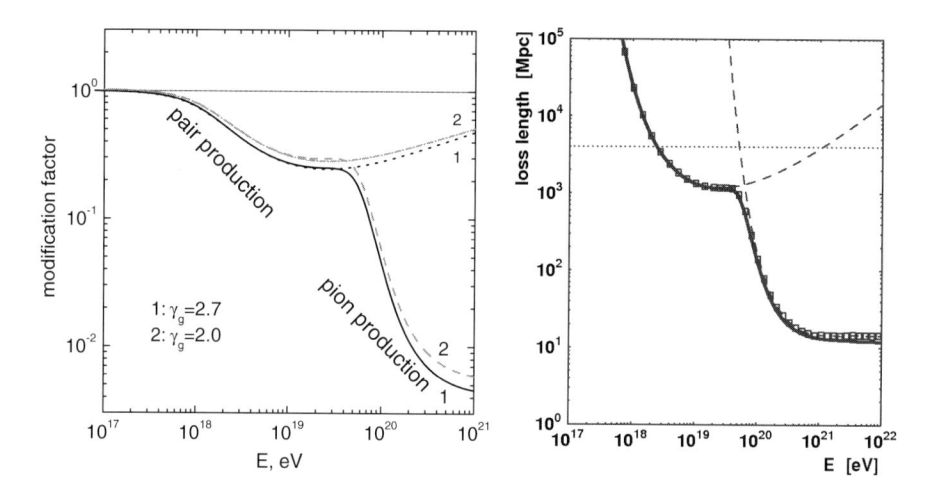

**Figure 2:** *Left*: Modification factor $f(E) = I_p(E)/I_0(E)$ of the cosmic-ray energy spectrum [23]. *Right*: Loss length for protons for pair production and pion photoproduction [51].

The energy loss length for pair production and pion photoproduction is depicted in Fig. 2 (*right*) [51]. Particles with energies above the threshold of the GZK effect can travel less than about 100 Mpc through the Universe, before their energy has decreased to $1/e$ of their initial value. Or, in other words, particles reaching the Earth at these energies have propagated less than 100 Mpc, see also [17], and their sources are inside a sphere with this radius

## 2.3 Multi Messenger Approach

It has been discussed that for both scenarios, acceleration and top-down models, hadronic cosmic rays are accompanied by high-energy photons and neutrinos. Also during the propagation of hadronic particles through the Universe high-energy photons and neutrinos are produced. To clarify the origin of the highest-energy particles in the Universe, simultaneous observations are desired of high-energy charged particles, photons, and neutrinos — a multi-messenger approach. Thus, the observation of high-energy charged particles, or charged particle astronomy, is complementary to observations in gamma ray astronomy [125, 126] and neutrino astronomy [36, 79, 112, 139].

Attention has to be paid on the 'simultaneous' observation: if a charged particle is deflected by an angle $\Theta$ in a (for this estimate simply homogeneous) magnetic field, its path $L_{ch}$ is somewhat longer as compared to the path of a massless neutral particle $L_\gamma$. Their relative difference can be approximated as

$$\mathcal{R} = \frac{L_{ch}}{L_\gamma} = \frac{2\pi\Theta}{360° \sin(\Theta)}. \tag{5}$$

If a charged particle from a source at a distance $L_\gamma = 100$ Mpc is deflected by $\Theta = 3°$, a value $\mathcal{R} = 1 + 4.6 \cdot 10^{-4}$ is obtained. This corresponds to a difference in the arrival time of a charged particle relative to a photon (both traveling at the speed of light) of about $\Delta T = 150 \cdot 10^3$ a. Thus, for a simultaneous detection the acceleration processes have to be stable over such a period in time.

## 3 Detection Method

The extremely steeply falling cosmic-ray energy spectrum ($\propto E^{-3}$) yields very low fluxes for the highest-energy particles. At the highest energies less than one particle is expected per square kilometer and century. This necessitates huge detection areas and large measuring times. At present, they are only realized in huge ground based installations, registering secondary particles produced in the atmosphere.

### 3.1 Extensive air showers

When high-energy cosmic-ray particles penetrate the Earths atmosphere they interact and generate a cascade of secondary particles, the extensive air showers. Hadronic particles interact and produce new hadronic particles or generate muons and photons through pion decays. Some of the muons may decay into electrons, while the photons and electrons/positrons regenerate themselves in an electromagnetic cascade. The by far dominant particles in a shower are electromagnetic particles (photons, electrons, and positrons). Most of the energy of the primary particle is absorbed in the atmosphere. However, a small fraction of the energy is transported to ground level and may be registered in detectors for electrons, muons, and hadrons. Particles traveling with relativistic speeds through the atmosphere (mostly electrons and positrons) emit Čerenkov light. The shower particles also excite nitrogen molecules in the air which in turn emit fluorescence light. While the Čerenkov light is collimated in the forward direction of the particle, the fluorescence light is emitted isotropically, thus, a shower can be "viewed from aside".

The objective of experiments observing extensive air showers is to determine the properties of the primary particle (energy $E_0$, mass $A$, arrival direction). In the energy regime of interest ($E > 10^{17}$ eV) mainly two methods are applied. Electrons (and positrons) as well as muons reaching ground level are observed in large arrays of detectors and the fluorescence light is viewed by imaging telescopes. An alternative technique, presently under investigation, is the detection of radio emission from air showers. Electrons and positrons are deflected in the Earths magnetic field and emit synchrotron radiation, which is detected in arrays of dipole antennae [57, 98, 149].

### 3.2 Measuring technique

The **direction** of air showers is inferred applying two techniques. The particles in a shower travel with nearly the speed of light through the atmosphere in a thin disc with a thickness of a few meters only. With detectors measuring the arrival time of the particles with a resolution of a few ns the angle of the shower front relative to the ground can be inferred, with the arrival direction being perpendicular to the shower plane. With imaging fluorescence telescopes the shower-detector plane is determined from the observed track in the camera. The orientation of the shower axis in this plane is then obtained by measurements of the arrival time of the photons at the detector. Using two (or more) telescopes to view the same shower allows a three-dimensional reconstruction of the shower axis.

The shower **energy** is proportional to the number of electrons $N_e$ and muons $N_\mu$ in the shower. A simple numerical model [96] yields the relations

$$E_0 = 3.01 \text{ GeV} \cdot A^{0.04} \cdot N_e^{0.96} \quad \text{and} \quad E_0 = 20 \text{ GeV} \cdot A^{-0.11} \cdot N_\mu^{1.11} \quad (6)$$

to estimate the primary energy. This illustrates that measuring $N_e$ or $N_\mu$ gives a good estimate for the energy almost independent of the particles mass.

With imaging fluorescence telescopes the amount of fluorescence light can be measured as function of depth in the atmosphere. The total amount of light collected is proportional to the shower energy. Using the number of electrons at shower maximum, the number of photons registered per square meter in a detector at a distance $r$ to the maximum of a shower with energy $E_0$ can be estimated as

$$\nu_\gamma = \frac{N_e X_0 N_\gamma}{4\pi r^2} \approx 790 \, \frac{\gamma}{\text{m}^2} \, A^{-0.046} \left(\frac{E_0}{\text{EeV}}\right)^{1.046} \frac{1}{(r/10 \text{ km})^2}, \quad (7)$$

where $N_\gamma \approx 4 \, \gamma/\text{m}$ is the fluorescence yield of electrons in air and $X_0 = 36.7 \text{ g/cm}^2$ (or 304 m at normal pressure) the radiation length. Absorption and scattering in the atmosphere have been neglected in this simple estimate, thus, the equation gives an upper limit for the registered photons.

Experimentally most challenging is the estimation of the **mass** of the primary particle. Showers induced by light and heavy particles develop differently in the atmosphere. The depth in the atmosphere $X_{max}$ at which the showers contain a maximum number of particles depends on the primary particles mass

$$X_{max}^A = X_{max}^p - X_0 \ln A, \quad (8)$$

where $X_{max}^p$ is the depth of the shower maximum for proton-induced showers [96, 115]. Experiments measuring the longitudinal shower profile by observations of fluorescence light estimate the mass by measurements of $X_{max}$.

If a shower develops higher in the atmosphere more particles (mostly electrons) are absorbed on the way to the ground. On the other hand, at high altitudes (with low air densities) charged pions are more likely to decay, thus, yielding more muons. Hence, the electron-to-muon ratio observed at ground level depends on the mass of the primary particle. A Heitler model of hadronic showers [96] yields the relation

$$\frac{N_e}{N_\mu} \approx 35.1 \left(\frac{E_0}{A \cdot 1 \text{ PeV}}\right)^{0.15}. \quad (9)$$

This implies that the registered electron-to-muon ratio depends on the energy per nucleon of the primary particle.

## 3.3 Cosmic-Ray Detectors

In the following we describe the most important recent detectors for ultra high-energy cosmic rays.

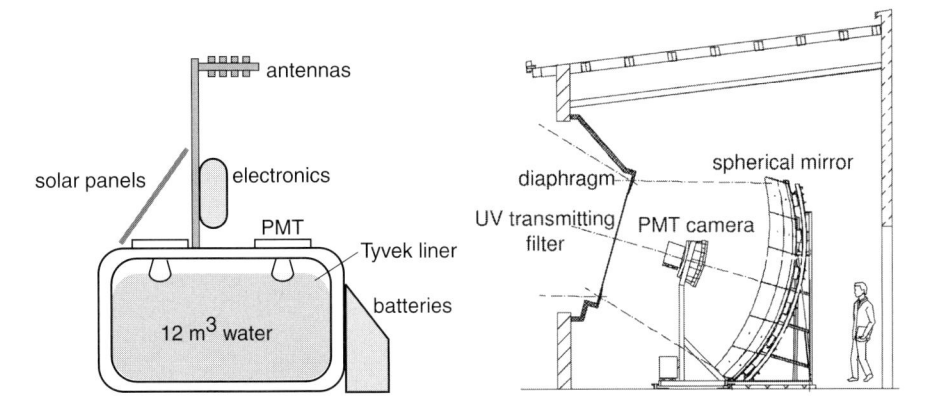

**Figure 3:** Schematic view of a water Čerenkov detector (*left*) and a fluorescence telescope (*right*) of the Pierre Auger Observatory [6].

**The AGASA experiment** The Akeno Giant Air Shower Array (AGASA) was a scintillator array located in Japan (35°N, 138°E), covering an area of 100 km$^2$ [144]. It consisted of 111 scintillation counters to register the electromagnetic shower component. Each station covered 2.2 m$^2$ in area. The scintillator blocks with a thickness of 5 cm were viewed by a 125 mm diameter photomultiplier tube. To register the muonic shower component, proportional counters were used with a cross section of $10 \times 10$ cm$^2$ and a length of 2 m or 5 m. The absorber consisted either of a 1 m thick concrete block, a 30 cm thick iron plate, or a 5 cm lead plate above a 20 cm thick iron plate. The threshold energy for muons is about 0.5 GeV. In total, 27 detector stations were installed with areas varying from 2.8 m$^2$ to 20 m$^2$.

**The HiRes experiment** The High Resolution Fly's Eye experiment (HiRes) was located in Utah, USA (40°N, 112°W) [13]. It was the successor of the Fly's Eye experiment [31], which pioneered the detection of fluorescence light from air showers. HiRes consisted of two detector sites (Hires I & II) separated by 12.6 km, providing almost 360° azimuthal coverage, each. Both telescopes were formed by an array of detector units. The mirrors consisted of four segments and formed a 5.1 m$^2$ spherical mirror. At its focal plane an array of $16 \times 16$ photomultiplier tubes was situated, viewing a solid angle of $16° \times 16°$. HiRes I consisted of 22 detectors, arranged in a single ring, overlooking between 3° and 17° in elevation. This detector used an integrating ADC read-out system, which recorded the photomultiplier tubes' pulse height and time information. HiRes II comprises 42 detectors, set up in two rings, looking between 3° and 31° in elevation. It was equipped with a 10 MHz flash ADC system, recording pulse height and timing information from its phototubes.

**The Pierre Auger Observatory** The observatory combines the observation of fluorescence light with imaging telescopes and the measurement of particles reaching ground level in a "hybrid approach" [6]. The southern site (near Malargue, Argentina, 35.2° S, 69.5° W, 1400 m above sea level) of the worlds largest air shower

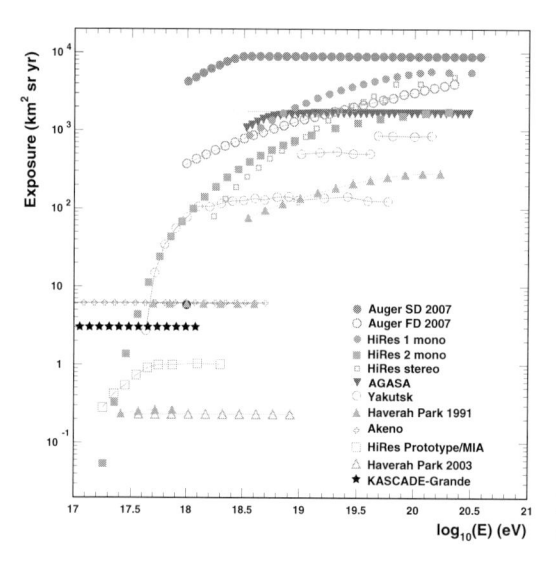

**Figure 4:** Accumulated exposures of various experiments [105].

detector is almost completed. It will consist of 1600 polyethylene tanks set up in an area covering 3000 km². Each water Čerenkov detector has 3.6 m diameter and is 1.55 m high, enclosing a Tyvak liner filled with 12 m³ of high purity water, see Fig. 3. The water is viewed by three PMTs (8 in or 9 in diameter). Signals from the PMTs are read by the electronics mounted locally at each station. Power is provided by batteries, connected to solar panels, and time synchronization relies on a GPS receiver. A radio system is used to provide communication between each station and a central data acquisition system.

Four telescope systems overlook the surface array. A single telescope system comprises six telescopes, overlooking separate volumes of air. A schematic cross-sectional view of one telescope is shown in Fig. 3. Light enters the bay through an UV transmitting filter. A circular diaphragm (1.7 m diameter), positioned at the center of curvature of a spherical mirror, defines the aperture of the Schmidt optical system. A 3.5 m × 3.5 m spherical mirror focuses the light onto a camera with an array of 22 × 20 hexagonal pixels. Each pixel has a photomultiplier tube, complemented by light collectors. Each camera pixel has a field of view of approximately 1.5°, a camera overlooks a total field of view of 30° azimuth × 28.6° elevation.

**Telescope Array** Like the Pierre Auger Observatory, the Telescope array is a hybrid detector, presently under construction in Mullard County, Utah, USA [63]. It covers an area of 860 km² and comprises 576 scintillator stations and three fluorescence detector sites on a triangle with about 35 km separation, each equipped with twelve fluorescence telescopes.

Accumulated exposures (i.e. experiment aperture times measuring time) for various high-energy experiments are presented in Fig. 4 [105]. For surface arrays the aperture is a function of the detector area and constant with energy. On the other hand, the aperture of fluorescence detectors depends on the shower energy, low energy showers can be seen up to a restricted distance only. This may be illustrated using the approximation (7): the fiducial volume to register $\nu_\gamma^{min}$ photons can be estimated as

$$V_{fid} \propto (\nu_\gamma^{min})^{-1.5} A^{-0.069} E_0^{1.569}. \quad (10)$$

This shows that the fiducial volume is a function of the primary energy. In this simple approximation there is a small dependence on the mass of the primary particle ($\approx 25\%$ difference between proton and iron induced showers) and an increase of almost a factor of 40 in the fiducial volume per decade in primary energy. A similar energy dependence can be recognized in Fig. 4 for the various fluorescence detectors. For fluorescence telescopes with a limited field of view in elevation an additional effect occurs: low energy showers penetrate less deep into the atmosphere and thus may have their maximum above the field of view of the telescopes, thus, reducing further the effective aperture.

The Pierre Auger Observatory, still under construction, is already the largest cosmic ray detector, the accumulated data exceed the data of all previous experiments. In particular, those of the largest scintillator array (AGASA) and the largest pure fluorescence detector (HiRes). Thus, the Pierre Auger Observatory is expected to measure the properties of ultra high-energy cosmic rays with unprecedented accuracy.

## 4 Energy Spectrum

Measurements of the energy spectrum provide important information about the origin of cosmic rays. Over a wide range in energy the all-particle differential energy spectrum is usually described by a power law $dN/dE \propto E^{-\gamma}$. For energies below $10^{15}$ eV a value for the spectral index $\gamma = -2.7$ has been established by many experiments. The most prominent feature in the all-particle spectrum is the so called knee at an energy of about $4 \cdot 10^{15}$ eV. At this energy the spectral index changes to $\gamma \approx -3.1$. The knee in the all-particle spectrum is caused by the subsequent cut-offs (or knees) of the spectra of individual elements, starting with protons at $E_k^p \approx 4.5 \cdot 10^{15}$ eV. However, this feature is below the focus of the current article, thus, the reader may be referred to e.g. [92, 94, 97] for a more detailed discussion about galactic cosmic rays and the knee. In the following we focus on energies above $10^{17}$ eV.

Recent energy spectra as obtained by the Pierre Auger Observatory are depicted in Fig. 5 [152]. The registered flux has been multiplied by $E^3$. Different methods are applied to reconstruct the spectra. The first method uses the data from the 3000 km$^2$ surface array. The detection efficiency reaches 100% for showers with zenith angles less than 60° for energies above $10^{18.5}$ eV and for inclined showers ($\Theta > 60°$) above $10^{18.8}$ eV. The signal at 1000 m from the shower axis is used to estimate the shower energy. To avoid a dependence on interaction models used in air shower

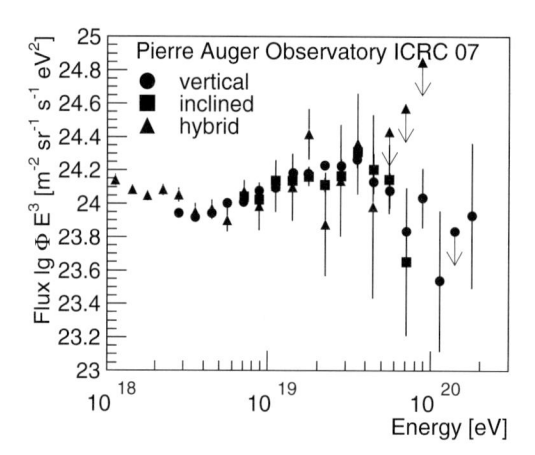

**Figure 5:** All-particle energy spectra measured by the Pierre Auger Observatory using different reconstruction methods [152].

simulation codes, an energy estimator is derived based on measured showers: a subset of showers contains so called hybrid events, seen simultaneously by the surface detector array and at least one fluorescence telescope. The fluorescence telescopes provide a nearly model independent calorimetric energy measurement of the showers in the atmosphere. Only a small correction for 'invisible energy' (high-energy muons and neutrinos carrying away energy) has to be applied. This factor amounts to about 10% and contributes with about 4% to the systematic error for the energy. The energy calibration thus obtained is applied to all events recorded with the surface detector array. Also inclined events with zenith angles exceeding 60° have been analyzed, yielding the second spectrum displayed. Finally, a set of showers which have been recorded by at least one surface detector tank and one fluorescence telescope have been analyzed. The resulting energy spectrum reaches energies as low as $10^{18}$ eV, as can be inferred from Fig. 5. It is interesting to point out that the different spectra have been analyzed independently and agree quite good with each other.

The all-particle energy spectra as obtained by various experiments are compiled in Fig. 6. The flux has been multiplied by $E^3$. The upper panel shows the original data. The different experiments yield absolute values which differ by almost an order of magnitude in this representation. Nevertheless, the overall shape of the energy spectrum seems to be reflected in all data, irrespective of the absolute normalization. This becomes more obvious when the energy scales are slightly readjusted. Typical systematic uncertainties for the energy scale are of order of 10% to 30% in the region of interest. When energy shifts are applied, the results have to be treated with care since the apertures of some experiments change as function of energy (see Fig. 4) and this effect has not been taken into account in the procedure used here.

In the lower panel of Fig. 6 the energy scales of the different experiments have been adjusted to fit the flux according to the poly-gonato model at $10^{18}$ eV. The latter has been obtained through a careful procedure extrapolating the measured spectra for individual elements at low energies [92]. Thus, the normalization applied provides a consistent description from direct measurements (10 GeV region) up to the highest energies. The corresponding energy shifts are listed in Table 1.

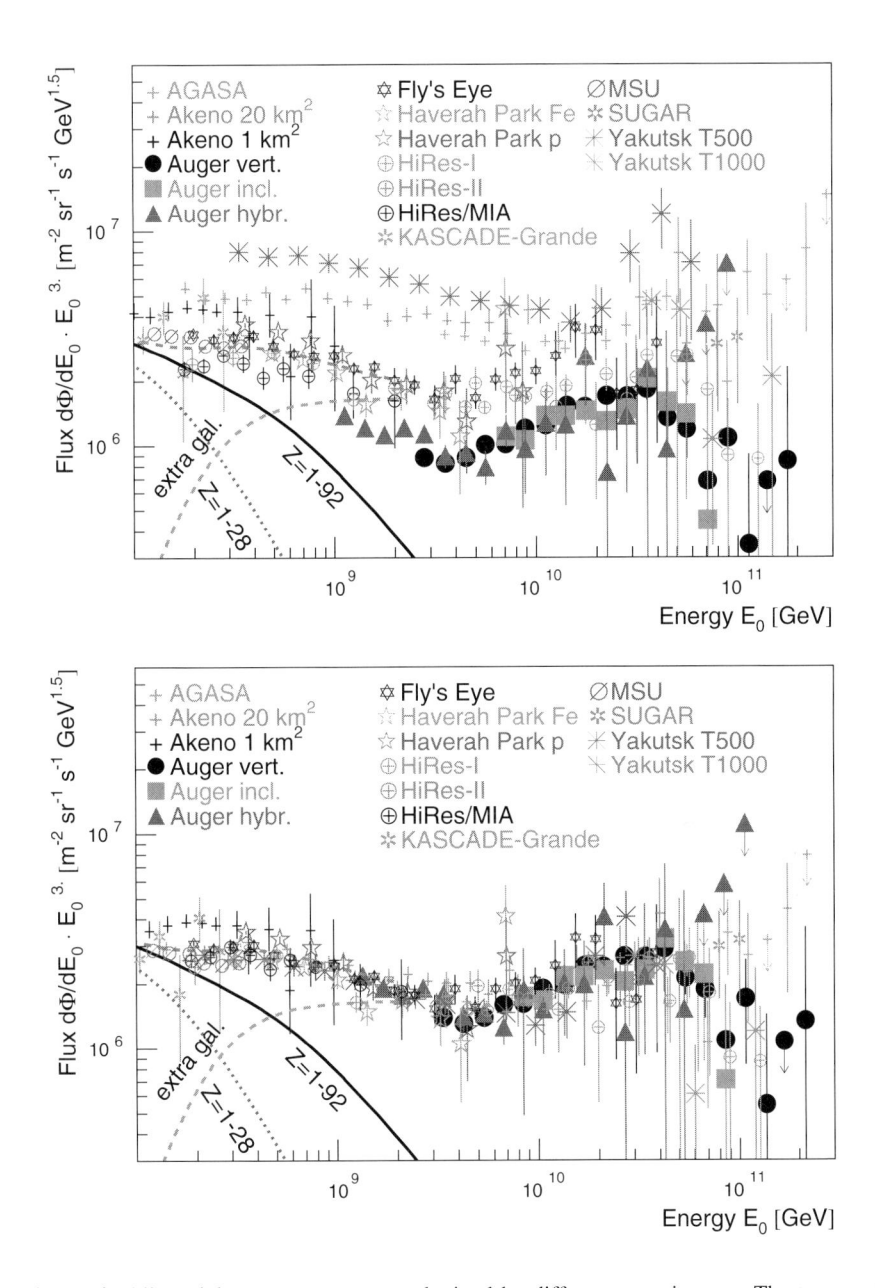

**Figure 6:** All-particle energy spectra as obtained by different experiments. The top panel shows the original values, in the bottom panel the energy scales of the individual experiments have been adjusted. For references and energy shifts, see Table 1. The lines indicate the end of the galactic component according to the poly-gonato model [92] and a possible contribution of extragalactic cosmic rays.

**Table 1:** Energy shifts applied to individual experiments as shown in Fig. 6.

| Experiment | Reference | Energy shift |
|---|---|---|
| AGASA | [144] | $-22\%$ |
| Akeno 1 km$^2$ | [119] | $-4\%$ |
| Akeno 20 km$^2$ | [120] | $-22\%$ |
| Auger | [152] | $+20\%$ |
| Fly's Eye | [44] | $-3\%$ |
| Haverah Park | [28] | $-2\%$ |
| HiRes-I | [1] | $0\%$ |
| HiRes-II | [3] | $0\%$ |
| HiRes-MIA | [14] | $+5\%$ |
| KASCADE-Grande | [84] | $-7\%$ |
| MSU | [60] | $-5\%$ |
| SUGAR | [25] | $0\%$ |
| Yakutsk T500 | [72] | $-35\%$ |
| Yakutsk T1000 | [72] | $-20\%$ |

The normalized spectra agree very well and seem to exhibit a clear shape of the all-particle energy spectrum. Some structures seem to be present in the spectrum. The second knee at about $4 \cdot 10^{17}$ eV, where the spectrum steepens to $\gamma \approx -3.3$ and the ankle at about $4 \cdot 10^{18}$ eV, above this energy the spectrum flattens again to $\gamma \approx -2.7$. Finally, above $4 \cdot 10^{19}$ eV the spectrum exhibits again a steepening with a spectral index $\gamma \approx -4$ to $-5$. The new Auger results help to clarify the situation in this energy region. While the AGASA experiment has reported events beyond the GZK threshold [142], the HiRes experiment has reported a detection of the GZK cut-off [5]. With the new results, a steeper falling spectrum above $4 \cdot 10^{19}$ eV is now confirmed.

The second knee possibly marks the end of the galactic component [92]. If the energy spectra for individual elements exhibit knees at energies proportional to their nuclear charge, the heaviest elements in galactic cosmic rays should fall off at an energy of about $92 \cdot E_k^p \approx 4 \cdot 10^{17}$ eV. An interesting coincidence with the position of the second knee. Different scenarios for the transition from galactic to extragalactic cosmic rays are discussed e.g. in [97, 104]. In the energy region around the ankle a depression is seen in the all-particle flux, also referred to as the dip. It is proposed that this dip is caused by interactions of ultra high-energy particles with the cosmic microwave background, resulting in electron-positron pair production, see Sect. 2.2. The steepening in the flux above $4 \cdot 10^{19}$ eV could be an indication of the GZK effect, i.e. photo-pion production of ultra high-energy cosmic rays with the microwave background, see Sect. 2.2. However, for a definite answer also other properties of cosmic rays have to be investigated.

# 5 Mass Composition

The elemental composition of galactic cosmic rays has been discussed elsewhere, e.g. [93, 97]. Above $10^{17}$ eV the situation is experimentally very challenging, since we are far away in parameter space from collider experiments, where the properties of high-energy interactions are studied in detail. Thus, the air shower models used to interpret the data have to extrapolate over a wide range in parameter space.

The fraction of iron nuclei in cosmic rays as deduced by many experiments has been investigated [52]. No clear conclusion can be drawn about the composition at highest energies. Tension in the interpretation of the measured data has been observed as well by the HiRes-MIA experiment [15]. The observed $X_{max}$ values exhibit a trend towards a lighter composition as function of energy in the range between $10^{17}$ and $10^{18}$ eV. On the other hand, measured muon densities indicate a very heavy composition in the same energy range.

Methods relying on the measured muon densities, the lateral distribution of Čerenkov light registered at ground level, or geometrically-based methods are rather indirect and depend on certain assumptions and/or interaction models. The most bias free mass estimator is probably a measurement of the depth of the shower maximum $X_{max}$, preferably with an imaging telescope such as fluorescence detectors. The best way to infer the mass is to measure $X_{max}$ distributions, rather than average values only. However, unfortunately, also the interpretation of the measured values depends on hadronic interaction models used in air shower simulations.

The average depth of the shower maximum registered by several experiments is plotted in Fig. 7 as function of energy. In the top panel the data are compared to predictions of air shower simulations for primary protons and iron nuclei, using different hadronic interaction models, namely QGSJET 01 [103], QGSJET II-3 [127], SYBILL 2.1 [54], and DPMJET 2.55 [133]. The models yield differences in $X_{max}$ of order of 30 g/cm² for iron nuclei and $\approx$ 50 g/cm² for proton induced showers. An overall trend seems to be visible in the data, the measured values seem to increase faster with energy as compared to the model predictions. This implies that the composition becomes lighter as function of energy. Through interactions with the cosmic microwave background heavy nuclei are expected to break up during their propagation through the Universe (GZK effect) and a light composition is expected at the highest energies. However, e.g. the Auger data at the highest energies correspond to a mixed composition for all models displayed.

In the bottom panel of Fig. 7 the measured values are compared to predictions of astrophysical models of the origin of high-energy cosmic rays.
The propagation of high-energy cosmic rays in extragalactic turbulent magnetic fields is considered in [71]. The average $X_{max}$ values are shown for a case, assuming a mixed source composition with an injection spectrum $\propto E^{-2.4}$, a continuous distribution of the sources, and no extragalactic magnetic field. Other cases studied deliver similar results in $X_{max}$, for details see [71].
Different scenarios for the transition from galactic to extragalactic cosmic rays are discussed in [24]. Two scenarios are distinguished, a 'dip' model in which the galactic and extragalactic fluxes equal at an energy below $10^{18}$ eV and an 'ankle' approach in which both components have equal fluxes at an energy exceeding $10^{19}$ eV.

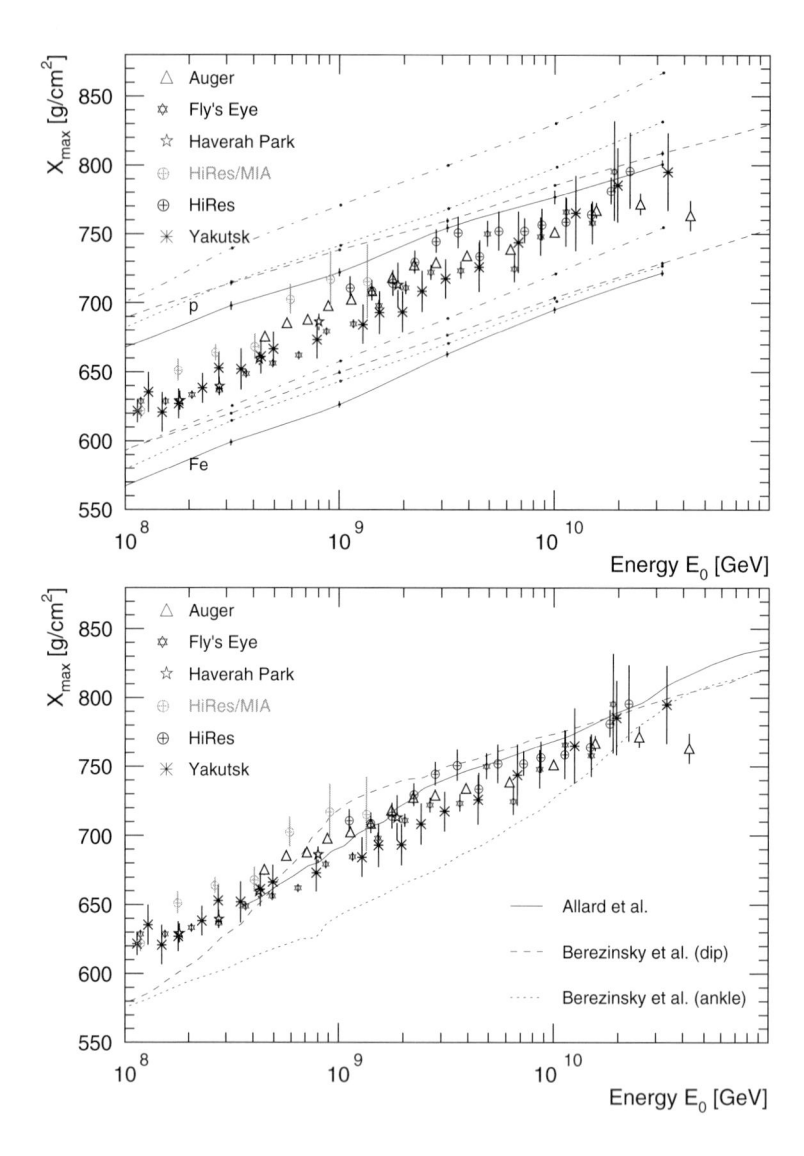

**Figure 7:** Average depth of the shower maximum $X_{max}$ as function of energy as measured by the Pierre Auger Observatory [147], as well as the Fly's Eye [44], Haverah Park [151], HiRes/MIA [15], HiRes [14], and Yakutsk [108] experiments. *Top*: measured values are compared to predictions for primary protons and iron nuclei for different hadronic interaction models QGSJET 01 [103] (—), QGSJET II-3 [127] (- - -), SYBILL 2.1 [54] (···), and DPMJET 2.55 [133] (· — ·). *Bottom*: comparison to astrophysical models according to [71] (—) as well as [24], for the latter a dip (- - -) and an ankle (· · ·) scenario are distinguished.

It is proposed that the dip is a consequence of electron-positron pair production, see Sect. 2.2. For the 'dip' model a source spectrum $\propto E^{-2.7}$ is assumed and a spectrum $\propto E^{-2}$ for the ankle approach. Cosmic rays have been propagated through an extragalactic magnetic field of 1 nG. The resulting average $X_{max}$ values, based on simulations using the interaction code QGSJET 01 are displayed in the figure.
The figure illustrates that we are entering an era where it should be possible to distinguish between different astrophysical scenarios.

Of great interest is also whether other species than atomic nuclei contribute to the ultra high-energy particle flux.

## 5.1 Photon Flux Limit

Air showers induced by primary photons develop an almost pure electromagnetic cascade. Experimentally they are identified by their relatively low muon content or their relatively deep shower maximum. Since mostly electromagnetic processes are involved in the shower development, the predictions are more reliable and don't suffer from uncertainties in hadronic interaction models. A compilation of recent upper limits on the contribution of photons to the all-particle flux is shown in Fig. 8 [7]. The best photon limits are the latest results of the Pierre Auger Observatory [7] setting rather strong limits on the photon flux. They are based on measurements with the

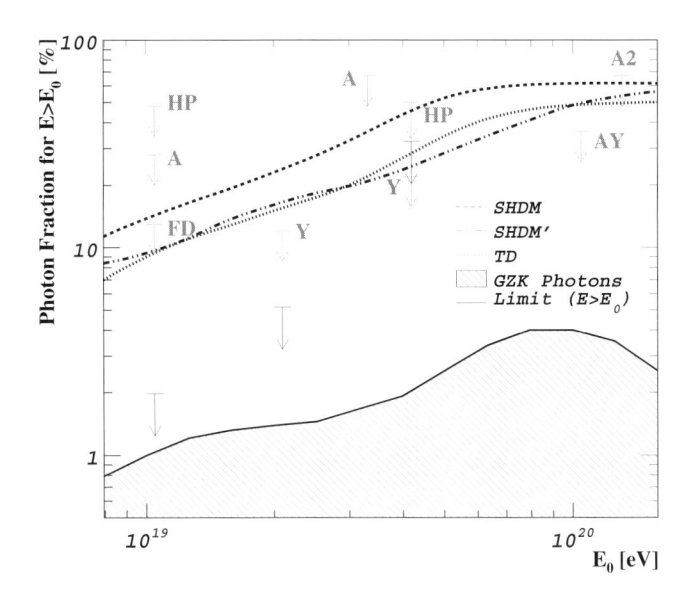

**Figure 8:** Upper limits on the fraction of photons in the integral cosmic-ray flux compared to predictions for GZK photons and top-down scenarios [7]. Experimental data are from the Auger surface detectors (arrows) [7] and a hybrid analysis (FD) [8], Haverah Park (HP) [27], AGASA (A) [134, 138], AGASA and Yakutsk (AY) [136], as well as Yakutsk (Y) [73].

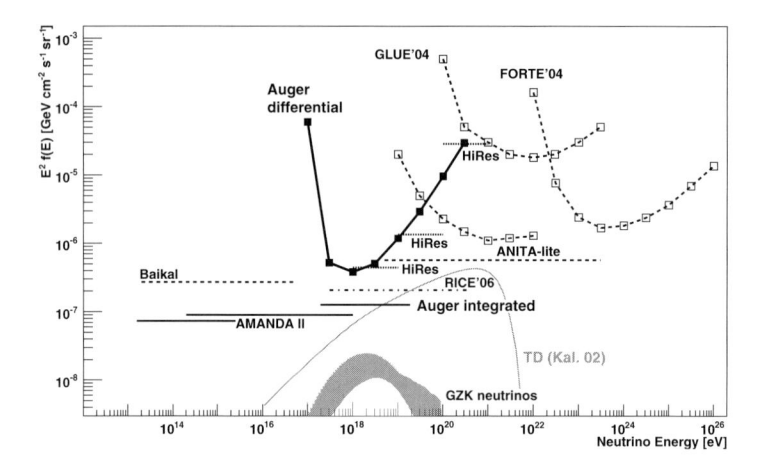

**Figure 9:** Limits at 90% confidence level for a diffuse flux of $\nu_\tau$ assuming a 1:1:1 ratio of the three neutrino flavors at Earth [9, 105]. The experimental results are compared to predictions for GZK neutrinos and a top-down model [101].

Auger surface detectors, taking into account observables sensitive to the longitudinal shower development, the signal rise time, and the curvature of the shower front. The photon fraction is smaller than 2%, 5.1%, and 31% above energies of $10^{19}$, $2 \cdot 10^{19}$, and $4 \cdot 10^{19}$ eV, respectively with 95% confidence level.

In top-down scenarios for high-energy cosmic rays, the particles are decay products of super-heavy particles. This yields relatively high-fluxes of photons predicted by such models. Several predictions are shown in the figure [22, 53]. These scenarios are strongly disfavored by the recent Auger results.

The upper limits are already relatively close to the fluxes expected for photons originating from the GZK effect [69], shown in the figure as shaded area.

## 5.2 Neutrino Flux Limit

The detection of ultra high-energy cosmic neutrinos is a long standing experimental challenge. Many experiments are searching for such neutrinos, and there are several ongoing efforts to construct dedicated experiments to detect them [56, 81]. Their discovery would open a new window to the Universe [32]. However, so far no ultra high-energy neutrinos have been detected. [2]

As discussed above (Sect. 2) ultra high-energy cosmic rays are expected to be accompanied by ultra high-energy neutrinos. The neutrinos are produced with different abundances for the individual flavors, e.g. pion decay leads to a ratio $\nu_e : \nu_\mu = 2 : 1$. However, due to neutrino oscillations the ratio expected at Earth is $\nu_\tau : \nu_\mu : \nu_e = 1 : 1 : 1$.

To discriminate against the huge hadronic background in air shower detectors, neutrino candidates are identified as nearly horizontal showers with a sig-

---

[2] Neutrinos produced in air showers (atmospheric neutrinos) [61], in the sun [21, 62], and during super nova 1987A [43, 91] have been detected, but are at energies much below our focus.

nificant electromagnetic component. The Pierre Auger Observatory is sensitive to Earth-skimming tau-neutrinos that interact in the Earth's crust. Tau leptons from $\nu_\tau$ charged-current interactions can emerge and decay in the atmosphere to produce a nearly horizontal shower with a significant electromagnetic component. Recent results from the Pierre Auger Observatory together with upper limits from other experiments are presented in Fig. 9. Assuming an $E_\nu^{-2}$ differential energy spectrum Auger derives a limit at 90% confidence level of $E_\nu^2 \, \mathrm{d}N_{\nu_\tau}/\mathrm{d}E_\nu < 1.3 \cdot 10^{-7}$ GeV cm$^{-2}$ s$^{-1}$ sr$^{-1}$ in the energy range between $2 \cdot 10^{17}$ and $2 \cdot 10^{19}$ eV.

According to top-down models for ultra high-energy cosmic rays a large flux of ultra high-energy neutrinos is expected. As an example, the predictions of a model [101] are shown in the figure as well. This model is disfavored by the recent upper limits. It should also be noted that the current experiments are only about one order of magnitude away from predicted fluxes of GZK neutrinos (cosmogenic neutrinos).

## 6 Arrival Direction

The arrival directions of cosmic rays provide an important observable to investigate the sources of these particles. Since charged particles are deflected in magnetic fields, the cosmic-ray flux observed at Earth is highly isotropic. A significant evidence for an anisotropy in the arrival directions would be the most direct hint towards possible cosmic-ray sources. Unfortunately, only limited experimental information is available about both, galactic [132, 148] and extragalactic [77, 110] magnetic fields. Selecting particles at the highest energies limits the field of view to distances less than 100 Mpc, see Fig. 2 (*right*). This implies two advantages: the number of source candidates is limited and the particles are only slightly deflected since they propagate a restricted distance only.

### 6.1 Galactic Center

The center of our galaxy is an interesting target for cosmic-ray anisotropy studies. It harbors a massive black hole associated with the radio source Sagittarius A* and a supernova remnant Sagittarius A East. Both are candidates to be powerful cosmic-ray accelerators. The importance is underlined by recent discoveries: the HESS experiment has reported the observation from TeV $\gamma$ rays near the location of Sagittarius A* [19] and discovered a region of extended emission from giant molecular clouds in the central 200 pc of the Milky Way [20].

Of particular interest to search for anisotropies in cosmic rays is the region around $10^{18}$ eV. At these energies the tail of the galactic component might still contribute significantly to the all-particle spectrum and neutrons from the galactic center can reach the Earth without decaying. Such neutrons would not be deflected by magnetic fields [18, 42, 47, 49, 76, 116].

The AGASA experiment has investigated anisotropies in the arrival directions of cosmic rays at energies around $10^{18}$ eV, see Fig. 10 [85]. The Galactic Center is just outside the field of view of the experiment. However, an excess in the Galactic-Center region has been detected. Also the SUGAR experiment, located in Australia

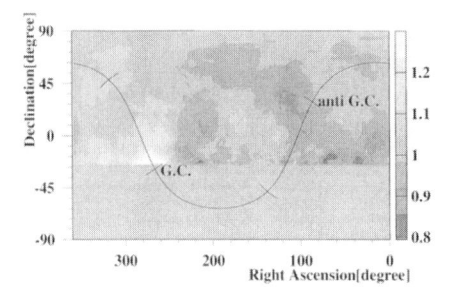

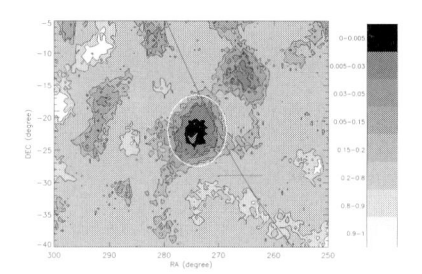

**Figure 10:** Significance maps of excess/deficit events in equatorial coordinates as measured by the AGASA [85] (*left*) and SUGAR [34] (*right*) experiments. The lines in both panels indicate the galactic plane. AGASA: events within a radius of 20° are summed up in each bin. SUGAR: The white circle with a radius of 5.5° indicates the error for a point source.

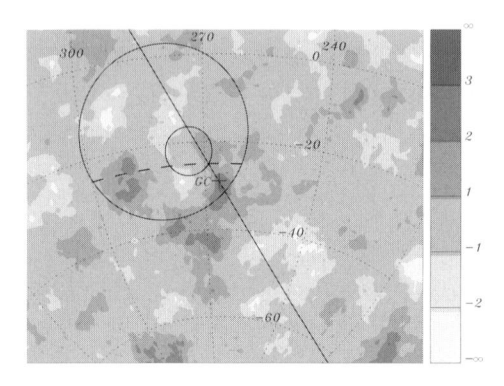

**Figure 11:** Map of cosmic-ray over-density significances near the Galactic Center as measured by the Pierre Auger Observatory [10]. The line indicates the galactic plane and "+" marks the Galactic Center. The circles represent the regions of excess events seen by the AGASA and SUGAR experiments.

has reported an excess of events from the region of the Galactic Center at $10^{18}$ eV, see Fig. 10 (*right*) [34]. It should be noted that both findings are on the 3 to $4\sigma$ level only.

Recently, data from the Pierre Auger Observatory have been searched for anisotropies in the region of the Galactic Center [10]. A map of resulting cosmic-ray over-density significances is displayed in Fig. 11. The regions were AGASA and SUGAR have found an excess are marked in the figure. With a statistics much greater than those of previous experiments, it has been searched for a point-like source in the direction of Sagittarius A. No significant excess has been found. Also searches on larger angular scales show no abnormally over-dense regions. These findings exclude recently proposed scenarios for a neutron source in the Galactic Center.

## 6.2 Clustering of Arrival Directions

The AGASA experiment has investigated small-scale anisotropies in the arrival directions of cosmic rays [143]. Above an energy of $4 \cdot 10^{19}$ eV they have found clusters of events coming from the same direction, see Fig. 12. One triplet and three doublets with a separation angle of 2.5° have been reported, the probability to observe these clusters by a chance coincidence under an isotropic distribution is smaller than 1%.

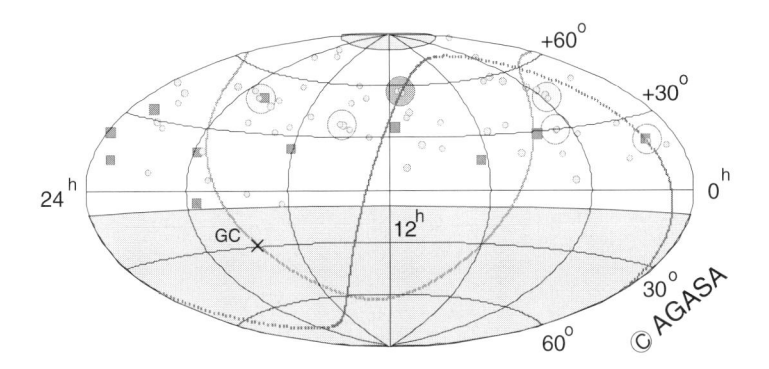

**Figure 12:** Arrival directions of cosmic rays with energies exceeding $4 \cdot 10^{19}$ eV in equatorial coordinates as observed by the AGASA experiment. Red squares and green circles represent cosmic rays with energies exceeding $10^{20}$ eV and $(4-10) \cdot 10^{19}$ eV, respectively. [16, 143]

The HiRes experiment has found no significant clustering at any angular scale up to 5° for energies exceeding 10 EeV [2]. Combining data from the AGASA, HiRes, SUGAR, and Yakutsk experiments at energies above 40 EeV a hint for a correlation has been found at angular scales around 25° [99].

Also the data of the Pierre Auger Observatory have been searched for clustering in the arrival directions [117]. The autocorrelation function has been analyzed adopting a method, in which a scan over the minimum energy $E$ and the separation angle $\Theta$ is performed [59]. For each value of $E$ and $\Theta$ a chance probability is calculated by generating a large number of isotropic Monte Carlo simulations of the same number of events, and computing the fraction of simulations having an equal or larger number of pairs than the data for those parameters. The result is depicted in Fig. 13, showing the probability as function of separation angle and threshold energy. A broad region with an excess of correlation appears at intermediate angular scales and large energies. The minimum is found at 7° for the 19 highest events ($E > 57.5$ EeV), where eight pairs were observed, while one was expected. The fraction of isotropic simulations with larger number of pairs at that angular scale and for that number of events is $P_{min} = 10^{-4}$. The chance probability for this value to arise from an isotropic distribution is $P \approx 2 \cdot 10^{-2}$.

## 6.3 Correlation with BL-Lacs

Interesting candidates as cosmic-ray sources are BL Lacertae objects. They are a sub class of blazars, active galaxies with beamed emission from a relativistic jet which is aligned roughly towards our line of sight. Several experiments have searched for correlations of the arrival directions of cosmic rays with the position of BL Lacs on the sky.

A correlation was found between a subset of BL Lac positions and arrival directions recored by AGASA with energies exceeding 48 EeV and by the Yakutsk experiment at energies above 24 EeV [145]. This correlation and further ones as

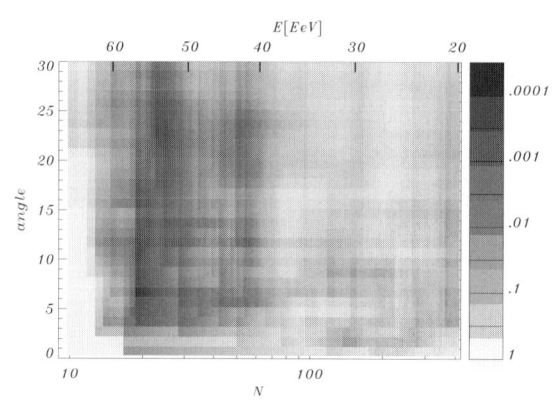

**Figure 13:** Autocorrelation scan for events with energies above $10^{20}$ eV recorded with the Pierre Auger Observatory [117]. The chance probability is shown as function of separation angle and threshold energy.

reported in [74, 146] between BL Lacs and ultra high-energy cosmic rays registered by the AGASA and Yakutsk experiments were not confirmed by data of the HiRes experiment [4]. On the other hand, an excess of correlations was found for a subset of BL Lacs and cosmic rays with energies above 10 EeV [4, 75].

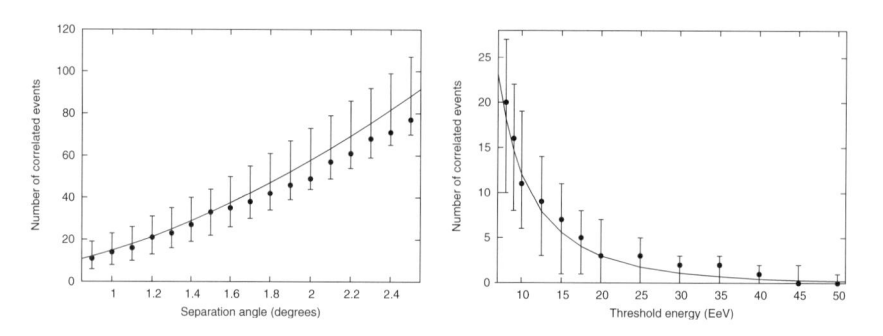

**Figure 14:** Number of events correlated with confirmed BL Lacs with optical magnitude $m < 18$ from the $10^{th}$ edition of the catalog of quasars and nuclei [150] (points) and average for an isotropic flux (solid line) along with dispersion in 95% of simulated isotropic sets (bars) [83]. As function of the angular separation (for $E > 10$ EeV, *left*) and as function of threshold energy ($\Theta < 0.9°$, *right*).

In spring 2007 the number of events recorded by the Pierre Auger Observatory above 10 EeV was six times larger than the data used in previous searches. The correlation hypotheses reported previously have been tested with the Auger data [83]. Since the southern detector of the Pierre Auger Observatory sees a different part of the sky as compared to the AGASA, HiRes, and Yakutsk experiments, only the 'recipes' could be tested but using different sources on the sky. Non of the previously reported hypotheses could be confirmed, the chance probabilities for the different approaches were found to be slightly smaller than 1%. The correlations search has been extended to a broader range of angular scales and energy thresholds, see Fig. 14. It shows the number of correlated events as function of separation angle (*left*) and energy threshold (*right*). The curves represent expectations for an isotropic flux.

**Figure 15:** Centaurus A as seen by the Hubble Space Telescope and the VLA radio telescope [http://hubblesite.org]. The radio lobes extend over a scale of about $10°$ along the super-galactic plane.

The error bars depict the dispersion within 95% of simulated isotropic sets. As can be inferred from the figure, the measured data are compatible with an isotropic distribution and they do not confirm earlier findings.

## 6.4 Correlation with AGN

Another interesting set of objects to serve as sources of ultra high-energy cosmic rays are Active Galactic Nuclei (AGN). The radiation from AGN is believed to be a result of accretion on to the super-massive black hole (with $10^6$ to $10^8$ solar masses) at the center of the host galaxy. AGN are the most luminous persistent sources of electromagnetic radiation in the Universe. An example of an AGN is shown in Fig. 15: Centaurus A is with a distance of 3.4 Mpc one of the closest AGN. The radio lobes are thought to be the result of relativistic jets emerging from the central black hole. Different scenarios related to AGN have been developed, which are supposed to accelerate particles to highest energies, e.g. [41, 114, 128, 131, 135].

The arrival directions of cosmic rays as measured by the Pierre Auger Observatory have been correlated with the positions of AGN [11, 12]. Data taken during the construction of the observatory since January 2004 have been analyzed, corresponding to slightly more than one year of data of the completed observatory. The angular resolution of the detector is better than $1°$ at energies above $10^{19}$ eV [29]. The positions of AGN according to the $12^{th}$ edition of the catalog of quasars and nuclei [150] within a distance $D$ have been used. A scan has been performed over the distance $D$, a threshold energy $E_{th}$, and the correlation angle $\Theta$. The best correlation has been found for events with energies exceeding $E_{th} = 57$ EeV, a maximum distance $D = 71$ Mpc, corresponding to a redshift $z = 0.017$, and a correlation angle $\Theta = 3.2°$. With these parameters 20 out of 27 cosmic rays correlate with at least one of the 442 selected AGN (292 in the field of view of the observatory). Only 5.6 are expected, assuming an isotropic flux. The 27 cosmic rays measured with the highest energies are shown in Fig. 16 together with the positions of the AGN. Many of the observed correlated events are aligned with the super-galactic plane. Two events have arrival directions less than 3° away from Centaurus A. These results indicate clearly that the arrival directions of cosmic rays at highest energies are not isotropic.

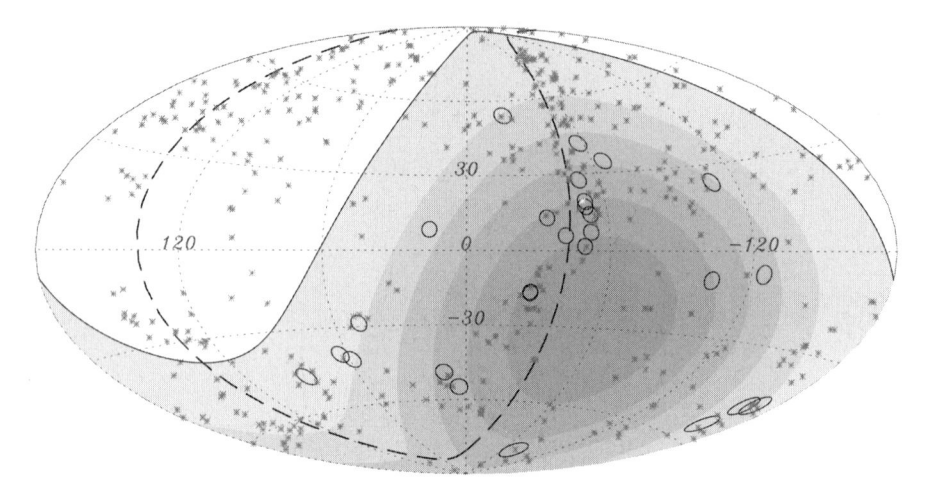

**Figure 16:** Aitoff projection of the celestial sphere in galactic coordinates with circles of 3.2° centered at the arrival directions of 27 cosmic rays detected by the Pierre Auger Observatory with energies $E > 57$ EeV [11]. The positions of AGN with redshift $z < 0.018$ ($D < 75$ Mpc) from the $12^{th}$ edition of the catalog of quasars and nuclei [150] are indicated by the asterisks. The solid line draws the border of the field of view of the southern observatory (for zenith angles $\Theta < 60°$). Darker color indicates larger relative exposure. The dashed line indicates the super-galactic plane. Centaurus A, one of the closest AGN is marked in white.

A cosmic ray with charge $Ze$ that travels a distance $D$ in a regular magnetic field $B$ is deflected by an angle [12]

$$\delta \approx 2.7° \frac{60 \text{ EeV}}{E/Z} \left| \int_0^D \left( \frac{dx}{\text{kpc}} \times \frac{B}{3 \text{ μG}} \right) \right|. \quad (11)$$

Assuming a coherence scale of order $\approx 1$ kpc [140] for the regular component of the galactic magnetic field, the deflection angle is a few degrees only for protons with energies larger than 60 EeV. This illustrates that the observed angular correlations are reasonable, but one has to keep in mind the limited knowledge about galactic magnetic fields. The angular scale of the observed correlations also implies that intergalactic magnetic fields along the line of sight to the sources do not deflect cosmic-ray trajectories by much more than a few degrees. The root mean square deflection of cosmic rays with charge $Ze$, traveling a distance $D$ in a turbulent magnetic field with coherence length $L_c$ is [12]

$$\delta_{rms} \approx 4° \frac{60 \text{ EeV}}{E/Z} \frac{B_{rms}}{\text{nG}} \sqrt{\frac{D}{100 \text{ Mpc}}} \sqrt{\frac{L_c}{1 \text{ Mpc}}}. \qquad (12)$$

As information on intergalactic magnetic fields is very sparse, the correlations observed can be used to constrain models of turbulent intergalactic magnetic fields. Within the observed volume they should be such that in most directions $B_{rms}\sqrt{L_c} \leq 10^{-9}$ G$\sqrt{\text{Mpc}}$. In the future the Pierre Auger Observatory will collect more data and more than one event per source should be detected. It should then be possible to use the data itself to set constraints on magnetic field models.

It should be noted that the findings by the Pierre Auger Observatory imply that the sources of the highest-energy-cosmic rays are spatially distributed like AGN. The actual acceleration sites could be the AGN itself or other candidates with the same spatial distribution as AGN.

## 7 Discussion and Outlook

"How do cosmic accelerators work and what are they accelerating?" is one of eleven science questions for the new century in physics and astronomy [121]. In the last few years important progress has been made in measuring the properties of ultra high-energy cosmic rays. In particular, the results of the Pierre Auger Observatory have significantly contributed to an improvement in understanding the origin of the highest-energy particles in the Universe. The discovery of correlations between the arrival directions of cosmic rays and the positions of AGN was among the most important scientific breakthroughs in 2007 for several science media organizations, see [www.auger.org/news/top_news_2007.html].

The must important findings discussed in this overview may be summarized as follows. The structures in the energy spectrum at highest energies seem to become more clear. In particular, there seems to be evidence for a steeper falling spectrum above $4 \cdot 10^{19}$ eV (Figs. 5 and 6). The question arises whether this steepening is due to the GZK effect or due to the maximum energy achieved during the acceleration processes. The most convincing evidence for the existence of the GZK effect is provided by the correlations of the arrival directions with AGN. They occur sharply above an energy of 57 EeV. At this energy, the flux measured by the Pierre Auger Observatory is about 50% lower than expected from a power law extrapolation from lower energies, see Fig. 5. Thus, there seems to be a connection between the steepening in the spectrum and the AGN correlation.

The correlations occur on an angular scale of about 3.2°. This indicates that the particles are deflected marginally only. In turn, this implies they should be light particles, with low $Z$, see (11) and (12). However, there is some tension between this expectation and the measurements of the average depth of the shower maximum $X_{max}$ (Fig. 7). The data at $4 \cdot 10^{19}$ eV are compatible with a mixed composition. But, since the correlations occur relatively sharp above 57 EeV, some dramatic change in composition above this energy can also not be excluded.

The correlation implies that the sources of ultra high-energy particles are in our cosmological neighborhood ($D < 75$). The GZK horizon, defined as the distance from Earth which contains the sources that produce 90% of the protons that arrive with energies above a certain threshold is 90 Mpc at 80 EeV and 200 Mpc at 60 EeV [82]. There seems to be a slight mismatch between these numbers and the Auger findings. Shifting upward the Auger energy scale by about 30%, as indicated by some simulations of the reconstruction procedures [55], a better agreement between the predicted GZK horizon and energy threshold with the observed data could be achieved [12].

The biggest uncertainty in the absolute energy scale of the Pierre Auger Observatory is the knowledge of the fluorescence yield. At present, intensive efforts are conducted by various groups to precisely determine the fluorescence yield of electrons in air [26]. Attention is paid to the dependence of the yield on atmospheric parameters, like pressure, temperature, and humidity. In particular, upcoming results from the AIRFLY experiment [30, 130] are expected to significantly reduce the uncertainties of the energy scale for fluorescence detectors.

The correlation between the arrival directions and the positions of AGN sets constraints on models for the acceleration of ultra high-energy particles. The results imply that the spatial distribution of sources is correlated with the distribution of AGN. Thus, already some scenarios are strongly disfavored. Ruled out are models proposing sources in our Galaxy, like neutron stars [46], pulsars [33], and black holes [50]. Models for sources in the galactic halo are also ruled out such as top-down scenarios with decaying super-heavy particles [37, 45, 111]. These models are also severely constraint by the upper limits on the photon flux (Fig. 8) and the neutrino flux (Fig. 9). Within the next years measurements of photons and neutrinos produced in the GZK effect seem to be in reach. Their detection would be an important and complementary information about the origin and propagation of ultra high-energy cosmic rays.

In summary, the acceleration of ultra high-energy particles in AGN seems to be very attractive, different scenarios have been proposed, e.g. [41, 114, 128, 131, 135]. However, other sources with a similar spatial distribution are not excluded.

With the energy density estimated in Sect. 2.1 we obtain a total cosmic ray power of about $9.7 \cdot 10^{43}$ erg/s within a sphere ($r = 75$ Mpc) seen by the Pierre Auger Observatory at the highest energies. The typical power in the jets of AGN is of order of $10^{44}$ to $10^{46}$ erg/s [109]. If we assume about 10% of this power is converted into cosmic rays, about 1 to 10 sources are needed to sustain the power of the observed extragalactic cosmic ray flux within a distance of 75 Mpc from Earth. If the efficiency is slightly smaller, the number of sources required is correspondingly slightly larger.[3] If the sources of the highest-energy particles are indeed related to AGN, the number of correlated events seen by the Pierre Auger Observatory seems to be of the right order of magnitude and one expects to see in future more events from the same sources.

When the number of correlated events found in the Auger data is compared to expectations for the AGASA and HiRes experiments, one has to be aware of the

---

[3] Based on statistical arguments a minimum number of sources $\geq 61$ has been estimated [12].

different energy scales, see Table 1. The energy scales of the AGASA and HiRes experiment are shifted relative to the Pierre Auger Observatory by about 42% and 20%, respectively. If the Auger prescription is applied to the data of these experiments, the energy threshold has to be adjusted correspondingly.

**In the next years** several experiments focus on the exploration of the energy region of the transition from galactic to extragalactic cosmic rays ($10^{17}$ to $10^{18}$ eV). The 0.5 km$^2$ KASCADE-Grande experiment [122] is taking data since 2004 [48]. The Ice Čerenkov detector Ice Cube [106] at the South Pole and its 1 km$^2$ surface air shower detector Ice Top [67] are under construction. In January 2008 40 Ice Cube strings and 40 surface detectors have been deployed, which implies the set-up is already 50% completed. Further experiments are the Telescope Array [63] and its low energy extension TALE, as well as extensions of the Pierre Auger Observatory to lower energies [102]. With this new high-quality data more detailed information will be available on the energy spectrum and the composition and it should be possible to distinguish between different scenarios for the transition from galactic to extragalactic cosmic rays.

A promising complementary detection method for high-energy cosmic rays is the measurement of radio emission from air showers. This method provides three-dimensional information about air showers, similar to the fluorescence technique, but with the advantage of a much higher duty cycle. In the next years air showers are expected to be detected with the LOFAR radio observatory [113]. An extensive research and development program is conducted in the Pierre Auger Collaboration with the goal to build a 20 km$^2$ radio antennae array [149].

The southern site of the Pierre Auger Observatory covers only a part of the whole sky, see Fig. 16. Since the distribution of matter in the Universe is different in the parts seen from the northern and southern hemispheres it is important to observe the whole sky. The Northern Auger Observatory is designed to complete and extend the investigations begun in the South [123]. To unambiguously identify the sources of the highest-energy cosmic rays requires collecting many more events in spite of the steeply falling energy spectrum. The planned Northern site will be located in Southeast Colorado, USA, having an instrumented area several times the area of Auger South.

The Northern Observatory needs unrestricted support now, it is the next step in exploring the high-energy Universe in the upcoming years. With the completed Pierre Auger Observatory, with its Southern and Northern sites operated simultaneously, an exciting future in astroparticle physics is ahead of us. It will establish charged particle astronomy on the whole sky and will provide high-accuracy data to test astrophysical models of the origin of ultra high-energy cosmic rays. In addition, it will improve our understanding of fundamental physics. The data will give insight into topics like the existence of vacuum Čerenkov radiation, the smoothness of space, and tests of Lorentz invariance [68, 107]. Already now, the existing Auger data set stringent limits on theories.

## Acknowledgment

I would like to thank the organizers of "Cosmic Matter" for the excellent meeting, bringing together the astroparticle physics and astronomy communities and for their invitation to give an overview talk. Many thanks to John Harton for critically reading the manuscript. I'm grateful to my colleagues from the Pierre Auger Observatory as well as the LOFAR, LOPES, KASCADE-Grande, and TRACER experiments for fruitful discussions.

## References

[1] Abbasi, R., et al., 2004. Phys. Rev. Lett. 92, 151101.
[2] Abbasi, R., et al., 2004. Astrophys. J. 610, L73.
[3] Abbasi, R., et al., 2005. Astropart. Phys. 23, 157.
[4] Abbasi, R., et al., 2006. Astrophys. J. 636, 680.
[5] Abbasi, R., et al., 2007. astro-ph/0703099.
[6] Abraham, J., et al., 2004. Nucl. Instr. & Meth. A 523, 50.
[7] Abraham, J., et al., 2007. arXiv:0712.1147, submitted to Astropart. Phys.
[8] Abraham, J., et al., 2007. Astropart. Phys. 27, 155.
[9] Abraham, J., et al., 2007. arXiv:0712.1909, submitted to Phys. Rev. Lett.
[10] Abraham, J., et al., 2007. Astropart. Phys. 27.
[11] Abraham, J., et al., 2007. Science 318, 938.
[12] Abraham, J., et al., 2007. arXiv:0712.2843.
[13] Abu-Zayyad, T., et al., 2000. Nucl. Instr. & Meth. A 450, 253.
[14] Abu-Zayyad, T., et al., 2000. Astrophys. J. 557, 686.
[15] Abu-Zayyad, T., et al., 2000. Phys. Rev. Lett. 84, 4276.
[16] AGASA, 2003. http://www-akeno.icrr.u-tokyo.ac.jp/AGASA/.
[17] Aharonian, F., Cronin, J., 1994. Phys. Rev. D 50, 1892.
[18] Aharonian, F., Neronov, A., 2005. Astrophys. J. 619, 306.
[19] Aharonian, F., et al., 2004. Astron. & Astroph. 425, L13.
[20] Aharonian, F., et al., 2006. Nature 439, 695.
[21] Ahmed, S., et al., 2004. Phys. Rev. Lett. 92, 181301.
[22] Aloisio, R., et al., 2004. PRD 69, 094023.
[23] Aloisio, R., et al., 2007. Astropart. Phys. 27, 76.
[24] Aloisio, R., et al., 2007. arXiv:0706.2834.
[25] Anchordoqui, L., Goldberg, H., 2004. Phys. Lett. B 583, 213.
[26] Arqueros, F., Hörandel, J., Keilhauer, B. (Eds.), 2008. Proceedings of the 5th Fluorescence Workshop, El Escorial, Spain. Nuclear Instr. and Meth., in press.
[27] Ave, M., et al., 2002. Phys. Rev. D 65, 063007.
[28] Ave, M., et al., 2003. Astropart. Phys. 19, 47.
[29] Ave, M., et al., 2007. Proc. 30th Int. Cosmic Ray Conf., Merida, in press (arXiv:0709.2125).
[30] Ave, M., et al., 2007. Astropart. Phys. 28, 41.
[31] Baltrusaitis, R., et al., 1988. Nucl. Instr. & Meth. A 264, 87.
[32] Becker, J., 2007. arXiv:0710.1557.
[33] Bednarek, W., 2003. Mon. Not. R. Astron. Soc. 345, 847.
[34] Bellido, J., et al., 2001. Astropart. Phys. 15, 167.
[35] Berezinsky, V., 2005. astro-ph/0509069.
[36] Berezinsky, V., 2006. Nucl. Phys. B (Proc. Suppl.) 151, 260.

[37] Berezinsky, V., et al., 1997. Phys. Rev. Lett. 79, 4302.
[38] Berezinsky, V., et al., 2004. Astropart. Phys. 21, 617.
[39] Bergman, D., Belz, J., 2007. J. Phys. G: Nucl. Part. Phys. 34, R359.
[40] Bhattacharjee, P., Sigl, G., 2000. Phys. Rep. 327, 109.
[41] Biermann, P., Strittmatter, P., 1987. Astrophys. J. 322, 643.
[42] Biermann, P., et al., 2004. Astrophys. J. 604, L24.
[43] Bionta, R., et al., 1987. Phys. Rev. Lett. 54.
[44] Bird, D., et al., 1994. Astrophys. J. 424, 491.
[45] Birkel, M., Sarkar, S., 1998. Astropart. Phys. 9, 297.
[46] Blasi, P., et al., 2000. Astrophys. J. 533, L123.
[47] Bossa, M., et al., 2003. J. Phys. G: Nucl. Part. Phys. 29, 1409.
[48] Chiavassa, A., et al., 2005. Proc. 29th Int. Cosmic Ray Conf., Pune 6, 313.
[49] Crocker, R., et al., 2005. Astrophys. J. 622, 892.
[50] Dar, A., Plaga, R., 1999. Astron. & Astroph. 349, 259.
[51] De Marco, D., Stanev, T., 2005. Phys. Rev. D 72, 081301.
[52] Dova, M., et al., 2005. Proc. 29th Int. Cosmic Ray Conf., Pune 7, 275.
[53] Ellis, J., et al., 2006. PRD 74, 115003.
[54] Engel, R., et al., 1999. Proc. 26th Int. Cosmic Ray Conf., Salt Lake City 1, 415.
[55] Engel, R., et al., 2007. Proc. 30th Int. Cosmic Ray Conf., Merida, in press (arXiv:0706.1921).
[56] Falcke, H., et al., 2004. New Astronomy Reviews 48, 1487.
[57] Falcke, H., et al., 2005. Nature 435, 313.
[58] Fichtner, H., 2005. Proc. 29th Int. Cosmic Ray Conf., Pune 10, 377.
[59] Finley, C., Westerhoff, S., 2004. Astropart. Phys., 359.
[60] Fomin, Y., et al., 1991. Proc. 22nd Int. Cosmic Ray Conf., Dublin 2, 85.
[61] Fukuda, Y., et al., 1998. Phys. Lett. B 436, 33.
[62] Fukuda, Y., et al., 1999. Phys. Rev. Lett. 82, 1810.
[63] Fukushima, M., et al., 2007. Proc. 30th Int. Cosmic Ray Conf., Merida, in press.
[64] Gaisser, T., 1997. astro-ph/9707283.
[65] Gaisser, T., 2006. astro-ph/0608553.
[66] Gaisser, T., Stanev, T., 2006. Nucl. Phys. A 777, 98.
[67] Gaisser, T., et al., 2003. Proc. 28th Int. Cosmic Ray Conf., Tsukuba 2, 1117.
[68] Galaverni, M., Sigl, G., 2007. arXiv:0708.1737.
[69] Gelmini, G., et al., 2005. astro-ph/0506128.
[70] Ginzburg, V., Syrovatskii, S., 1964. The origin of cosmic rays. Pergamon Press.
[71] Globus, N., et al., 2007. arXiv:0709.1541.
[72] Glushkov, A., et al., 2003. Proc. 28th Int. Cosmic Ray Conf., Tsukuba 1, 389.
[73] Glushkov, A., et al., 2007 85, 163.
[74] Gorbunov, D., et al., 2002. Astrophys. J. 577, L93.
[75] Gorbunov, D., et al., 2004 80, 145.
[76] Grasso, D., Maccione, L., 2005. Astropart. Phys. 24, 273.
[77] Grasso, D., Rubinstein, H., 2001. Phys. Rep. 348, 163.
[78] Greisen, K., 1966. Phys. Rev. Lett. 16, 748.
[79] Halzen, F., 2005. astro-ph/0506248.
[80] Halzen, F., 2006. astro-ph/0604441.
[81] Halzen, F., et al., 2002. Reports on Progress in Physics 65, 1025.
[82] Harari, D., et al., 2006. J. Cosmol. & Astropart. Phys. 11, 012.
[83] Harari, D., et al., 2007. Proc. 30th Int. Cosmic Ray Conf., Merida, in press (arXiv:0706.1715).
[84] Haungs, A., et al., 2008. Nucl. Phys. B (Proc. Suppl.) 175–176, 354.

[85] Hayashida, N., et al., 1999. Astropart. Phys. 10, 303.
[86] Heber, B., 2005. Int. J. Mod. Phys. A 20, 6621.
[87] Hess, V., 1912. Phys. Zeitschr. 13, 1084.
[88] Hill, C., Schramm, D., 1983. Phys. Lett. B 131, 247.
[89] Hill, C., Schramm, D., 1985. Phys. Rev. D 31, 564.
[90] Hillas, A., 1984. Ann. Rev. of Astron. and Astroph. 22, 425.
[91] Hirata, K., et al., 1987. Phys. Rev. D 58, 1490.
[92] Hörandel, J., 2003. Astropart. Phys. 19, 193.
[93] Hörandel, J., 2003. J. Phys. G: Nucl. Part. Phys. 29, 2439.
[94] Hörandel, J., 2004. Astropart. Phys. 21, 241.
[95] Hörandel, J., 2007. arXiv:0710.4909 (Nuclear Instruments and Methods A in press).
[96] Hörandel, J., 2007. Mod. Phys. Lett. A 22, 1533.
[97] Hörandel, J., 2008. Adv. Space Res. 41, 442.
[98] Huege, T., Falcke, H., 2005. Astropart. Phys. 24, 116.
[99] Kachelriess, M., Semikoz, D., 2006. Astropart. Phys. 26, 10.
[100] Kahler, S., et al., 2005. Proc. 29th Int. Cosmic Ray Conf., Pune 10, 367.
[101] Kalashev, O., et al., 2002. Phys. Rev. D 66, 063004.
[102] Kalges, H., et al., 2007. Proc. 30th Int. Cosmic Ray Conf., Merida, in press.
[103] Kalmykov, N., et al., 1997. Nucl. Phys. B (Proc. Suppl.) 52B, 17.
[104] Kampert, K.-H., 2007. Nucl. Phys. B (Proc. Suppl.) 165, 294.
[105] Kampert, K.-H., 2008. arXiv:0801.1986.
[106] Kestel, M., et al., 2004. Nucl. Instr. & Meth. A 535, 139.
[107] Klinkhammer, F., Risse, M., 2008. Phys. Rev. D 77, 016002.
[108] Knurenko, S., et al., 2001. Proc. 27th Int. Cosmic Ray Conf., Hamburg 1, 177.
[109] Körding, E., et al., 2008. Mon. Not. R. Astron. Soc. 383, 277.
[110] Kronberg, P., 1994. Reports on Progress in Physics, 325.
[111] Kuz'min, V., Rubakov, V., 1998. Phys. Atom. Nucl. 61, 1028.
[112] Lipari, P., 2006. Nucl. Instr. & Meth. A 567, 405.
[113] LOFAR, 2007. www.lofar.org.
[114] Lyutikov, M., Ouyed, R., 1999. Astropart. Phys. 27, 473.
[115] Matthews, J., 2005. Astropart. Phys. 22, 387.
[116] Medina Tanco, G., Watson, A., 2001. Proc. 27th Int. Cosmic Ray Conf., Hamburg 2, 531.
[117] Mollerarch, S., et al., 2007. Proc. 30th Int. Cosmic Ray Conf., Merida, in press (arXiv:0706.1749).
[118] Nagano, M., Watson, A., 2000. Rev. Mod. Phys. 72, 689.
[119] Nagano, M., et al., 1984. J. Phys. G: Nucl. Part. Phys. 10, 1295.
[120] Nagano, M., et al., 1984. J. Phys. G: Nucl. Part. Phys. 18, 423.
[121] National Research Council, 2003. Connecting quarks with the cosmic - Eleven science questions for the new century. National Academy Press, Washington, D.C.
[122] Navarra, G., et al., 2004. Nucl. Instr. & Meth. A 518, 207.
[123] Nitz, D., et al., 2007. Proc. 30th Int. Cosmic Ray Conf., Merida, in press (arXiv:0706.3940).
[124] Olinto, A., 2000. Phys. Rep. 333, 329.
[125] Ong, R., 1998. Phys. Rep. 305, 93.
[126] Ong, R., 2005. Proc. 29th Int. Cosmic Ray Conf., Pune 10, 329.
[127] Ostapchenko, S., 2005. astro-ph/0412591.
[128] Ostrowski, M., 1999. Astron. & Astroph., 134.
[129] Ostrowski, M., 2002. Astropart. Phys. 18, 229.
[130] Privitera, P., et al., 2007. Proc. 30th Int. Cosmic Ray Conf., Merida, in press.

[131] Rachen, J., Biermann, P., 1993. Astron. & Astroph. 272, 161.
[132] Rand, R., Kulkarni, S., 1989. Astrophys. J. 343, 760.
[133] Ranft, J., 1995. Phys. Rev. D 51, 64.
[134] Risse, M., et al., 2005. Phys. Rev. Lett. 95, 171102.
[135] Romero, G., et al., 1996. Astropart. Phys. 5, 279.
[136] Rubtsov, G., et al., 2006. Phys. Rev. D 73, 063009.
[137] Ryan, J., 2005. Proc. 29th Int. Cosmic Ray Conf., Pune 10, 357.
[138] Shinozaki, K., et al., 2002. Astrophys. J. 571, L117.
[139] Spiering, C., 2003. J. Phys. G: Nucl. Part. Phys. 29, 843.
[140] Stanev, T., 1997. Astrophys. J. 479, 290.
[141] Strong, A., et al., 2007. Ann. Rev. Nucl. Part. Sci. 57, 285.
[142] Takeda, M., et al., 1998. Phys. Rev. Lett. 81, 1163.
[143] Takeda, M., et al., 1999. Astrophys. J. 522, 225.
[144] Takeda, M., et al., 2003. Astropart. Phys. 19, 447.
[145] Tinyakov, P., Tkachev, I., 2001 74, 445.
[146] Tinyakov, P., Tkachev, I., 2002. Astropart. Phys. 18, 165.
[147] Unger, M., et al., 2007. Proc. 30th Int. Cosmic Ray Conf., Merida.
[148] Vallée, J., 2004. New Astronomy Reviews 48, 763.
[149] van den Berg, A., et al., 2007. Proc. 30th Int. Cosmic Ray Conf., Merida, in press.
[150] Véron-Cetty, M.-P., Véron, P., 2006. Astron. & Astroph. 455, 773.
[151] Watson, A., 2000. Phys. Rep. 333–334, 309.
[152] Yamamoto, T., et al., 2007. Proc. 30th Int. Cosmic Ray Conf., Merida, in press, arXiv:0707.2638.
[153] Zatsepin, G., Kuz'min, V., 1966. JETP Lett. 4, 78.

# Hydrodynamical Simulations of the Bullet Cluster

Chiara Mastropietro & Andreas Burkert

Universitäts Sternwarte München
Scheinerstr.1, 81679 München, Germany
chiara@usm.uni-muenchen.de, http://www.usm.uni-muenchen.de/people/chiara/

## Abstract

*We present high resolution N-body/SPH simulations of the interacting cluster 1E0657−56. The main and the sub-cluster are modeled using extended cuspy $\Lambda CDM$ dark matter halos and isothermal $\beta$-profiles for the collisional component. We investigate the X-ray morphology and derive the most likely impact parameters, mass ratios and initial relative velocities. We find that the observed displacement between the X-ray peaks and the associated mass distribution, the morphology of the bow shock, the surface brightness and projected temperature profiles across the shock discontinuity can be well reproduced by offset 1:6 encounters where the sub-cluster has initial velocity (in the rest frame of the main cluster) close to 2 times the virial velocity of the main cluster dark matter halo. We find that a relatively high concentration ($c = 6$) of the main cluster dark matter halo is necessary in order to prevent the disruption of the associated X-ray peak. For a selected sub-sample of runs we perform a detailed three dimensional analysis following the past, present and future evolution of the interacting systems.*

## 1 Introduction

The "bullet" cluster 1E0657-56 represents one of the most complex and unusual large-scale structures ever observed. Located at a redshift $z = 0.296$ it has the highest X-ray luminosity and temperature of all known clusters as a result of overheating due to a recent supersonic Mach $M \sim 3$ (Markevitch et al. 2006) central encounter of a sub-cluster (the bullet) with its main cluster. The 500 ks Chandra ACIS-I image of 1E0657-56 (Fig. 1 in Markevitch 2006) shows two plasma concentrations with the bullet sub-cluster on the right of the image being deformed in a classical bow shock on the western side as a result of its motion through the hot gas of the main cluster. The analysis of the shock structure leads to the conclusion that the bullet is now moving away from the main cluster with a velocity of $\sim 4700 \, \mathrm{km \, s^{-1}}$. The line-of-sight velocity difference between the two systems is only $600 \, \mathrm{km \, s^{-1}}$ suggesting that the encounter is seen nearly in the plane of the sky (Barrena et al. 2002). As the core passage must have occurred $\sim 0.15$ Gyr ago we have the unique opportunity to study this interaction in a very special short-lived stage, far away from thermal and dynamical equilibrium. As a result of the encounter, the collision-dominated hot plasma and the collisionless stellar and dark matter components have been separated. The galaxy components of both clusters are clearly offset from the associated X-ray emitting cluster gas (Barrena et al. 2002). In addition, weak and strong lensing maps (Clowe et al. 2006, Bradač et al. 2006) show that the gravitational potential does not trace the distribution of the hot cluster gas that dominates the baryonic mass but follows approximately the galaxy distribution as expected for a collisionless dark matter component. Recent numerical works (Milosavljević et al. 2007, Springel & Farrar 2007) have demonstrated that the relative velocity of the dark matter components associated to the main and the sub-cluster are not necessarily coincident with the bullet velocity inferred from the shock analysis. In details, Milosavljević et al. 2007 using a 2-D Eulerian code well reproduced the observed increase in temperature across the shock front with a dark halo velocity $\sim 16\%$ lower than that of the shock, while Springel & Farrar 2007 found even a larger difference between the shock velocity ($\sim 4500$ km s$^{-1}$ in their best model) and the speed of the halo (only $\sim 2600$ km s$^{-1}$). Moreover, according to Milosavljević et al. 2007, due to a drop in ram-pressure after the cores' interaction the gas component of the sub-cluster can eventually be larger than that of its dark matter counterpart. The simulations of Springel & Farrar 2007 represent the most complete three dimensional numerical modeling of the 1E0657-56 system so far. Nevertheless they focus preferentially on the speed of the bullet but fail in reproducing the observed displacement of the X-ray peaks which represent an important indicator of the nature of the interaction. In particular, they do not obtain any displacement in the X-ray distribution associated with the main cluster suggesting that the baryonic component is suffering too little ram-pressure. Moreover, the concentrations used for the main cluster (and obtained by modeling the lensing data) are much smaller than those suggested by $\Lambda$CDM (Macciò et al. 2007) for halos of similar masses.

The aim of this paper is to investigate the evolution of the bullet cluster in details using high resolution SPH simulations. We quantify the initial conditions that are required in order to better reproduce its observed state and predict its subsequent evolution.

## 2 Models

Both the main and the sub-cluster are two components spherical systems modeled assuming a cuspy dark matter halo and a distribution of hot gas in hydrostatic equilibrium within the global potential of the cluster. The dark halo has a NFW profile and the concentration parameter assumed to be dependent on the halo mass (Macciò et al. 2007). The distribution of hot gas follows an isothermal $\beta$-model of the form:

$$\rho(r) = \rho_0 (1 + (r/r_c)^2)^{-3/2\beta}, \qquad (1)$$

We take the asymptotic slope parameter $\beta = 2/3$ (Jones & Forman 1984) and $r_c = 1/2 r_s$ (Ricker & Sarazin 2001). The adopted gas fraction ranges from a mini-

**Table 1:** Initial conditions of the simulations. For each run dark matter virial mass ($M_{vir}$) and concentration $c$ of the main and the sub-cluster are indicated. $v_i$ and $v$ are the initial velocities of the sub-cluster in the system of reference where the main-cluster is at rest and in the center of mass rest-frame, respectively.

| Run | $M_{vir}$ main [$10^{14} M_\odot$] | $c$ main | $M_{vir}$ bullet [$10^{14} M_\odot$] | $c$ bullet | $v_i$ [km/s] | $v$ [km/s] |
|---|---|---|---|---|---|---|
| 1:6b0 | 7.13 | 6 | 1.14 | 8 | 5000 | 4286 |
| 1:6v3000b0 | 7.13 | 6 | 1.14 | 8 | 3000 | 2571 |
| 1:6 | 7.13 | 6 | 1.14 | 8 | 5000 | 4286 |
| 1:6v3000 | 7.13 | 6 | 1.14 | 8 | 3000 | 2571 |
| 1:6v2000 | 7.13 | 6 | 1.14 | 8 | 2000 | 1714 |
| 1:3 | 7.13 | 6 | 2.4 | 7 | 5000 | 3750 |
| 1:8 | 7.13 | 6 | 0.91 | 8 | 5000 | 4445 |
| 1:6v3000big | 14.2 | 6 | 2.4 | 7 | 3000 | 2571 |
| 1:6c4 | 7.13 | 4 | 1.14 | 8 | 5000 | 4286 |
| 1:6lfg | 7.13 | 6 | 1.14 | 8 | 5000 | 4286 |
| 1:3lfg | 7.13 | 6 | 2.4 | 7 | 5000 | 3750 |
| 1:6c | 7.13 | 6 | 1.14 | 8 | 5000 | 4286 |
| 1:3clfg | 7.13 | 6 | 2.4 | 7 | 5000 | 3750 |

mum value of 12%, comparable with the gas mass fraction provided by X-ray observations of galaxy clusters (McCarthy et al. 2007) to 17%, consistent with the recent WMAP results (Spergel et al. 2007). Assuming a spherically symmetric model, the temperature profile is determined by the condition of hydrostatic equilibrium by the cumulative total mass distribution and the density profile of the gas (see Mastropietro et al. 2005). We assumed a molecular weight $\mu = 0.6$ for a gas of primordial composition, which appears to be a reasonable approximation since the mean temperature of 1E0657-56 is $T \sim 14$ keV according to Markevitch et al. 2006 and cooling is dominated by bremsstrahlung and almost independent of the metallicity.

Masses are assigned to the models according to the weak and strong lensing mass reconstruction of Bradac et al.2006. In particular we assume that the inferred mass enclosed within the field of the HST Advanced Camera for Surveys (ACS) is comparable with the total projected mass of our simulated system (calculated when the two centers of the mass distribution are at a distance similar to the observed one) within the same area. Since the ACS field represents only the central fraction of the area covered by the entire system, this mass constraint is strongly influenced by the concentration of the dark matter halos (Nusser 2007). With a cosmologically motivated choice of $c = 6$ (Macciò et al. 2007) for the main cluster initial halo model, we can reproduce the lensing mass reasonably well adopting a main cluster total mass (within the virial radius) of $\sim 8.34 \times 10^{14} M_\odot$ (Table 1), almost a factor 1.8 smaller than the mass obtained by fitting lensing data with extremely low concentrated ($c < 2$) NFW halos where the inner density profile is much flatter than the one suggested by $\Lambda$CDM simulations. One of the simulations presented in this paper (run 1:6c4) adopts a main halo with a lower ($c = 4$) concentration value. We

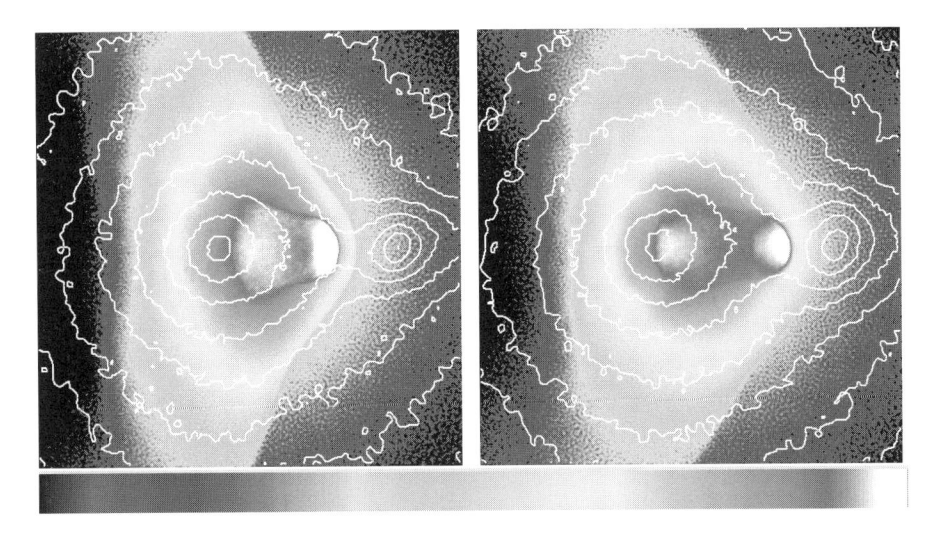

**Figure 1:** 0.8–4 keV surface brightness maps of runs 1:6vb0 (left panel) and 1:6v3000b0 (right). Logarithmic colour scaling is indicated by the key at the bottom of the figure with violet corresponding to $10^{38}$ erg s$^{-1}$ kpc$^{-2}$ and white to $2 \times 10^{41}$ erg s$^{-1}$ kpc$^{-2}$. White contours trace the total surface mass density of the system within $2.3 \times 10^3$ and $2.3 \times 10^8$ M$_\odot$ kpc$^{-2}$. The box size is 1800 kpc.

will see that in this case the X-ray intensity peak associated with the main cluster is easily destroyed during the interaction. We model encounters with mass ratios 1:3, 1:6 and 1:8 between the sub and the main cluster in order to investigate the effects of tidal and ram-pressure stripping, which significantly reduce the mass associated to the sub-cluster and lead to values closer to the 1:10 ratio inferred from lensing observations (Clowe et al. 2006, Bradac et al. 2006).

The main cluster is initially at rest and the sub-cluster moves in the $x$ direction with a velocity which ranges from 2000 to 5000 km s$^{-1}$. The initial conditions of the different runs are summarized in Table 1. The velocities of the sub-cluster relative to the center of mass of the system are listed in the last but one column of the Table. All the simulations were carried out using GASOLINE, a parallel SPH tree-code with multi-stepping (Wadsley, Stadel & Quinn 2004). Most of the runs are adiabatic, with $\gamma = 5/3$. Radiative cooling for a primordial mixture of hydrogen and helium in collisional equilibrium is implemented in 1:6c and 1:3lfgc. The consequences of assuming a lower concentration for the main dark halo are investigated in run 1:6c4, where $c = 4$. The main cluster is modeled with $1.8 \times 10^6$ particles, $10^6$ SPH and $8 \times 10^5$ collisionless. The sub-cluster, with the exception of run 1:8 (where the number of gas particles in the bullet is $4 \times 10^5$), has $9 \times 10^5$ particles, $5 \times 10^5$ collisional and the remainder dark matter particles. The gravitational spline softening is set equal to 5 kpc for the gaseous and dark component.

**Table 2:** Present time. $\Delta$ is the projected (perpendicular to the plane of the encounter) distance between the peaks of the total mass distributions, associated with the two clusters. The third and fourth columns represent the projected offset between each X-ray peak and the associated mass density peak. $v_{gas}$ and $v_{dark}$ are the sub-cluster gas and dark matter velocity calculated in the center of mass system of reference.

| Run | $\Delta$ kpc | offset bullet kpc | offset main kpc | $v_{gas}$ km/s | $v_{dark}$ km/s |
|---|---|---|---|---|---|
| 1:6b0 | 753 | 278 | 188 | 3215 | 4715 |
| 1:6v3000b0 | 741 | 213 | 66 | 3131 | 3134 |
| 1:6 | 742 | 237 | 128 | 3609 | 4756 |
| 1:6v3000 | 729 | 185 | 172 | 2893 | 3137 |
| 1:6v2000 | 721 | 126 | 230 | 2849 | 2425 |
| 1:3 | 784 | 162 | 223 | 3908 | 4076 |
| 1:8 | 737 | 228 | 117 | 3647 | 4858 |
| 1:6v3000big | 725 | 139 | 92 | 3927 | 3528 |
| 1:6c4 | 735 | 151 | 192 | 4168 | 4799 |
| 1:6lfg | 718 | 200 | 189 | 3811 | 4804 |
| 1:3lfg | 779 | 197 | 242 | 3746 | 4145 |
| 1:6c | 736 | 234 | 127 | 3497 | 4806 |
| 1:3clfg | 780 | 228 | 272 | 3595 | 4205 |

## 3 Projected analysis

A first indication about the validity of a model arises from the qualitative comparison of our simulated X-ray surface brightness maps with the X-ray 500 ks *Chandra* ACIS-I images provided by Markevitch 2006.

The impact parameter $b$ is not strictly constrained by observations. Nevertheless a head-on merger, with $b = 0$, seems to be excluded by comparing deep X-ray observations and weak lensing maps. In particular in Fig. 1b of Clowe et al. 2006 the brightest gas associated with the main cluster is not located along the line which connects the centers of the two total mass distributions. Moreover, the X-ray emission from the main cluster is asymmetric, with a peak in the north of the image and an extended tail of less bright material pointing south. These features are hardly associable with a zero impact parameter interaction, as shown in Fig. 1 which illustrate two simulated 1:6 head-on encounters where the sub-cluster moves from the left to the right of the image ($x$ axis of the simulation) with decreasing initial velocities (the left panel corresponds to run 1:6vb0 of Table 1, the right one to 1:6v3000b0). Images are projected along an axis perpendicular to the collision plane (the encounter is seen face-on) and the selected snapshot is the one which most closely matches the observed distance between the centers of the total mass distributions, associated with the two clusters – about 720 kpc from Bradac et al. 2006 – once the sub-cluster has passed through the core of the main system (hereafter the present time). In Fig. 1 colours represent X-ray maps in the *Chandra* energy band (0.8–4 keV) generated using the Theoretical Image Processing System (TIPSY), which produces projected

X-ray surface brightness maps with the appropriate variable SPH kernel applied individually to the flux represented by each particle. The cooling function is computed using a Raymond-Smith code (Raymond & Smith 1977) for a gas of primordial composition. The energy band (0.8–4 keV) is chosen in such a way to reproduce the 500 ks *Chandra* ACIS-I image of the bullet cluster (Markevitch 2006). The entire energy band, used to calculate bolometric X-ray luminosities in the following of the paper, goes from 5 eV to $5 \times 10^4$ keV. Fig. 1 shows that after an encounter with zero impact parameter the displacement of the main cluster's X-ray peak is aligned with the $x$ axis. A large relative velocity (central panel) induces a significant offset between the dark and baryonic component of the main cluster (see Table 2 for details) but it also leads to substantial disruption of the main cluster gaseous core. Moreover, the displacement of the bullet from its dark halo (278 kpc) is much larger than observed. Decreasing the relative velocity between the two clusters (bottom panel) two X-ray peaks are clearly visible but the displacement of the main cluster gas is now negligible due to the lower ram-pressure experienced by the main cluster core. Springel & Farrar 2007 (Fig. 7 in their article) provide further examples of head-on encounters with even lower mass ratios and relative velocities ($v = 2600 \, \text{km s}^{-1}$ in the center of mass rest frame). Even assuming extremely low concentrations ($c = 2$) for the main halo, the authors never reproduce the displacement observed in the two systems. Increasing the concentration strongly increases the luminosity of the main cluster, which appears much brighter than the bullet, contrary to what is observed. A bow shock is clearly visible on the right of each image. The shape of the shock front is only marginally dependent on the kinematics of the model while the distance between the edge of the bullet (the so called contact discontinuity) and the shock front becomes larger for decreasing bullet velocities. The contact discontinuity itself is much flatter in the case of 1:6b0 than in the low velocity encounter 1:6v3000b0 and clearly not comparable with observations, which show a more narrow structure.

The rest of the runs listed in Table 1 have an impact parameter $b$ equal to 150 kpc, comparable with the core radius $r_c$ of the main cluster gas distribution for most of the models. In Fig. 2 and 3 we illustrate the projected X-ray surface brightness maps of some interesting models with impact parameter $b = 150$ kpc. The encounters are shown face-on. The box size and the surface mass density contours are the same as in Fig. 1. In order to underline the morphological details of the high emission regions the upper limit of the surface brightness scale varies in the different images. Individual values are indicated in the captions. Among this sub-sample of runs, 1:3 produces the largest displacement of the X-ray peak associated with the main cluster, but the X-ray map differs from the observations. In particular a large strip of strongly emitting gas still connects the two X-ray peaks while the morphology of the main cluster peak is much more elongated than observed. Decreasing the mass ratio between the two interacting systems (runs 1:6 and 1:8) the displacement in the bullet becomes larger than that in the main cluster. The run 1:6 is characterized by an initial sub-cluster velocity of 5000 km s$^{-1}$ (as well as runs 1:3 and 1:8) in the system of reference where the main cluster is at rest, that corresponds to a present time velocity of $\sim 4300$ km s$^{-1}$ in the center of mass rest frame. The same model is simulated assuming lower relative velocities (1:6v3000 and 1:6v2000). With decreasing velocities the offset in the sub-cluster becomes smaller and the X-ray emission from

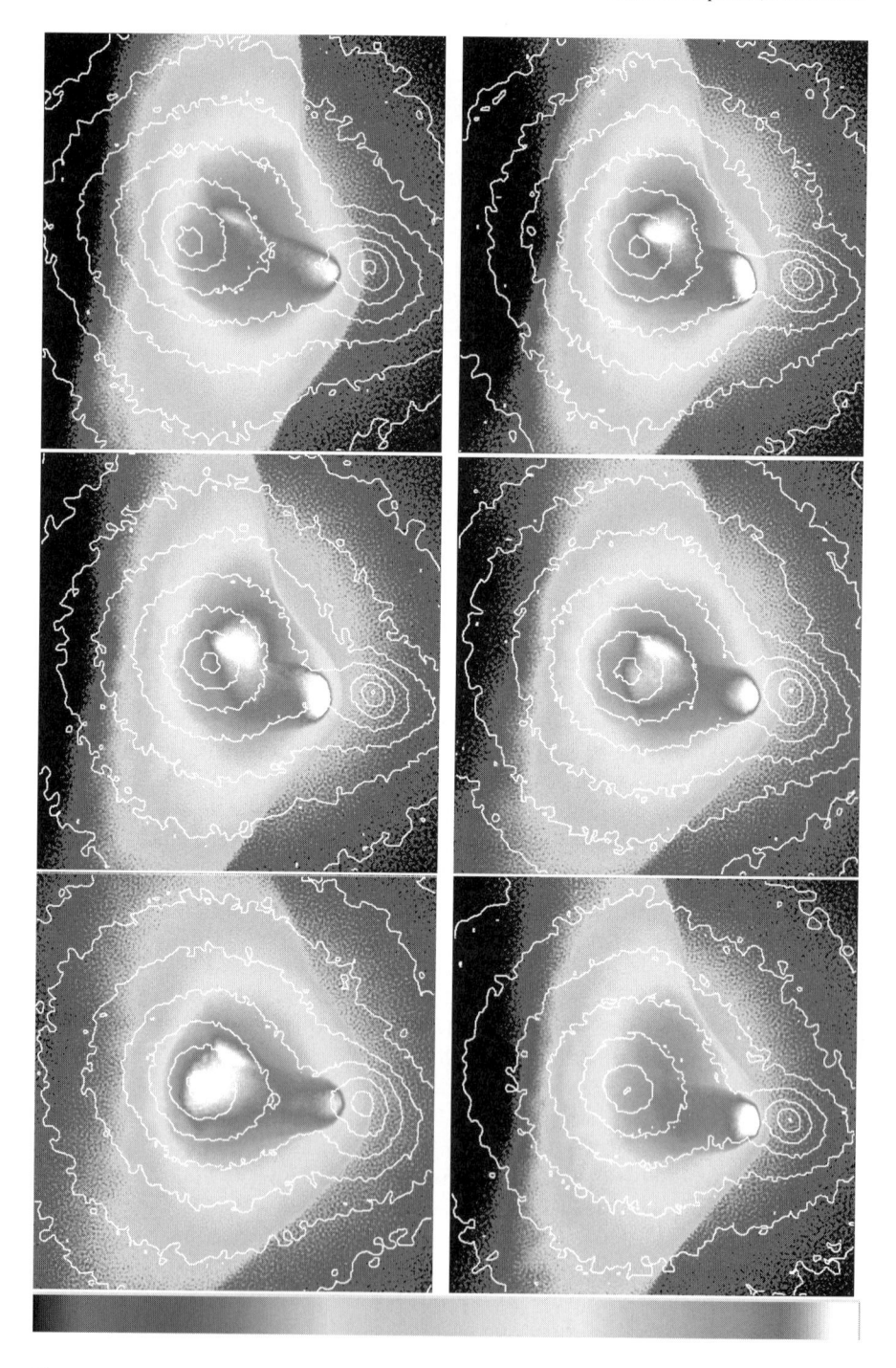

**Figure 2:** 0.8–4 keV surface brightness maps of runs (from the top left to the bottom right) 1:3, 1:8, 1:6, 1:6v3000, 1:6v2000 and 1:6c4. Logarithmic colour scaling is indicated by the key at the bottom of the figure, with violet corresponding to $10^{38}$ erg s$^{-1}$ kpc$^{-2}$ and white to $1.8 \times 10^{41}$ erg s$^{-1}$ kpc$^{-2}$ in runs 1:8 and 1:6c4, to $2.34 \times 10^{41}$ erg s$^{-1}$ kpc$^{-2}$ in run 1:3 and to $2 \times 10^{41}$ erg s$^{-1}$ kpc$^{-2}$ in the remaining cases. Projected isodensity contours of the total mass distribution are shown. Limits are $2.3 \times 10^3$ and $2.3 \times 10^9$ M$_\odot$ kpc$^{-2}$. Each box size is 1800 kpc.

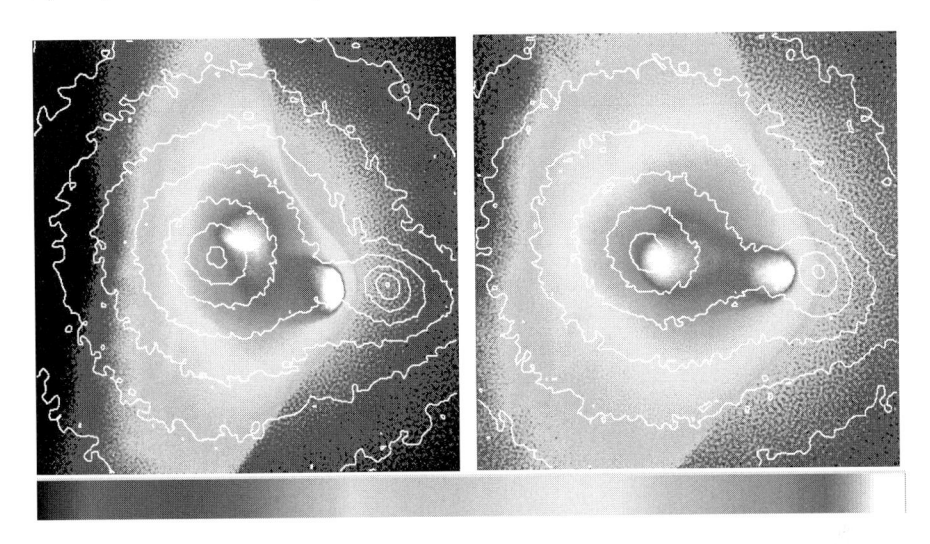

**Figure 3:** Same as in Fig. 2 for the cooling run 1:6c (the X-ray upper limit is $1.8 \times 10^{41}$ erg s$^{-1}$kpc$^{-2}$) and run 1:6v3000big ($3.9 \times 10^{41}$ erg s$^{-1}$kpc$^{-2}$).

the bullet less pronounced with respect to the bright X-ray emitting region at the center of the main cluster. At the same time the shape of the contact discontinuity changes, getting progressively more narrow while the distance between the contact discontinuity and the shock front increases as will be shown more quantitatively in the next section. The displacement associated with the main-cluster is determined by the distance of closest approach between the centers of the two clusters, which becomes smaller – assuming the same initial impact parameter $b$ – with decreasing bullet velocities. Indeed, the separation between the main cluster X-ray emission peak and its dark matter counterpart is maximum in the case of the low velocity run 1:6v2000. A not negligible fraction of the X-ray emission visible at the present time near the center of the main-cluster is actually associated with hot gas stripped from the external regions of the bullet. Figure 4 refers to run 1:6v3000. It shows the individual distribution of gas originating from the main (top) and sub-cluster (bottom) and lying at the present time within 1 Mpc from the center of the system. Comparing these images with the middle right panel of Fig. 2 it appears evident that the bright elongated X-ray feature crossing the second innermost isodensity contour is associated with the displaced gaseous center of the main cluster while the surrounding more diffuse region hosts a significant amount of sub-cluster gas. Indeed, as it shown in the upper panel of Fig. 4, the motion of the bullet across the inner regions of the main system creates a low density "tunnel" in the main cluster gas distribution. At the same time the sub-cluster looses a large amount of baryonic material during the phase of core-core interaction. This material, which fills the tunnel, falls back into the gravitational center of the main cluster and resides now at more than 500 kpc distance from the X-ray bullet. As we will show in the next section, the amount of gas deposited by the sub-cluster in the central regions of the main system increases with decreasing relative velocities. This trend explains the relative increase in luminosity

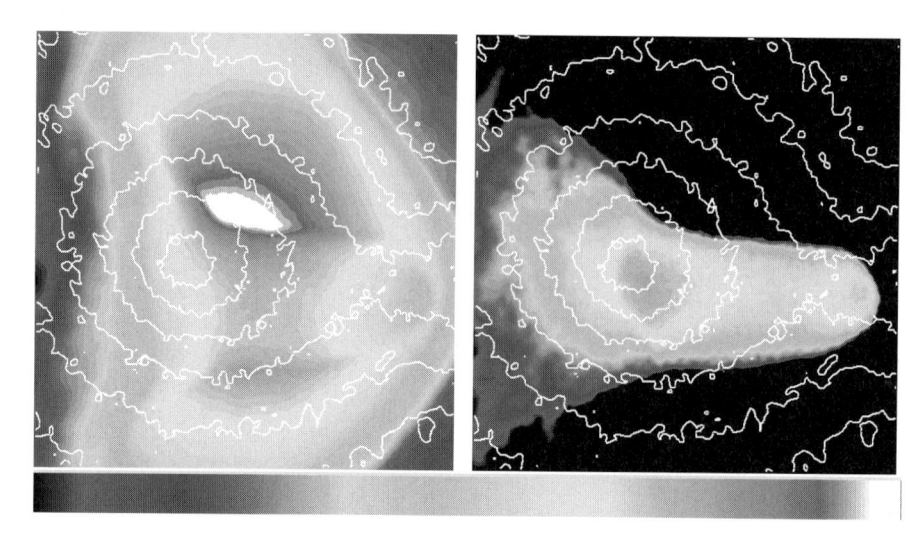

**Figure 4:** Run 1:6v3000. Gas originating from the main (top panel) and the sub-cluster (bottom) is projected individually along the $z$-axis perpendicular to the plane of the encounter. Violet corresponds to a surface density of $2.3 \times 10^5 M_\odot$ kpc$^{-2}$ and white to $2.3 \times 10^8 M_\odot$ kpc$^{-2}$. Projected isodensity contours of the total mass distribution are drawn on top of the image. The box size is 1 Mpc.

of the diffuse strongly emitting component if compared to the peak associated to the main cluster core gas (always in the upper right region with respect to the center of the mass density distribution), when we compare run 1:6 with the low velocity run 1:6v2000 where it becomes the primary peak of X-ray emission. As shown in the bottom right panel of Fig. 2 a main halo with low concentration ($c = 4$) does not survive a 1:6 sub-cluster encounter with velocity $v = 5000$ km s$^{-1}$ and its X-ray peak is destroyed. The choice of a lower gas fraction ($f_g = 0.12$ in three of the last four runs of Table 1) does not affect significantly the X-ray map morphology although the displacement of the two luminosity peaks with respect to their dark matter counterparts slightly changes.

Finally, we tested the consequences caused by including radiative cooling and choosing a larger main halo model (Fig. 3). Cooling makes the contact discontinuity more narrow and the amount of diffuse X-ray gas around the peak associated with the main cluster smaller. The offset in the main and sub-cluster remains unaltered. Run 1:6v3000big is characterized by a main cluster total mass of $\sim 1.64 \times 10^{15} M_\odot$, which is closer to the value adopted by Springel & Farrar 2007 and predicted by fitting the large field weak lensing data with extremely low concentrated ($c \sim 2$) NFW halos. Although the initial relative velocity is only 3000 km s$^{-1}$, due to the large mass of the host halo the present time velocity of the bullet in the center of mass rest frame is much higher than in the corresponding 1:6v3000 run. Consequently the offset of the X-ray peak associated with the main cluster is less than 100 kpc and the amount of gas lost by the bullet in the core of the main system closer to that observed in 1:6 than in 1:6v3000.

Summarizing, run 1:6v3000, with mass ratio 1:6, initial relative velocity $v = 3000 \text{ km s}^{-1}$ and present time sub-cluster velocity $v \sim 3100 \text{ km s}^{-1}$ in the center of mass rest frame, best reproduces the main features observed in X-ray maps, in particular the peculiar morphology of the X-ray emission associated with the main cluster, the relative surface brightness between the main and the sub-cluster, the shape of the shock front and of the contact discontinuity. Although the low velocity run 1:6v2000 leads to a X-ray displacement closer to that observed by Clowe et al. 2006 both in the main and in the sub-cluster, this model seems to be excluded on the basis of a pure morphological comparison with the observational data. Indeed, the bullet seems to be much less bright than the center of the main cluster.

In order to calculate projected temperatures we need to define a weighting function. We used the spectroscopic-like temperature $T_{sl}$ defined by Mazzotta et al. 2004, which gives a good approximation (within few percent) of the spectroscopic temperature obtained from data analysis of Chandra when applied to clusters hotter than 2–3 keV.

Figure 5 illustrates the projected spectroscopic-like temperature profiles across the shock for the different runs of Table 1. All the values refer to the present time. In particular, the upper left panel shows the temperature jumps associated with different relative velocities of the two clusters. Decreasing the initial relative velocity from 5000 to 2000 km s$^{-1}$ (in the system of reference where the main cluster is at rest) reduces the temperature peak by $\sim$ 7 keV while the peak itself becomes broader since the thickness of the shock increases by almost a factor of two due to the lower pressure exercised by the pre-shock gas. Both the 1:6v3000 and 1:6v2000 models seem to fit quite nicely the observed height ($\sim$ 27–30 keV) and thickness (150–200 kpc) of the shock front (Markevitch 2006) while the 1:6 run produces a peak which is too narrow ($\sim$ 100 kpc) and pronounced ($\sim$ 35 keV). The upper right panel refers to different mass ratios. If the encounter is characterized by the same initial relative velocity and gas fraction the strongest shock is associated with the most massive sub-cluster. A 1:3 adiabatic encounter produces a $\sim$ 45 keV temperature peak with thickness $\sim$ 150 kpc while the 1:8 temperature profile is not substantially different from the one of the 1:6 run. In the same plot we also show the shock created in the massive run 1:6v3000big. Clearly the maximum temperature is much higher than the observed one. Including radiative cooling (bottom left panel) has the effect of reducing the peak in temperature but does not influence the thickness of the shock region. The adoption of a simple cooling model is actually questionable. Indeed, although the estimate of the temperature jump is reasonable, the entire temperature profile drops by 5 keV due to the fact that once cooling is activated the main cluster gas component becomes thermally unstable in the early phases of the interaction and overcools in the central regions. If cooling is important, models with higher mass ratios and relative velocities (like the 1:3lfg run in plot) have still to be taken in account and can not be excluded as the high temperature peaks could actually cool significantly. The choice of a different line of sight (bottom right plot) does not affect significantly the temperature profile along the shock. Even decreasing the baryonic fraction in the clusters and assuming a much less concentrated main halo, the height and thickness of the temperature peak do not change.

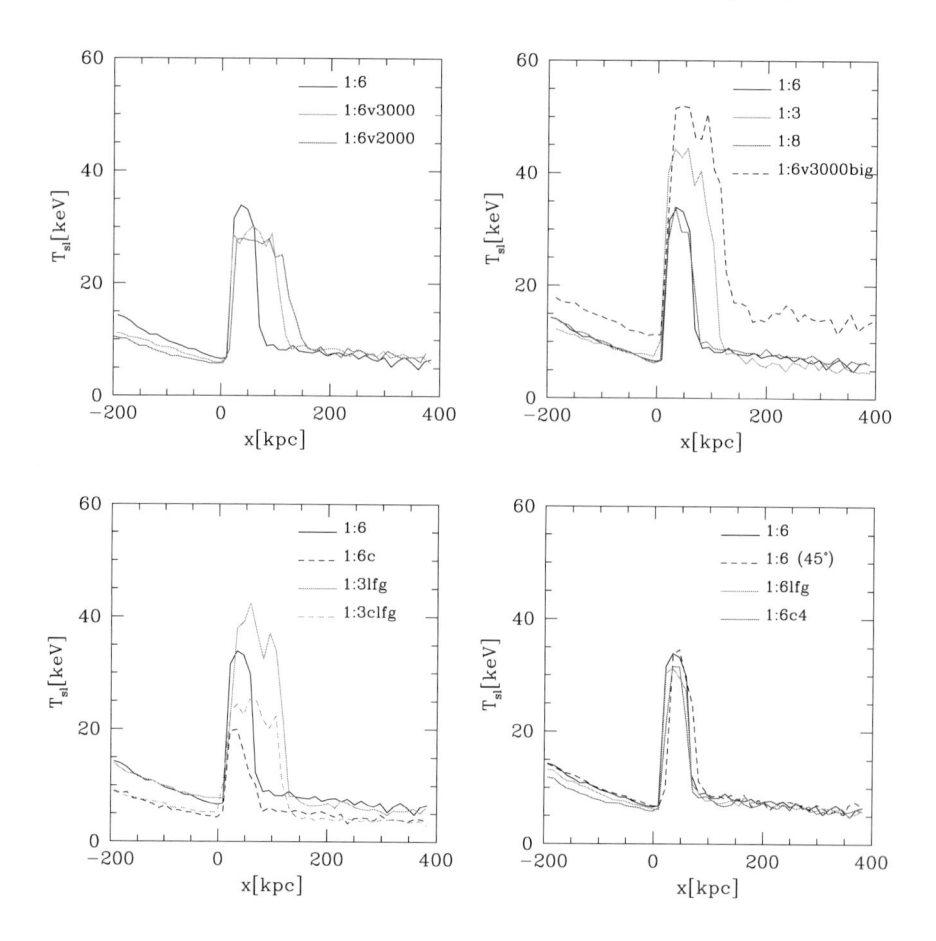

**Figure 5:** Spectroscopic-like temperature profiles measured in a narrow slit (20 kpc) across the shock. The bullet is located at $x = 0$ kpc with $x$ increasing towards the pre-shock region. Upper left panel: 1:6 runs with different relative velocities are compared. Upper right: different mass ratios. Bottom left: comparison between adiabatic and cooling 1:6 and 1:3 encounters. Bottom right: effects of inclination, lower gas fraction and lower concentration of the main halo.

In Fig. 6 we compare the spectroscopic-like temperature profile across the shock region with the emission weighted one for our favourite run 1:6v3000. While the projected temperature profile calculated according to the two definitions is similar ($T_{ew}$ is only slightly higher than $T_{sl}$) in the pre and post-shock regions, the emission weighted temperature $T_{ew}$ in the region $0 \leq x \leq 150$ kpc is $\sim 20\%$ higher than $T_{sl}$. $T_{shock}$ represents the actual temperature along the $x$-axis through the shock. Indeed the blue curve in Fig. 6 gives the exact temperature jump across the shock (the "true temperature" of the shock) which is characterized by an even higher peak with respect to the projected ones. This deprojected temperature profile is actually the one

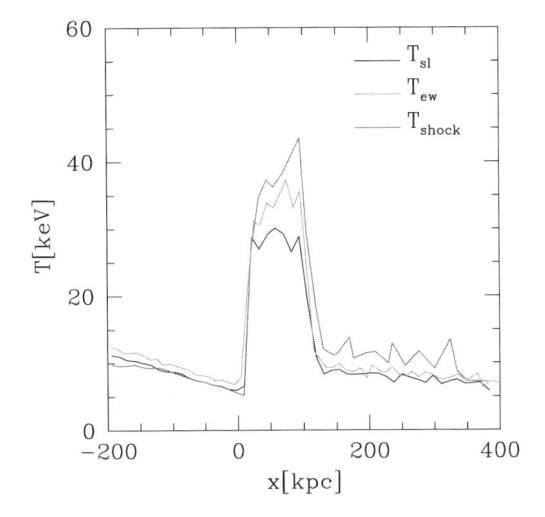

**Figure 6:** Run 1:6v3000: spectroscopic-like ($T_{sl}$), emission weighted ($T_{ew}$) and true ($T_{shock}$) temperature profiles across the shock.

shown by Springel & Farrar 2007 in their Fig. 9 where they compare their model with observations which on the other hand refer to projected quantities (Markevitch 2006). As the calculated local temperatures of Springel & Farrar2007 fit the projected observed temperature profile very well we conclude that their projected temperature profiles are actually inconsistent with the observations by Markevitch 2006.

In Fig. 7 and 8 we show for our selected adiabatic run 1:6v3000 the evolution with time of the 0.8–4 keV X-ray surface brightness and spectroscopic-like temperature during the central phases of the interaction. The quantities are projected along an axis perpendicular to the plane of the collision. On top of the color maps we draw in black the projected isodensity contours of the total mass distribution, which is dominated by the dark matter component. Time increases from the left to the right and from the top to the bottom. The sequence of six panels covers an interval in time of 400 Myr, between 1.3 and 1.7 Gyr from the beginning of the simulation. The snapshot corresponding to the present time is the one on the bottom left. The bullet approaches the main cluster from the left, with an initial velocity of 3000 km s$^{-1}$ in the main cluster rest-frame. The shock front has an arc-like shape and becomes progressively more asymmetric as the bullet moves closer to the core of the main cluster. When the centers of the two clusters are less than 250 kpc separated, ram-pressure becomes effective in producing a displacement (visible in the projected density maps as well as in surface brightness) between the gaseous core of the bullet and the peak of its associated mass distribution. The offset in the main cluster is evident only when the bow shock passes through its core. In the central panels of Fig. 7 the X-ray luminosity saturates in order to distinguish features in the maps at later times. Nevertheless, it is visible how the main cluster core gets compressed and displaced

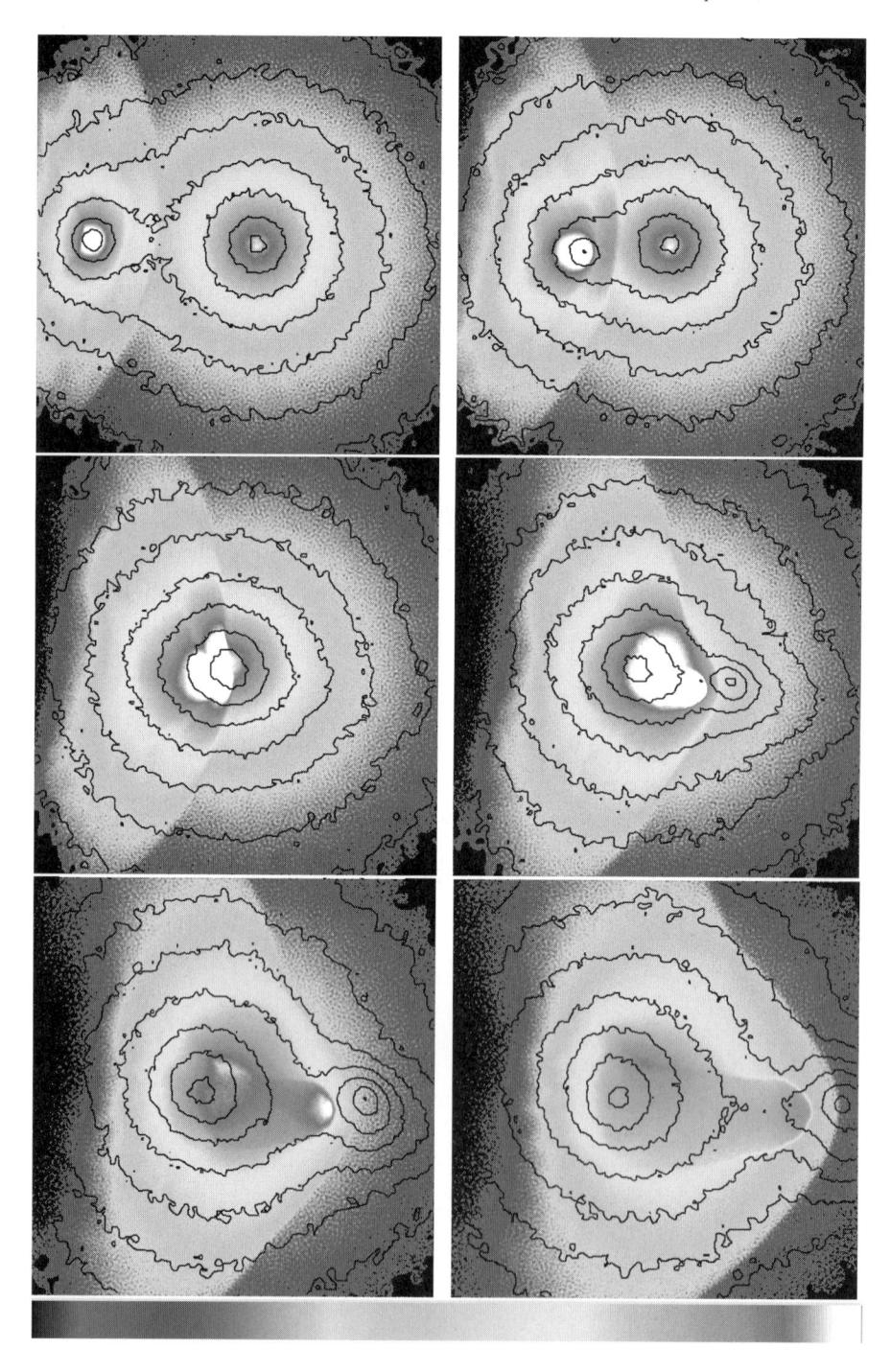

**Figure 7:** Time evolution of run 1:6v3000. X-ray (0.8–4 keV) surface brightness maps. Logarithmic colour scaling is indicated by the key to the bottom of the figure, with violet corresponding to $10^{38}$ ergs$^{-1}$ kpc$^{-2}$ and white to $2.5 \times 10^{41}$ erg s$^{-1}$kpc$^{-2}$. Projected isodensity contours of the total mass distribution are drawn on top of temperature maps.

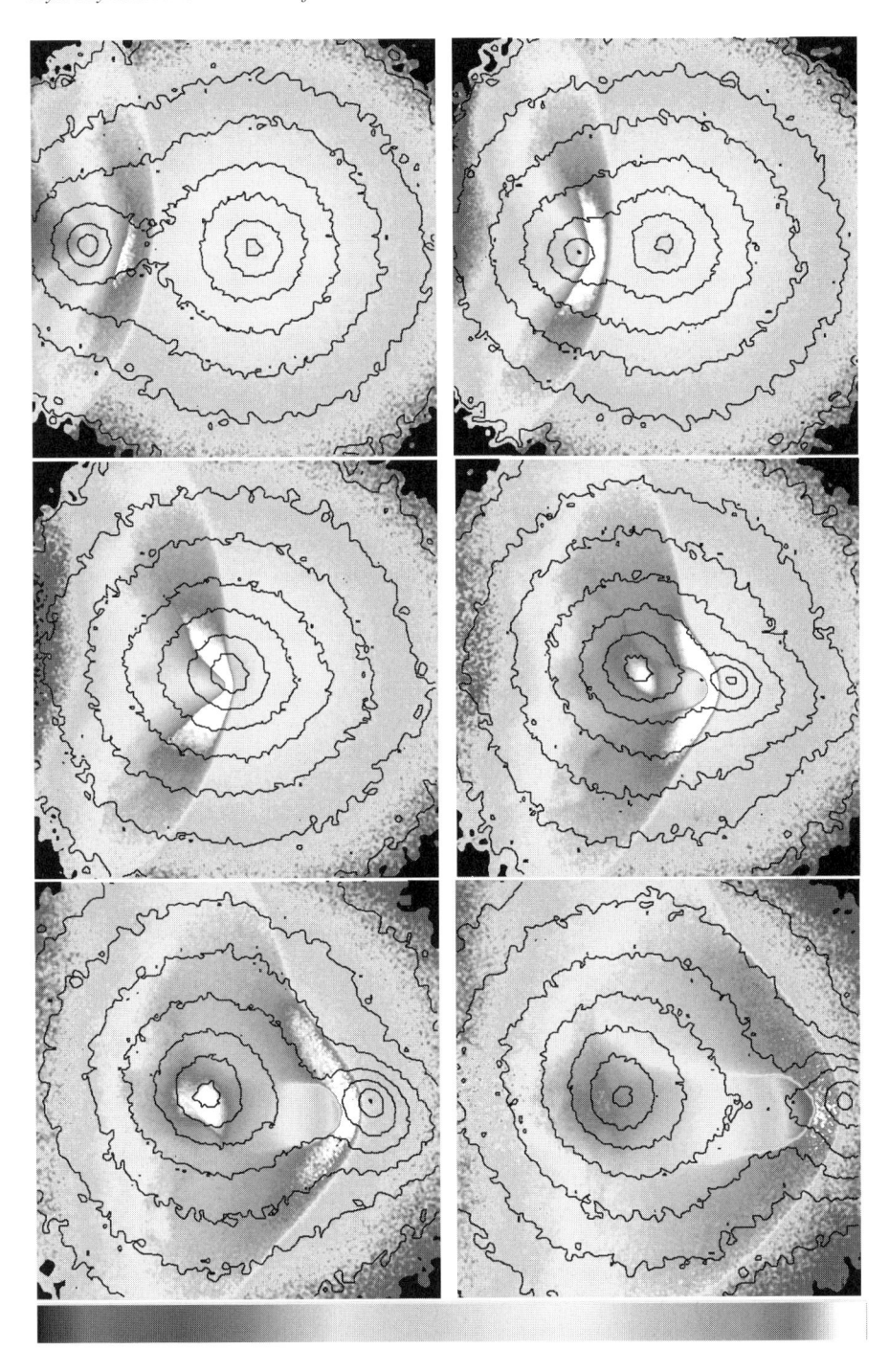

**Figure 8:** Time evolution of run 1:6v3000. Spectroscopic-like X-ray temperature maps. Logarithmic colour scaling is indicated by the key to the bottom of the figure, with violet corresponding to 0.9 keV and white to 86 keV. Projected isodensity contours of the total mass distribution are drawn on top of temperature maps.

from the center of the potential towards the top right of the image and appears in the X-ray maps (last two panels) as an elongated structure, characterized by high surface brightness.

Temperature maps better describe the evolution of the shock region, which gets compressed and hotter during the core passage while in a later phase it cools down and becomes thicker due to lower pressure of the pre-shock gas. The bullet itself expands as it leaves the central regions of the main cluster. Despite of its strong X-ray emission (it is the brightest feature in the post core-core interaction X-ray maps) the bullet remains relatively cold. Even if radiative cooling is not activated, the core of the sub-cluster is heated only to a maximum temperature of $10^8$ K, while the shock front is much hotter ($5 \times 10^8$ K). The projected temperature associated with the bullet is higher when it passes through the core of the main system (central panels of Fig. 8) due to the line of sight overlap with the hot gas from the main cluster and the shock heated material stripped from the bullet itself which surrounds it. As soon as the sub-cluster moves out into the cool external regions of the main cluster and loses part of the hot envelope of stripped gas, its projected temperature decreases to values comparable to the observed ones. Another peculiar feature in the temperature maps is the high temperature region next to the innermost total density contour of the main cluster and visible in the middle right and bottom left panels of Fig. 8. This area could be associated with the southeastern high temperature region observed by Markevitch et al. 2002 (regions $f$, $i$ and $l$ in their Fig. 2) in the main cluster X-ray map and assumed to be coincident with the main merger site. In our simulations the high temperature region is filled with hot gas stripped from the external regions of the sub-cluster and deposited within the core of the main system (compare with Fig. 4). This high temperature material, combined with the main cluster gas which lies in the same projected region, produces the diffuse X-ray high emission feature visible at the present time in the main cluster below the primary peak (bottom left panel of Fig. 7 or middle right panel of Fig. 2 for a better colour contrast). The eastern side of the high temperature region is less bright in X-ray emission and lies beyond the luminosity peak associated with the main cluster. As the sub-cluster moves to larger radii, the high temperature region expands and cools. Shortly ($\sim 70$ Myr) after the present time this gas is not anymore clearly distinguishable in the X-ray maps.

## 4  3D analysis

In this Section we will focus on a sample of runs from Table 1 – namely 1:6, 1:6v3000, 1:6v2000 and 1:3clfg – and investigate in details their three-dimensional characteristics. Three of these simulations are adiabatic models characterized by different initial relative velocities and will permit us to study the effects of the sub-cluster speed on the present and future state of the encounter. The remaining run 1:3clfg is the one which better reproduces – together with 1:6v3000 and 1:6v2000 – the observed jump in temperature across the shock front.

The Mach number $\mathcal{M}$ in our simulations is determined from the temperature jump – which shows a better defined discontinuity compared with the density jump – using the Rankine-Hugoniot conditions. We find values in good agreement with observations (Markevitch 2006) for 1:6v3000 and the cooling run 1:3clfg (both with $\mathcal{M} \sim 3$). 1:6v2000 is characterized by a slightly lower value of $\mathcal{M}$ ($\sim 2.9$) while the high velocity run 1:6 has a stronger shock, with $\mathcal{M} = 3.2$.

As seen in the previous Section, the pre-shock temperature ($\sim$11–12 keV for the adiabatic runs) is slightly higher than the one obtained by projection along the line of sight (Fig. 6). Assuming $T = 11$ keV and $\mathcal{M} = 3$ we predict a pre-shock sound speed $c_S = 1700 \, \text{km s}^{-1}$ and a shock velocity $v_S = \mathcal{M} c_S \sim 5100 \, \text{km s}^{-1}$. This value reduces to the observed shock velocity $v_S \sim 4700 \, \text{km s}^{-1}$ if we use the projected average pre-shock temperature ($T \sim 9$ keV) adopted by Markevitch 2006.

As previously noticed by other authors (Springel & Farrar 2007) the velocity jump shown in the last panel of each plot is much smaller than the theoretically inferred shock velocity. Actually the pre-shock gas is not at rest but shows a negative velocity along the $x$-axis which however can only be partially explained by the fact that the center of mass of the system is moving in the positive direction of the $x$-axis following the bullet. Indeed the upstream velocity maintains a negative sign even with respect to the rest frame of the parent cluster, indicating a pre-shock infall towards the bullet. This effect is explained in Springel & Farrar 2007 by studying the dynamical evolution of the system's global potential, which becomes deeper after the core-core interaction and induces infall of material from the region ahead of the shock. Nevertheless the infall velocity which characterizes our models is a significantly smaller than the sub-cluster velocity in contrast to the values found by Springel & Farrar 2007 and ranges only between 500 and 900 km s$^{-1}$ in the different runs.

Figures 9 and 10 illustrate the characteristic velocities of the sub-cluster in the orbital plane for the three runs 1:6v3000, 1:6 and 1:6v2000. All the velocities are calculated in the center of mass rest frame and the time corresponding to the present position is indicated by a vertical line. The velocity of the dark matter component peaks at the moment of closest approach between the two cores and then decreases faster than for a ballistic orbit as a result of dynamical friction. The escape velocity at a given sub-cluster position is calculated assuming a spherical unperturbed host potential and is indicated by a black dotted curve. All the runs have initially unbound sub-clusters. Due to the effects of dynamical friction after the phase of core-core interaction the 1:6v2000 sub-cluster is actually bound to the main system, while sub-clusters with initial velocities $v = 3000$ km s$^{-1}$ and 5000 km s$^{-1}$ have velocities slightly or much larger than the escape velocity from the main cluster. The gaseous bullet initially follows its dark matter counterpart but before the point of closest approach it is slowed down by ram-pressure. At the same time the morphology of the sub-cluster gas distribution changes. The contact discontinuity assumes an arc-like shape (partially reducing the effect of ram-pressure) and the bow shock forms. It is interesting to notice that after the point of closest approach the evolution of the relative velocity between gas and dark matter in the sub-cluster strongly depends on the intensity of the ram-pressure force. In particular for relatively low ram-pressure values (1:6v2000 and 1:6v3000) the gaseous bullet is accelerated towards its dark matter counterpart as soon as it leaves the core of the host cluster where it experienced the largest external densities and ram-pressure. As a result the gravitational acceleration relative velocity between the gaseous and dark component

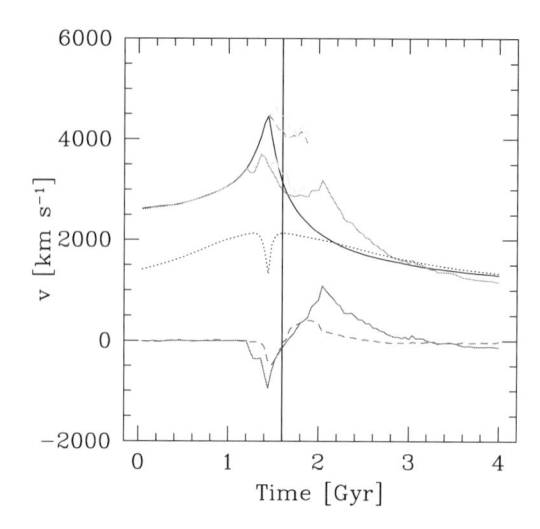

**Figure 9:** Run 1:6v3000. Characteristic bullet velocities plotted as a function of time. The black and red solid curves represent the velocity of the dark matter and gaseous component of the sub-cluster in the plane of sky, while the black dotted line indicates the escape velocity from the main cluster. The blue curves show the relative velocity of the two components of the sub-cluster in the two directions perpendicular to the line of sight. The red solid and dashed curves show the velocity of the edge of the bullet in the center of mass and pre-shock gas rest-frame, respectively. Finally, the green curves represent the shock velocity obtained by differentiating the shock position. The vertical line indicates the present time. More details are given in the text.

of the sub-cluster (whose two components in the orbital plane are represented for 1:6v3000 by the blue solid and dashed curves in Fig. 9) is larger than zero. At the time corresponding to the present position the two velocities look comparable in the case of 1:6v3000 while in 1:6v2000, where the ram-pressure acting on the bullet is lower, the acceleration starts earlier and the gaseous bullet is already $\sim 500\,\mathrm{km\,s^{-1}}$ faster than its dark counterpart. For larger impact velocities – as in the case of 1:6 – ram-pressure is effective in slowing down the gaseous bullet even at large distances from the center of the main-cluster. The velocity of the gaseous bullet is therefore smaller than that of its dark counterpart until when the sub-cluster is well outside of the virial radius of the main system.

The green solid curve in each plot represents the velocity of the front shock obtained by differentiating the positions of the shock front at increasing times. The shock velocity rapidly increases after the core-core interaction and at the present time it is always larger than the velocity of the gaseous bullet. In order to calculate the bullet and shock velocities in the system of reference of the pre-shock gas (red and green dashed curve, respectively) the infall velocity of the upstream gas is calculated at different times before and after the present one.

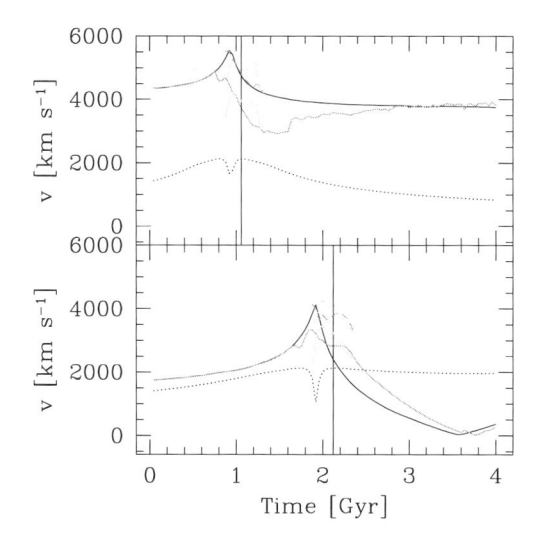

**Figure 10:** Run 1:6 (top panel) and 1:6v2000 (bottom panel). Same as in Fig. 9.

The shock velocity $v_S$ at the present position is 4100, 4500 and 5100 km s$^{-1}$ in the case of 1:6v2000, 1:6v3000 and 1:6, respectively, consistently with the shock velocity $v_S = Mc_S$ inferred from the Rankine-Hugoniot jump conditions and with the value provided by observations. Only 1:6v2000 shows a two-dimensional shock velocity well below the observational uncertainties. As found by Springel & Farrar 2007, after the point of closest interaction the shock velocity is always larger than the velocity of the sub-cluster mass centroid, but the amount of the difference strongly depends on the model. In particular, in the case of the two low velocity runs 1:6v2000 and 1:6v3000 $v_S$ is $\sim 65\%$ and $\sim 40\%$ larger than the velocity of the dark matter component, while in 1:6 the difference is almost negligible (only 6%).

The cluster collision produces a drastic increase in luminosity and temperature. Figures 11 and 12 show the bolometric X-ray luminosity $L_{Xbol}$ and average spectroscopic-like temperature $T_{sl}$ as functions of time. Both quantities are calculated for the entire simulated box and scaled to their initial values. The first phase of the interaction – which involves only the external low density regions of the two clusters – is characterized by an identical slow increment of luminosity and temperature in all the adiabatic runs. The jump is associated with the phase of core-core interaction and peaks right after the time of closest approach. The present time is indicated with an empty squared and sits on the downturning curve. The high velocity run 1:6 is associated with the largest increase in temperature ($T_{spec}/T_{spec}(0) \sim 3.5$) and with the smallest jump in luminosity ($L_{bol}/L_{bol}(0) \sim 4$). For decreasing sub-cluster velocities the peak in temperature becomes slightly less pronounced while the luminosity jump rises by a factor of 1.5.

The loss of baryonic material from the sub-cluster within the high density core of the main system is indeed larger for low velocity encounters and leads to higher

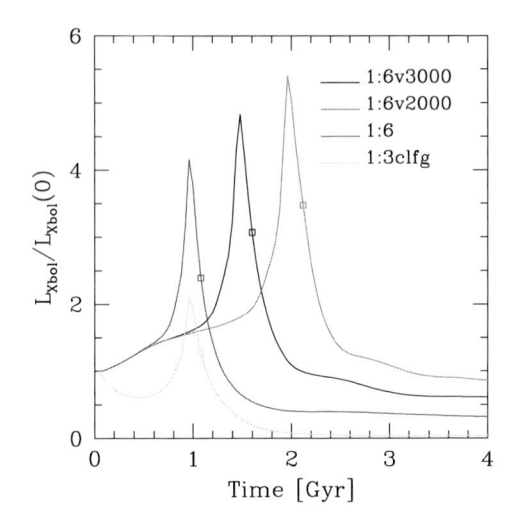

**Figure 11:** Bolometric X-ray luminosity $L_{bol}$ as function of time for four selected runs. Luminosity is scaled to its initial value and calculated for the entire simulated volume.

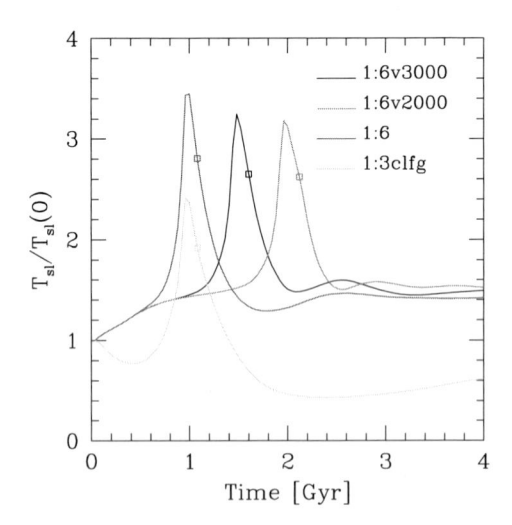

**Figure 12:** Average spectroscopic-like temperature $T_{sl}$ as function of time for four selected runs. Temperature is scaled to its initial value and averaged for the entire simulated volume.

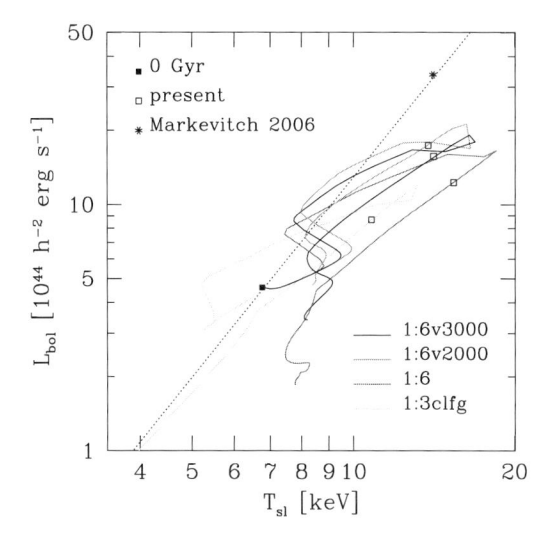

**Figure 13:** Evolution of the interacting system along the $L_X - T$ diagram. Here $L_{Xbol}$ and $T_{sl}$ are the bolometric X-ray luminosity and the spectroscopic-like temperature within the virial radius of the main system. The dotted line represents the $L_X - T$ relation by Markevitch 1998 for local clusters. The initial and present time of each simulation, as well as the observed position of the bullet cluster on the $L_X - T$ relation (Markevitch 2006) are indicated.

luminosities even if the increment in temperature is smaller with respect to the high velocity runs. Excluding the bound run 1:6v2000, the amount of gas stripped from the bullet and lying within the virial radius of the main cluster is $1.6 \times 10^{13} M_\odot$ in 1:6 and $1.8 \times 10^{13} M_\odot$ in 1:6v3000, which in both cases corresponds to almost $\sim 60\%$ of the initial baryonic content of the sub-cluster. The difference becomes more pronounced if we consider only the core ($r < r_s$) of the main cluster, where the mass of sub-cluster gas is $1.4 \times 10^{12} M_\odot$ in the 1:6v3000 run and one order of magnitude less in the case of 1:6.

At 4 Gyrs after the beginning of the simulation the total luminosity of the system is similar to the initial one in the case of the bound system 1:6v2000 where the center of mass of the bullet does not move out to distances beyond the virial radius of the main system. The luminosity drops to $50\%$ or even less of the initial luminosity for the runs 1:6v3000 and 1:6 respectively. This decrease in luminosity is motivated by the fact that at the final stages of the simulations a significant fraction of the gas is unbound and very extended. Due to its low density it does not contribute to the luminosity of the system despite the high temperature. In particular, for the same intracluster distance, the high velocity encounter 1:6 is associated to the highest fraction of unbound material, as will be shown later in this Section. On the other hand, the final temperature of the system is higher than the one associated to the initially isolated clusters and converges to a value $T_{sl}/T_{sl}(0) \sim 1.5$ almost independently of the sub-cluster velocity. The cooling run 1:3clfg shows a somehow different behaviour: both, luminosity and temperature profiles have an initial

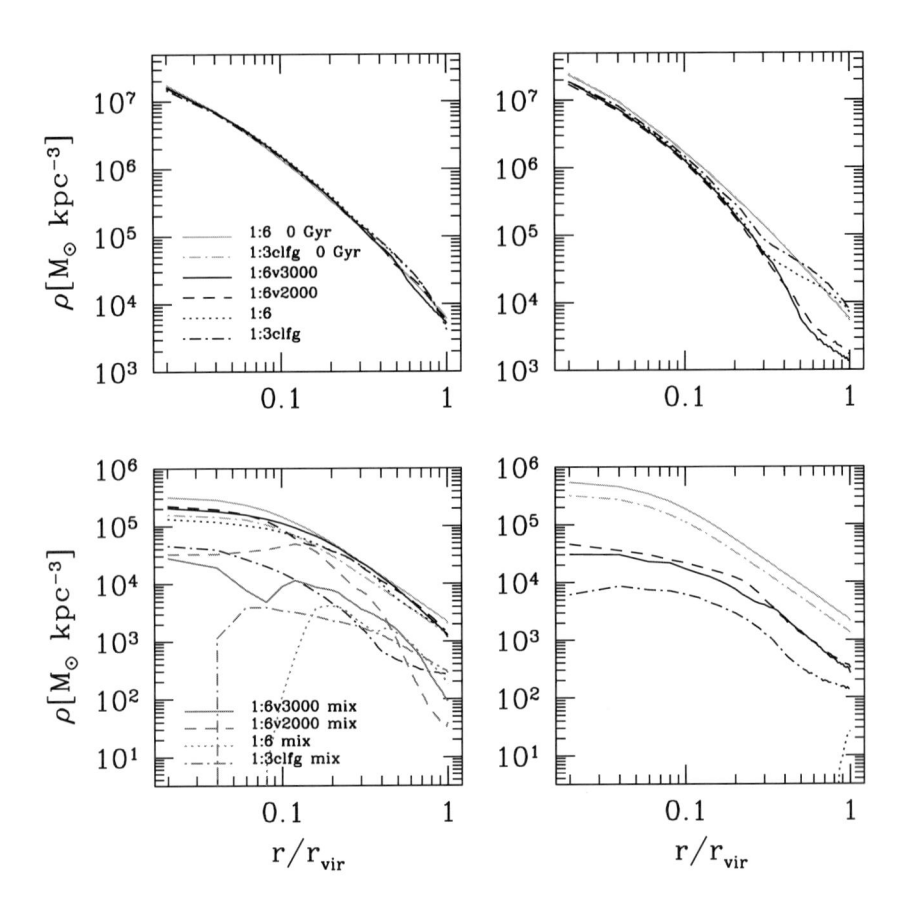

**Figure 14:** Dark matter (upper panels) and gas (bottom) density profiles of the main (left panels) and sub-cluster (right). Initial values and profiles at the final time (see text) are shown. Blue curves in the bottom left image refer to gas stripped from the sub-cluster and lying in the potential of the main system. Radius is scaled to the virial radius $r_{vir}$ of the dark matter distribution.

decrement due to the cooling of the central regions of the two approaching clusters. Already during the early phases of the interaction the cooling run moves out of the equilibrium. The peaks in luminosity and temperature are much smaller than the corresponding adiabatic ones (not illustrated). The final luminosity approaches zero. Markevitch 2006 found that the bullet cluster lies exactly on the $L_X - T$ relation for nearby clusters but its temperature is much higher than the one expected according to weak lensing mass estimates. Figure 13 illustrates the drift of the simulated systems along the $L_X - T$ diagram. The position of 1E0657−−560 is indicated by a star. $L_{Xbol}$ and $T_{sl}$ are calculated in a cylindrical region centered on the center of mass of the main cluster and radius equal to its initial virial radius. The luminosity of each model is normalized in such a way that the initial main cluster lies on the $L_X - T$ relation for local clusters despite of the different initial gas fractions. The starting time of the simulations is indicated with a black solid square. All the adiabatic runs present a similar evolution and move roughly parallel to the $L_X - T$ relation. During the early stages of the encounter the cluster moves along a curve which is flatter than the observed $L_X - T$ relation: the compression of the low density gas at the outskirts of the cluster produces an increase in temperature which is only marginally accompanied by a luminosity growth. The time when the core of the sub-cluster enters the virial radius of the main system represents an inversion point in the diagram: despite the formation of a bow-shock the temperature decreases due to the expansion of the main-cluster gas and the presence of low temperature baryons belonging to the sub-cluster within the virial radius of the main system. At the same time the luminosity rises as a result of the shock and the cluster moves perpendicularly towards the $L_X - T$ relation. This phase is actually quite short (on average $\sim 0.12$ Gyr) but it characterizes all the adiabatic runs. Later on the cluster moves almost parallel to the $L_X - T$ relation toward larger values of $T_{sl}$, with the peak in temperature being reached at the point of closest approach. Most of the runs show a small delay ($\sim 40$ Myr) between the time characterized by the highest temperature and the time with highest luminosity with the curve making a knot in the diagram. The branch of the curve associated with the post core-core interaction is parallel to the increasing one but shifted to smaller luminosities: during the strongest phase of the interaction some of the hot material is lost beyond the virial radius of the main cluster and indeed the largest shift in luminosity is observable in the high velocity run 1:6. Both, luminosity and temperature now decrease until they reach a second inversion point in the curve (in the case of 1:6 it is only a change in slope), associated with the egress of the bullet from the virial radius of the main system. The cooling run 1:3clfg is characterized by a first decrease in temperature which corresponds to the initial phase of thermal instability and central cooling of the main cluster. Later on it moves in the $L_X - T$ diagram similarly to the adiabatic runs although the peaks in luminosity and temperature are much less prominent. Nevertheless in a pure cooling model the main cluster does not return to a state of thermal equilibrium at the end of the interaction since nothing prevents the central regions from cooling and the system moves toward extremely low values of luminosity and temperature.

Fig. 14 illustrates the final structure of the remnants. In the case of the two highest velocities the final time is chosen in such a way that the distance between the centers of the two systems is about $1000$ kpc larger than the sum of the virial radii at $Time = 0$. This occurs at $Time = 2.2$ and $4$ Gyr in the case of 1:6 and 1:6v3000, respectively. The low velocity run 1:6v2000 is analyzed at the time corresponding to the first apocenter when the core of the sub-cluster is close to the virial radius of the main cluster. The collisionless component of the main cluster (top left panel) in all cases is not substantially affected by the interaction while the sub-cluster dark matter halo appears to be strongly perturbed and retains the original spherical symmetry only within its scale radius. The sub-cluster central density profile (top right) is vertically shifted downward without a significant change of slope while for $r > 0.2 r_{vir}$ the loss of material becomes more significant and in the case of low velocity encounters ($v = 2000$ and $3000$ km s$^{-1}$) with mass ratio 1:6 the profile shows a large jump of more than one order of magnitude between $0.3r_{vir}$ and $0.5r_{vir}$. Beyond the scale radius the isodensity contours appear elongated and show a large plateau associated to tidally stripped material. The evolution of gas density profiles is represented by the two panels on the bottom row of Fig. 14. In the 1:6 runs the interaction affects the central slope of the main cluster which becomes shallower while the 1:3 encounter produces the largest deviation from the initial values, with the final density profile shifted down by $\sim 25\%$. As mentioned earlier, a not negligible part of the gas within the virial radius of the main cluster originally belonged to the sub-cluster and was subsequently stripped by ram-pressure during the central phases of the interaction. The density profiles of the stripped gas are drawn in blue for the different runs. In the case of 1:6v2000 only the gas outside the virial radius of the sub-cluster is considered. Both 1:6v3000 and 1:6v2000 are characterized by a large fraction of sub-cluster gas lost to the core of the main system, with a flat (1:6v2000) or even positive (1:6v3000) central slope, while the high velocity run 1:6, despite the larger ram-pressure values, has less time to deposit gas in the central regions and shows a clear cut-off for $r < 0.1r_{vir}$.

At larger radii, baryonic material is still accreting onto the remnant. Indeed, part of the main cluster gas has been pushed out by the bow shock and is now falling back into the cluster potential together with a fraction of the material lost by the sub-cluster. Curves in the bottom right panel of Fig. 14 represent the sub-cluster gas density profiles. The low velocity sub-clusters 1:6v3000 and 1:6v2000 retain less than one tenth of their initial gas within $0.1r_{vir}$ while the density profiles in the external regions drop by a factor $\geq 5$. In general, the encounter flattens the gas density profile of the bullet core and this effect is evident in the cooling simulation as well. The 1:6 bullet at $Time = 2.2$ Gyr has lost almost all the gas within $r_{vir}$. Half of this baryonic material was stripped from its dark halo by ram-pressure during the central phases of the interaction and now accelerates towards its dark matter counterpart. It will be accreted by the sub-cluster halo at later times.

Figure 14 compares dark matter and gas density profiles at times when the main and sub-cluster are close to a state of virial equilibrium. Despite the changes observable in the gas density profiles and the tidal stripping affecting the bullet beyond its core radius, the central slope of the collisionless component seems to be not strongly perturbed by the interaction (Kazantzidis et al. 2004). The situation changes comparing the density profiles of the initial systems with those of the main and sub-cluster at the present time, when the bullet is located at almost one third of the host virial radius. As illustrated in the two upper panels of Fig. 15, the density profile of the interacting dark matter halos increases in the inner regions and shows a decrement beyond $0.4r_{vir}$. This behavior is similar for different models and therefore independent of the orbital details, of the mass ratios and is seen for both, the main and the sub-cluster. The two-body relaxation time scale beyond the softening radius is longer than $10^4$ Gyr (Arad & Johansson 2005), thus implying that dark matter density profiles are not affected by numerical relaxation. The loss of gas in the central regions of the sub-cluster is expected as a result of ram-pressure stripping. But interestingly also the inner 200 kpc of the main cluster are completely devoid of gas originally belonging to the main cluster. Later on, after the displacement of the main cluster gaseous core, the baryonic material stripped

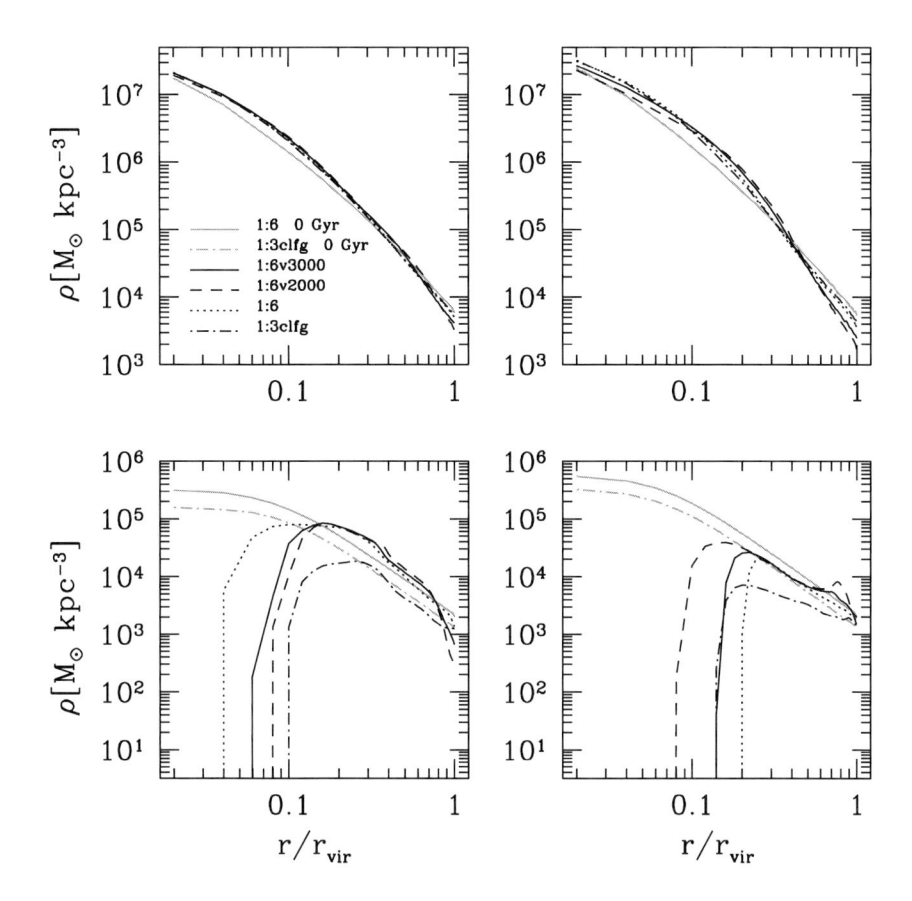

**Figure 15:** Dark matter (upper panels) and gas (bottom) density profiles of the main (left panels) and sub-cluster (right). Initial values and profiles at the present time are shown. Radius is scaled to the virial radius $r_{vir}$ of the dark matter distribution.

from the bullet replenishes the central regions of the host system. As soon as the bullet moves toward the outskirts of the main system, the main cluster gas has time to collapse again into the center of the potential (bottom left panel of Fig. 14).

# 5 Conclusions

We used high resolution N-body/SPH simulations to perform an extensive parameter study of the "bullet" cluster system 1E0657-56. We have shown that:

- Most of the main features in the observed X-ray maps are not well reproduced by encounters with zero impact parameter. An impact parameter corresponding to the core radius of the main cluster gas distribution provides enough ram-pressure to produce a displacement comparable to observations and introduces asymmetries in the main cluster emissivity map, similar to those detected in X-ray.

- A low concentrated ($c = 4$) main cluster does not survive the collision and its X-ray emissivity peak is destroyed.

- Encounters with mass ratios as large as 1:3 do not match the observed X-ray morphology and the size of the projected temperature jump across the shock discontinuity. Introducing cooling in the simulations, the temperature peak is cooled to a value comparable with the observed one.

- A pure cooling model neglecting energetic feedback still does not give a realistic description of the interacting system. What we are witnessing is overcooling in the central regions following a thermal instability in the early phases of the interaction. Nevertheless, cooling simulations provide a significant reduction of the temperature peak across the shock indicating that a more realistic treatment of the gas physics make models with higher mass ratios more realistic.

- The choice of a different gas fraction does not affect significantly the results.

- The morphology of X-ray maps is best simulated by adiabatic 1:6 encounters. In particular the run 1:6v3000, with initial velocity $v = 5000 \,\text{km}\,\text{s}^{-1}$ ($v \sim 2570 \,\text{km}\,\text{s}^{-1}$ in the center of mass system of reference) and present time dark matter velocity $v \sim 3100 \,\text{km}\,\text{s}^{-1}$ (again in the center of mass rest-frame), reproduces most of the main X-ray features. Decreasing the relative velocity the bullet becomes much less bright with respect to the center of the host system in disagreement with observations, although the displacement in the two clusters is better reproduced. The high velocity run 1:6 produces a contact discontinuity much broader than observed.

- A significant fraction of the X-ray emission next to the center of the main cluster is associated with gas stripped from the external regions of the bullet.

- The projected temperature jump across the shock discontinuity gives important indications about the nature of the encounter. Indeed, both, the height and the thickness of the peak change, with the peak becoming broader for decreasing bullet velocities.

- Temperature maps reveal some interesting features. The bullet remains relatively cold despite of its dominant X-ray emission. For a short time, a high temperature region appears next to the center of the main cluster mass distribution immediately after the central phases of the interaction. Its location partially corresponds to the X-ray diffuse peak, while its eastern component could be related to the high temperature region observed by Markevitch et al. 2002 southeast of the main-cluster peak.

For a selected sub-sample of runs we performed a detailed three dimensional analysis following their past, present and future evolution for four Gyr.

- Only the low velocity bullet 1:6v2000 is actually bound to the main cluster at the end of the simulation while in 1:6v3000 and 1:6 the sub-cluster has a velocity larger than the escape velocity from the host system.

- The velocity of the gaseous component of the bullet starts diverging from that of its dark matter counterpart before the point of closest approach between the two clusters, when the gas is slowed down by ram-pressure. The behavior of the sub-cluster gas after the point of closest approach strongly depends on the initial velocity of the bullet. For relatively low velocities (and ram-pressure values, run 1:6v3000 and 1:6v2000) the gaseous bullet is accelerated towards its dark matter counterpart as it leaves the core of the main cluster. For larger encounter velocities (1:6) ram-pressure is more effective in slowing down the bullet even beyond the core radius of host cluster and the gaseous bullet is always slower than dark matter.

- It has been already noticed by Milosavljevic et al. 2007 and Springel & Farrar 2007 that the sub-cluster velocities do not coincide with the shock velocity $v_s$ as measured by observers. In the case of the three runs (1:6v3000, 1:6v2000 and 1:6clfg) which best reproduce the projected temperature jump, the Mach number and shock velocity determined using the Rankine-Hugoniot conditions across the shock discontinuity show a good agreement with the values provided by Markevitch 2006. The two dimensional shock velocity calculated by tracking the shock position as function of time is consistent with these values. We find that, although after the point of closest approach $v_s$ is always larger than the velocity of the sub-cluster dark matter halo, the difference depends on the model and becomes less significant for higher velocities

- The collision produces a drastic increase in luminosity and temperature. The highest velocity impacts are associated with the largest increase in temperature (and shocks) and the smallest peaks in luminosity. We followed the evolution of the main cluster in the $L_X - T$ diagram during the different phases of the interaction and find that it moves roughly parallel to the $L_X - T$ relation for nearby clusters.

- After the encounter, as soon as the bullet is close or beyond the virial radius of its host system, the dark matter density profile of the main cluster does not deviate anymore significantly from the original one, while the gas profile becomes shallower in the central regions. The situation changes drastically if we compare the density profiles of the initial systems with those at the present time, when the bullet is still located well within the virial radius of the main cluster. In this case the interacting systems are not in virial equilibrium and the dark matter densities of both, the main and sub-cluster, increase in the inner regions and show a decrement beyond 0.4 $r_{vir}$. At a time corresponding to the present configuration the center of the host system is completely devoid of main cluster gas as a result of the displacement of the main cluster gaseous core by the bullet. The final sub-cluster dark matter density profiles seem significantly affected by the interaction beyond their scale radius, where the isodensity contours are elongated and show a plateau.

# 6 Acknowledgements

The work was partly supported by the DFG Sonderforschungsbereich 375 "Astro-Teilchenphysik". The numerical simulations were performed on a local SGI-Altix 3700 Bx2 which was partly funded by the cluster of excellence "Origin and Structure of the Universe".

## References

Arad, I., Johansson, P. H. 2005, MNRAS, 362, 252

Barrena, R., Biviano, A., Ramella, M., Falco, E. E., Seitz, S. 2002, A&A, 386, 816

Bradač, M., et al. 2006, ApJ, 652, 937

Clowe, D., Bradač, M., Gonzalez, A. H., Markevitch, M., Randall, S. W., Jones, C., Zaritsky, D. 2006, ApJL, 648, L109

Farrar, G. R., Rosen, R. A. 2007, Physical Review Letters, 98, 171302 Kazantzidis, S., Mayer, L., Mastropietro, C., Diemand, J., Stadel, J., Moore, B. 2004, ApJ, 608, 663

Macciò A. V., Dutton, A. A., van den Bosch, F. C., Moore, B., Potter, D., Stadel, J. 2007, MNRAS, 378, 55

Markevitch, M., Gonzalez, A. H., David, L., Vikhlinin, A., Murray, S., Forman, W., Jones, C., Tucker, W. 2002, ApJL, 567, L27

Markevitch, M. 2006, Proceedings of the The X-ray Universe 2005, 604, 723

Mastropietro, C., Moore, B., Mayer, L., Wadsley, J., Stadel, J. 2005, MNRAS, 363, 509

Mazzotta, P., Rasia, E., Moscardini, L., Tormen, G. 2004, MNRAS, 354, 10

Milosavljević, M., Koda, J., Nagai, D., Nakar, E., Shapiro, P. R. 2007, ApJL, 661, L131

Nusser, A. 2007, astro-ph/07093572

Springel, V., & Farrar, G. 2007, MNRAS, 380, 911

Wadsley, J. W., Stadel, J., Quinn, T. 2004, New Astronomy, 9, 137

# Pulsar Timing – From Astrophysics to Fundamental Physics

Michael Kramer

University of Manchester
Jodrell Bank Centre for Astrophysics
Alan-Turing Building
Oxford Road, Manchester M13 9PL, UK
Michael.Kramer@manchester.ac.uk

### Abstract

*Since their discovery 40 years ago, pulsars have been remarkable tools for a large variety of applications in astrophysics and fundamental physics. These range from the study of superdense matter to cosmology and precision tests of theories of gravity. This use of pulsars is possible due their nature as very stable cosmic clocks and a technique known as pulsar timing. I will present some of the achievements enabled by pulsar timing, which are exemplified by the rich laboratory provided by the unique double pulsar system, and explain how pulsars also contribute to the study of the nature and sources of gravitational waves.*

## 1 Introduction

In order to understand the Universe and the fundamental laws of physics that govern it, studies of "extreme physics" are essential. The questions that we try to answer range from the small scales ruled by quantum mechanics to the origin and evolution of the cosmos, cosmology. The quest to answer these questions as diverse as "What is the equation of state of super-dense matter?", "What is Dark Energy'?" or "Can we combine relativistic gravity with quantum mechanics?" requires input from a wide range of experiments and observational data. Fortunately, nature has provided us with a tool that can help us with the exploration of all these questions: pulsars!

Pulsars are fascinating objects. They represent not only the most extreme matter in the observable Universe, but their gravitational field is also strong enough, and their clock-like character precise enough, to allow for the best tests of theories of gravity in strong gravitational fields. Indeed, to date the theory of general relativity (GR) has passed all observational tests with flying colours. But GR may not be the last word in our understanding of gravitational physics. Still, because GR has been so successful, it forms the basis of most cosmological models and is also often the starting point in the search of quantum gravity. It is therefore important to experimentally confront the theory, as well as alternative theories of gravity, with new observations at ever more extreme conditions.

*Reviews in Modern Astronomy 20.* Edited by S. Röser
Copyright © 2008 WILEY-VCH Verlag GmbH & Co. KGaA, Weinheim
ISBN: 978-3-527-40820-7

In the following, I will summarise how objects with the size of a city can be used to understand the origin and fate of the Universe as a whole, by putting GR and other theories of gravity to the test.

## 2 Pulsars – fascinating tools

Pulsars are highly magnetised, rotating neutron stars which emit a narrow radio beam along the magnetic dipole axis. As the magnetic axis is inclined to the rotation axis, the pulsar acts like a cosmic light-house emitting a radio pulse that can be detected once per rotation period when the beam is directed toward Earth. For some very fast rotating pulsars, the so-called millisecond pulsars, the stability of the pulse period is similar to that achieved by the best terrestrial atomic clocks. Using these astrophysical clocks by accurately measuring the arrival times of their pulses, a wide range of experiments is possible. For most of these it is not necessarily important *how* the radio pulses are actually created.

Pulsars are born in supernova explosions of massive stars. Created in the collapse of the stars' core, neutron stars are the most compact objects next to black holes. From timing measurements of binary pulsars (see Sect. 5.2), we determine the masses of pulsars to be typically around $1.35\pm0.04 M_\odot$ (e.g. Thorsett & Chakrabarty 1999) although this range has been expanded recently from $\sim 1.2 M_\odot$ well over $2 M_\odot$. Modern calculations for different equations of state produce results for the size of a neutron star which are quite similar to the very first calculations by Oppenheimer & Volkov (1939), i.e. a diameter of about 20 km. Such sizes are consistent with independent estimates derived from X-ray light-curves and luminosities of pulsars (e.g. Zavlin & Pavlov 1998).

Pulsars emit electromagnetic radiation and, in particular, magnetic dipole radiation as they essentially represent rotating magnets. Assuming that this is the dominant process of loss in rotational energy and hence responsible for the observed increase in rotation period, $P$, described by $\dot{P}$, we can equate the corresponding energy output of the dipole to the loss rate in rotational energy. We obtain an estimate for the magnetic field strength at the pulsar (equatorial) surface from

$$B_S = 3.2 \times 10^{19} \sqrt{P\dot{P}} \text{ Gauss,} \qquad (1)$$

with $P$ measured in s and $\dot{P}$ in s s$^{-1}$. Millisecond pulsars (MSPs) have lower field strengths of the order of $10^8$ to $10^{10}$ Gauss which appear to be a result of their evolutionary history. These magnetic fields are consistent with values derived from X-ray spectra of neutron stars where we observe cyclotron lines (Bignami et al. 2003).

The notion that MSPs represent a separate evolutionary phase of pulsars is supported by an estimate of their age. This is achieved by comparing the current pulse period (and hence rotation period) with the spin-down rate. Assuming that the initial spin frequency was infinite, and that the spin-down is steady and indeed given by magnetic dipole radiation, we obtain the *characteristic* of a pulsar

$$\tau = \frac{P}{2\dot{P}}. \qquad (2)$$

Even though this quantity is not a very reliable estimator, it indicates that normal pulsars have ages from about 1,000 years to about 100 Million years, while MSPs have ages of typically 1 Billion years and more. As the vast majority of MSPs (about $\sim 80\%$) is also in binary systems (in contrast to less than $\sim 1\%$ for normal pulsars), MSPs are therefore believed to be the end product of an evolutionary process which only occurs in binary pulsars.

## 3  Evolution of Pulsars

Before we describe the evolution of MSPs in particular, we review the evolution of pulsars in general. We can describe this evolution in by systematic changes in period, $P$, and slow-down, $\dot{P}$, which we can display in a logarithmic $P$-$\dot{P}$-diagram as shown in Figure 1 where we plot all known pulsars for which $P$ and $\dot{P}$ have been measured.[1] Since the estimates for both magnetic field (Eq. (1)) and characteristic age (Eq. (2)) depend only $P$ and $\dot{P}$, we can draw lines of constant magnetic field and constant characteristic age. Accordingly, young pulsars should be located in the upper left area of Fig. 1. Pulsars are generally considered to be young if their characteristic age is less than 100 kyr. Specifically, pulsars with characteristic ages of less than 10 kyr appear in the cross-hatched area, whilst pulsars with ages between 10 and 100 kyr are located in the hatched area. The latter pulsars are often compared to the Vela pulsar if they match its *spin-down luminosity*, i.e. $\dot{E} > 10^{36}$ erg s$^{-1}$. The spin-down luminosity is simply given by the loss in rotational energy which can be measured from the observed period and period derivative,

$$\dot{E} = 4\pi^2 I \dot{P} P^{-3} \text{ erg s}^{-1} \qquad (3)$$

where a neutron star moment of inertia of $I = 10^{45}$ g cm$^2$ is assumed. Obviously, $\dot{E}$ represents the maximum energy output available for spin-powered pulsars across the *whole* electromagnetic spectrum. A line of a constant, Vela-like $\dot{E} = 10^{36}$ erg s$^{-1}$ is shown in Fig. 1 together with a line for $\dot{E} = 10^{33}$ erg s$^{-1}$.

When pulsars age, they move into the central part of the $P$–$\dot{P}$-diagram where they spend most of their lifetime. Consequently, most known pulsars have spin periods between 0.1 and 1.0 s with period derivatives of typically $\dot{P} = 10^{-15}$ s s$^{-1}$. Selection effects are only partly responsible for the limited number of pulsars known with very long periods, the longest known period being 8.5 s (Young et al. 1999). The dominant effect is due to the "death" of pulsars when their slow-down has reached a critical state. This state seems to depend on a combination of $P$ and $\dot{P}$ which can be represented in the $P$–$\dot{P}$-diagram as a *pulsar death-line*. To the right and below this line (see Figure 1) the electric potential above the polar cap may not be sufficient to produce the particle plasma that is responsible for the observed radio emission. While this model can indeed explain the lack of pulsars beyond the death-line, the truth may be more complicated as the position of the 8.5-sec pulsar deep in the *pulsar graveyard* indicates. Nevertheless, it is clear that the normal life of radio pulsars

---

[1] We do not plot MSPs in globular clusters for which the observed $\dot{P}$ is severely affected by an extrinsic acceleration in the gravitational potential of the cluster.

is limited and that they die eventually after tens to a hundred million years. The position the MSPs in the lower left part of Fig. 1 cannot be explained by the above picture of normal pulsar life. Their evolution is closely linked to their nature as members of binary systems. For reasons explained in the following, they are also called *recycled pulsars*.

## 4 Recycled Pulsars

The MSPs located in lower-left part of the $P - \dot{P}$ diagram are clearly different from the majority of pulsars. Firstly, they exhibit much shorter pulse periods. The first discovered MSP, PSR B1937+21 (Backer et al. 1982), has has a period of only 1.56 ms and remained the pulsar with the shortest period known for more than 20 years. Only recently, a millisecond pulsar, PSR J1748-2446ad, was discovered which has a slightly shorter period of 1.40 ms (Hessels et al. 2006). Secondly, MSPs also have very small period derivatives, $\dot{P} \lesssim 10^{-18}$ s s$^{-1}$, making them much older (see Eq. (2)) than normal pulsars as discussed earlier. In the standard scenario, which finds its ultimate confirmation in the discovery of the Double Pulsar, MSPs are recycled from a dead binary pulsar via an X-ray binary accretion phase. The pulsars' millisecond periods are obtained when mass and thereby angular momentum is transferred from an evolving binary companion while it overflows its Roche lobe (e.g. Alpar et al. 1982).

Even though most ordinary stars are in binary systems, most pulsars do not evolve into a MSP. The birth of the pulsar usually disrupts the system, preventing the access to a mass donor and explaining why most pulsars are isolated. In the binary systems that survive the supernova explosion, the pulsar will eventually cease to emit radio emission, before the system evolves into a X-ray binary phase during which mass accretion onto the pulsar occurs. The pulsar spins up and is recycled into a radio millisecond pulsar when $P$ and $\dot{P}$ have been altered such that the pulsar has crossed the death-line again in the other direction. The final spin period of such a recycled pulsar depends on the mass of the binary companion. A more massive companion evolves faster, limiting the duration of the accretion process and hence the angular momentum transfer.

The majority of millisecond pulsars will have had a low-massive companion. These systems evolve into low-mass X-ray binaries (LMXBs) and will result into a fast-spinning millisecond pulsar with period of $P \sim 1-10$ ms with a low-mass white-dwarf companion. Systems with a more massive companion evolve into high-mass X-ray binaries (HMXBs) which represent the progenitors for double neutron star systems (DNSs). DNSs are rare since these systems need to survive a total of two supernova explosions. If this happens, the millisecond pulsar is only mildly recycled with a period of tens of millisecond.

The properties of MSPs and X-ray binaries are consistent with the described picture, such as the relative fraction of pulsars with a binary companion. For MSPs with a low-mass white dwarf companion the orbit is nearly circular due to a circularisation of the orbit during the recycling process. In case of DNS systems, the orbit is affected by the unpredictable nature of the kick imparted onto the newly born neutron star in

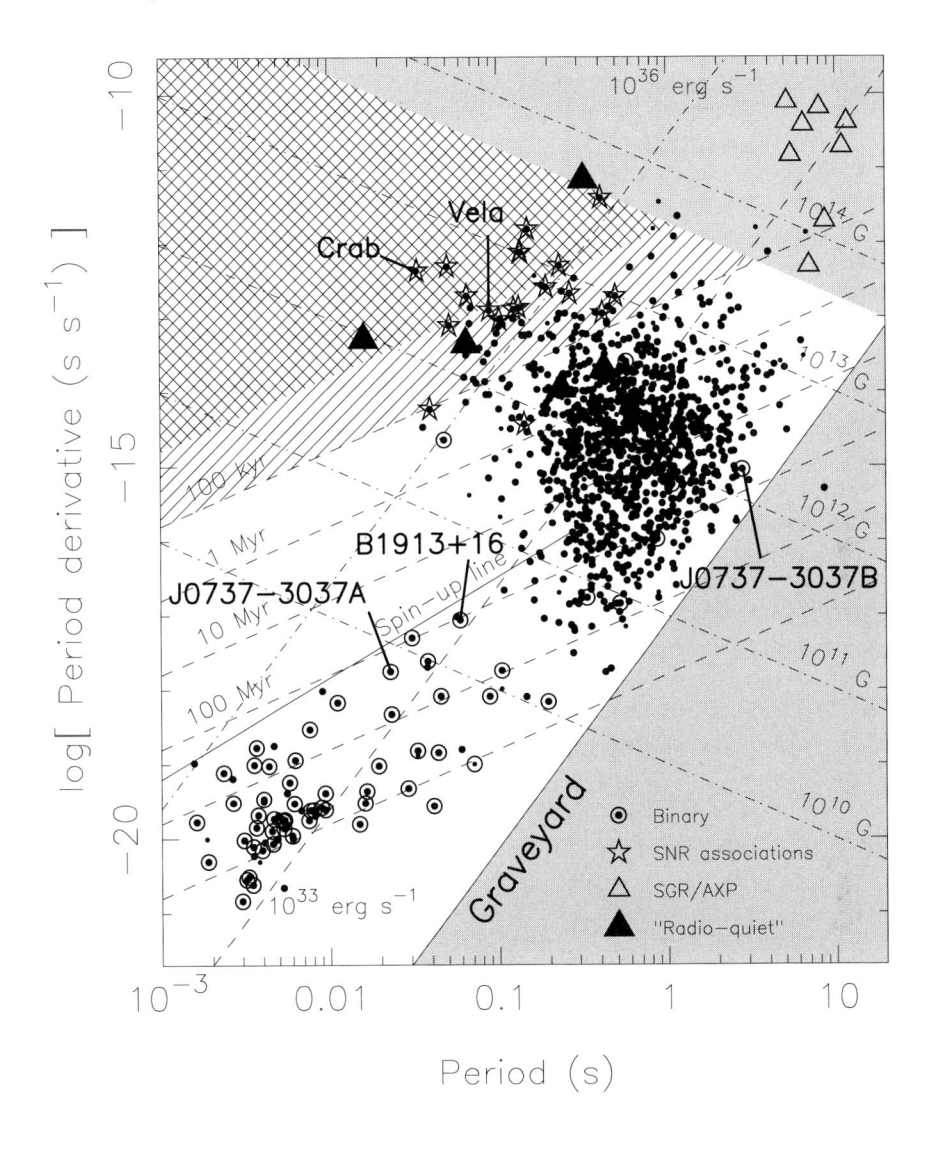

**Figure 1:** The $P$–$\dot{P}$-diagram for the known pulsar population. Lines of constant characteristic age, surface magnetic field and spin-down luminosity are shown. Binary pulsars are marked by a circle. The lower solid line represents the pulsar "death line" enclosing the "pulsar graveyard" where pulsars are expected to switch off radio emission. The grey area in the top right corner indicates the region where the surface magnetic field appears to exceed the quantum critical field of $4.4 \times 10^{13}$ Gauss. For such values, some theories expect the quenching of radio emission in order to explain the radio-quiet "magnetars" (i.e. Soft-gamma ray repeaters, SGRs, and Anomalous X-ray pulsars, AXPs). The upper solid line is the "spin-up" line which is derived for the recycling process as the period limit for millisecond pulsars. The two pulsars of the Double Pulsar are marked along with the Hulse-Taylor pulsar, PSR B1913+16, and the Vela and the Crab pulsar.

the asymmetric supernova explosion of the companion. If the system survives, the result is typically an eccentric orbit with an orbital period of a few hours.

## 5 Applications to fundamental physics

Most applications of pulsars in fundamental physics involve a technique called *pulsar timing*. In this procedure we make use of the clock-like stability of pulsars by a precise monitoring of the pulsars' rotations. In other words, rather than measuring simply variations in pulse period, we measure the exact arrival of a pulse at the telescope on Earth. In a procedure outlined below, we therefore count every single rotation of the neutron star, i.e. we connect the rotational phases in a coherent manner. We obviously have to take into account that the revolving Earth is moving around the Sun, and that the pulsar itself may be in a binary system. However, by having a phase-coherent "solution" for the pulsar's rotational model, corresponding variations in arrival time can be detected and small effects (for instance, affecting the binary motion) can be detected.

While the basic spin and astrometric parameters can be derived for essentially all pulsars, millisecond pulsars are the most useful objects for more exotic applications. Their pulse arrival times can be measured much more precisely than for normal pulsars (scaling essentially with the pulse period) and their rotation is also much smoother, making them intrinsically better clocks. Specifically, they usually do not exhibit rotational instabilities such as *timing noise* and *glitches* known for normal pulsars. The latter phenomena are important observables for the study of neutron star dynamics and the behaviour of super-dense matter. However, they also mean that young pulsars can, usually, not be used for high precision tests of theories of gravity.

### 5.1 Pulsar timing

The key quantity of interest is the *time of arrival* (TOA) of pulses at the telescope. However, since individual pulses are usually too weak to be detected, and since they also show a jitter in arrival time within a window given by the extend of the pulse profile, it is the latter which is used for timing. The stability of pulse profiles allows us to compare the observed profile with a high signal-to-noise ratio template that is constructed from previous observations. The time-offset between template and profile determines the TOA. Because we use pulse profiles rather than individual pulses, the TOA is defined usually as the arrival time of the nearest pulse to the mid-point of the observation. As the pulses have a certain width, the TOA refers to some *fiducial point* on the profile. Ideally, this point coincides with the plane defined by the rotation and magnetic axes of the pulsar and the line of sight to the observer which is defined geometrically and independent of observing frequency or propagation effects.

The aim of pulsar timing is to count the number of neutron star rotations between two observations. Each TOA can therefore be assigned with a pulse number $N$ which depends on rotation frequency $\nu$ and TOA $t$ as

$$N = N_0 + \nu_0(t - t_0) + \frac{1}{2}\dot{\nu}(t - t_0)^2 + \frac{1}{6}\ddot{\nu}(t - t_0)^3 + \cdots \qquad (4)$$

where $N_0$ is the pulse number at the reference epoch $t_0$. If $t_0$ coincides with the arrival of a pulse and the pulsar spin-down (i.e. $\nu$ and $\dot{\nu}$) is known accurately, the pulses should appear at integer values of $N$ when observed in an inertial reference frame. However, our observing frame is not inertial: we are using telescopes that are located on a rotating Earth orbiting the Sun. Before analysing TOAs measured with the observatory clock (topocentric arrival times), we need to transfer them to the centre of mass of the Solar System (solar system barycentre, SSB). To a very good approximation, the SSB is an inertial reference frame.

The time transformation also corrects for any relativistic time delay that occurs due to the presence of masses in the Solar System. An additional advantage of analysing these barycentric arrival times is that they can easily be combined with other TOAs measured at different observatories at different times.

Given a minimal set of starting parameters, a least squares fit is needed to match the measured arrival times to pulse numbers according to Equation (4). We minimise the expression

$$\chi^2 = \sum_i \left( \frac{N(t_i) - n_i}{\sigma_i} \right)^2 \qquad (5)$$

where $n_i$ is the nearest integer to $N(t_i)$ and $\sigma_i$ is the TOA uncertainty in units of pulse period (turns).

As outlined earlier, the aim is to obtain a phase-coherent solution that accounts for every single rotation of the pulsar between two observations. One starts off with a small set of TOAs that were obtained sufficiently close in time so that the accumulated uncertainties in the starting parameters do not exceed one pulse period. Gradually, the data set is expanded, maintaining coherence in phase. When successful, post-fit residuals expressed in pulse phase show a Gaussian distribution around zero with a root mean square that is comparable to the TOA uncertainties (see Fig. 2).

After starting with fits for only period and pulse reference phase over some hours and days, longer time spans slowly require fits for parameters like spin frequency derivative(s) and position. Incorrect or incomplete timing models cause systematic structures in the post-fit residuals identifying the parameter that needs to be included or adjusted (see Figure 2). The precision of the parameters improves with length of the data span and the frequency of observation, but also with orbital coverage in the case of binary pulsars.

## 5.2 Using binary pulsars

Observations of pulsars in binary orbits show a periodic variation in pulse arrival time. The timing model therefore needs to incorporate the additional motion of the pulsar as it orbits the common centre of mass of the binary system. For non-relativistic binary systems, the orbit can be described using Kepler's laws. For a

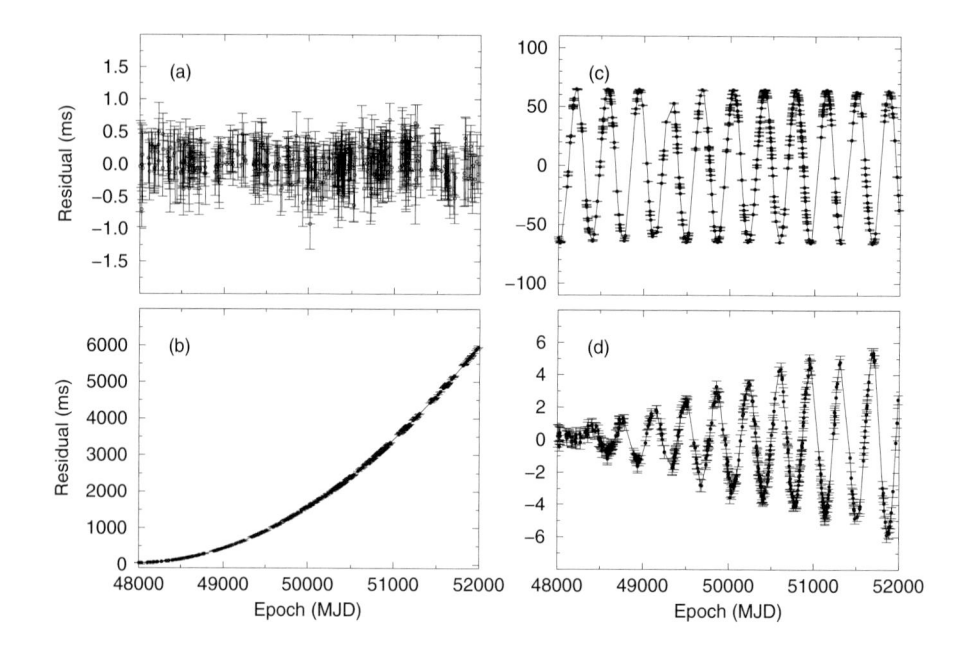

**Figure 2:** (a) Timing residuals for the 1.19 s pulsar B1133+16. A fit of a perfect timing model should result in randomly distributed residuals. (b) A parabolic increase in the residuals is obtained if $\dot{P}$ is underestimated, here by 4 per cent. (c) An offset in position (in this case a declination error of 1 arcmin) produces sinusoidal residuals with a period of 1 yr. (d) The effect of neglecting the pulsar's proper motion, in this case $\mu_{\rm T} = 380$ mas yr$^{-1}$. In all plots we have set the reference epoch for period and position to the first TOA at MJD 48000 to show the development of the amplitude of the various effects. Note the different scales on each of the vertical axes.

number of binary systems however, the Keplerian description of the orbit is not sufficient and relativistic corrections need to be applied.

Kepler's laws can be used to describe a binary system in terms of the five Keplerian parameters, shown schematically in Figure 3. These five parameters are required to refer the TOAs to the binary barycentre: (a) orbital period, $P_{\rm b}$; (b) projected semi-major orbital axis, $a_{\rm p} \sin i$ (see below); (c) orbital eccentricity, $e$; (d) longitude of periastron, $\omega$; (e) the epoch of periastron passage, $T_0$.

For pulsars in close binary systems about white dwarfs, other neutron stars, or perhaps eventually black holes, relativistic effects due to strong gravitational fields and high orbital velocities produce observable signatures in the timing residuals. Even though GR appears to be the best description of the strong-field regime to date (Kramer et al. 2006), alternative theories of gravity should be considered and tested against it. A straightforward means of comparison is to parametrise the timing model in terms of the theory-independent "post-Keplerian" (PK) parameters. While the measurement of these PK parameters can indeed be performed without assuming a

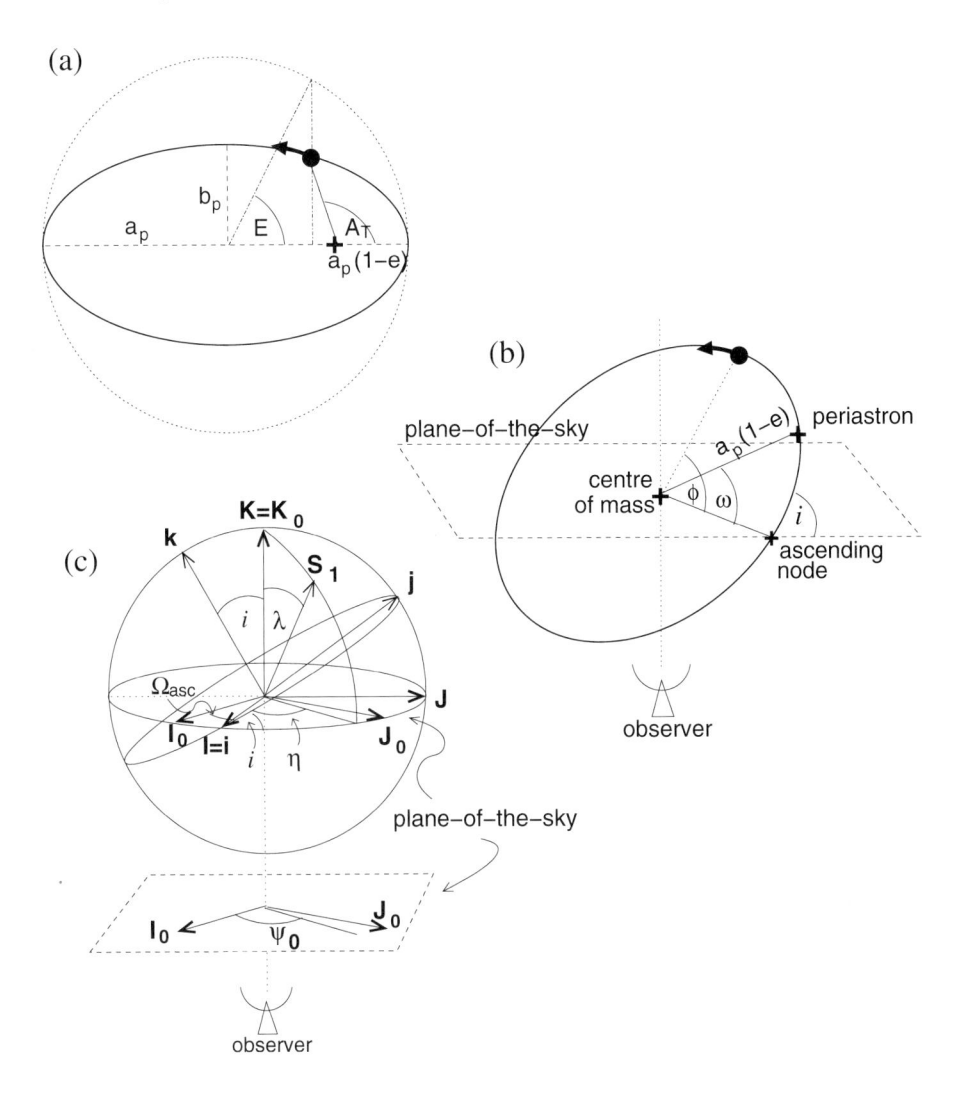

**Figure 3:** Definition of the orbital elements in a Keplerian orbit and the angles relating both the orbit and the pulsar to the observer's coordinate system and line of sight. (a) is drawn in the plane of the orbit; (b) shows the orbit inclined to the plane of the sky. The closest approach of the pulsar to the centre of mass of the binary system marks periastron, given by the longitude $\omega$ and a chosen epoch $T_0$ of its passage. The distance between centre of mass and periastron is given by $a_p(1-e)$ where $a_p$ is the semi-major axis of the orbital ellipse and $e$ its eccentricity. (b) Usually, only the projection on the plane of the sky, $a_p \sin i$, is measurable, where $i$ is the orbital inclination defined as the angle between the orbital plane and the plane of the sky. The *true anomaly*, $A_T$, and *eccentric anomaly*, $E$, are related to the *mean anomaly* by Kepler's law. The orbital phase of the pulsar $\Phi$ is measured relative to the ascending node. (c) The spatial orientation of the pulsar's spin-vector, $\mathbf{S_1}$, is given by the angles $\lambda$ and $\eta$ in the coordinate system shown as defined by Damour and Taylor (1992). The angle $\Omega_{\mathrm{asc}}$ gives the longitude of ascending node in the plane of the sky.

particular theory of gravity, we can compare the observed values with the predictions made by a theory to be tested.

For point masses with negligible spin contributions, the PK parameters in each theory should only be functions of the a priori unknown pulsar and companion mass, $M_p$ and $M_c$, and the easily measurable Keplerian parameters (Damour & Taylor 1992). With the two masses as the only free parameters, an observation of two PK parameters will already determine the masses uniquely in the framework of the given theory. The measurement of a third or more PK parameters then provides a consistency check for the assumed theory.

## 5.3 Relativistic corrections

The PK parameters are additional parameters in a theory independent timing model which describes the pulse arrival times in a phenomenological way, using the combination of astrometric parameters and the Keplerian and PK parameters. The best timing model for describing relativistic binary pulsars is the Damour-Deruelle (DD) timing model (Damour & Deruelle 1985, 1986). They take different forms in different theories of gravity. In general relativity, the five most important PK parameters are given by (e.g. Damour & Taylor 1992):

$$\dot{\omega} = 3T_\odot^{2/3} \left(\frac{P_b}{2\pi}\right)^{-5/3} \frac{1}{1-e^2} (M_p + M_c)^{2/3}, \tag{6}$$

$$\gamma = T_\odot^{2/3} \left(\frac{P_b}{2\pi}\right)^{1/3} e \frac{M_c(M_p + 2M_c)}{(M_p + M_c)^{4/3}}, \tag{7}$$

$$\dot{P_b} = -\frac{192\pi}{5} T_\odot^{5/3} \left(\frac{P_b}{2\pi}\right)^{-5/3} \frac{\left(1 + \frac{73}{24}e^2 + \frac{37}{96}e^4\right)}{(1-e^2)^{7/2}} \frac{M_p M_c}{(M_p + M_c)^{1/3}}, \tag{8}$$

$$r = T_\odot M_c, \tag{9}$$

$$s = T_\odot^{-1/3} \left(\frac{P_b}{2\pi}\right)^{-2/3} x \frac{(M_p + M_c)^{2/3}}{M_c}, \tag{10}$$

where the masses $M_p$ and $M_c$ of pulsar and companion, respectively, are expressed in solar masses ($M_\odot$). We define the constant $T_\odot = GM_\odot/c^3 = 4.925490947$ μs where $G$ denotes the Newtonian constant of gravity and $c$ the speed of light. The first PK parameter, $\dot{\omega}$, is the easiest to measure and describes the relativistic advance of periastron. It provides an immediate measurement of the total mass of the system, $(M_p + M_c)$. The parameter $\gamma$ denotes the amplitude of delays in arrival times caused by the varying effects of the gravitational redshift and time dilation (second order Doppler) as the pulsar moves in its elliptical orbit at varying distances from the companion and with varying speeds. The decay of the orbit due to gravitational wave damping is expressed by the change in orbital period, $\dot{P_b}$. The other two parameters, $r$ and $s$, are related to the Shapiro delay caused by the gravitational field of the companion. These parameters are only measurable, depending on timing precision, if the orbit is seen nearly edge-on.

The PK parameters listed above are those which have been measured in binary systems to date. However, the list can be extended (see Damour & Taylor 1992) if the binary system is extreme enough. We expect the Double Pulsar to be such a system where never before measured PK parameters will be needed to describe the observations adequately. In the following, we summarise the properties of the Double Pulsar and then demonstrate how the technique described above can be used to test theories of gravity and general relativity in particular.

## 6 The Double Pulsar

While DNSs were known since the discovery of the Hulse-Taylor pulsar, the Double Pulsar in the only known system were both neutron stars are active radio pulsars that can be detected on Earth. We discovered the first of the two pulsars, the 22.8-ms pulsar then-called J0737−3039 in April 2003 (Burgay et al. 2003) in an extension to the hugely successful Parkes Multi-beam survey (Manchester et al. 2001). It was soon found to be a member of the most extreme relativistic binary system ever discovered: its short orbital period ($P_b = 2.4$ hrs) is combined with a remarkably high value of periastron advance ($\dot{\omega} = 16.9 \deg \mathrm{yr}^{-1}$, i.e. four times larger than for the Hulse-Taylor pulsar PSR B1913+16). This large precession of the orbit was measurable after only a few days of observations. The system parameters predict that the two members of the binary system will coalesce on a short time scale of only $\sim 85$ Myr. This boosts the hopes for detecting a merger of two neutron stars with first-generation ground-based gravitational wave detectors by a factor of 5 to 10 compared to previous estimates based on only the double neutron stars B1534+12 and B1913+16 (Burgay et al. 2003, Kalogera et al. 2004).

In October 2003, we then also detected radio pulses from the second neutron star (Lyne et al. 2004). The reason why signals from the 2.8-s pulsar companion (now called PSR J0737−3039B, hereafter "B") to the millisecond pulsar (now called PSR J0737−3039A, hereafter "A") had not been found earlier, became clear when it was realized that B was only bright for two short parts of the orbit. For the remainder of the orbit, the pulsar B is extremely weak and only detectable with the most sensitive equipment. The detection of a young companion B around an old millisecond pulsar A and their position in the $P - \dot{P}$-diagram (see Figure 1) confirms the evolution scenario proposed for recycled pulsars (see Section 4) and provides a truly unique testbed for relativistic gravity and also plasma physics.

### 6.1 System properties

Indeed, the double pulsar provides an unprecedented opportunity to probe the workings of pulsars. The pulse emission from B is strongly modulated with orbital phase, most probably as a consequence of the penetration of the A's wind into B's magnetosphere. Figure 4 shows the pulse intensity for B as a function of pulse phase and orbital longitude for three radio frequencies. The first burst of strong emission, centred near orbital longitude 210 deg, covers about 13 min of the orbit, while the second burst, centred near longitude 280 deg, is shorter and last only for about 8 min. This pattern is stable over successive orbits and obviously frequency independent over the range probed. Deep integrations reveal other orbital phases, where B

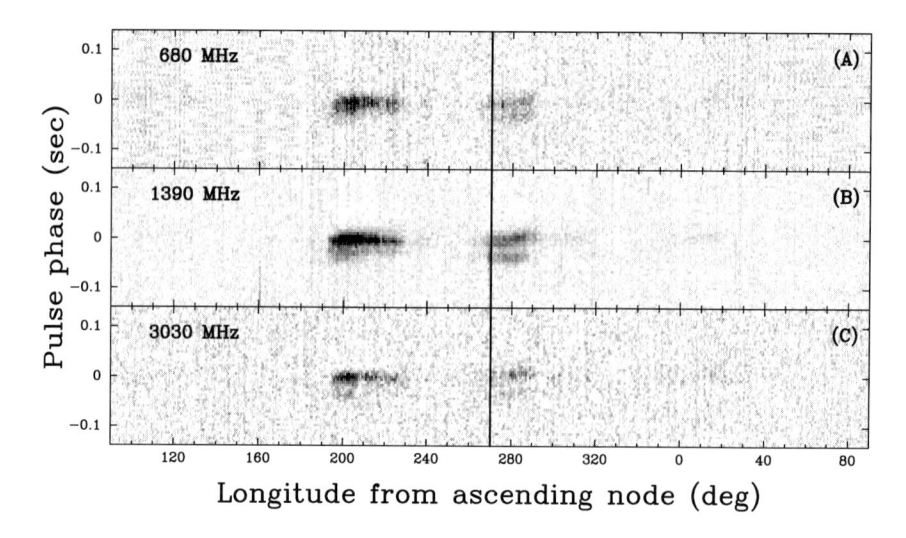

**Figure 4:** Grey-scale images showing the pulse of PSR J0737–3039B as a function of orbital phase at three observing frequencies (Lyne et al. 2004).

is visible but much weaker than during the two main burst periods. The figure also shows that not only does the pulse intensity change with orbital phase, but that the pulse shape changes as well. At the start of the first burst the pulse has a strong trailing component and a weaker leading component which dies out in the later phases of the burst. In the second burst, there are two components of more equal amplitude. Decoding this pattern as the orbit precesses due to relativistic effects and the system is viewed from different directions, offers a unique chance to probe the magnetosphere. Indeed, the "light-curve" of B is changing with time, probably due to the effects of geodetic precession.

It is important to note that by simply seeing B functioning as a radio pulsar, albeit with orbital phases of rather weak emission, confirms our ideas about the location of the origin of pulsar radio emission in general: The fact that B is still emitting, despite the loss of most of its magnetosphere due to A's wind, indicates that the fundamental processes producing radio emission are likely to occur close to the neutron star surface – in accordance with emission heights determined for normal radio pulsars (see e.g. Lorimer & Kramer 2005).

The quenching or attenuation of B's radio emission for most of its orbit is only part of the interaction between A and B that is observed. For about 27 seconds of the orbit, A's emission is eclipsed when A is lined up behind B at superior conjunction (Fig. 5). At that moment, the pulses of A pass in about 30,000km distance to the surface of B. It appears that the magnetospheric transmission for A's emission is modulated during the rotation of B, depending on the relative orientation of the spin-axis of B to A and our line-of-sight. Indeed, a modulation of the light-curve inside the eclipse region consistent with B's (full and half) rotation period is observed (McLaughlin et al. 2004a, Fig. 6).

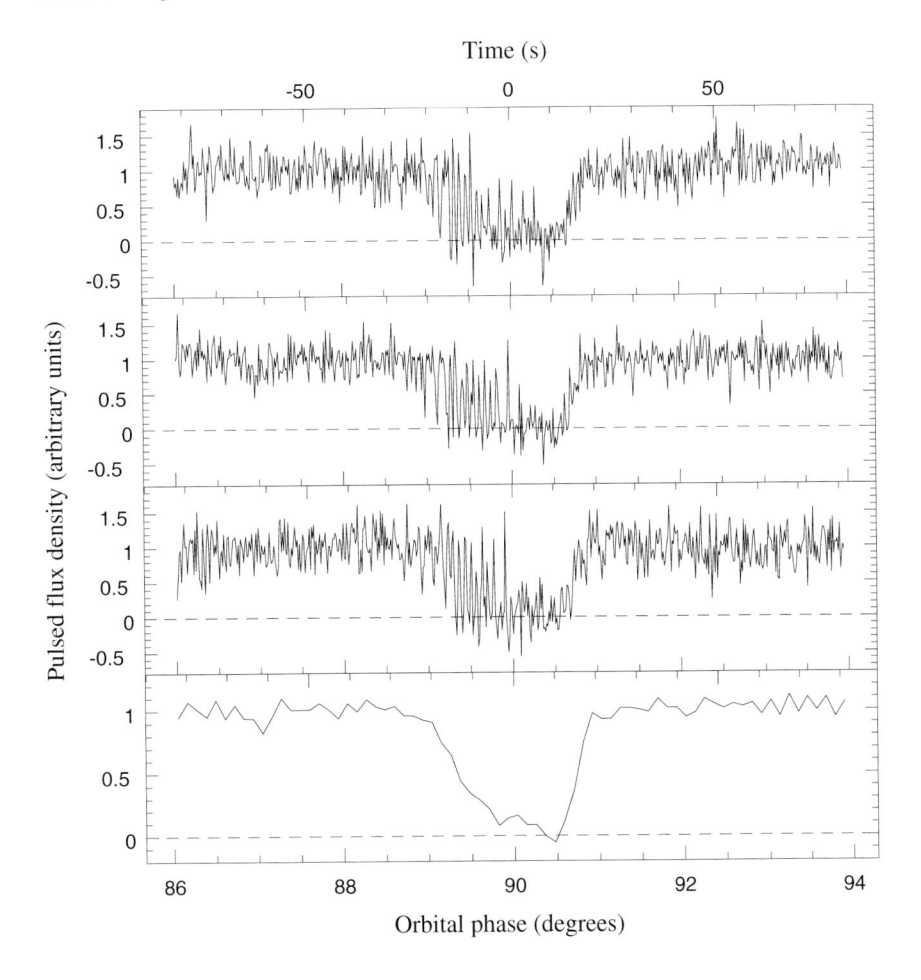

**Figure 5:** The pulsed flux density of A versus time (with respect to superior conjunction) and orbital phase for (top three panels) the three eclipses in the 820-MHz observation and (bottom panel) all three eclipses summed (McLaughlin et al. 2004b). In the individual eclipse light curves, every 12 pulses have been averaged for an effective time resolution of $\sim 0.27$ s. Every 100 pulses have been averaged to create the lower, composite light curve for an effective time resolution of $\sim 2.3$ s. Pulsed flux densities have been normalised such that the pre-eclipse average flux density is unity.

Perhaps even more exciting is the discovered evidence that A's radiation has some direct impact on the radiation pattern of B. Figure 7 shows a blow-up of B's emission at orbital phases where B is strongest. At the right orientation angles, a drifting sub-pulse pattern emerges that coincides with the arrival times of A's pulses at B (McLaughlin et al. 2004b). This is the first time pulsar emission is observed to be triggered by some external force, and it is likely that this will help us to understand the conditions and on-set of pulsar emission in general.

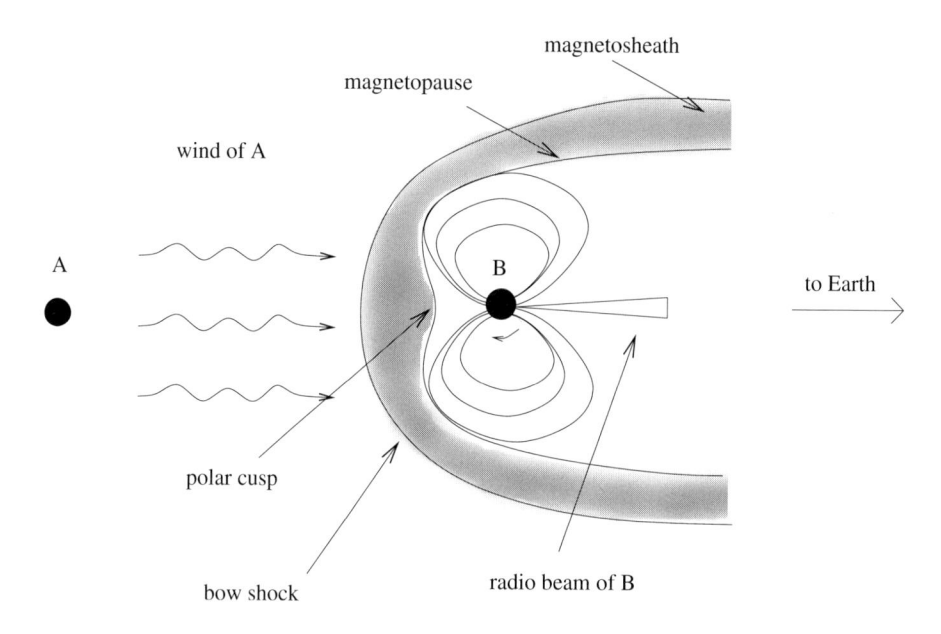

**Figure 6:** Cartoon (not to scale) showing the interaction between the relativistic wind of A and the magnetosphere of B when the radio beam of B is pointing towards the Earth (from McLaughlin et al. 2004b).

## 6.2 Tests of general relativity

Since neutron stars are very compact massive objects, the Double Pulsar (together with other DNSs) can be considered as almost ideal point sources for testing theories of gravity in the strong-gravitational-field limit. We have conducted timing observations of PSR J0737−3039A/B since its discovery with all major radio-telescopes that enable us to observe this system. The published results (Kramer et al. 2006) are based on a total of 131,416 pulse times-of-arrival (TOAs) measurements, leading to the pulsar and binary system parameters listed in Table 1. Because of its narrower and more stable pulse profile, TOAs from A have a much higher precision than those from B and hence are used to determine the position, proper motion and main orbital parameters of the system. For B, the only fitted parameters were the pulse phase, the pulsar spin frequency, $\nu \equiv 1/P$, its first time-derivative $\dot\nu$ and the projected semi-major axis, $x_{\rm B} \equiv (a_{\rm B}/c)\sin i$.

The timing observations allowed us to measure a total of five PK parameters: the rate of periastron advance $\dot\omega$, the gravitational redshift and time dilation parameter $\gamma$, and the Shapiro-delay parameters $r$ and $s$, and the orbital decay, $\dot P_{\rm b}$, due to gravitational wave emission (Kramer et al. 2006). When trying to see whether these PK parameter measurements are in agreement with the predictions of a theory of gravity, we can construct these tests in a very elegant way (Damour & Taylor 1992): We expect every theory of gravity to predict a unique relationship between the two masses of the system and each PK parameter. These relationships can be drawn in a diagram

*Pulsar Timing* 269

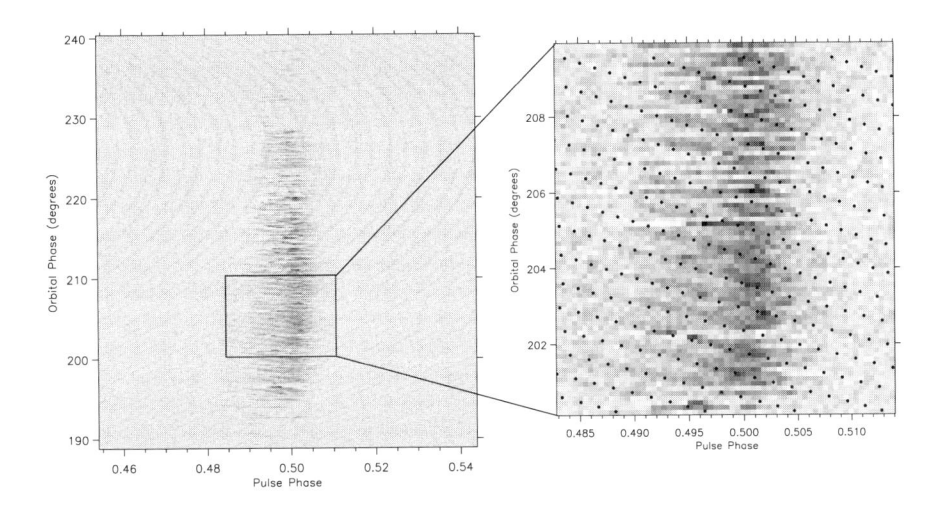

**Figure 7:** Observations of single pulses of B at 820 MHz for orbital phases 190 – 240 deg (only 10% of the pulse period is shown). Drifting features are present through most of these data, but are particularly obvious from orbital phases ∼ 200 – 210deg which is enlarged on the right. Single pulses of A can be seen in the background of the left figure, where differential Doppler shifts from the orbital motion result in different apparent pulse periods and hence drifting patterns. The expanded view on the right is overlaid with dots marking the arrival of pulses of A at the centre of B, coinciding with the observed drift pattern in B. (McLaughlin et al. 2004a)

showing the mass of A on one axis and that of B on the other. While the curves will be different for each PK parameter, all curves must intersect in a single point if the chosen theory is a valid description of the nature of this system, i.e. only one unique pair of neutron star masses should exist that satisfies all constraints. Figure 8 displays such a mass-mass diagram where we draw the relationships predicted by GR. We can see that all measured constraints are indeed consistent with this theory of gravity, intersecting in a point which allows us to obtain a precision measurement of the masses of the two pulsars, i.e. $m_A = 1.3381 \pm 0.0007$ M$_\odot$ and $m_B = 1.2489 \pm 0.0007$ M$_\odot$.

Every theory of gravity predicts different relationships between the masses and the PK parameters, usually leading to intersection points located at different positions in the mass-mass diagram. This degree-of-freedom is, however, uniquely removed in the Double Pulsar. The possibility to measure the orbit of both A and B provides a new, qualitatively different constraint in such an analysis. Indeed, with a measurement of the projected semi-major axes of the orbits of both A and B, we obtain a precise measurement of the mass ratio simply from Kepler's third law, via $R \equiv M_A/M_B = x_B/x_A$ where $M_A$ and $M_B$ are the masses and $x_A$ and $x_B$ are the (projected) semi-major axes of the orbits of both pulsars, respectively. We can expect the mass ratio, $R$, to follow this simple relationship to at least the first Post-Newtonian (1PN or $(v/c)^2$ order) level. In particular, the $R$ value is not only

**Table 1:** Parameters for PSR J0737−3039A (A) and PSR J0737−3039B (B) as measured by Kramer et al. (2006). The values were derived from pulse timing observations using the DD and DDS models of the timing analysis program TEMPO and the Jet Propulsion Laboratory DE405 planetary ephemeris (Standish 1998). Estimated uncertainties, given in parentheses after the values, refer to the least significant digit of the tabulated value and are twice the formal 1-$\sigma$ values given by TEMPO. The positional parameters are in the DE405 reference frame which is close to that of the International Celestial Reference System. Pulsar spin frequencies $\nu \equiv 1/P$ are in barycentric dynamical time (TDB) units at the timing epoch quoted in Modified Julian Days. The five Keplerian binary parameters ($P_b, e, \omega, T_0$, and $x$) are derived for pulsar A. The first four of these (with an offset of 180° added to $\omega$) and the position parameters were assumed when fitting for B's parameters. Five post-Keplerian parameters have now been measured. An independent fit of $\dot{\omega}$ for B yielded a value (shown in square brackets) that is consistent with the much more precise result for A. The value derived for A was adopted in the final analysis. The dispersion-based distance is based on a model for the interstellar electron density (Cordes & Lazio 2002) and has an uncertainty of order 20%.

| Timing parameter | PSR J0737−3039A | PSR J0737−3039B |
|---|---|---|
| Right Ascension $\alpha$ | $07^h37^m51^s.24927(3)$ | − |
| Declination $\delta$ | $-30°39'40''.7195(5)$ | − |
| Proper motion in the RA direction (mas yr$^{-1}$) | −3.3(4) | − |
| Proper motion in Declination (mas yr$^{-1}$) | 2.6(5) | − |
| Parallax, $\pi$ (mas) | 3(2) | − |
| Spin frequency $\nu$ (Hz) | 44.054069392744(2) | 0.36056035506(1) |
| Spin frequency derivative $\dot{\nu}$ (s$^{-2}$) | $-3.4156(1) \times 10^{-15}$ | $-0.116(1) \times 10^{-15}$ |
| Timing Epoch (MJD) | 53156.0 | 53156.0 |
| Dispersion measure DM (cm$^{-3}$pc) | 48.920(5) | − |
| Orbital period $P_b$ (day) | 0.10225156248(5) | − |
| Eccentricity $e$ | 0.0877775(9) | − |
| Projected semi-major axis $x = (a/c)\sin i$ (s) | 1.415032(1) | 1.5161(16) |
| Longitude of periastron $\omega$ (deg) | 87.0331(8) | 87.0331 + 180.0 |
| Epoch of periastron $T_0$ (MJD) | 53155.9074280(2) | − |
| Advance of periastron $\dot{\omega}$ (deg/yr) | 16.89947(68) | [16.96(5)] |
| Gravitational redshift parameter $\gamma$ (ms) | 0.3856(26) | − |
| Shapiro delay parameter $s$ | 0.99974(−39, +16) | − |
| Shapiro delay parameter $r$ (μs) | 6.21(33) | − |
| Orbital period derivative $\dot{P_b}$ | $-1.252(17) \times 10^{-12}$ | − |
| Timing data span (MJD) | 52760 − 53736 | 52760 − 53736 |
| RMS timing residual $\sigma$ (μs) | 54 | 2169 |
| Total proper motion (mas yr$^{-1}$) | 4.2(4) | |
| Distance $d$(DM) (pc) | ∼ 500 | |
| Distance $d(\pi)$ (pc) | 200 − 1000 | |
| Transverse velocity ($d = 500$ pc) (km s$^{-1}$) | 10(1) | |
| Orbital inclination angle (deg) | 88.69(−76,+50) | |
| Mass function ($M_\odot$) | 0.29096571(87) | 0.3579(11) |
| Mass ratio, $R$ | 1.0714(11) | |
| Total system mass ($M_\odot$) | 2.58708(16) | |
| Neutron star mass ($m_\odot$) | 1.3381(7) | 1.2489(7) |

*Pulsar Timing* 271

theory-independent, but also independent of strong-field (self-field) effects which is not the case for PK-parameters. Any combination of masses derived from the PK-parameters *must* be consistent with the mass ratio derived from Kepler's 3rd law, i.e. the intersection point mass fall on the mass-ratio line in the diagram. With five PK parameters already available, this additional constraint makes the Double Pulsar the most overdetermined system to date where the most relativistic effects can be studied in the strong-field limit.

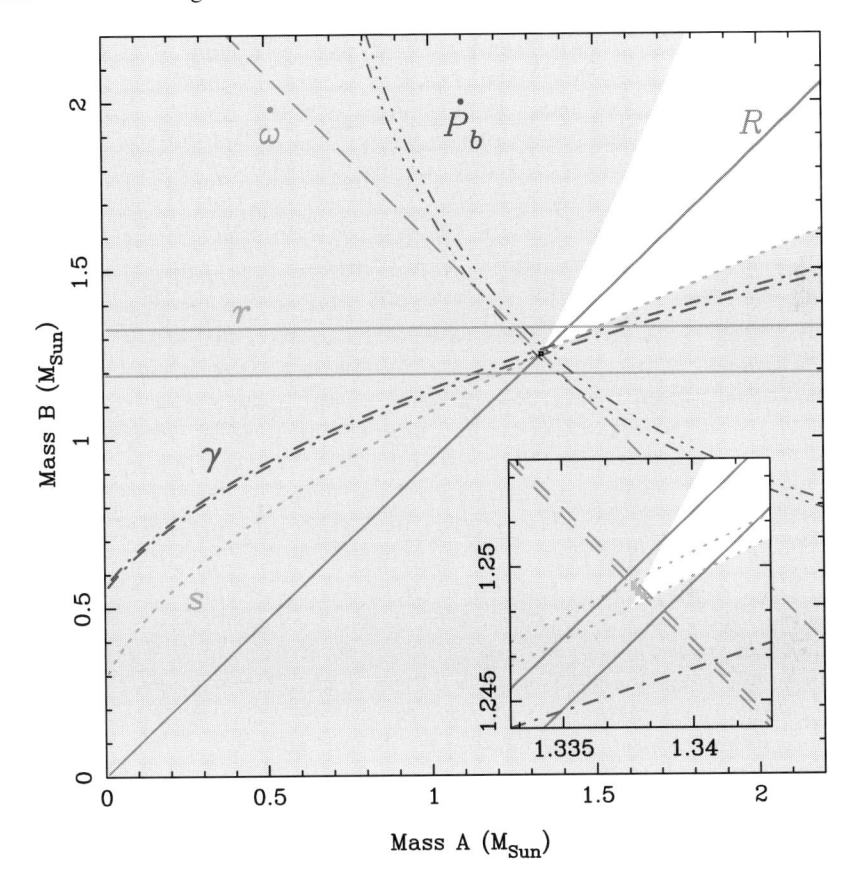

**Figure 8:** 'Mass–mass' diagram showing the observational constraints on the masses of the neutron stars in the double pulsar system J0737–3039. The shaded regions are those that are excluded by the Keplerian mass functions of the two pulsars. Further constraints are shown as pairs of lines enclosing permitted regions as given by the observed mass ratio, $R$, and the PK parameters shown here as predicted by general relativity (see text). Inset is an enlarged view of the small square encompassing the intersection of these constraints. See Kramer et al. (2006) for details.

Assuming GR, we can define the intersection point using the theory-independent mass ratio and the most precise PK parameter $\dot{\omega}$. This leaves four PK parameters, each of which has to pass the very same intersection point. This represents four independent tests of GR which are shown in Table 2, where we list the expected and observed value for each remaining PK parameters. The Shapiro delay gives the most precise test, with $s_{\exp}/s_{\rm obs} = 1.000 \pm 0.0005$. This is by far the best test of GR in the strong-field limit, having a higher precision than the test based on the observed orbit decay in the PSR B1913+16 system with a 30-year data span (Weisberg & Taylor 2005). As for the PSR B1534+12 system (Stairs et al. 2002), the PSR J0737−3039A/B Shapiro-delay test is complementary to that of B1913+16 since it is not based on predictions relating to emission of gravitational radiation from the system (Taylor et al. 1992). Most importantly, the four tests of GR presented here are qualitatively different from all previous tests because they include one constraint ($R$) that is independent of the assumed theory of gravity at the 1PN order. GR also passes this additional constraint with the best precision so far.

The precision of the measured timing system parameters increases continuously with time as further and better observations are made. Soon, we expect the measurement of additional PK parameters, allowing more and new tests of theories of gravity. Some of these parameters arise from a relativistic deformation of the pulsar orbit and those which find their origin in aberration effects and their interplay with geodetic precession. In a few years, we will measure the decay of the orbit so accurately, that we can put limits on alternative theories of gravity which should even surpass the precision achieved in the solar system. On somewhat longer time scales, we will even achieve a precision that will require us to consider post-Newtonian terms that go beyond the currently used description of the PK parameters. Indeed, we already achieve a level of precision in the $\dot{\omega}$ measurement where we expect corrections and contributions at the 2PN level. One such effect involves the prediction by GR that, in contrast to Newtonian physics, the neutron stars' spins affect their orbital motion via spin-orbit coupling. This effect modifies the observed $\dot{\omega}$ by an amount that depends on the pulsars' moment of inertia, so that a potential measurement of this effect would allow the moment of inertia of a neutron star to be determined for the very first time (Damour & Schäfer 1988, Lyne et al. 2004). We do not expect this measurement to be easy, but we will certainly try! Indeed, we only just started to study and exploit the relativistic phenomena that can be investigated in great detail in this wonderful cosmic laboratory.

## 6.3 Alternative theories of gravity

In the previous section, we have concentrated on the comparison of the observations with the predictions of GR. However, alternative theories of gravity exist, all of which must pass the tests provided by the Double Pulsar if they are to be considered further. Some theories, for instance, predict that the Universe's global matter distribution selects a preferred rest frame for local gravitational physics, so that in contrast to GR, the outcome of gravitational experiments depends on the motion of the laboratory with respect to this preferred frame. In particular, theories in which gravity is partially mediated by a vector field or a second tensor field are known to exhibit such preferred-frame effects (PFEs) whose strength is determined by cosmological matching parameters.

**Table 2:** Four independent tests of GR provided by the double pulsar as presented by Kramer et al. (2006). The second column lists the observed PK parameters obtained by fitting a DDS timing model to the data. The third column lists the values expected from general relativity given the masses determined from the intersection point of the mass ratio $R$ and the periastron advance $\dot{\omega}$. The last column gives the ratio of the observed to expected value for each test. Uncertainties refer to the last quoted digit and were determined using Monte Carlo methods.

| PK parameter | Observed | GR expectation | Ratio |
|---|---|---|---|
| $\dot{P}_\mathrm{b}$ | 1.252(17) | 1.24787(13) | 1.003(14) |
| $\gamma$ (ms) | 0.3856(26) | 0.38418(22) | 1.0036(68) |
| $s$ | 0.99974(−39,+16) | 0.99987(−48,+13) | 0.99987(50) |
| $r$(μs) | 6.21(33) | 6.153(26) | 1.009(55) |

In Wex & Kramer (2007), we present a consistent, theory-independent methodology to measure PFEs in binary pulsars that exhibit a high rate of periastron advance. We show that the existence of a preferred frame for gravity would lead to characteristic periodic changes in the orbital parameters, with a period determined by the timescale for the orbital precession. Based on the work by Damour & Deruelle (1985, 1986), we develop a new timing model that describes this signature and demonstrated how timing observations can be used to either measure or constrain the parameters related to a violation of the local Lorentz invariance of gravity in the strong internal fields of neutron stars. In particular, if PFEs indeed exist, two pairs of their newly introduced timing parameters will have a unique relationship that depends only on the orbital parameters of the binary system. This has far-reaching consequences: using the system parameters of the Double Pulsar, we can compute the two numbers that we should expect to measure in the presence of PFEs. If these values were to be significantly measured and were to deviate from these two numbers, the existence of preferred frames and a violation of gravitational Lorentz invariance could be ruled out.

The Double Pulsar is indeed particularly suitable for testing PFEs. As the characteristic PFE signatures are periodic with the orbital precession period, periastron has to have advanced sufficiently far to separate the PFE timing amplitudes from other orbital parameters. After applying our timing model, we however find that the current data span is insufficient, so that we can only derive upper limits for the PFE amplitudes. Currently the periastron has not advanced far enough to separate the PFE amplitudes from other orbital parameters, in particular the decrease in the orbital period due to the emission of gravitational waves. Still, we can translate the current timing results into quantitative limits within the generalised Einstein-Infeld-Hoffmann (EIH) formalism (see Wex & Kramer 2007 for details). The derived limits are currently not very tight (see Fig. 9), but are in fact several orders of magnitude worse than the limits for the corresponding weak-field PPN parameters. However, these limits are qualitatively different as they probe strong-field effects, and they will also improve considerably during the next couple of years. By 2013 the periastron will have advanced by nearly $\pi$ since the pulsar discovery, providing us with measurements of further strong-field parameters.

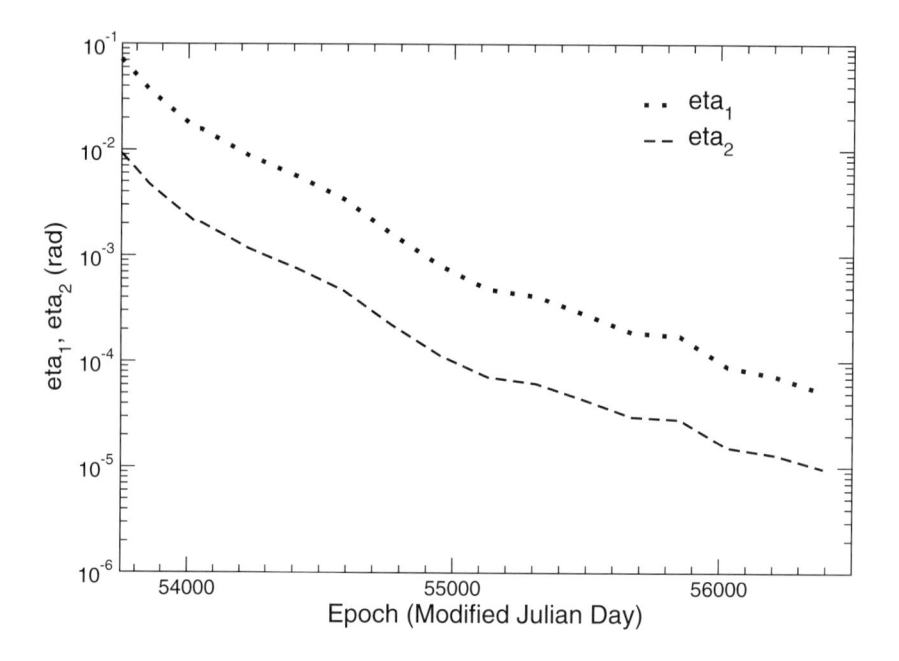

**Figure 9:** Limits on the existence of preferred-frame effects present in the orbital motion of the Double Pulsar as measured by the two parameters $\eta_1^{(\omega)}$ and $\eta_2^{(\omega)}$ as a function of time. See text and Wex & Kramer (2007) for details.

## 7 The Future: The Square-Kilometre-Array

Whilst the currently possible tests of GR are exciting, they are only the prelude to what will be possible once the Square-Kilometre-Array (SKA) comes on-line. The SKA project is a global effort to built a radio telescope interferometer with a total collecting area of $10^6 m^2$. It will be about 100 times more sensitive than the VLA, GBT or Effelsberg and about 200 times more sensitive than the Lovell telescope. Pulsar surveys with the SKA will essentially discover all active pulsars in the Galaxy that are beamed toward us. In addition to this complete Galactic Census, pulsars will be discovered in external galaxies as far away as the Virgo cluster. Most importantly for probing relativistic gravity is the prospect that the SKA will almost certainly discover the first pulsar orbiting a black hole (BH). Strong-field tests using such unprecedented probes of gravity have been identified as one of the key science projects for the SKA (Kramer et al. 2004). The SKA will enable us to measure both the BH spin and the quadrupole moment using the effects of classical and relativistic spin-orbit coupling – impossible with the timing precision affordable with present-day telescopes (Wex & Kopeikin 1999). Having extracted the dimensionless spin and quadrupole parameters, $\chi$ and $q$,

$$\chi \equiv \frac{c}{G}\frac{S}{M^2} \quad \text{and} \quad q = \frac{c^4}{G^2}\frac{Q}{M^3}, \tag{11}$$

where $S$ is the angular momentum and $Q$ the quadrupole moment, we can use these measured properties of a BH to confront the predictions of GR (Wex & Kopeikin et al. 1999, Kramer et al. 2004) such as the "Cosmic Censorship Conjecture" and the "No-hair theorem".

In GR the curvature of space-time diverges at the centre of a BH, producing a singularity, which physical behaviour is unknown. The Cosmic Censorship Conjecture was invoked by Penrose in 1969 (Hawking & Penrose 1970) to resolve the fundamental concern that if singularities could be seen from the rest of space-time, the resulting physics may be unpredictable. The Cosmic Censorship Conjecture proposes that singularities are always hidden within the event horizons of BHs, so that they cannot be seen by a distant observer. Whether the Cosmic Censor Conjecture is correct remains an unresolved key issue in the theory of gravitational collapse. If correct, we would always expect $\chi \leq 1$, so that the complete gravitational collapse of a body always results in a BH rather than a naked singularity (Wald 1984). In contrast, a value of $\chi > 1$ would imply that the event horizon has vanished, exposing the singularity to the outside world. Here, the discovered object would not be a BH as described by GR but would represent an unacceptable naked singularity and hence a violation of the Cosmic Censorship Conjecture (Hawkings & Ellis 1973).

One may expect a complicated relationship between the spin of the BH, $\chi$, and its quadrupole moment, $q$. However, for a rotating Kerr BH in GR, both properties share a simple, fundamental relationship (Thorne 1980, Thorne et al. 1986),

$$q = -\chi^2. \tag{12}$$

This equation reflects the "no-hair" theorem of GR which implies that the external gravitational field of an astrophysical (uncharged) BH is fully determined by its mass and spin (Shapiro & Teukolsky 1983). Therefore, by determining $q$ and $\chi$ from timing measurements with the SKA, we can confront this fundamental prediction of GR for the very first time.

Finally, about 1000 millisecond pulsars to be discovered with the SKA can also be used to directly detect gravitational radiation in contrast to the indirect measurements from orbital decay in binaries. Pulsars discovered and timed with the SKA act effectively as the endpoints of arms of a huge, cosmic gravitational wave detector which can measure a stochastic background spectrum of gravitational waves predicted from energetic processes in the early Universe. This "device" with the SKA at its heart promises to detect such a background, at frequencies that are below the band accessible even to LISA (Kramer et al. 2004).

**Figure 10:** Artistic impression of the SKA as conceived in its reference design. Parabolic dishes shown in the background are complemented by aperture arrays monitoring the whole sky at once visible in the foreground. (Figure provided by the International SKA Project Office)

# 8 Conclusions

Pulsars are versatile objects that find their applications in a wide range of fundamental physics, with implications for diverse topics such as the properties of super-dense matter, cosmology and quantum gravity. Often this is related to the fact, that pulsars provide some of the most stringent and in many cases the only constraints for theories of relativistic gravity in the strong-field limit. Being precise clocks, moving in deep gravitational potentials, they are a physicist's dream-come-true. With the discoveries of the first pulsar, the first binary pulsar, the first millisecond pulsar, and now recently also the first double pulsar, a wide range of parameter space can be probed. The SKA will provide yet another leap in our understanding of relativistic gravity and hence in the quest for quantum gravity.

## Acknowledgements

The author thanks the organizers of the conference and expresses his gratitude to his colleagues working on the double pulsar and in particular to Maura McLaughlin and Marta Burgay for providing figures and to Ingrid Stairs and Norbert Wex for many valuable discussions.

## References

Alpar M. A., Cheng A. F., Ruderman M. A., Shaham J., 1982, Nature, 300, 728

Backer D. C., Kulkarni S. R., Heiles C., Davis M. M., Goss W. M., 1982, Nature, 300, 615

Bignami G. F., Caraveo P. A., de Luca A., Mereghetti S., 2003, in F. Combes D. Barret T. C., Pagani L., eds, SF2A-2003: Semaine de l'Astrophysique Francaise. p. 381

Burgay M. et al., 2003, Nature, 426, 531

Cordes J. M., Lazio T. J. W., 2002, astro-ph/0207156

Damour T., Deruelle N., 1985, Ann. Inst. H. Poincaré (Physique Théorique), 43, 107

Damour T., Deruelle N., 1986, Ann. Inst. H. Poincaré (Physique Théorique), 44, 263

Damour T., Schäfer G., 1988, Nuovo Cim., 101, 127

Damour T., Taylor J. H., 1992, Phys. Rev. D, 45, 1840

Hawking S. W., Penrose R., 1970, Royal Society of London Proceedings Series A, 314, 529

Hawkings S. W., Ellis G. F. R., 1973, The Large Scale Structure of Space-Time. Cambridge University Press, Cambridge

Hessels J. W. T., Ransom S. M., Stairs I. H., Freire P. C. C., Kaspi V. M., Camilo F., 2006, Science, 311, 1901

Kalogera V. et al., 2004, ApJ, 601, L179

Kramer M., Backer D. C., Cordes J. M., Lazio T. J. W., Stappers B. W., Johnston S. ., 2004, New Astronomy Reviews, 48, 993

Kramer M. et al., 2006, Science, 314, 97

Lorimer, D. R. and Kramer, M., 2005, Handbook of Pulsar Astronomy. Cambridge University Press

Lyne A. G. et al., 2004, Science, 303, 1153

Manchester R. N. et al., 2001, MNRAS, 328, 17

McLaughlin M. A. et al., 2004a, ApJ, 613, L57

McLaughlin M. A. et al., 2004b, ApJ, 616, L131

Oppenheimer J. R., Volkoff G., 1939, Phys. Rev., 55, 374

Shapiro S. L., Teukolsky S. A., 1983, Black Holes, White Dwarfs and Neutron Stars. The Physics of Compact Objects. Wiley–Interscience, New York

Stairs I. H., Thorsett S. E., Taylor J. H., Wolszczan A., 2002, ApJ, 581, 501

Standish E. M., 1998, A&A, 336, 381

Taylor J. H., Wolszczan A., Damour T., Weisberg J. M., 1992, Nature, 355, 132

Thorne K. S., 1980, Reviews of Modern Physics, 52, 299

Thorne K. S., Price R. H., Macdonald D. A., 1986, Black Holes: The Membrane Paradigm. New Haven: Yale Univ. Press

Thorsett S. E., Chakrabarty D., 1999, ApJ, 512, 288

Wald R. M., 1984, General relativity. Chicago: University of Chicago Press, 1984

Weisberg J. M., Taylor J. H., 2005, in Rasio F., Stairs I. H., eds, Binary Radio Pulsars. Astronomical Society of the Pacific, San Francisco, p. 25

Wex N., Kopeikin S., 1999, ApJ, 513, 388

Wex N., Kramer M., 2007, MNRAS, 380, 455

Young M. D., Manchester R. N., Johnston S., 1999, Nature, 400, 848

Zavlin V. E., Pavlov G. G., 1998, A&A, 329, 583

# The Assembly of Present-Day Galaxies as Witnessed by Deep Surveys

Klaus Meisenheimer

Max Planck Institute for Astronomy
Königstuhl 17, 69117 Heidelberg, Germany
meise@mpia.de

### Abstract

*This review emphasizes the strength of deep look-back surveys in understanding the formation and assembly of present-day galaxies. Its main aim is to introduce the* non-expert *into strengths and limitations of survey-based observational cosmology. To this end, commonly used observational techniques and methods for the analysis of galaxy surveys are discussed. The present article focusses on optically selected surveys, the completeness of which is limited to $z \lesssim 1.2$ (i.e. eight billion years ago). It is a sign of maturity, that these surveys have agreed on the separation between the two main constituents of the galaxy population – the* red sequence *of passive galaxies and the* blue cloud *of actively star-forming galaxies – and reach converging results on the evolution of their integrated properties, like luminosity and stellar mass functions. More detailed studies are required to understand the global picture in terms of evolutionary tracks of individual galaxies. Here subtle selection effects and systematic errors play an important role. Five examples of state-of-the-art investigations are discussed to demonstrate the enormous progress made recently. I conclude with a summary of established results and the most pressing open questions.*

## 1 Introduction

Surveys which reach objects at redshift $z > 1$ and therefore *look back* over a considerable fraction of the age of the universe date back into the late 1960s when the first quasar surveys based on the 3C radio catalogue identified a handful of objects at redshifts $z > 1$ (Schmidt 1968). A few years later it became clear that quasars without radio emission are even more common and that they can be identified either *photometrically* by their extremely blue UV continuum or *spectroscopically* by their very broad and strong emission lines.

In the course of the 1970s and 1980s many optical quasar surveys were carried out (for an overview about methods an results see Hewett & Foltz 1994). They reached the consensus that at redshift $z \simeq 2$ bright quasars have been $100\times$ more common than today, although no agreement could be reached whether this is caused by pure luminosity evolution (*i.e.* the average quasar was more luminous in the past) or whether also the number density was higher then. However, in any case, QSOs that are the Active Nuclei of Galaxies (AGN), which are powered by accretion on a super-massive black hole, at that time could not tell us much about the evolution of the general galaxy population.[1]

Thus, when in the 1990s Steidel et al. (1996) demonstrated, that a simple color selection in the $(u-g)-(g-R)$–plane is able to identify thousands of galaxies per $\Box°$ at $z \simeq 3$ the interest in quasars ceased. At the same time, surveys for Lyman–$\alpha$ galaxies were pushing the barrier for detecting galaxies out to $z = 5.7$ (Hu et al. 1999), overtaking the highest redshift quasar known at that time. But recently interest in quasars has been revitalized by the Sloan Digital Sky Survey (SDSS) which soon after its commissioning detected the first $z = 5.8$ quasar (Fan et al. 2000) and meanwhile has detected more than a dozen QSOs at $z > 5.7$ (Fan et al. 2006). Their number density and black hole masses (in excess of $10^9 M_\odot$) indicate that they reside in the most pronounced overdensities of the universe and thus mark locations which evolve into the most massive galaxy clusters seen today.

The topic of the present review – deep surveys for distant galaxies – could not prosper before the late 1980s, when the first multi-object spectrographs at 4m class telescopes became available (Low Dispersion Survey Spectrograph – LDSS; see Colless et al. 1990). The first look-back survey for galaxies in our modern sense was the Canadian-French Redshift Survey (CFRS, see Lilly et al. 1995a and subsequent papers in the same issue of ApJ). It contained about 600 galaxies to $I_{AB} \lesssim 22$, which had been sparsely selected in five independent fields. Based on this – by modern standards – tiny sample the CFRS was able for the first time to study how the luminosity function of *normal* galaxies evolved since $z = 1$ (Lilly et al. 1995b). Most notably, already an early version of the "Lilly-Madau diagram" (*c.f.* Fig. 2) could be derived, which clearly indicated that the cosmic Star Formation Rate (SFR) dropped by about a factor of ten between $z = 1$ and today (Lilly et al. 1996). Since then the advent of new observational methods and multi-object spectrographs at 10 m class telescopes have caused an exponential growth in the field of deep surveys: recent studies of the evolution of galaxies are typically based on several ten thousands of galaxies.

Here I will give an overview over the methods and recent results of deep *optical* galaxy surveys. The optical selection restricts the redshift range to $z \lesssim 1.2$, since further out the spectral imprint of an old stellar population is shifted beyond 900 nm and the inclusion of near-infrared (NIR) observations is mandatory to get fair samples of galaxies. Until recently, however, the field of view of NIR cameras was tiny compared to the optical ones and NIR spectroscopy from the ground is still extremely time-consuming even with 8 m telescopes. At the time of writing we witness a new revolution, as the InfraRed Array Camera (IRAC) on board of the Spitzer Space Observatory offers unprecedented sensitivity in a wide range of mid-infrared wavelengths (3.6 to 8 µm). This opens up entirely new opportunities for galaxy surveys at

---

[1] This has changed: today's consensus is that AGN and galaxy evolution are closely linked. See section 5.5.

$z > 1$. But currently samples of IRAC-selected galaxies with accurate redshifts are sparse. Thus, it seems premature to include these efforts here.

My contribution will be centered around the COMBO-17 survey[2] (Wolf et al. 2004) and its high-resolution imaging complement GEMS carried out with the *Advanced Camera for Surveys* (ACS) on board of the Hubble Space Telescope (Rix et al. 2004) for two reasons: *first,* I am deeply involved in those projects and can, therefore, give first-hand account on their methods and results, and *second,* the course of investigations in COMBO-17 and GEMS can serve as leitmotif of the scientific progress which has been achieved by look-back surveys over the past couple of years.

The structure of this review is as follows: In *Section 2* the main goals of look-back surveys will be outlined. *Sections 3 & 4* will present the basic observational techniques and discuss common methods for analyzing the statistical properties of the galaxy populations. The central *Section 5* will demonstrate with five examples, how profound our understanding of galaxy assembly has become. The concluding *Section 6* summarizes established results and points to the most pressing open issues.

The "concordance cosmology" ($H_0 = 73 \,\mathrm{km\,s^{-1} Mpc^{-1}}$, $\Omega_m = 0.27$, $\Omega_v = 0.73$) will be used throughout this article.[3]

## 2 Scientific goals of lock-back surveys

Galaxy surveys out to redshifts $z \lesssim 1.2$ (that is over the period of the last 8 billion years) essentially try to understand how galaxies have evolved into the present day population. In the local universe we can distinguish three main populations of galaxies: spheroidal/elliptical galaxies, spiral galaxies and irregular (dwarf) galaxies. The spheroids are supported by random motion of the stars and are characterized by very red colors, indicating an old stellar population. Spirals (and most irregulars) are supported by rotation and differ in size and regularity of their stellar distribution; most of them contain many young stars which make their overall colors blue. As Fig. 1 demonstrates, this dichotomy existed already six billion years ago.

Today more than 2/3 of the stellar mass is concentrated in the red galaxies which do not form stars any more ("red and dead" galaxies). But recent studies show that this has not always been the case (see lower left panel of Fig. 2): about 7 billion years ago only 40% of the stellar mass resided in red galaxies. Interestingly, the integrated stellar mass density in blue star forming galaxies has not changed over the same period (see upper panel of Fig. 2). On the other hand, we know that the star forming activity of the universe has dropped dramatically (by about a factor of 10) over the last 8 billion years (right panel of Fig. 2). These global trends have been found during the first decade of look-back surveys. But how can they be understood? Obviously, the population of blue, star-forming galaxies has not retained their stars. Somehow they must have been transferred to the red population.

Integrated properties can be measured with reasonable accuracy from rather small samples of galaxies, as long as the systematic effects are well under control.

---

[2] COMBO-17 = Classifying Objects by Multi-Band Observations in 17 filters.
[3] Alternative cosmological parameters would essentially change the look-back times $t_{back}$ assumed here, according to $t'_{back} \simeq t'_0/13.7 \,\mathrm{Gyr} \times t_{back}$, where $t'_0$ is the alternative age of the universe.

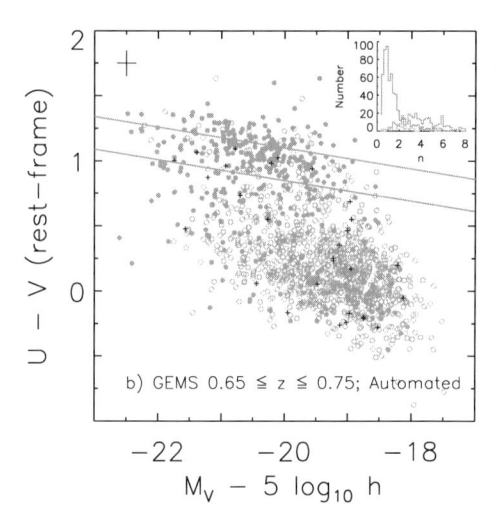

**Figure 1:** Color dichotomy of the galaxy population. Like today, galaxies at $z \simeq 0.7$ (6 Gyrs ago) can be separated by their rest-frame $(U - V)_{\rm rest}$ -colors into a red sequence (($U - V)_{\rm rest} \simeq 1.2$) and a blue cloud (($U-V)_{\rm rest} < 1.0$; Bell et al. 2004). The color-bimodality is almost perfectly mirrored by morphology: red galaxies are dominated by a spheroidal stellar distribution (red •), while blue galaxies are of spiral or irregular morphology (blue ○).

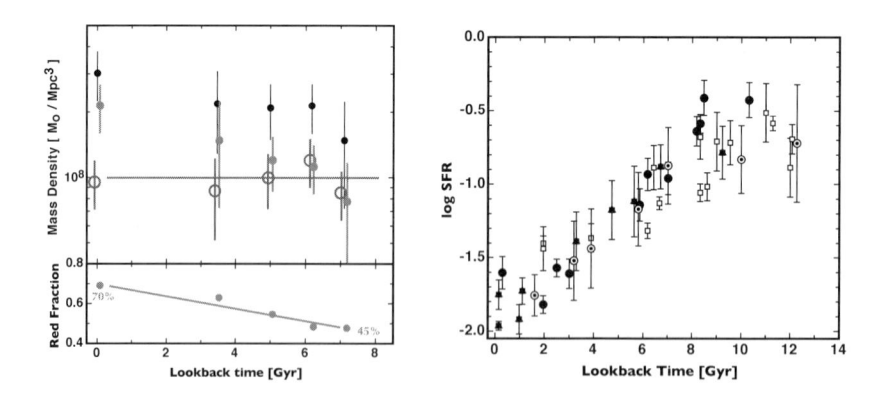

**Figure 2:** The build-up of stellar mass over the last 8 billion years. *Left panel:* Evolution of the integrated stellar mass density in *all* (uppermost, black dots), *red* (red dots) and *blue* galaxies (blue circles). The error bars represent field-to-field variations among three COMBO-17 fields. The *fraction* of red galaxies (lower subpanel) is more robust since the field-to-field variance cancels out (from Borch et al. 2006). *Right panel:* Evolution of the average star formation density ("Lilly-Madau diagram"). The exponential decline of the SFR is established robustly, despite the scatter between different estimators.

*Present-Day Galaxies* 283

But a detailed understanding of the galaxy population including transformation from one class to another requires much larger samples which allow to split them into many bins of varying parameters (such as mass or SFR). Great efforts were undertaken in the past couple of years to collect suitably large samples of galaxies out to $z \simeq 1.2$ in order to find answers to the following questions (see Section 5):

a) In which type of galaxies have the stars been formed?
   For instance, is the drop of star formation caused by the fact that a special population of highly star forming galaxies has died out?

b) How have spiral disks evolved with cosmic time?
   Hierarchical models of galaxy formation predict a substantial growth of the disk size over cosmic time. Can this be verified observationally?

c) How were elliptical galaxies built up?
   Stars predominantly form in blue galaxies but accumulate in red elliptical galaxies. How does the transformation occur?

d) What is the role of the environment of a galaxy in these processes?

e) What is the role of black holes and nuclear activity?
   The tight relation between the mass of the stellar bulge and the mass of its embedded super-massive black hole indicates that their formation is related. An active nuclear phase (AGN) in the life of every galaxy might establish this relation.

As the reader might note, these questions mainly address the assembly of *single* galaxies which is predominately governed by interactions between stars and gas (baryonic matter). Deliberately, I will leave out all questions of *structure formation* in the universe here, as it is dominated by the Dark Matter and Dark Energy content of the universe and galaxies merely act as bright tracers of the underlying invisible structure.

## 3 Methods of deep look-back surveys

The methods that are used in deep surveys can naturally be split as follows. On one hand, one needs to determine *integrated* properties of the galaxies. Most important is the redshift, which allows us to determine the distance (and thus the conversion from observed flux to luminosity and rest-frame colors) as well as the look-back time to an object. But in order to understand the assembly of galaxies from gas and stars or pre-existing building blocks, also the *morphology* and *kinematics* of the galaxies has to be studied over cosmic time. Here, I will discuss the current methods to obtain both the integrated and spatially resolved properties in turn.

### 3.1 Redshift, luminosity, SED

The classical method for determining the redshift of a galaxy is to obtain a *spectrum*. It has been used in deep galaxy surveys since the early days of the LDSS and the

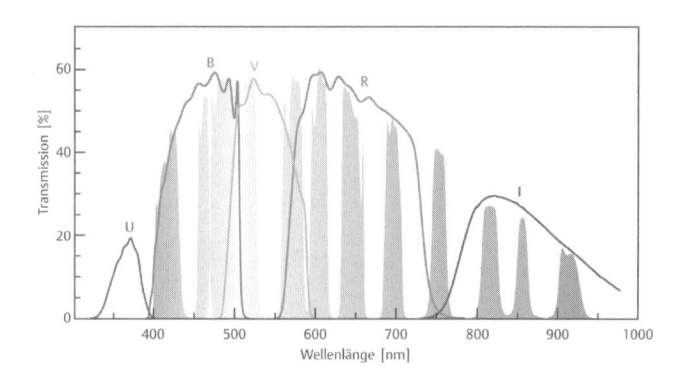

**Figure 3:** COMBO-17 filter set. Photometry in five broad- and 12 medium-band filters results in low-resolution "spectra" for each object.

CFRS (see section 1) which were carried out with the first multi-object spectrographs at 4 m telecopes. With the advent of huge multi-object spectrographs at the VLT 8 m telescope (VIMOS) and the Keck 10 m telescope (DEIMOS) it has become feasible to collect large amounts of galaxy spectra down to $R \simeq 24$ within a reasonable observing time. Three of the most important look-back surveys are or have been carried out at these facilities: The DEEP2 survey (Davis et al. 2003), the VVDS (Le Fevre et al. 2005), and zCOSMOS – the spectroscopic follow up of the multi-band imaging survey COSMOS (Scoville et al. 2007, Lilly et al. 2007).

The huge advantage of "spectroscopic redshifts" is their high redshift accuracy ($\sigma_z/(1+z) \simeq 0.001$ or better). This allows to determine the location of a galaxy (in a cluster, group or close pair) very reliably. Moreover, line diagnostics provides the best tool for constraining age, metallicity and formation history of a stellar population. But even spectrographs with enormous multiplexing capability face problems to sample the surface density of distant galaxies ($\sim 10/\square'$ down to $R = 23.5$) in "one shot". Thus, either sparse sampling of the galaxy population or – for complete surveys – very substantial observing time are still required on 8–10 m telescopes. In addition, slit losses often require photometric calibration of the spectroscopic SEDs.

The alternative method for getting redshifts is to use a set of medium- and broad-band filters. It has been developed for the Calar Alto Deep Imaging Survey (CADIS, for the methodology see Wolf et al. 2001) and successfully applied to investigations of the faint galaxy population out to $z = 1$ (Fried et al. 2001). The first systematic look-back survey with the multi-color technique is COMBO-17 that uses the filter-set shown in Fig. 3 to obtain redshifts with an accuracy $\sigma_z/(1+z) = 0.01$ (at the bright end), deteriorating with photon statistics to $\sigma_z/(1+z) \simeq 0.03$ (at $R = 23.5$; see Hildebrandt, Wolf, Benítez 2008). COMBO-17 has been carried out with the Wide Field Imager (WFI) at the – comparatively small – MPG/ESO 2.2 m telescope.

The multi-color technique has the obvious advantage, that to *every* galaxy above a certain flux limit a redshift will be assigned. No pre-selection or sparse sampling is needed. Since such "photometric redshifts" require to have very accurate photometry in the first place, very well calibrated SEDs are delivered "for free". However, even

*Present-Day Galaxies* 285

the best photometric redshifts are not accurate enough to identify any specific galaxy as member of a group or a cluster: a redshift accuracy of $\sigma_z = 0.01$ corresponds to an uncertainty of the position along the line-of-sight of $> 40$ Mpc!

In the past five years the photometric redshift techniques have matured considerably: The FORS Deep Field (FDF) project (Appenzeller et al. 2004) demonstrated that photometric redshifts as good as $\sigma_z/(1 + z) = 0.03$ can be obtained with broad-band filters only, as long as the wavelength coverage is sufficiently wide (by including NIR-bands) and the relative calibration of data and template spectra is well under control (Bender et al. 2001). Similar results have been reached in the GOODS survey (Mobasher et al. 2004). A redshift accuracy $\sigma_z/(1 + z) \lesssim 0.01$ has recently been reached by the COSMOS-team on the basis of very deep medium- and broad-band observations obtained with the *SuprimeCam* at the Subaru 8 m telescope.

When the redshift of a galaxy is known and good photometry over a wide wavelength range exists, determination of *rest-frame* luminosities and colors in given bands is straight-forward: the simple approach interpolates the flux $F_\lambda$ at some rest-frame $\lambda_0$ between adjacent observed bands $A, B$: $\lambda_A < (1+z)\lambda_0 < \lambda_B$. A more accurate and robust approach is to adjust an appropriate galaxy *template* spectrum to the observed SED and then integrate this template over the desired band transmission $T(\lambda_0)$ in redshifted wavelengths $\lambda = (1+z)\lambda_0$. The latter approach even allows for moderate extrapolations beyond the observed wavelength coverage (by 10–20%).

## 3.2 Images and morphology

At redshifts $0.7 < z < 1.2$ the typical scale factor is $0\rlap{.}''13/\text{kpc}$. Thus, ground-based, seeing limited images can hardly resolve the largest galaxies at that distance. Despite the huge progress in using *adaptive optics* in the near-infrared on large telescopes, the workhorse for imaging studies of distant galaxies has been the Advanced Camera for Surveys (ACS) on board of the HST. Its Wide Field Camera provides a field of view of $3.5 \times 3.5 \,\square'$ and delivers an image quality of $0\rlap{.}''04$ ($u$-band) to $0\rlap{.}''1$ ($i$-band). Unfortunately, in January 2007 the electronics of the wide field camera failed, and we now have to wait for the next *HST servicing mission* before further imaging surveys will be possible. In Tab. 1 the most important imaging surveys with the ACS are summarized.

**Table 1:** Recent ACS imaging surveys. The number of orbits refers to the total on each pointing ($3\rlap{.}'5 \times 3\rlap{.}'5$ FoV).

| Survey | Area | Filters | Orbits | Reference |
|---|---|---|---|---|
| UDF | $3\rlap{.}'4 \times 3\rlap{.}'4$ | $b\,v\,i\,z$ | 400 | Beckwith et al. 2006 |
| GOODS | $2 \times 10' \times 16'$ | $b\,v\,i\,z$ | 15 | Giavalisco et al. 2004 |
| AEGIS | $12' \times 59'$ | $v\,i$ | 2 | Davis et al. 2007 |
| GEMS | $30' \times 30'$ | $v\,z$ | 2 | Rix et al. 2004 |
| STAGES | $30' \times 30'$ | $i$ | 1 | Gray et al. 2007 |
| COSMOS | $85' \times 85'$ | $i$ | 1 | Scoville et al. 2007 |

# 4 Analysis of look-back surveys

Obviously, a look-back survey cannot follow the evolution of any *individual* galaxy over cosmic time. So, the scientific analysis of these surveys have to follow the evolution of the galaxy *population*. In order to understand how galaxies evolve, one needs to build models for galaxy evolution, the parameters of which have to be adjusted such, that the properties of the galaxy population at any time in the past is reproduced. A generic difficulty in this process is that galaxies evolve *forward* in time, while the surveys become less and less accurate when looking *backwards*. That is, one normally cannot start from a complete set of observed *initial conditions*, but most of the initial conditions are free or pre-determined by other means (like observations of the cosmic microwave background). The defining *boundary condition* is provided by the galaxy population in the local universe. As examples for the analysis of the galaxy population I discuss the evolution of stellar luminosity and stellar mass.

## 4.1 The luminosity function and its evolution

The most common way in describing the luminosity distribution of the population of galaxies is the *luminosity function*. It represents the number density of galaxies in an *absolute* magnitude bin $M$, $M + \Delta M$ per unit *comoving* volume. The top row in Fig. 4 clearly demonstrates an important limitation for following the luminosity function over large periods (*i.e.* large distances): in a flux-limited survey the absolute magnitude limit for the faintest galaxies changes by 4 magnitudes when going from $z \simeq 0.3$ (leftmost panel in Fig. 4) to $z \simeq 1.1$ (rightmost panel). In this way the faint end slope of the luminosity function becomes less and less determined. Since in the standard description of galaxy luminosity functions as *Schechter functions* (that is a power-law, which cuts off exponentially towards the bright end), the faint end slope $\alpha$ and the other two parameters $\phi^*$, the normalization at some characteristic magnitude $M^*$ all depend on each other, this uncertainty in $\alpha$ can also render the formal values of the fitted $\phi^*$, $M^*$ meaningless. The normal way around this problem is to fix $\alpha$ at the value which is well determined at the least distant redshift bin. For understanding the galaxy evolution in an *hierarchical scenario*, in which luminous galaxies are built up from small (faint) ones, this is very bad news: the initial distribution of faint galaxies remains rather undetermined. A combination of shallow, wide surveys with narrow but very deep surveys can partly solve the problem (see upper left panels in Fig. 4).

## 4.2 The stellar mass function and its evolution

The luminosity of a galaxy at (rest-frame) optical wavelengths strongly depends on its star formation rate. Even if only a minor fraction of its stars have been formed very recently, the very luminous, massive young stars will outshine any older stellar population. Therefore, sorting galaxies by their luminosity could be very misleading. A much better measure for the bigness of a galaxy is its mass. Although it

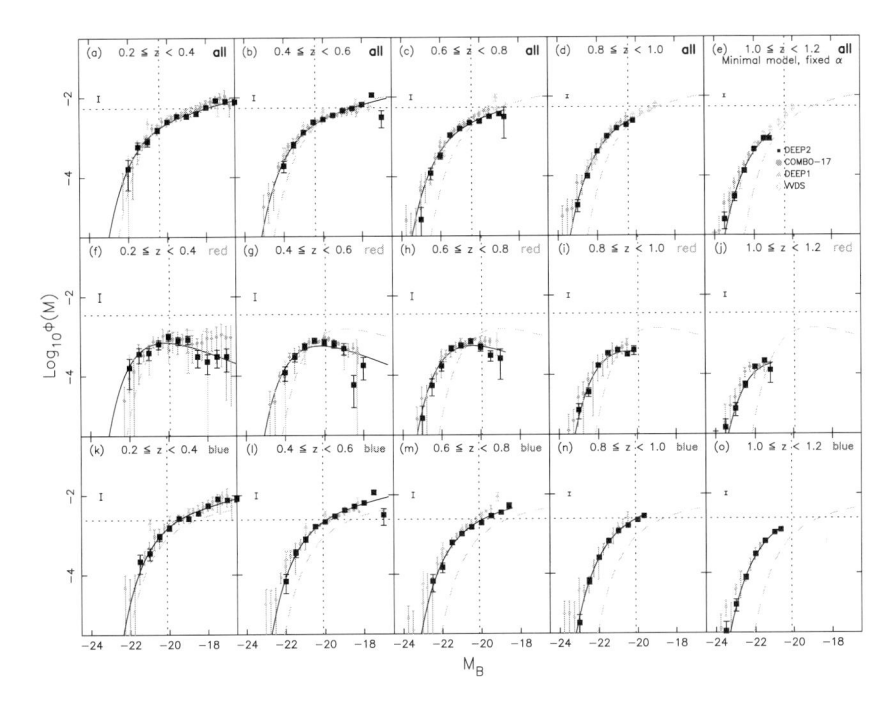

**Figure 4:** Evolution of the $B$-band Luminosity Function (LF) since redshift $z = 1.2$ (from Faber et al. 2007). The *top row* shows the LF for *all* galaxies. The *middle* and *bottom row*, respectively, separate *red* and *blue* galaxies. The data from three large surveys (DEEP2, COMBO-17 and VVDS) are in good agreement. Compared with the most recent LF (at $0.2 < z < 0.4$, long dashed) both the brightest red and blue galaxies were brighter on average in the past. The population of red galaxies seems to have increased over time. Note, how the deep part of the VVDS ($I_{AB} \leq 24$) constrains the faint end slope at $z > 0.8$ (two rightmost panels in *top row*).

would be desirable to use *dynamical* masses, this can be achieved currently only for a few hundred galaxies beyond $z = 0.5$. Therefore, it is common practice to work with the *stellar mass* of galaxies as a proxy. *In principle,* the stellar mass $M_\star$ can be derived from the luminosity $L$ in some band and the mass-to-light ratio $M_\star/L$ which depends on the stellar population mix of the galaxy. The latter has to be determined from the SED. *In practice,* however, even with spectroscopically observed SEDs it is far from trivial to dissect the stellar population of a distant galaxy according to stellar *age, metallicity* or to estimate the average *extinction*, to name the most important parameters which influence $M_\star/L$. Photometric SEDs, which need to be interpreted with the help of spectral templates, generated by population synthesis codes (*e.g.* Fioc & Rocca-Volmerage et al. 1997, Bruzual and Charlot 2003, Maraston et al. 2006), are subject to even larger uncertainties, since the ambiguities between *e.g.* older age, higher metallicity or stronger extinction can hardly be distinguished. Fortunately, when determining the stellar mass, the errors in those parameters do not aggregate but rather tend to cancel each other (Bell & de Jong 2001).

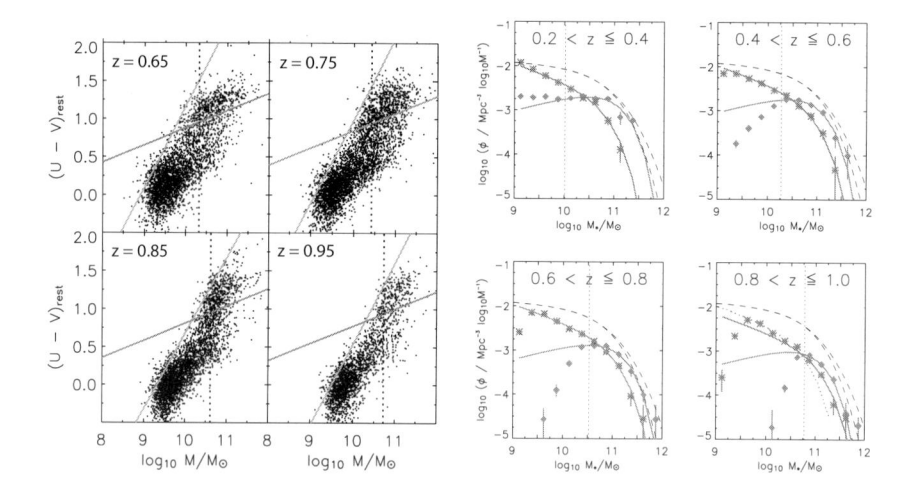

**Figure 5:** Stellar Mass Function derived from COMBO-17 (Borch et al. 2006): *Left panel:* Color-mass diagram for the four highest redshift bins ($\Delta z = 0.1$). The inclined dotted line shows that the optical selection $R < 24$ introduces a color-dependent mass cut. The completeness limit is set by the reddest galaxies (dashed vertical line). An inclined continuos line shows the separation between red and blue galaxies adopted in COMBO-17. *Right panel:* Mass functions for red ($\diamond$) and blue ($*$) galaxies. The Schechter function fits assume a redshift-independent low-mass slope. The dashed lines represent the total mass function of local galaxies.

Nevertheless, UV- or blue-band $M_\star/L$ ratios are more strongly affected by extinction and age (more precisely: Star Formation History – SFH) of the stellar population and thus most uncertain. But since young, luminous stars dominate the luminosity in any optical-to-NIR band, similar uncertainties exist also at longer wavelengths (see Maraston et al. 2006). A rest-frame $I$–band ($0.8 < \lambda_0 < 1.0\,\mu$m) seems to offer a good compromise.

For optically selected surveys a further complication arises for sampling galaxies by mass and determining their mass function: the observed selection band shifts into bluer and bluer rest-frame wavelength, thus more and more favoring the detection of young stars (with low $M_\star/L$). Accordingly, the limiting stellar mass of the survey is a strong function of the rest-frame color (see left panel of Fig. 5).

The current mass estimation in the COMBO-17 survey (Borch et al. 2006) uses a one dimensional sequence of population synthesis templates generated by PEGASE2 for a three-component SFH (very old and young burst, some continuous star formation) to interpret the observed SED. We assume a Kroupa IMF (Kroupa 2001). The mass estimate always refers to the flux in the reddest deep medium band of COMBO-17 ($\lambda/\Delta\lambda = 815/16$ nm), thus using a redshift dependent $M_\star/L(\lambda_0)$ with $\lambda_0 = 815\,\mathrm{nm}/(1+z)$. The requirement to sample the old stellar population in the filter limits this method to $z < 1$. The main results are that the most massive galaxies have been red at all times since $z \simeq 1$ and the characteristic mass of blue star forming galaxies moves to lower mass with cosmic time, indicating that active star formation becomes confined to smaller galaxies (sometimes referred to as "down-sizing").

Integrating the mass functions in Fig. 5 for the three observed COMBO-17 fields separately, results in the stellar mass density evolution (and its variance due to field-to-field variations) given in Fig. 2.

## 4.3 Stellar mass functions from DEEP2 and VVDS

In order to get a feeling for the uncertainties of current determinations of stellar mass function, it is instructive to compare results obtained from different surveys with alternative techniques. Conceptionally, the approach used by Bundy et al. (2006) is similar to the COMBO-17 method, but using the observed flux in the $K_s$-band as reference. The sample selection and redshifts are taken from DEEP2. The template set assumes an exponentially declining SFH (of various age) and a wide variety in metallicity and extinction. The grid of templates is compared with the observed BRIK colors and the resulting probability distribution is binned according to stellar mass. The median of this probability distribution is chosen as best mass estimate. As Fig. 6 (left panel) shows, the results of Bundy et al. are in excellent agreement with the COMBO-17 results. Minor discrepancies can easily be understood by differences in sample selection and the separation between red and blue galaxies.

Type-dependent mass functions for 4048 galaxies at $0.5 < z < 1.3$ from the VVDS have been presented by Vergani et al. (2007; right panel of Fig. 6). Their methods differ in two aspects: first, galaxy templates are fitted to a combination of the *spectroscopic* and *photometric* data. The grid of templates is derived by assuming a single population of stars with exponentially declining SFH by varying the initial age and the decline time. A Salpeter IMF (which typically results in $\sim 1.8\times$ higher stellar masses than the Kroupa IMF) is assumed. Second, the authors divide their sample not in red and blue galaxies but in *early* and *late type* galaxies according to the $D_{4000}$ parameter (strength of the 4000 Å break as *e.g.* used in Kauffmann et al. 2003). Despite these differences, the mass function for the late type galaxies seems to agree reasonably well with those for the blue galaxies from DEEP2 and COMBO-17. However, especially at the highest redshifts (lower panels of Fig. 6) the normalization of the early type galaxies is more than a factor of 2 lower than reported by Bundy et al. (2006). The reason for this disagreement is unclear.

## 4.4 Morphological classification

Traditionally, galaxy morphologies are classified by *Hubble type*. This is a human eye-based scheme which takes many aspects like light concentration, bulge-to-disk ratio, strength and opening angle of spiral arms, occurrence of a bar and others into consideration. Modern surveys (*e.g.* SDSS or the various ACS surveys listed in Tab. 1) need to classify many thousands of galaxies. Thus, more and more efforts are spend to resort to machine-based classifications. As a first step, the light profile of a galaxy is fitted by a more or less complicated model, taking the effects of a finite point spread function into account (*e.g.* GALFIT, Peng et al. 2002; GIM2D, Simard et al. 2002). As second step the parameters of the fit are sorted in various categories

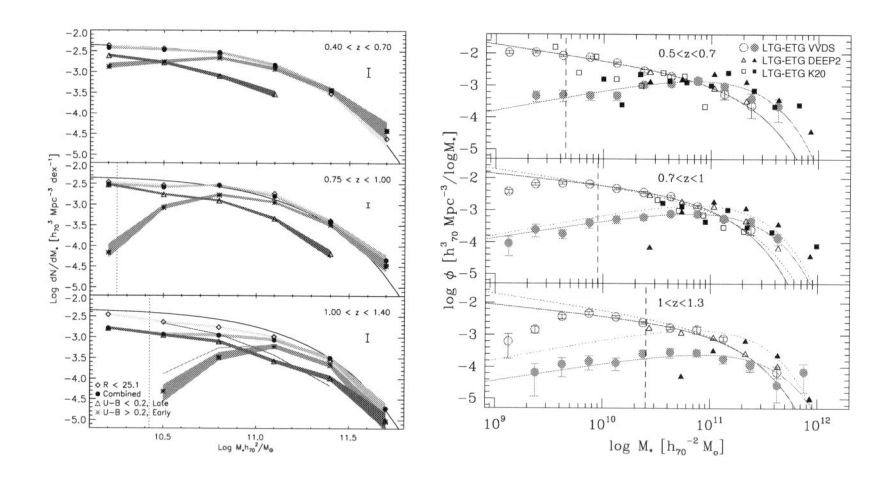

**Figure 6:** Evolution of stellar mass function for red and blue galaxies as derived from DEEP2 (Bundy et al. 2006; *left panel*) and VVDS (Vergani et al. 2007; *right panel*).

(like bulge-to-disk ratio, isophotal twist, asymmetry). The challenge in getting a consistent morphological classification in look-back surveys is two-fold: first, the observed band ($\lambda_{\rm obs}$) will correspond to a shifting rest-frame band $\lambda_0 = \lambda_{\rm obs}/(1+z)$, thus increasing the influence of young (blue) stars with redshift. Second, the surface brightness dimming $\propto (1+z)^{-4}$ will confine the detected area above the background noise to smaller and smaller physical radii.

For these reasons, unbiases studies of the evolution of higher-order moments (like the bulge-to-disk ratio) are extremely tough. Instead many investigations limit themselves to a generalize description of the radial light profile $i(r)$, the Sérsic profile:

$$\sigma(r) = \sigma_e \exp\left\{-\kappa(n)\left[(r/r_e)^{1/n} - 1\right]\right\},$$

which contains essentially one free parameter[4], the Sérsic index $n$: a pure exponential disk has $n = 1$, the de Vaucouleur profile of elliptical galaxies corresponds to $n = 4$ (Sérsic 1968). Commonly a critical value $n_c = 2.5$ or 2.0 is used to decide between spheroid-(bulge-) and disk-dominated galaxies. In practice, the inclination of the galaxy, which results in an ellipticity of the 2D profile, is taken into account as a second free parameter.

## 5 The assembly of galaxies – selected results

In this section, I will demonstrate with a few selected examples that look-back surveys have allowed us to make considerable progress in understanding how the properties of present-day galaxies have evolved over cosmic time. Its is organized along the five key questions which have been posed in Section 2.

---

[4]The parameter $\kappa(n)$ secures the normalization $\sigma(r_e) \equiv \sigma_e$

## 5.1 In which type of galaxies have the stars been formed?

It is well established, that over the last eight billion years the global star formation rate of the universe has dropped by a factor of ten (see right panel of Fig. 2). Is this drop caused by a general decline of star formation activity throughout all types of galaxies or is it rather a sign that the most actively star forming galaxies (*i.e.* Ultra-Luminous InfraRed Galaxies – ULIRGs) have been much more abundant in the past?

In todays universe $> 90\%$ of new stars are formed in spiral disks and irregular galaxies (Fig. 7), while major mergers (with their ULIRG-type SFR) are so rare that they contribute $< 5\%$ to the overall budget.

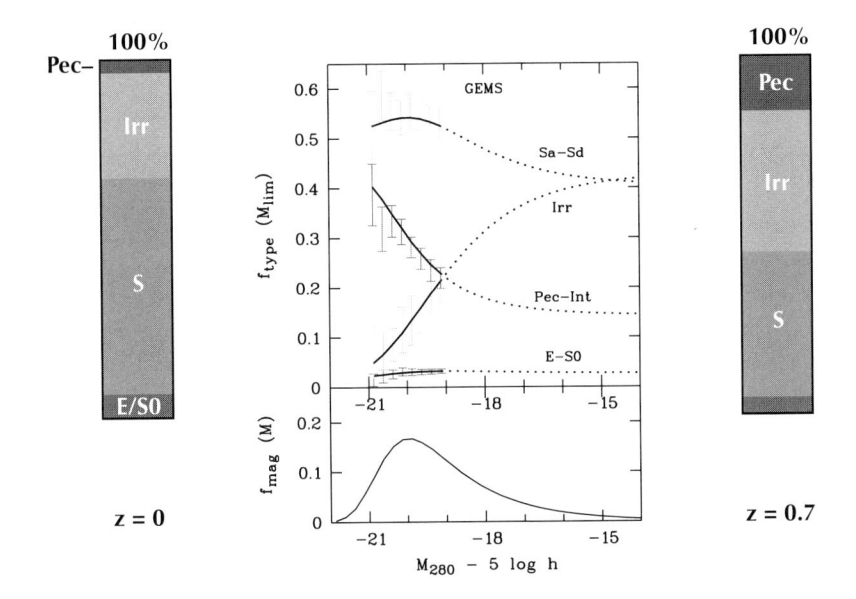

**Figure 7:** Fractional contribution to the total SFR by different types of galaxies today (*left* panel) and at $z \simeq 0.7$ (*right* panel). The *middle* panel shows the fractional contribution to the total UV-luminosity at 280 nm, as derived from integration over the type-dependent luminosity functions.

Two studies based on the combined COMBO-17/GEMS data have addressed this question by comparing the type-dependent SFR at $0.65 < z < 0.75$ with the present situation. In both studies, galaxies are classified into four morphological types: E/S0, Sa-Sd, Irr, and "Peculiar-Interacting". The latter type includes very asymmetric or disturbed morphologies, indicative of ongoing mergers. The morphological classification – by the eyes of three "observers" – was obtained on the ACS images through the F850LP filter (*i.e.* rest-frame $\lambda_0 \simeq 540$ nm at $z = 0.7$). The first study (Wolf et al. 2005) uses the UV-luminosity at rest-frame wavelength $\lambda_0 = 280$ nm as proxy for the SFR. Integrating over the fractional UV luminosity functions of the four morphological types then gives the contribution to the total SFR (Fig. 7, middle panel). Although there is some uncertainty in the morphological classification between Irr and Pec-Int at the faint observed end ($M_{280} > -20$) and the extrapolation towards fainter galaxies, it is established that no more than 25% of the SF at $z \simeq 0.7$ happens in merging systems. The rightmost panel of Fig. 7 summarizes this result and shows that, like today, $> 75\%$ of the star formation was occurring in spiral and irregular galaxies. The second study (Bell et al. 2005) uses the MIR luminosity at $\lambda_0 \simeq 14\,\mu m$ derived from MIPS 24 μm observations with the Spitzer Space Observatory as proxy for FIR luminosity, which in turn is a very good measure of the total SFR. Although most sensitive to obscured star formation (as in ULIRGs), the essential result is the same: $> 50\%$ of the stars are formed in spirals and $< 25\%$ in Pec-Int systems. The 24 μm observations are not sensitive enough to detect a fair fraction of the irregular galaxies. Thus their contribution remains undetermined. It is worth to note, that the spiral galaxies at $z = 0.7$ are substantially more MIR luminous than their local relatives – moving them at least into the class of Luminous IR Galaxies (LIRGs). This is consistent with earlier claims that at higher redshift an increasing fraction of star formation occurs in IR-luminous galaxies but without supporting the original interpretation of an enhanced merger rate.

## 5.2 How have spiral disks evolved with time?

Our interest in understanding the assembly of spiral galaxies is twofold: in the last section we have seen, that they dominate the global SFR at least during the second half of the life of the universe. Second, since we live in a large spiral galaxy, it is part of our own pre-history to understand how the Milky Way evolved into its present structure.

There are various geometrical parameters which could give hints to disk evolution: first the disk size characterized by the effective radius $r_e$, second the bulge-to-disk ratio which might increase due to mergers, and third the fraction of barred spirals.

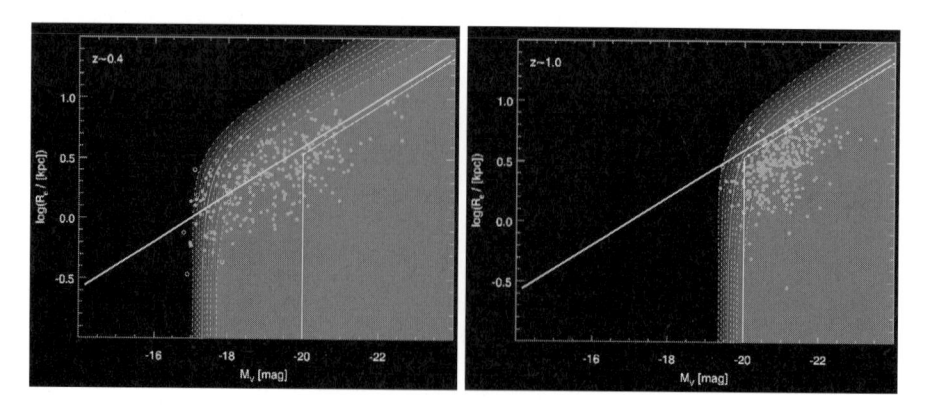

**Figure 8:** Magnitude-size relation of disk galaxies in GEMS at $0.3 < z < 0.5$ *(left)* and $0.9 < z < 1.1$ *(right)*. The grey shaded areas represent the completeness maps for the center of the redshift bin. A thick (orange) line shows the local relation from SDSS.

The large area of the GEMS mosaic allowed a very detailed study of the disk size evolution on the basis of 8000 galaxies at $z < 1.1$ with well-determined properties (Barden et al 2005). The Sérsic index $n$ (see Section 4.4) determined with GALFIT (Peng et al. 2002) on the ACS F850LP images was used to identify disk-dominated galaxies by $n < 2.5$. In the end we were left with 5664 disk galaxies with well-determined $r_e$. The resulting magnitude-size relations for two of our five redshift bins are shown in Fig. 8. Monte-Carlo simulations were used to determine redshift-dependent selection effects both in absolute magnitude (flux limit of the sample) and surface brightness (dimming by $(1+z)^{-4}$) and show that even in the highest redshift bin the distribution is hardly affected by the surface brightness limit (Fig. 8). A comparison sample of local disk galaxies ($z \simeq 0.05$) was constructed from the SDSS and treated in a similar manner. As at all redshifts the slope of the magnitude-size relation is not far from that of constant surface brightness, we chose to represent our results by one parameter only – the average surface brightness $\langle \mu_V \rangle$ at $M_V = -20$ as derived from the rest-frame V luminosity and the effective radius $r_e$. The left panel of Fig. 9 demonstrates that $\langle \mu_V \rangle$ is dimming by about 1 mag between $z = 1$ and today, that is disk galaxies have either been smaller or brighter in the past.

A more conclusive picture emerges when one plots the evolution of the average (stellar) mass surface density within $r_e$ by using the stellar mass $M_*$ derived by COMBO-17 (see Section 4.2): as the right panel in Fig. 9 shows this surface mass density stays amazingly constant since $z = 1$. Since disk galaxies are actively forming stars throughout that period, there must be a moderate increase in disk size. However, naive CDM models, in which the disk size scales with the size of the dark matter halo (Mo, Mao & White 1998) predict a much more pronounced size evolution which effectively would decrease the surface mass density by about a factor of 3 since $z = 1$ (see Fig. 9).

Does this imply that the moderate disk size evolution is in conflict with the hierarchical picture of galaxy assembly? Somerville et al. (2008) have inspected this issue more closely: in fact they can demonstrate that the on-going concentration of the baryonic matter in the center of the DM halo leads to a substantial deepening of the central potential thus invalidating the scaling between halo and disk size. Although their model still predicts a slight decrease in average surface mass density (Fig. 10), it is consistent with the observed evolution when considering the statistical errors and systematic effects of only 25% which certainly cannot be excluded.

An independent confirmation that the surface mass density in disk galaxies stayed roughly constant over the last 7 billion years comes from recent investigations of the evolution of the Tully-Fisher (TF) relation. Based on 135 disk galaxies in the GEMS field, for which good (central) rotation curves could be obtained with VIMOS at the VLT, Koposov et al. (2008) find that the $B$-band TF relation was shifted by $\Delta m \simeq 1.5$ at $z = 1$ with respect to the local relation. When using stellar mass instead of luminosity, there is little evidence for changes in the slope or the normalization of the mass-TF relation (Fig. 11). This result agrees well with the study of the mass TF relation in the AEGIS/DEEP2 project (Kassin et al 2007) which also found no evidence for evolution, when disordered motions of the gas are properly taken into account. It should be noted, however, that based on early DEEP2 data, Vogt et al. (1997) and Vogt (2001) did not see the apparent brightening in the $B$-band TF

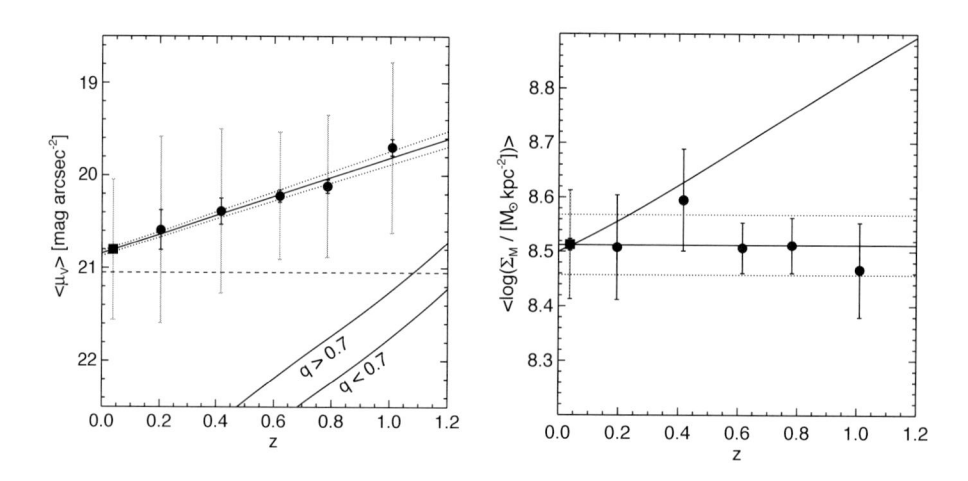

**Figure 9:** Evolution of disk galaxies. *Left panel:* apparent surface brightness. Large error bars refer to the RMS scatter, short error bars to the error of the mean value. *Right panel:* surface mass density and error of its average. The prediction by Mo et al. (inclined line) is inconsistent with the observations.

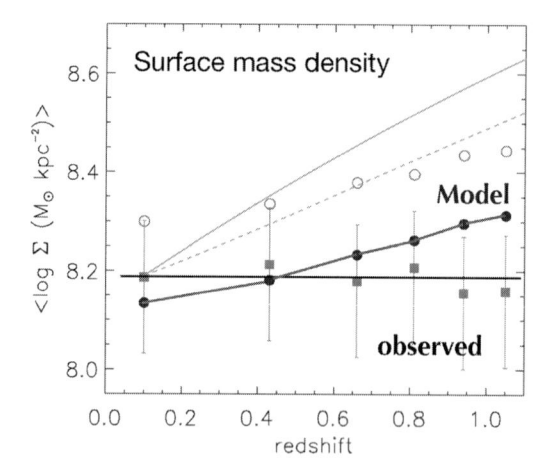

**Figure 10:** Surface mass density evolution for disk galaxies (corrected for inclination). Observations from Barden et al. (shown as ■ with error bars) are compared with predictions by Somerville et al. (connected, blue ●). Other symbols and lines show alternative models.

relation out to $z = 1$. Moreover, based on data from the FDF, Ziegler et al. (2002) reported a change in slope near the faint end of the TF relation. Some indication for that might indeed be present in Fig. 11.

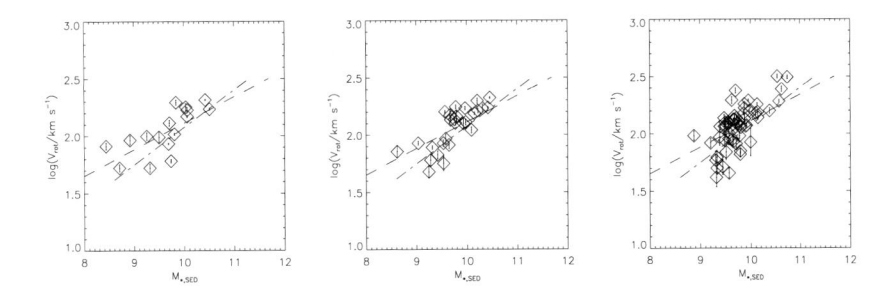

**Figure 11:** Mass Tully-Fisher relation for redshift bins around $z = 0.3$, $z = 0.5$ and $z = 0.7$ (from left to right). The dashed and dashed-dotted lines, resp., indicate two determinations of the mass TF relation in the local universe.

## 5.3 How have elliptical galaxies built up?

From the evolution of the integrated stellar mass density (Fig. 2) and the finding of Section 5.1 that almost all stars form in blue disk and irregular galaxies an apparent contradiction is evident: most of the stars accumulate in elliptical galaxies although the cannot form there. A graphic sketch of the dilemma is given in Fig. 12: constant SFR would move a disk galaxy right and slightly up in the diagram but never onto the red sequence. In order to reach the red sequence star formation has to be turned off completely (vertical arrow). But it is evident that blue progenitors, which could produce the most massive ellipticals ($M_\star > 3 \times 10^{11} M_\odot$) in this way, do not exist. If one carefully considers the different tracks along which galaxies could evolve through the diagram, it turns out that the most likely track to reach the red sequence at $M_\star > 10^{11} M_\odot$ is along the arrows parallel to the red sequence, indicating the merger between two elliptical galaxies of similar mass. Is there evidence in the surveys that such "dry mergers" occur and do they happen often enough to build up the most massive elliptical galaxies?

Bell et al. (2006) have used the GEMS/COMBO-17 data base to investigate this question. Among 379 red galaxies they found six pairs (12 galaxies) which show strong signs of an ongoing merger (top row in Fig. 13). In order to determine the merger rate (per unit time) one needs to estimate for how long obvious merger signatures could be visible on the GEMS images. To this end they performed N-body simulations of major mergers between spheroidal systems, assigned a proper luminosity and color to them and simulated their appearance in the GEMS survey (lower row in Fig. 13). From the comparison between observed mergers and the models, they derived a typical timescale of 200 Myrs during which a merger event should be evident on the GEMS images. From this they conclude, that each or every second massive elliptical galaxy today has experienced a major dry merger over the past seven billion years – a rate which seems (just) sufficient to expain the build-up of the most massive part of the red sequence.

A similar study in the AEGIS survey by Lotz et al. (2008) reaches very similar conclusions about the abundance of red pairs (see Fig. 14).

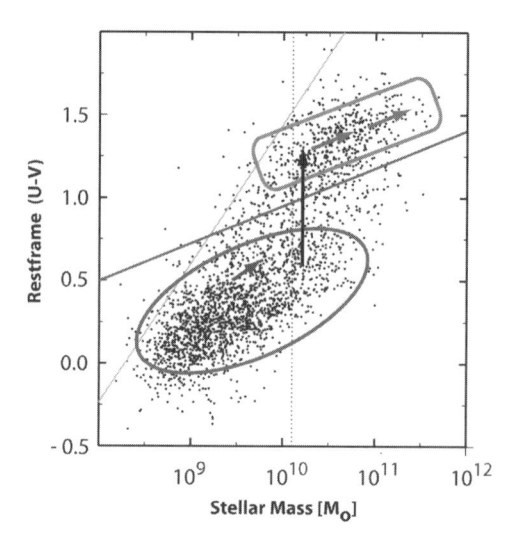

**Figure 12:** The population of galaxies at intermediate redshift, $z \simeq 0.45$ displayed as a color – stellar mass diagram. The red sequence is clearly separated from the blue cloud (dividing line at $(U - V)_{\rm rest} = 0.225 \log(M_\star/M_\odot) - 1.3$). See text for description of the possible evolutionary tracks (arrows).

## 5.4 What is the role of environment?

Early hints to the role of environment in shaping the galaxy population came from the morphology-density relation (Dressler 1980) or the equivalent color-density relation and from the Butcher-Oemler effect (Butcher & Oemler 1984). It is a long-standing matter of debate, whether the transformation from blue to red galaxies mostly happens in a cluster environment or already in smaller groups which later agglomerate into the clusters.

As an example how look-back surveys can address this issue, I discuss our finding of hidden star formation in red spirals in the outskirts of the super-cluster Abell 901/902 located at $z = 0.17$ (Wolf, Gray & Meisenheimer 2005). This work is based on the current COMBO-17 library of galaxy spectra which adds a screen of extinction as second dimension to a age sequence of stellar populations (exponentially declining SFH). Starting from a redshift selection of probable cluster members, a cut at "Age> 3000 Myr" provides a distinction between blue cloud and the red sequence (see Fig. 15). A second cut $E_{B-V} > 0.1$ isolates those galaxies on the red sequence, the spectra of which are not consistent with a purely old stellar population. Indeed, the observed color-color diagrams show that the reddest galaxies at $V - I > 1.5$ split into a tight "true" red sequence (upper left panel of Fig. 15) and a branch of galaxies which are bluer in color indices of filters straddling the 4000 Å break at $\lambda_{\rm obs} = 470$ nm (*i.e.* these galaxies show a relative small $D_{4000}$, compared to their color at longer wavelengths). Enhanced dust extinction indeed provides a

*Present-Day Galaxies*

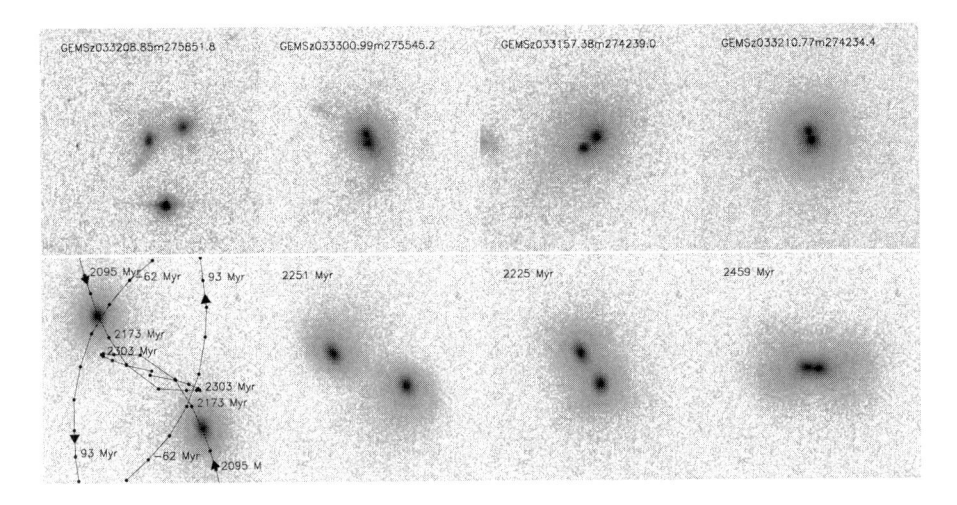

**Figure 13:** Major "dry mergers" between two red galaxies of comparable luminosity. *Upper row:* Four out of the six pairs observed in GEMS. *Lower row:* Simulations of mergers between two equal mass gas-free speroidal galaxies.

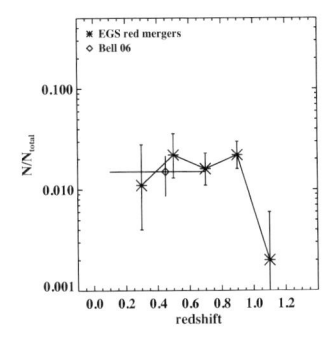

**Figure 14:** Fraction of red galaxy pairs in the Extended Groth Strip (EGS) which seem to form a dry merger system (Lotz et al. 2007). There is good agreement with the GEMS average over $0.1 < z < 0.7$ (Bell et al. 2006).

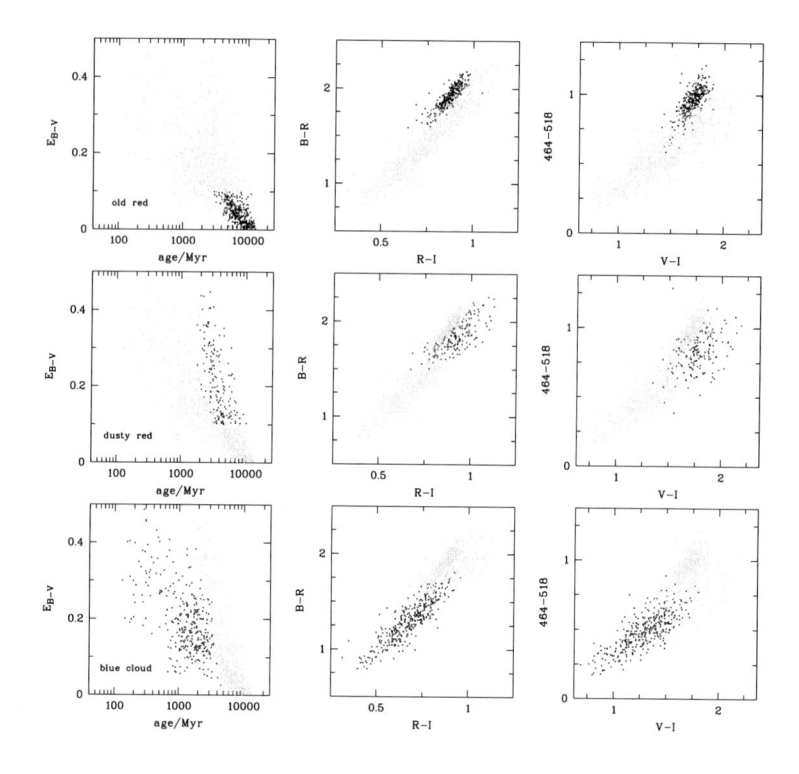

**Figure 15:** The distinction between old red *(top row)*, dusty red *(middle)* and blue galaxies *(bottom)* in the COMBO-17 data of Abell 901. The three types selected in the *age-extinction* plot *(left column)* can clearly be distinguished on various *color-color* diagrams *(middle and right columns)*.

natural explanation for he redder $R - I$ or $V - I$ colors. The presence of [OII]372.7 emission in stacked spectra of these "dusty red galaxies" prove that they are actively star-forming. In addition, most of them are detected at 24 µm with MIPS on board of the Spitzer satellite (Gallazzi et al. 2008), further supporting the case for enhanced star formation.

Interestingly, the dusty red galaxies show a projected distribution across the $32' \times 31'$ A901/902 field which is quite special: they are neither as concentrated towards the cluster cores as the old red galaxies nor do they appear as wide-spred as the blue galaxies. Figure 16 demonstrates this behavior by using $\Sigma_{10}$, that is the projected number density (per Mpc$^2$) of surrounding galaxies within the radius to the 10th neighbor.[5] In each $\Sigma_{10}$ bin, the fraction of blue, old red and dusty red galaxies is calculated. The field value is taken from other COMBO-17 fields and plotted at the mean field density $\log \Sigma_{10} = 0.7$. The behavior of the blue and old red galaxies is a perfect demonstration of the color-density relation. The fraction of the dusty

---

[5] In a strong overdensity like A901, the projected density estimator $\Sigma_{10}$ is a monotonic function of volume density.

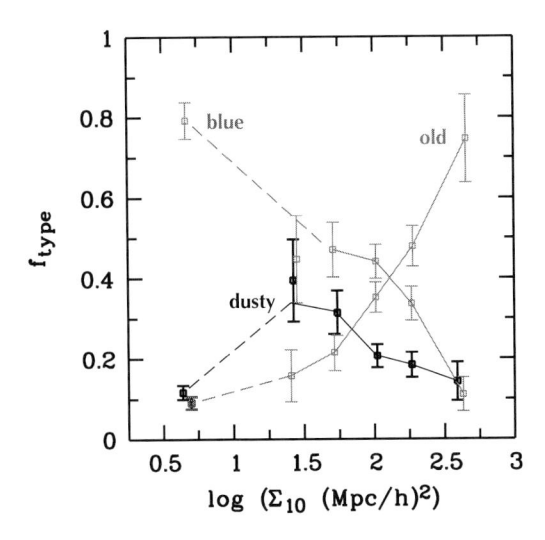

**Figure 16:** Fractional contribution of old red, dusty red and blue galaxies, sorted by the local density of their environment, $\Sigma_{10}$.

red galaxies, however, increases outwards from the cluster cores (at $\log \Sigma_{10} > 2.5$) reaching $f = 40\%$ in the outskirts of the A901/902 complex. As their fraction is only about 12% in the comparison fields, we conclude that there exists some maximum not far outside our field-of-view (projected radius from the cores: $\sim 10' = 2.0\,\mathrm{Mpc}$). As a word of caution, I would like to emphasize that a few more cluster fields at various redshifts need to be studied before general conclusion about the location of the galaxy transition should be drawn, although the dusty red galaxies share many properties with the "optically passive cluster infall spirals" identified by Poggianti et al. (1999).

## 5.5 The host galaxies of quasars

Two lines of evidence point to the possibility that Active Galactic Nuclei (AGN) and their most luminous members – the quasars – play an important role in the lives of most galaxies: *first,* the so-called $M - \sigma$ relation (Farrarese & Merritt 2000) or its alternative formulation, the $M_{\rm bh} - M_{\rm bulge}$ relation (Häring & Rix 2004) in nearby galaxies indicate that the formation of the bulge stars and the central super-massive black hole are related. *Second,* the abrupt "shut-down" of star formation which is needed to move a blue galaxy onto the red sequence is hard to understand since in hierarchical galaxy formation scenarios a steady infall of gas from the halo or satellite galaxies should continuously replenish the reservoir for star formation. A violent super-wind from an AGN could blow away and/or heat this gas to temperatures from which it is not able to cool within a Hubble time.

The essential question here is, whether we could observe the "smoking guns" which show direct evidence that an AGN affects its host galaxy. One problem is

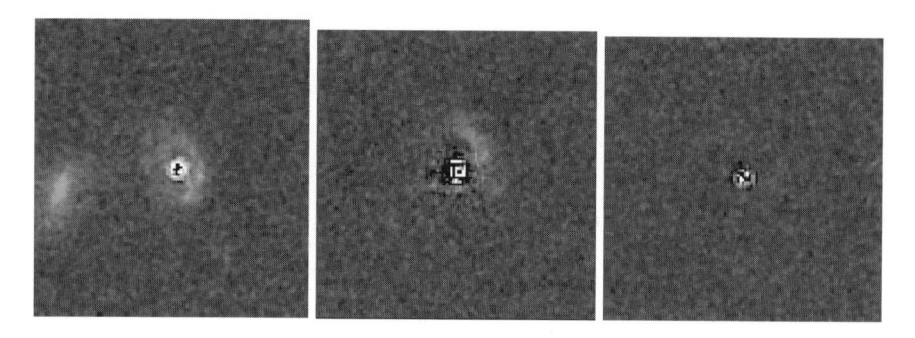

**Figure 17:** Quasar host galaxies in the COSMOS survey. The point source subtracted images show a clearly resolved spiral host at $z = 1.53$, asymmetric "fuzz" around a quasar at $z = 2.16$, and an example for the typical, unresolved quasar at $z = 2.2$ (from *left to right*).

that the AGN phenomenon might be rather short-lived (several $10^7$ yr). Another complication is due to the fact that only *type 1* quasars (characterized by a blue continuum and broad emission lines) can easily be detected in optical or NIR surveys, while *type 2* AGN are rather inconspicuous in their optical-NIR spectra. Deep X-ray surveys which are available for most of the surveys will provide help in getting an unbiased AGN sample.

The large look-back surveys with ACS imaging data (GEMS, COSMOS, STAGES) offer the unique opportunity to detect a statistically meaningful number of AGN out to $z \simeq 2.5$ and to provide the comparison sample of normal galaxies at the same redshift. Thus several investigations have started to compare the properties of the quasar host galaxies (containing the smoking gun) with the general galaxy population. Unfortunately, the notorious problems of look-back surveys – the $(1 + z)^{-4}$ dimming and the shift of the observable rest-frame wavelengths plague these studies in a specific manner: The dimming of the extended galaxy component tends to increase the huge contrast between QSO core and host even further, while the shift in rest-frame wavelength biases the detection of hosts towards actively star forming galaxies (note that the $i$−band images of the $z \simeq 2$ quasars in Fig. 17 sample $\lambda_0 \simeq 270$ nm).

The most recent study of quasar hosts is based on the COSMOS survey. It contains several hundred type 1 quasars at $0.3 < z \simeq 2.5$ (K. Jahnke, *priv. comm.*). Representative examples for the highest redshift QSOs in the sample are displayed in Fig. 17. Not surprisingly, most of the high-redshift quasars remain unresolved. Since COSMOS has observed only one band, F810W, with the ACS, no color information of the host galaxies is available. We will have to wait for the installation of the Wide Field Camera 3 (WFC3) during the next *HST servicing mission* to get colors of these host galaxies. The WFC3 observations of quasar hosts will be superior to any previous studies since its longest wavelength (1.7 μm) can reach the rest-frame $V$-band out to $z \gtrsim 2$.

Nevertheless, a first glimpse on the properties of quasar host could already be obtained with GEMS and its ACS observations in the $V$- and $z$-bands (Jahnke et al 2004, Sánchez et al. 2004). The colors of the host galaxies detected by GEMS are

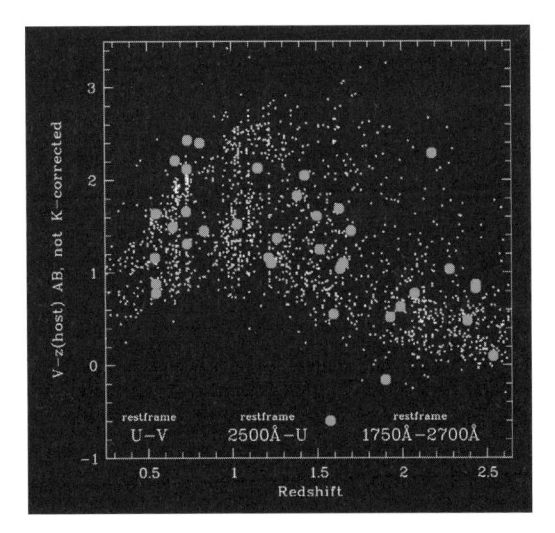

**Figure 18:** Observed $V - z$ colors of quasar host galaxies (large red •) from GEMS. The comparison with typical field galaxies at the same redshift (white dots, from GOODS) show that quasar hosts populate the full color range.

compared with those of the field population of galaxies at the same redshift in Fig. 18: it is obvious that quasar hosts populate the entire range of colors found in normal galaxies. However, one might get the impression that quasar hosts are somewhat over-abundant in the "green valley" which separates the red sequence from the blue cloud (note the six quasar hosts at $(V - z)_{\rm obs} > 1.4$ and $1.1 < z < 1.7$ in Fig. 18). This could be regarded as first hint, that some quasar host galaxies might indeed undergo a transition from the blue cloud to the red sequence. On the other hand, one should not forget the aforementioned problem to detect red host galaxies at $z > 1.2$ with the $z$-band.

# 6 Conclusions and Prospects

This review introduces the observational and analytical methods of deep lock-back surveys which aim to follow the assembly of present-day galaxies over the period of the last eight billion years. Hopefully, I could convince the reader with some selected examples of recent studies that look-back surveys indeed have made exciting progress. Most noteworthy, the following results seem firmly established now:

- The dichotomy between passively evolving galaxies – the red sequence – and actively star forming galaxies – the blue cloud – observed today has existed already eight billion years ago. The evolution of the integrated properties of these two populations has been well determined: red galaxies increase mainly in number with cosmic time while their characteristic luminosity $L^*$ stays roughly constant. This and the inevitable increase of the mass-to-light ratio in their aging stellar population leads to an overall increase of the stellar mass trapped in red systems by about a factor of 2 since $z = 1$. The characteristic luminosity $L^*$ of the blue population (defined by spirals) has decreased substantially during the same period while their stellar mass function seems to undergo little evolution.

- At all times since $z = 1$, stars formation predominantly happens in spiral disk galaxies and (less luminous) irregular galaxies. Earlier claims that the decrease in the integrated SFR by a factor of 10 since $z = 1$ (c.f. right panel in Fig. 2) might be caused by mergers becoming less frequent are ruled out. It is mainly the decline in star formation in the disks and irregulars which leads to the decrease of the global SFR in the last half of the universe' life.

- As expected from the hierarchical model, galactic disks evolve inside-out. But the observed size evolution is more moderate than predicted by the growth of the dark matter halos. Obviously, the concentration of the baryons counter-acts the halo growth.

Despite this impressive progress, several of the most interesting and important questions still remain to be solved:

- The formation of the most massive, "red and dead" elliptical galaxies is a puzzle. There is not only the questions of possible *progenitors* in the blue cloud population. In hierarchical models one also would expect that the most massive systems assemble late ($z < 0.5$) and and that ongoing (minor) merger events could provide new fuel for a persistent star formation activity. "Dry mergers" between gas-free red galaxies provide a possible evolutionary track, but the question remains why and how the restart of star formation activity is prohibited. Current hypotheses of AGN "feedback" or environmental influence (most massive ellipticals live in clusters today) could not yet be verified by look-back surveys.

- In general, the role of environment in shaping the present-day galaxy population is not well understood. There are 20 years of observational evidence that red galaxies accumulate in the cores of galaxy clusters. But, since it is the destiny of galaxies to end in a cluster, it is by no means clear where and by which process the transformation from a blue to a red galaxy occurs. This question can only be tackled by studying the full range of environmental densities between the field and cluster cores. So far, the areas of look-back surveys have been too small (typically $\sim$ 1–2 Mpc projected diameter at $z \simeq 0.5$) to investigate density-dependent transition processes. In addition, only spectroscopic redshifts provide the necessary accuracy for establishing membership in galaxy groups. Here we will need to wait for the next generation of surveys targeting many $\Box°$ by photometric and spectroscopic observations.

- The recent discovery that the mass of the stellar bulge of a galaxy and the mass of its central super-massive black hole are correlated led to the idea that black hole evolution and star formation need to be related somehow. Moreover, the complete "shut-down" of star formation in the most massive galaxies seems to require some "magic" process which is capable of preventing (hot or cold) circum-galactic material to rain into the galaxy and form new stars. Semi-analytic models *ad hoc* introduce various kinds of "AGN feedback" to explain this shut-down (*e.g.* Croton et al. 2006). The verification of the "AGN feedback paradigm" seems one of the toughest goals for observational cosmology to me: first, the AGN or quasar phase may be rather short-lived ($10^7$ years) compared to cosmological timescales ($10^{10}$ years). Second, considerable time-delays might exist between the active AGN phase and its influence on the host galaxy as a whole. Progress in this field might require to identify post-AGN signatures and very huge statistical samples.

In the near future, the most substantial progress can be expected in the redshift range $1 < z \lesssim 3$ (7.5 to 11 billion years ago), that is the pre-history of the epoch addressed in this review. The near-infrared (NIR) facilities, which are mandatory to sample the old stellar population in galaxies at those redshifts are rapidly evolving: wide-field NIR cameras are operating at UKIRT and CFHT on Hawaii (WFCAM, WIRCam) and the infrared survey telescope VISTA with its 67 Megapixel camera is planned be commissioned in fall 2008. Deep surveys which target galaxies at $z > 1$ are either under way (UKIDSS with WFCAM) or shortly before their start (VIDEO and Ultra-VISTA at VISTA). A new suite of NIR spectrographs which will provide multi-object capabilities over substantial fields of view are currently under construction: LUCIFER for the LBT on Mount Graham, FMOS for the Subaru telescope on Mauna Kea and KMOS for the VLT on Cerro Paranal.

In addition, the unprecedented depth reached in the thermal infrared ($\lambda > 2.5\,\mu m$) with the InfraRed Array Camera (IRAC) on board of the Spitzer Space Telescope allows us to sample the spectra of high-redshift galaxies out to the rest-frame infrared.

Our ongoing program COMBO-17+4 extends the multi-color technique of COMBO-17 into the NIR by adding three medium-band filters and the standard $H$-band to the 17 optical bands. Three COMBO-17 fields will be observed. The observations are obtained with the wide-field camera Omega2000 at the 3.5 m telescope on Calar Alto. We expect to finish the complete analysis of the first field in summer 2008. Its catalogue of $H$-band selected galaxies ($H \lesssim 21.0$) will contain more than thousand galaxies at $1 < z < 2$ with accurate photometric redshifts and well sampled SEDs out to at least the rest-frame visual ($\lambda_0 = 550\,nm$). On this basis, the color-dichotomy of the galaxy population will be followed into the era when the star formation activity in the universe peaked, and the early formation of the red sequence should be witnessed. Moreover, the current uncertainties in the stellar mass functions at $0.8 < z < 1.3$ should be solved. The large sample of galaxies at $z > 1$ will provide an input catalogue for detailed studies with the next generation of multi-object NIR spectrographs.

**Acknowledgements**

This review would not have been possible without hundreds of discussions about methods, astrophysical results and their interpretation within the teams of the COMBO-17 and GEMS surveys. COMBO-17 is led by Christian Wolf (now at University of Oxford) and myself; GEMS is led by Hans-Walter Rix in Heidelberg. Most of the results have been obtained by the devoted effort of the team members: Marco Barden, Steve Beckwith, Eric Bell, Andrea Borch, Michael Brown, John Caldwell, Simon Dye, Isabel Franco, Meghan Gray, Boris Häußler, Catherine Heymans, Knud Jahnke, Sharda Jogee, Martina Kleinheinrich, Sergej Koposov, Zoltan Kovacs, Dan McIntosh, Marie-Hélène Nicol, Chien Peng, Sebastian Sánchez, Rachel Somerville, Christian Tapken, Andy Taylor, Lutz Wisotzki, Xianzhong Zheng. The author would like to thank Christian Wolf for his critical reading of the manuscript.

# References

Appenzeller, I., Bender, R., Böhm, A. et al. 2004, ESO Messenger 116, 18

Barden, M., Rix, H.-W., Somerville, R. S. et al. 2005, ApJ 635, 959

Beckwith, S. V. W. B., Stiavelli, M., Koeckemoer, A. M. et al. 2006, AJ 132, 1729

Bell, E. F. & De Jong, R. S. 2001, ApJ 550, 212

Bell, E. F., Wolf, C., Meisenheimer, K. et al. 2004, ApJ 608, 752

Bell, E. F., Papovich, C., Wolf, C. et al. 2005, ApJ 625, 23

Bell, E. F., Naab, T., McIntosh, D. H. et al. 2006, ApJ 640, 241

Bender, R., Appenzeller, I., Böhm, A. et al. 2001, in ESO proceedings: "Deep Fields" (eds. S. Christiani, A. Renzini, R. Williams), Springer 2001, p.96

Borch, A., Meisenheimer, K., Bell, E. F. et al. 2006, A&A 453, 869

Bruzual, G. & Charlot, S. 2003, MNRAS 344, 1000

Bundy, K., Ellis, R. S., Conselice, C. J. et al. 2006, ApJ 651, 120

Butcher, H. R. & Oemler, Jr. A. 1984, ApJ 285, 426

Colless, M., Ellis, R. S., Taylor, K., Hook, R. N. 1990, MNRAS 244, 408

Croton, D. J., Springel, V., White, S. D. M. et al. 2006, MNRAS 365, 11

Davis, M., Faber, S. M., Newman, J. et al. 2003, SPIE 4834, 161

Davis, M., Guhathakurta, P., Konidaris, N. P. et al. 2007, ApJL 660, L1

Dressler, A. 1980, ApJ 236, 351

Faber, S. M., Willmer, C. N. A., Wolf, C. et al. 2007, ApJ 665, 265

Fan, X., White, R. L., Davis, M. et al. 2000, AJ 120, 1167

Fan, X., Strauss, M. A., Richards, G. T. et al. 2006, AJ 131, 1203

Farrarese, L. & Merritt, D. 2000, ApJL 539, L9

Fioc, M. & Rocca-Volmerage, B. 1997, A&A 326, 950

Fried, J. W., von Kuhlmann, B., Meisenheimer, K. et al. 2001, A&A 367, 788

Gallazzi, A., Bell, E. F. et al. 2008, to appear in ApJ

Giavalisco, M., Ferguson, H. C., Koeckemoer, A. M. et al. 2004, ApJL 600, L93

Gray, M., Aragon-Salamanca, A., Bacon, D. et al. 2007, AAS Abstracts 211, 132.20

Häring, N. & Rix, H.-W. 2004, ApJL 604, L89

Hewett, P. C., Foltz, C. B. 1994, PASP 106, 113

Hildebrandt, H., Wolf, C., Benítez, N. 2008, submitted to A&A, arXiv:0801.2975

Hu, E. M., McMahon, R. G., Cowie, L. L. 1999, ApJL 522, L9

Jahnke, K., Sánchez, S., Wisotzki, L. et al. 2004, ApJ 614, 568

Kassin, S. A., Weiner, B. J., Faber, S. M. et al. 2007, ApJL 660, L35

Kauffmann, G., Heckman, T. M., White, S. D. M. et al. 2003, MNRAS 341, 33

Koposov, S., Barden, M., Rix, H.-W. et al. 2008, submitted to ApJ

Kroupa, P. 2001, MNRAS 322, 231

Le Fevre, O., Vettolani, G., Garilli, B. et al. 2005, A&A 439, 863

Lilly, S. J., Le Fevre, O., Crampton, D., Hammer, F., Tresse, L. 1995a, ApJ 455, 50. *(see also following papers on pages 60, 75, 88, 96 and 108 in the same issue)*

Lilly, S. J., Tresse, L., Hammer, F. et al. 1995b, ApJ 455, 108.

Lilly, S. J., Le Fevre, O., Hammer, F., Crampton, D. 1996, ApJL 460, L1

Lilly, S. J., Le Fevre, O., Renzini, A. et al. 2007, ApJS 172, 70

Lotz, J. M., Davis, M., Faber, S. M. et al. 2008, ApJ 672, 177

Maraston, C., Daddi, E., Renzini, A. et al. 2006, ApJ 564, 69

Mo, H. J., Mao, S. and White, S. D. M. 1998, MNRAS 295, 319

Mobasher, B., Idzi, R., Benítez, N. et al. 2004, ApJL 600, L187

Peng, C. Y., Ho, L. C., Impey, C. D., Rix, H.-W. 2002, AJ 124, 266

Poggianti, B. M., Smail, I., Dressler, A. et al. 1999, ApJ 518, 576

Rix, H.-W., Barden, M., Beckwith, S. V. W. B. et al. 2004, ApJS 152, 163

Sánchez, S., Jahnke, K., Wisotzki, L. et al. 2004, ApJ 614, 586

Scoville, N., Aussel, H., Brusa, M. et al. 2007, ApJS 172, 1

Schmidt, M. 1968, ApJ 151, 393

Sérsic, J. L. 1968, Atlas de Galaxias Australes, Obs. Astron. Cordoba 1968

Simard, L., Willmer, C. N. A., Vogt, N. P. et al. 2002, ApJS 142, 1

Somerville R. S., Barden, M., Rix, H.-W. et al. 2008, ApJ 672, 776

Steidel C. C., Giavalisco, M., Pettini, M. et al. 1996, ApJL 462, L17

Vergani, D., Scodeggio, M., Pozzetti, L. et al. 2007, arXiv:0705.3018

Vogt, N. P., Phillips, A. C., Faber, S. M. et al. 1997, ApJ 479, 121

Vogt, N. P. 2001, ASP Conf. Ser. 240, 89

Wolf, C., Meisenheimer, K., Röser, H.-J. 2001, A&A 365, 660

Wolf, C., Meisenheimer, K., Kleinheinrich, M. et al. 2004, A&A 421, 913

Wolf, C., Bell, E. F., McIntosh, D. H. et al. 2005, ApJ 630, 771
Wolf, C., Gray, M., Meisenheimer, K. 2005, A&A 443, 435
Ziegler, B. L., Böhm, A., Fricke, K. J. et al. 2002, ApJ 564, 69

# The First Stars

Volker Bromm

Department of Astronomy
University of Texas at Austin
2511 Speedway, Austin, TX 78712, USA
vbromm@astro.as.utexas.edu

### Abstract

*The formation of the first stars at redshifts $z \geq 15 - 20$ signaled the transition from the simple initial state of the universe to one of ever increasing complexity. I here review recent progress in understanding the assembly process of the first galaxies, starting with cosmological initial conditions and modelling the detailed physics of star formation. In particular, I discuss the role of HD cooling in ionized primordial gas, the impact of UV radiation produced by the first stars, and the propagation of the supernova blast waves triggered at the end of their brief lives. I conclude by discussing how the chemical abundance patterns observed in extremely low-metallicity stars allow us to probe the properties of the first stars.*

## 1 Introduction

One of the key goals in modern cosmology is to study the formation of the first generations of stars and to understand the assembly process of the first galaxies. With the formation of the first stars, the so-called Population III (Pop III), the universe was rapidly transformed into an increasingly complex, hierarchical system, due to the energy and heavy element input from the first stars and accreting black holes (Barkana & Loeb 2001; Bromm & Larson 2004; Ciardi & Ferrara 2005; Miralda-Escudé 2003). Currently, we can directly probe the state of the universe roughly a million years after the Big Bang by detecting the anisotropies in the cosmic microwave background (CMB), thus providing us with the initial conditions for subsequent structure formation. Complementary to the CMB observations, we can probe cosmic history all the way from the present-day universe to roughly a billion years after the Big Bang, using the best available ground- and space-based telescopes. In between lies the remaining frontier, and the first stars and galaxies are the sign-posts of this early, formative epoch.

To simulate the build-up of the first stellar systems, we have to address the feedback from the very first stars on the surrounding intergalactic medium (IGM), and the formation of the second generation of stars out of material that was influenced by this feedback. There are a number of reasons why addressing the feedback from the first stars and understanding second-generation star formation is crucial:

(i) The first steps in the hierarchical build-up of structure provide us with a simplified laboratory for studying galaxy formation, which is one of the main outstanding problems in cosmology.

(ii) The initial burst of Pop III star formation may have been rather brief due to the strong negative feedback effects that likely acted to self-limit this formation mode (Greif & Bromm 2006; Yoshida et al. 2004). Second-generation star formation, therefore, might well have been cosmologically dominant compared to Pop III stars.

(iii) A subset of second-generation stars, those with masses below $\simeq 1\ M_\odot$, would have survived to the present day. Surveys of extremely metal-poor Galactic halo stars therefore provide an indirect window into the Pop III era by scrutinizing their chemical abundance patterns, which reflect the enrichment from a single, or at most a small multiple of, Pop III SNe (Beers & Christlieb 2005; Frebel et al. 2007). Stellar archaeology thus provides unique empirical constraints for numerical simulations, from which one can derive theoretical abundance patterns to be compared with the data.

Existing and planned observatories, such as HST, Keck, VLT, and the *James Webb Space Telescope (JWST)*, planned for launch around 2013, yield data on stars and quasars less than a billion years after the Big Bang. The ongoing *Swift* gamma-ray burst (GRB) mission provides us with a possible window into massive star formation at the highest redshifts (Bromm & Loeb 2002, 2006; Lamb & Reichart 2000). Measurements of the near-IR cosmic background radiation, both in terms of the spectral energy distribution and the angular fluctuations provide additional constraints on the overall energy production due to the first stars (Dwek et al. 2005; Fernandez & Komatsu 2006; Kashlinsky et al. 2005; Magliocchetti et al. 2003; Santos et al. 2002). Understanding the formation of the first galaxies is thus of great interest to observational studies conducted both at high redshifts and in our local Galactic neighborhood.

## 2 Primordial Star Formation

The first stars in the universe likely formed roughly 150 Myr after the Big Bang, when the primordial gas was first able to cool and collapse into dark matter minihalos with masses of the order of $10^6\ M_\odot$ (e.g., Haiman et al. 1996; Tegmark et al. 1997; Bromm et al. 1999, 2002; Abel et al. 2000, 2002; Yoshida et al. 2003). These stars are believed to have been very massive, with masses of the order of $100\ M_\odot$, owing to the limited cooling properties of primordial gas, which could only cool in minihalos through the radiation from $H_2$ molecules. While the initial conditions for the formation of these stars are, in principle, known from precision measurements of cosmological parameters (e.g. Spergel et al. 2007), Pop III star formation may have occurred in different environments which may allow for different modes of star formation. Indeed, it has become evident that Pop III star formation might actually consist of two distinct modes: one where the primordial gas collapses into a DM

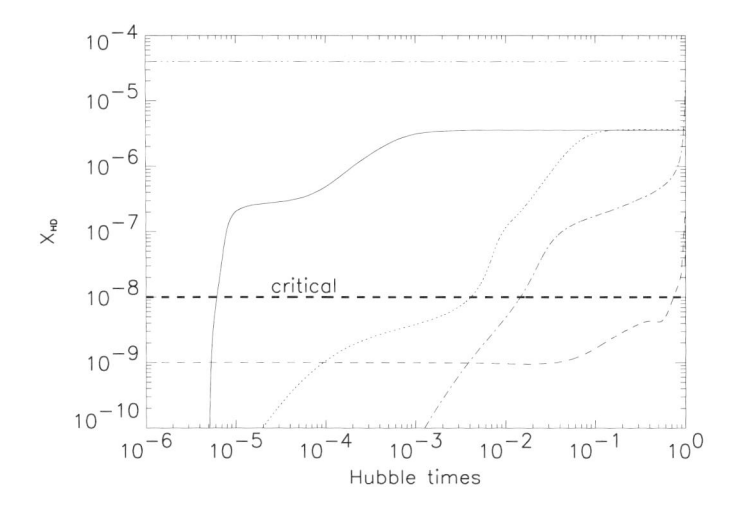

**Figure 1:** Evolution of the HD abundance, $X_{\rm HD}$, in primordial gas which cools in four distinct situations. The solid line corresponds to gas with an initial density of 100 cm$^{-3}$, which is compressed and heated by a SN shock with velocity $v_{\rm sh} = 100$ km s$^{-1}$ at $z = 20$. The dotted line corresponds to gas at an initial density of 0.1 cm$^{-3}$ shocked during the formation of a $3\sigma$ halo at $z = 15$. The dashed line corresponds to unshocked, un-ionized primordial gas with an initial density of 0.3 cm$^{-3}$ collapsing inside a minihalo at $z = 20$. Finally, the dash-dotted line shows the HD fraction in primordial gas collapsing from an initial density of 0.3 cm$^{-3}$ inside a relic H II region at $z = 20$. The horizontal line at the top denotes the cosmic abundance of deuterium. Primordial gas with an HD abundance above the critical value, $X_{\rm HD,crit}$, denoted by the bold dashed line, can cool to the CMB temperature within a Hubble time.

minihalo (see below), and one where the metal-free gas becomes significantly ionized prior to the onset of gravitational runaway collapse (Johnson & Bromm 2006). We had termed this latter mode of primordial star formation 'Pop II.5' (Greif & Bromm 2006; Johnson & Bromm 2006; Mackey et al. 2003). To more clearly indicate that both modes pertain to *metal-free* star formation, we here follow the new classification scheme suggested by Chris McKee (see McKee & Tan 2007; Johnson et al. 2008). Within this scheme, the minihalo Pop III mode is now termed Pop III.1, whereas the second mode (formerly 'Pop II.5') is now called Pop III.2. The hope is that McKee's terminology will gain wide acceptance.

While the very first Pop III stars (so-called Pop III.1), with masses of the order of 100 $M_\odot$, formed within DM minihalos in which primordial gas cools by H$_2$ molecules alone, the HD molecule can play an important role in the cooling of primordial gas in several situations, allowing the temperature to drop well below 200 K (Abel et al. 2002; Bromm et al. 2002). In turn, this efficient cooling may lead to the formation of primordial stars with masses of the order of 10 $M_\odot$ (so-called Pop III.2) (Johnson & Bromm 2006). In general, the formation of HD, and the concomi-

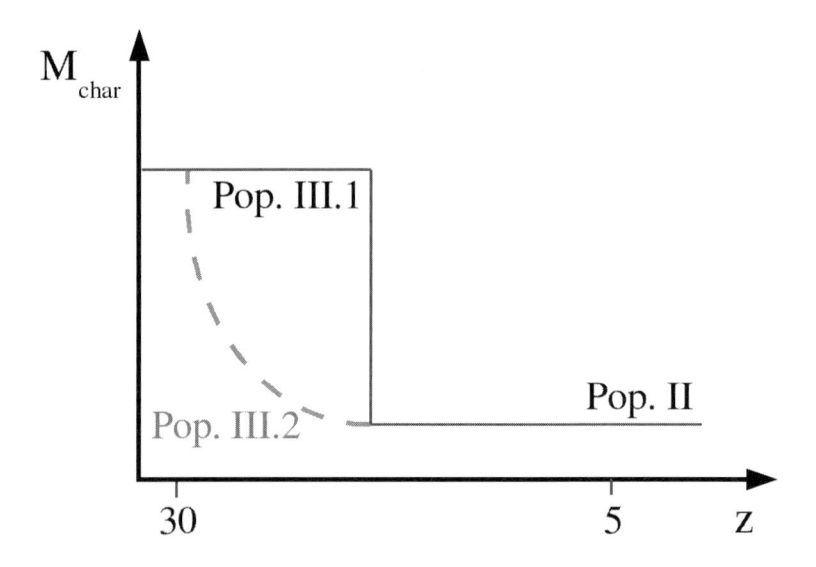

**Figure 2:** Characteristic stellar mass as a function of redshift. Pop III.1 stars, formed from unshocked, un-ionized primordial gas are characterized by masses of the order of 100 $M_\odot$. Pop II stars, formed in gas which is enriched with metals, emerged at lower redshifts and have characteristic masses of the order of 1 $M_\odot$. Pop III.2 stars, formed from ionized primordial gas, have characteristic masses reflecting the fact that they form from gas that has cooled to the temperature of the CMB. Thus, the characteristic mass of Pop III.2 stars is a function of redshift, but is typically of the order of 10 $M_\odot$.

tant cooling that it provides, is found to occur efficiently in primordial gas which is strongly ionized, owing ultimately to the high abundance of electrons which serve as catalyst for molecule formation in the early universe (Shapiro & Kang 1987).

Efficient cooling by HD can be triggered within the relic H II regions that surround Pop III.1 stars at the end of their brief lifetimes, owing to the high electron fraction that persists in the gas as it cools and recombines (Johnson et al. 2007; Nagakura & Omukai 2005; Yoshida et al. 2007). The efficient formation of HD can also take place when the primordial gas is collisionally ionized, such as behind the shocks driven by the first SNe or in the virialization of massive DM halos (Greif & Bromm 2006; Johnson & Bromm 2006; Machida et al. 2005; Shchekinov & Vasiliev 2006). In Figure 1, we show the HD fraction in primordial gas in four distinct situations: within a minihalo in which the gas is never strongly ionized, behind a 100 km s$^{-1}$ shock wave driven by a SN, in the virialization of a $3\sigma$ DM halo at redshift $z = 15$, and in the relic H II region generated by a Pop III.1 star at $z \sim 20$ (Johnson & Bromm 2006). Also shown is the critical HD fraction necessary to allow the primordial gas to cool to the temperature floor set by the CMB at these redshifts. Except for the situation of the gas in the virtually un-ionized minihalo, the fraction of HD becomes large quickly enough to play an important role in the cooling of the gas, allowing the formation of Pop III.2 stars.

Figure 2 schematically shows the characteristic masses of the various stellar populations that form in the early universe. In the wake of Pop III.1 stars formed in DM minihalos, Pop III.2 star formation ensues in regions which have been previously ionized, typically associated with relic H II regions left over from massive Pop III.1 stars collapsing to black holes, while even later, when the primordial gas is locally enriched with metals, Pop II stars begin to form (Bromm & Loeb 2003; Greif & Bromm 2006). Recent simulations confirm this picture, as Pop III.2 star formation ensues in relic H II regions in well under a Hubble time, while the formation of Pop II stars after the first SN explosions is delayed by more than a Hubble time (Greif et al. 2007; Yoshida et al. 2007a,b; but see Whalen et al. 2008).

## 3 Radiative Feedback from the First Stars

Due to their extreme mass scale, Pop III.1 stars emit copious amounts of ionizing radiation, as well as a strong flux of $H_2$-dissociating Lyman-Werner (LW) radiation (Bromm et al. 2001b; Schaerer 2002). Thus, the radiation from the first stars dramatically influences their surroundings, heating and ionizing the gas within a few kpc (physical) around the progenitor, and destroying the $H_2$ and HD molecules locally within somewhat larger regions (Alvarez et al. 2006; Abel et al. 2007; Ferrara 1998; Johnson et al. 2007; Kitayama et al. 2004; Whalen et al. 2004). Additionally, the LW radiation emitted by the first stars could propagate across cosmological distances, allowing the build-up of a pervasive LW background radiation field (Haiman et al. 2000).

### 3.1 Local Radiative Effects

The impact of radiation from the first stars on their local surroundings has important implications for the numbers and types of Pop III stars that form. The photoheating of gas in the minihalos hosting Pop III.1 stars drives strong outflows, lowering the density of the primordial gas and delaying subsequent star formation by up to $100$ Myr (Johnson et al. 2007; Whalen et al. 2004; Yoshida et al. 2007a). Furthermore, neighboring minihalos may be photoevaporated, delaying star formation in such systems as well (Ahn & Shapiro 2007; Greif et al. 2007; Shapiro et al. 2004; Susa & Umemura 2006; Whalen et al. 2007). The photodissociation of molecules by LW photons emitted from local star-forming regions will, in general, act to delay star formation by destroying the main coolants that allow the gas to collapse and form stars.

The photoionization of primordial gas, however, can ultimately lead to the production of copious amounts of molecules within the relic H II regions surrounding the remnants of Pop III.1 stars (Johnson & Bromm 2007; Nagakura & Omukai 2005; Oh & Haiman 2002; Ricotti et al. 2001). Recent simulations tracking the formation of, and radiative feedback from, individual Pop III.1 stars in the early stages of the assembly of the first galaxies have demonstrated that the accumulation of relic H II regions has two important effects. First, the HD abundance that develops in relic H II regions allows the primordial gas to re-collapse and cool to the temperature of the

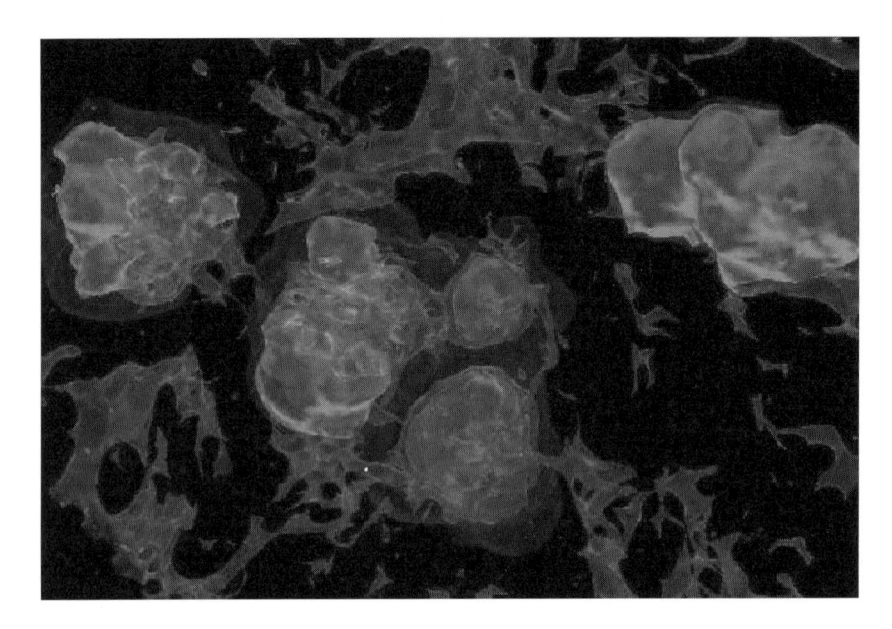

**Figure 3:** The chemical interplay in relic H II regions. While all molecules are destroyed in and around active H II regions, the high residual electron fraction in relic H II regions catalyzes the formation of an abundance of $H_2$ and HD molecules. The light and dark shades of blue denote regions with a free electron fraction of $5 \times 10^{-3}$ and $5 \times 10^{-4}$, respectively, while the shades of green denote regions with an $H_2$ fraction of $10^{-4}$, $10^{-5}$, and $3 \times 10^{-6}$, in order of decreasing brightness. The regions with the highest molecule abundances lie within relic H II regions, which thus play an important role in subsequent star formation, allowing molecules to become shielded from photodissociating radiation and altering the cooling properties of the primordial gas.

CMB, possibly leading to the formation of Pop III.2 stars in these regions (Johnson et al. 2007; Yoshida et al. 2007b). Second, the molecule abundance in relic H II regions, along with their increasing volume-filling fraction, leads to a large optical depth to LW photons over physical distances of the order of several kpc. The development of a high optical depth to LW photons over such short length-scales suggests that the optical depth to LW photons over cosmological scales may be very high, acting to suppress the build-up of a background LW radiation field, and mitigating negative feedback on star formation.

Figure 3 shows the chemical composition of primordial gas in relic H II regions, in which the formation of $H_2$ molecules is catalyzed by the high residual electron fraction. Figure 4 shows the average optical depth to LW photons across the simulation box, which rises with time owing to the increasing number of relic H II regions.

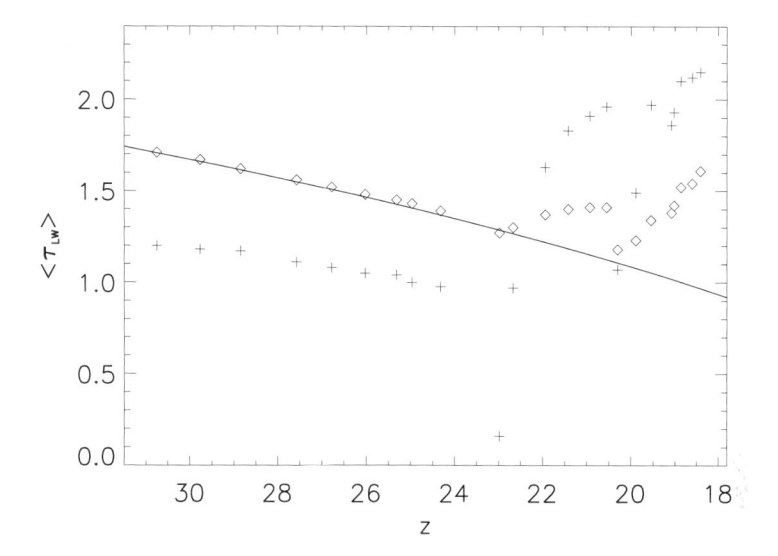

**Figure 4:** Optical depth to LW photons due to self-shielding, averaged over two different scales, as a function of redshift. The diamonds denote the optical depth averaged over the entire cosmological box of comoving length 660 kpc, while the plus signs denote the optical depth averaged over a cube of 220 kpc (comoving) per side, centered on the middle of the box. The solid line denotes the average optical depth that would be expected for a constant $H_2$ fraction of $2 \times 10^{-6}$ (primordial gas), which changes only due to cosmic expansion.

## 3.2 Global Radiative Feedback

While the reionization of the universe is likely to have occurred at later times, as inferred from the *WMAP* third year results (Spergel et al. 2007), the process of primordial star formation can be affected by the build up of a LW background very soon after the formation of the first stars. This LW radiation, which acts to destroy $H_2$, the very coolant that enables the formation of the first stars, can, in principle, dramatically lower the formation rate of Pop III stars in minihalos (e.g. Haiman et al. 2000; Machacek et al. 2001; Yoshida et al. 2003; Mackey et al. 2003).

While star formation in more massive systems may proceed relatively unimpeded, through atomic line cooling, during the earliest epochs of star formation these atomic-cooling halos are rare compared to the minihalos which host individual Pop III stars. While the process of star formation in atomic-cooling halos is not well understood, for a broad range of models the dominant contribution to the LW background is from Pop III.1 stars formed in minihalos at $z \geq 15$–$20$ (Johnson et al. 2008). Therefore, at these redshifts the LW background radiation may be largely self-regulated, with Pop III.1 stars producing the very radiation which, in turn, suppresses their formation. Johnson et al. (2008) argue that there is a critical value for the LW flux, $J_{\rm LW,crit} \sim 0.04$, at which Pop III.1 star formation occurs self-

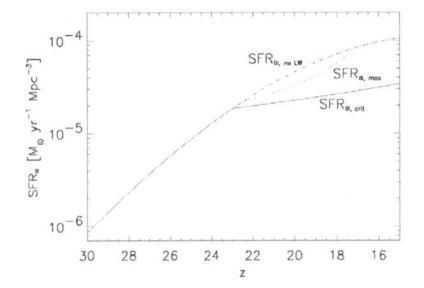

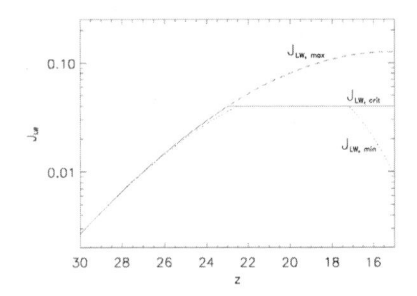

**Figure 5:** The Pop III star formation rates (*left panel*) and the corresponding LW background fluxes (*right panel*) for three models of the build-up of the LW background by Pop III.1 stars formed in minihalos. The maximum possible LW background, $J_{\rm LW,max}$, is generated for the case that every minihalo with a virial temperature $\geq 2 \times 10^3$ K hosts a Pop III star, without the LW background in turn diminishing the SFR, labeled here as $\rm SFR_{III,noLW}$. The self-regulated model considers the coupling between the star formation rate, $\rm SFR_{LW,crit}$, and the critical LW background that it produces, $J_{\rm LW,crit}$. The minimum value for the LW background, $J_{\rm LW,min}$, is produced for the case of a high opacity through the relic H II regions left by the first stars, in which case the self-consistent SFR, $\rm SFR_{III,max}$ can approach the undiminished $\rm SFR_{III,noLW}$.

consistently, with the implication that the PopIII.1 star formation rate in minihalos at $z > 15$ is decreased by only a factor of a few, as shown in Fig. 5. The star formation rate may be even higher if the cosmological average optical depth to LW photons through the relic H II regions left by the first stars is sufficiently high. An analytical model of this effect shows that the SFR may be only negligibly reduced once the volume-filling factor of relic H II regions becomes large, as is also shown in Fig. 5 (Johnson et al. 2008).

Simulations of the formation of a dwarf galaxy at $z \geq 10$ which take into account the effect of a LW background at $J_{\rm LW,crit}$ show that Pop III.1 star formation takes place before the galaxy is fully assembled, suggesting that the formation of metal-free galaxies may be a rare event in the early universe (Johnson et al. 2008). Figure 6 shows the temperature and density of the protogalaxy simulated by these authors at $z \sim 12.5$.

## 4 The First Supernova Explosions

Recent numerical simulations have indicated that primordial stars forming in DM minihalos typically attain 100 $M_\odot$ by efficient accretion, and might even become as massive as 500 $M_\odot$ (Bromm & Loeb 2004; O'Shea & Norman 2007; Omukai & Palla 2003; Yoshida et al. 2006). After their main-sequence lifetimes of typically 2–3 Myr, stars with masses below $\simeq 100\ M_\odot$ are thought to collapse directly to black holes without significant metal ejection, while in the range 140–260 $M_\odot$ a pair-instability supernova (PISN) disrupts the entire progenitor, with explosion energies

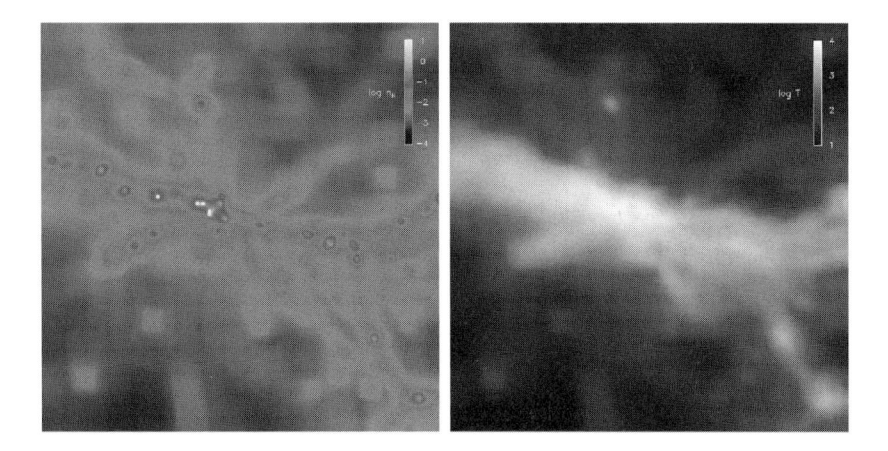

**Figure 6:** The hydrogen number density and temperature of the gas in the region of the forming galaxy at $z \sim 12.5$. The panels show the inner $\sim 10.6$ kpc (physical) of our cosmological box. The cluster of minihaloes harboring dense gas just left of the center in each panel is the site of the formation of the two Pop III.1 stars which are able to form in our simulation including the effects of the self-regulated LW background. The remaining minihaloes are not able to form stars by this redshift, largely due to the photodissociation of the coolant $H_2$. The main progenitor DM halo, which hosts the first star at $z \sim 16$, by $z \sim 12.5$ has accumulated a mass of $9 \times 10^7$ $M_\odot$ through mergers and accretion. Note that the gas in this halo has been heated to temperatures above $10^4$ K, leading to a high free electron fraction and high molecule fractions in the collapsing gas. The high HD fraction that is generated likely leads to the formation of Pop III.2 stars in this system.

ranging from $10^{51} - 10^{53}$ ergs, and yields up to 0.5 (Heger et al. 2003; Heger & Woosley 2002).

The significant mechanical and chemical feedback effects exerted by such explosions have been investigated with a number of detailed calculations, but these were either performed in one dimension (Kitayama & Yoshida 2005; Machida et al. 2005; Salvaterra et al. 2004; Whalen et al. 2008), or did not start from realistic initial conditions (Bromm et al. 2003; Norman et al. 2004). The most realistic, three-dimensional simulation to date took cosmological initial conditions into account, and followed the evolution of the gas until the formation of the first minihalo at $z \simeq 20$ (Greif et al. 2007). After the gas approached the 'loitering regime' at $n_H \simeq 10^4$ cm$^{-3}$, the formation of a primordial star was assumed, and a photoheating and ray-tracing algorithm determined the size and structure of the resulting H II region (Johnson et al. 2007). An explosion energy of $10^{52}$ ergs was then injected as thermal energy into a small region around the progenitor, and the subsequent expansion of the SN remnant was followed until the blast wave effectively dissolved into the IGM. The cooling mechanisms responsible for radiating away the energy of the SN remnant, the temporal behavior of the shock, and its morphology could thus be investigated in great detail (see Figure 7).

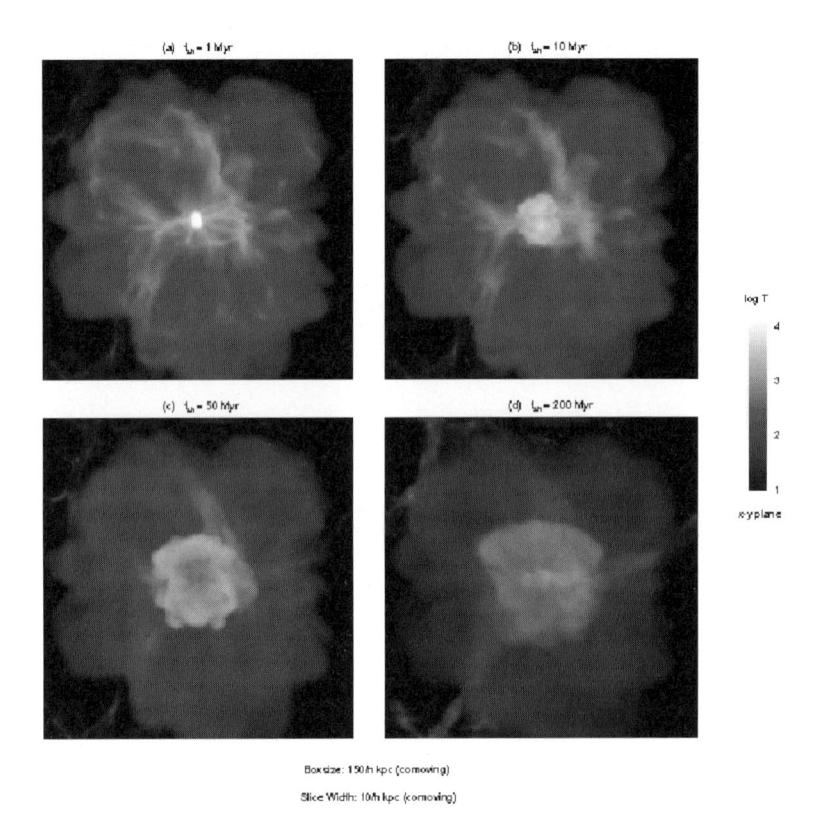

**Figure 7:** Temperature averaged along the line of sight in a slice of $10/h$ kpc (comoving) at 1, 10, 50, and 200 Myr after a Pop III PISN (Greif et al. 2007). In all four panels, the H II region and SN shock are clearly distinguishable, with the former occupying almost the entire simulation box, while the latter is confined to the central regions. *(a)*: The SN remnant has just left the host halo, but temperatures in the interior are still well above $10^4$ K. *(b)*: After 10 Myr, the asymmetry of the SN shock becomes visible, while most of the interior has cooled to well below $10^4$ K. *(c)*: The further evolution of the shocked gas is governed by adiabatic expansion. *(d)*: After 200 Myr, the shock velocity approaches the local sound speed and the SN remnant stalls. By this time the post-shock regions have cooled to roughly $10^3$ K.

The dispersal of metals by the first SN explosions transformed the IGM from a simple primordial gas to a highly complex medium in terms of chemistry and cooling, which ultimately enabled the formation of the first low-mass stars. However, this transition required at least a Hubble time, since the presence of metals became important only after the SN remnant had stalled and the enriched gas re-collapsed to high densities (Greif et al. 2007; but see Whalen et al. 2008). Furthermore, the metal distribution was highly anisotropic, as the post-shock gas expanded into the voids in the shape of an 'hour-glass', with a maximum extent similar to the final mass-weighted mean shock radius (Greif et al. 2007).

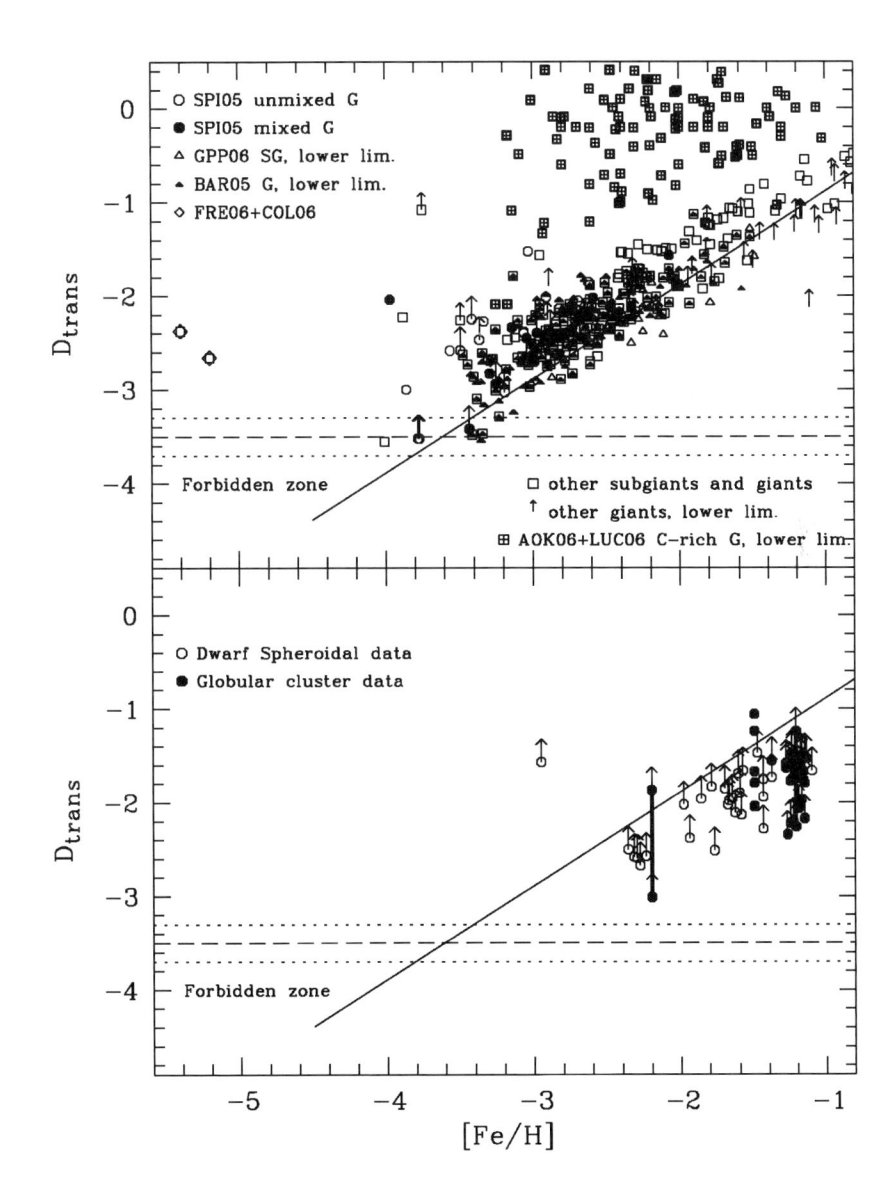

**Figure 8:** Transition discriminant, $D_{\rm trans}$, for metal-poor stars collected from the literature as a function of [Fe/H]. *Top panel*: Galactic halo stars. *Bottom panel*: Stars in dSph galaxies and globular clusters. G indicates giants, SG subgiants. The critical limit is marked with a dashed line. The dotted lines refer to the uncertainty. The detailed references for the various data sets can be found in Frebel et al. (2007).

To efficiently mix the metals with all components of the swept-up gas, a DM halo of at least $M_{\rm vir} \simeq 10^8\ M_\odot$ had to be assembled (Greif et al. 2007), and with an initial yield of 0.1, the average metallicity of such a system would accumulate to $Z \simeq 10^{-2.5} Z_\odot$, well above any critical metallicity (Bromm & Loeb 2003; Bromm et al. 2001a; Schneider et al. 2006; see also Wise & Abel 2007). Thus, if energetic SNe were a common fate for the first stars, they would have deposited metals on large scales before massive galaxies formed and outflows were suppressed. Hints to such ubiquitous metal enrichment have been found in the low column density Ly$\alpha$ forest (Aguirre et al. 2005; Songaila & Cowie 1996; Songaila 2001), and in dwarf spheroidal satellites of the Milky Way (Helmi et al. 2006).

## 5  The Chemical Signature of the First Stars

The discovery of extremely metal-poor stars in the Galactic halo has made studies of the chemical composition of low-mass Pop II stars powerful probes of the conditions in which the first low-mass stars formed. While it is widely accepted that metals are required for the formation of low-mass stars, two general classes of competing models for the Pop III – Pop II transition are discussed in the literature: *(i)* atomic fine-structure line cooling (Bromm & Loeb 2003; Santoro & Shull 2006); and *(ii)* dust-induced fragmentation (Schneider et al. 2006). Within the fine-structure model, C II and O I have been suggested as main coolants (Bromm & Loeb 2003), such that low-mass star formation can occur in gas that is enriched beyond critical abundances of $[\text{C/H}]_{\rm crit} \simeq -3.5 \pm 0.1$ and $[\text{O/H}]_{\rm crit} \simeq -3 \pm 0.2$. The dust-cooling model, on the other hand, predicts critical abundances that are typically smaller by a factor of 10–100.

Based on the theory of atomic line cooling (Bromm & Loeb 2003), a new function, the 'transition discriminant' has been introduced:

$$D_{\rm trans} \equiv \log_{10}\left(10^{[\text{C/H}]} + 0.3 \times 10^{[\text{O/H}]}\right), \qquad (1)$$

such that low-mass star formation requires $D_{\rm trans} > D_{\rm trans,crit} \simeq -3.5 \pm 0.2$ (Frebel et al. 2007). Figure 9 shows values of $D_{\rm trans}$ for a large number of the most metal-poor stars available in the literature. While theories based on dust cooling can be pushed to accommodate the lack of stars with $D_{\rm trans} < D_{\rm trans,crit}$, it appears that the atomic-cooling theory for the Pop III – Pop II transition naturally explains the existing data on metal-poor stars. Future surveys of Galactic halo stars will allow to further populate plots such as Figure 8, and will provide valuable insight into the conditions of the early universe in which the first low-mass stars formed.

The abundance patterns observed in the most metal-poor stars can also provide information about the types of SN that ended the lives of the first stars, as the metals that are emitted in these explosions will become incorporated into later generations of stars, some of which are observed in the halo of the Galaxy. Interestingly, while detailed numerical simulations of the formation of the first stars suggest that they were often massive enough to explode as PISN, no clear signature of such a PISN has yet been detected in a metal-poor star. Does this imply that the first stars did not explode as PISN, or that they were not very massive (Ekström et al. 2008)?

*The First Stars* 319

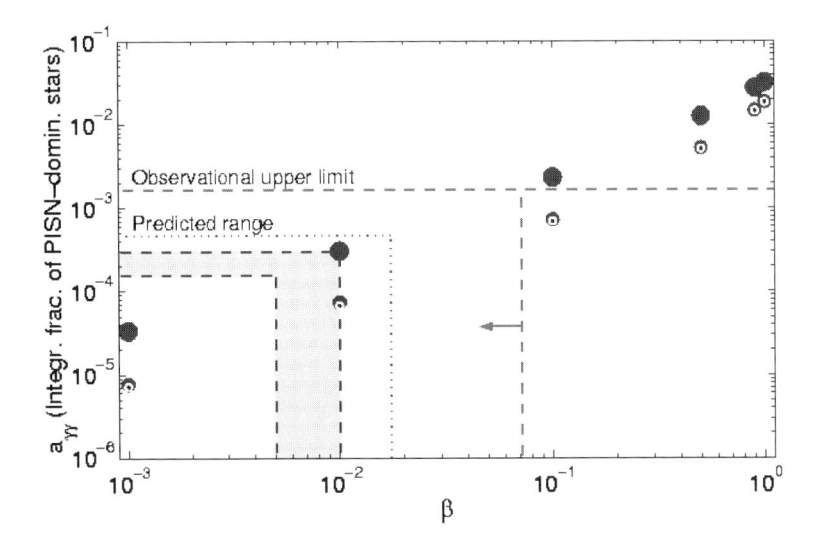

**Figure 9:** The predicted integrated (total) fraction of PISN-dominated stars below [Ca/H] = −2 as a function of $\beta$, corresponding to > 90% (big blue dots), > 99% (medium sized, blue dots), and > 99.9% (small, dark blue dots) PISN-enrichment. The dashed (red) lines indicate the observational upper limit of $\beta$, assuming that none of the ∼ 600 Galactic halo stars with [Ca/H] ≤ −2 for which high-resolution spectroscopy is available show any signature of PISNe (N. Christlieb, priv. comm.). The dotted (blue) line and shaded (blue) area denote the predicted range of $a_{\gamma\gamma}$ anticipated from the calculations of Padoan et al. (2007) and Greif & Bromm (2006), respectively.

PISNe may have ejected enough mass in metals to enrich the IGM to a metallicity well above those of the most metal-poor stars. Therefore, one possible explanation for the apparent lack of Pop III PISNe is that the few stars which might have formed from gas dominantly enriched by a PISN may have relatively high metallicities, and so may have eluded surveys seeking such true second generation stars at lower metallicity (Karlsson et al. 2008). Karlsson et al. (2008) developed a model for the inhomogeneous chemical enrichment of the gas collapsing to become a dwarf galaxy at z ∼ 10 in which the formation of both Pop III stars from metal-free gas and the formation of Pop II stars from metal-enriched gas were tracked self-consistently. These authors find that the lack of the discovery of a metal-poor star showing signs of enrichment dominated by PISN yields in the existing catalog of metal poor stars is not inconsistent with theories predicting that the first stars were very massive, as shown in Figure 9. It is hoped that future surveys of stars in the Galactic halo will test this model by searching for PISN-enriched stars with metallicities [Fc/H] ≥ −2.5.

# 6 Conclusion

Understanding the formation of the first galaxies marks the frontier of high-redshift structure formation. It is crucial to predict their properties in order to develop the optimal search and survey strategies for the *JWST*. Whereas *ab-initio* simulations of the very first stars can be carried out from first principles, and with virtually no free parameters, one faces a much more daunting challenge with the first galaxies. Now, the previous history of star formation has to be considered, leading to enhanced complexity in the assembly of the first galaxies. One by one, all the complex astrophysical processes that play a role in more recent galaxy formation appear back on the scene. Among them are external radiation fields, comprising UV and X-ray photons, and possibly cosmic rays produced in the wake of the first SNe (Stacy & Bromm 2007). There will be metal-enriched pockets of gas which could be pervaded by dynamically non-negligible magnetic fields, together with turbulent velocity fields built up during the virialization process. However, the goal of making useful predictions for the first galaxies is now clearly drawing within reach, and the pace of progress is likely to be rapid.

## Acknowledgments

My sincere thanks to the Astronomische Gesellschaft for the kind invitation to give this review talk. It is my pleasure to acknowledge the many contributions to the work presented here made by my young collaborators Anna Frebel, Thomas Greif, Jarrett Johnson, Athena Stacy and Torgny Karlsson. Support from the U.S. National Science Foundation under grant AST-0708795 is gratefully acknowledged. The simulations presented here were carried out at the Texas Advanced Computing Center (TACC).

## References

Abel, T., Bryan, G. L., & Norman, M. L. 2000, ApJ, 540, 39

Abel, T., Bryan, G. L., & Norman, M. L. 2002, Science, 295, 93

Abel, T., Wise, J. H., & Bryan, G. L. 2007, ApJ, 659, L87

Aguirre, A., Schaye, J., Hernquist, L., Kay, S., Springel, V., & Theuns, T. 2005, ApJ, 620, L13

Ahn, K., & Shapiro, P. R. 2007, MNRAS, 375, 881

Alvarez, M. A., Bromm, V., & Shapiro, P. R. 2006, ApJ, 639, 621

Barkana, R., & Loeb, A. 2001, Phys. Rep., 349, 125

Beers, T. C., & Christlieb, N. 2005, ARA&A, 43, 531

Bromm, V., Coppi, P. S., & Larson, R. B. 1999, ApJ, 527, L5

Bromm, V., Coppi, P. S., & Larson, R. B. 2002, ApJ, 564, 23

Bromm, V., Ferrara, A., Coppi, P. S., & Larson, R. B. 2001a, MNRAS, 328, 969

Bromm, V., Kudritzki, R. P., & Loeb, A. 2001b, ApJ, 552, 464

Bromm, V., & Larson, R. B. 2004, ARA&A, 42, 79

Bromm, V., & Loeb, A. 2002, ApJ, 575, 111

—. 2003b, Nature, 425, 812

—. 2004, New Astronomy, 9, 353

—. 2006, ApJ, 642, 382

Bromm, V., Yoshida, N., & Hernquist, L. 2003, ApJ, 596, L135

Ciardi, B., & Ferrara, A. 2005, Space Science Reviews, 116, 625

Dwek, E., Arendt, R. G., & Krennrich, F. 2005, ApJ, 635, 784

Ekström, S., Meynet, G., & Maeder, A. 2008, to appear in proceedings of IAU Symposium 250, "Massive Stars as Cosmic Engines", eds. F. Bresolin, P. A. Crowther & J. Puls (arXiv: 0801.3397)

Fernandez, E. R., & Komatsu, E. 2006, ApJ, 646, 703

Ferrara, A. 1998, ApJ, 499, L17

Frebel, A., Johnson, J. L., & Bromm, V. 2007, MNRAS, 380, L40

Greif, T. H., & Bromm, V. 2006, MNRAS, 373, 128

Greif, T. H., Johnson, J. L., Bromm, V., & Klessen, R. S. 2007, ApJ, 670, 1

Haiman, Z., Abel, T., & Rees, M. J. 2000, ApJ, 534, 11

Haiman, Z., Thoul, A. A., & Loeb, A. 1996, ApJ, 464, 523

Heger, A., Fryer, C. L., Woosley, S. E., Langer, N., & Hartmann, D. H. 2003, ApJ, 591, 288

Heger, A., & Woosley, S. E. 2002, ApJ, 567, 532

Helmi, A. et al. 2006, ApJ, 651, L121

Johnson, J. L., & Bromm, V. 2006, MNRAS, 366, 247

—. 2007, MNRAS, 374, 1557

Johnson, J. L., Greif, T. H., & Bromm, V. 2007, ApJ, 665, 85

Johnson, J. L., Greif, T. H., & Bromm, V. 2008, MNRAS, in press (arXiv:0711.4622)

Karlsson, T., Johnson, J. L., & Bromm, V. 2008, ApJ, 679, 6

Kashlinsky, A., Arendt, R. G., Mather, J., & Moseley, S. H. 2005, Nature, 438, 45

Kitayama, T., & Yoshida, N. 2005, ApJ, 630, 675

Kitayama, T., Yoshida, N., Susa, H., & Umemura, M. 2004, ApJ, 613, 631

Lamb, D. Q., & Reichart, D. E. 2000, ApJ, 536, 1

Machacek, M. E., Bryan, G. L., Abel, T. 2001, ApJ, 548, 509

Machida, M. N., Tomisaka, K., Nakamura, F., & Fujimoto, M. Y. 2005, ApJ, 622, 39

Mackey, J., Bromm, V., & Hernquist, L. 2003, ApJ, 586, 1

Magliocchetti, M., Salvaterra, R., & Ferrara, A. 2003, MNRAS, 342, L25

McKee C. F., & Tan, J. C. 2007, ApJ, in press (arXiv:0711.1377)

Miralda-Escudé, J. 2003, Science, 300, 1904

Nagakura, T., & Omukai, K. 2005, MNRAS, 364, 1378

Oh, S. P., & Haiman, Z. 2002, ApJ, 569, 558

Omukai, K., & Palla, F. 2003, ApJ, 589, 677

O'Shea, B. W., & Norman, M. L. 2007, ApJ, 654, 66

Ricotti, M., Gnedin, N. Y., & Shull, J. M. 2001, ApJ, 560, 580

Salvaterra, R., Ferrara, A., & Schneider, R. 2004, New Astronomy, 10, 113

Santoro, F., & Shull, J. M. 2006, ApJ, 643, 26

Santos, M. R., Bromm, V., & Kamionkowski, M. 2002, MNRAS, 336, 1082

Schaerer, D. 2002, A&A, 382, 28

Schneider, R., Omukai, K., Inoue, A. K., & Ferrara, A. 2006, MNRAS, 369, 1437

Shapiro, P. R., Iliev, I. T., & Raga, A. C. 2004, MNRAS, 348, 753

Shapiro, P. R., & Kang, H. 1987, ApJ, 318, 32

Shchekinov, Y. A., & Vasiliev, E. O. 2006, MNRAS, 368, 454

Songaila, A. 2001, ApJ, 561, L153

Songaila, A., & Cowie, L. L. 1996, AJ, 112, 335

Spergel, D. N. et al. 2007, ApJS, 170, 377

Stacy, A., & Bromm, V. 2007, MNRAS, 382, 229

Susa, H., & Umemura, M. 2006, ApJ, 645, L93

Tegmark, M., Silk, J., Rees, M. J., Blanchard, A., Abel, T., & Palla, F. 1997, ApJ, 474, 1

Whalen, D., Abel, T., & Norman, M. L. 2004, ApJ, 610, 14

Whalen, D., O'Shea, B. W., Smidt, J., & Norman, M. L. 2007, to appear in "First Stars III", eds. B. O'Shea, A. Heger & T. Abel (arXiv:0708.3466)

Whalen, D., van Veelen, B., O'Shea, B. W., & Norman, M. L. 2008, ApJ, in press (arXiv:0801.3698)

Wise, J. H., & Abel, T. 2007, ApJ, in press (arXiv:0710.3160)

Yoshida, N., Abel, T., Hernquist, L., & Sugiyama, N. 2003, ApJ, 592, 645

Yoshida, N., Bromm, V., & Hernquist, L. 2004, ApJ, 605, 579

Yoshida, N., Oh, S. P., Kitayama, T., & Hernquist, L. 2007a, ApJ, 663, 687

Yoshida, N., Omukai, K., & Hernquist, L. 2007b, ApJ, 667, L117

Yoshida, N., Omukai, K., Hernquist, L., & Abel, T. 2006, ApJ, 652, 6

# Massive Stars as Tracers for Stellar and Galactochemical Evolution

Norbert Przybilla

Dr. Remeis-Sternwarte Bamberg
Astronomisches Institut der Universität Erlangen-Nürnberg
Sternwartstr. 7, 96049 Bamberg, Germany
przybilla@sternwarte.uni-erlangen.de,
http://www.sternwarte.uni-erlangen.de/∼ai32/

## Abstract

*Hot, massive stars are key drivers of the cosmic cycle of matter. They are the dominant contributors to the energy and momentum budget of the interstellar medium, via ionizing radiation, stellar winds and supernova explosions, and they are important sites of nucleosynthesis. Because of their high luminosities these stars are primary targets for spectroscopy over large distances. With the present generation of 8–10m-class telescopes and modern instruments they can be used as powerful tracers of chemical composition throughout the Milky Way, other star-forming galaxies of the Local Group, and beyond. The most frequent massive stars, OB-type stars in the mass range from about 8 to 30 $M_\odot$, are of particular interest to us, which are studied in a homogeneous way from the main sequence to the evolved stage as BA-type supergiants.*

*In the past, quantitative studies of early-type stars were not accurate enough (e.g. only to within a factor ∼2 at most for elemental abundances) to derive conclusive results. Now, advances in quantitative spectroscopy allow stellar parameters and chemical abundances to be constrained with much higher precision than previously possible: to $\lesssim$1–2% uncertainty in effective temperature, 0.05–0.10 dex in (logarithmic) surface gravity, and ∼10–25% (random) plus ∼25% (systematic $1\sigma$-errors) in abundance. This is achieved by the use of improved atomic data and the identification and consequent reduction of systematic errors. Massive stars are thus turned into universal tools for modern astrophysics.*

*The improved model atmosphere analyses have a strong impact on our understanding of the evolution of massive stars and the chemical evolution of galaxies, as thorough conclusions can be drawn for the first time. We discuss massive stars as tracers for: I) abundances of the light elements, which make possible the study of the effects of complex (magneto)hydrodynamic mixing processes in the stellar interior; II) the abundances of the heavier elements, which allow the present-day spatial distribution of metals in the Milky Way to be investigated. Finally, first results from studies of massive supergiants in other galaxies within the Local Group and beyond are discussed. Attention is drawn in particular to galactic abundance gradients and a novel spectroscopic distance determination technique.*

# 1 Introduction

Cosmology, the formation and evolution of galaxies, and the cosmic cycle of matter pose major challenges for modern astronomy. Stars have been the driving force of evolution ever since the Cosmic Dark Ages. Massive stars are of particular importance as sites of nucleosynthesis and as dominant contributors to the energy and momentum budget of the interstellar medium (ISM), via ionizing radiation, stellar winds and supernova explosions (SNe).

One cornerstone in the wide field of galaxy evolution is the investigation of the chemical evolution of the Milky Way (e.g. Pagel 1997; Matteucci 2001). Here, the predictive power of different model scenarios (among the more recent work e.g. Hou et al. 2000; Chiappini et al. 2001, 2003; Oey 2003; Cescutti et al. 2007) can be verified in great detail by comparison with observation. The solar neighbourhood[1] is of particular interest. Long-lived low-mass stars in the solar neighbourhood play an important rôle, as they allow the evolution of the Galaxy to be studied over time (e.g. Edvardsson et al. 1993; Fuhrmann 2004, and references therein). In order to sample larger distances, more luminous indicators are required, like H II regions or massive early-type stars. These short-lived objects trace present-day abundances only. However, they provide important information complementary to cool stars: they allow elemental abundances to be sampled spatially over the entire disk of the Milky Way.

The drivers behind the evolution of the Galaxy are the stars. Massive ones determine the pace because of their short life cycles. Only the massive stars pass through all (static) fusion cycles up to silicon burning, preparing the seed for the synthesis of the heaviest elements (Burbidge et al. 1957; Cameron 1957) in SNe type II. A quantitative understanding of the evolution of massive stars is therefore essential for the deeper study of the whole picture.

Models of massive star evolution have reached a high degree of sophistication recently. Models accounting for the effects of mass loss and rotation (Heger & Langer 2000; Maeder & Meynet 2000) succeeded in reproducing observed abundance patterns in massive stars. In particular, enriched helium and nitrogen, and depleted carbon are indicative for the mixing of nuclear-processed matter from the stellar core to the atmosphere, which can be observed in fast-rotating stars even on the main sequence. More recent developments concentrated on the interplay of rotation and magnetic fields (Heger et al. 2005; Maeder & Meynet 2005), which – depending on the input physics – may result in quantitatively different predictions on the amount of chemical mixing. Systematic investigations of massive stars of different mass and metallicity can provide the necessary observational constraints to distinguish between competing treatments of the complex (magneto)hydrodynamic processes, which are not fully understood from first principles. Moreover, such observational constraints may guide further refinements of theory.

However, deriving accurate observational constraints on stellar and Galacto-chemical evolution is not simple. Most of the quantitative information in astronomy comes from spectroscopy. The information cannot be inferred directly from

---

[1] In the following we consider the region at distances shorter than $\sim$1 kpc from the Sun as solar neighbourhood.

observation. One has to rely on the interpretation of radiation from light-emitting plasmas, and its interaction with matter, i.e. *quantitative spectroscopy*. Accurate physical modelling is crucial, with systematic uncertainties often dominating the error budget.

Well-understood astrophysical indicators that can contribute to all the fields mentioned above in equal measure are most valuable. Such indicators must meet several criteria: I) they have to be highly luminous in order to be observable over large distances to sample different environments in the universe; II) they should exhibit the spectral features required to address the key questions above; III) the analysis methods for these indicators should be essentially free of systematic uncertainties. Dwarfs and giants of late O and early B-type (OB-type stars) are in principle such indicators, and even more so their evolved progeny, the supergiants of spectral types B and A.

**OB-type Stars.** The advantage of OB-type dwarfs and giants lies in their simple atmospheres in radiative equilibrium, which, in combination with insignificant mass-loss, makes the modelling in principle rather easy. Their spectra exhibit signatures of the light elements, most of the $\alpha$-process elements, and iron, i.e. many of the astrophysically interesting metal species are accessible. Pristine elemental abundances (unaffected by mixing with nuclear-processed matter or by dust formation and subsequent depletion on dust grains as in the case of H II regions) can be derived only from these unevolved objects. However, large bolometric corrections limit ground-based high-resolution spectroscopy with the available telescopes to objects in the Milky Way and the Magellanic Clouds (MCs).

Numerous abundance studies of early-type dwarfs and giants in the solar neighbourhood have been performed (among the more recent e.g. Gies & Lambert 1992; Kilian 1992; Cunha & Lambert 1994; Daflon et al. 1999, 2001a,b; Lyubimkov et al. 2005). The early-type stars imply sub-solar present-day abundances[2] on the mean and a large range in abundance of $\sim$1 order of magnitude (see Sect. 4 for more details). Even within individual star clusters/associations the range in derived abundances can reach a factor $\sim$3–5, e.g. in the Orion association and its sub-groups (Cunha & Lambert 1994). A high degree of *chemical inhomogeneity* is indicated for the matter in the solar neighbourhood. This is in contrast to other observational indicators which imply *chemical homogeneity*, like the gas-phase ISM (e.g. Sofia & Meyer 2001), and to theory, which predicts efficient homogenisation through turbulent mixing (Edmunds 1975; Roy & Kunth 1995). Some of the discrepancy is certainly related to differences in atmospheric parameters derived by the various authors for individual objects. These can amount to more than 15% in effective temperature $T_{\mathrm{eff}}$ and $\sim$0.5 dex in surface gravity $\log g$.

The literature on present-day abundances from H II regions and early-type stars throughout the Galactic disk is too vast to be discussed in detail. A selective but illustrative summary on the state-of-the-art is given in Fig. 1 (for oxygen; similar trends are derived for other elements). Observation indicates the presence of a relatively

---

[2] A reduction of the discrepancies is found if the revised solar abundance standard from hydrodynamic 3-D modelling of the solar atmosphere (Asplund et al. 2005) is adopted instead of earlier work (e.g. Grevesse & Sauval 1998).

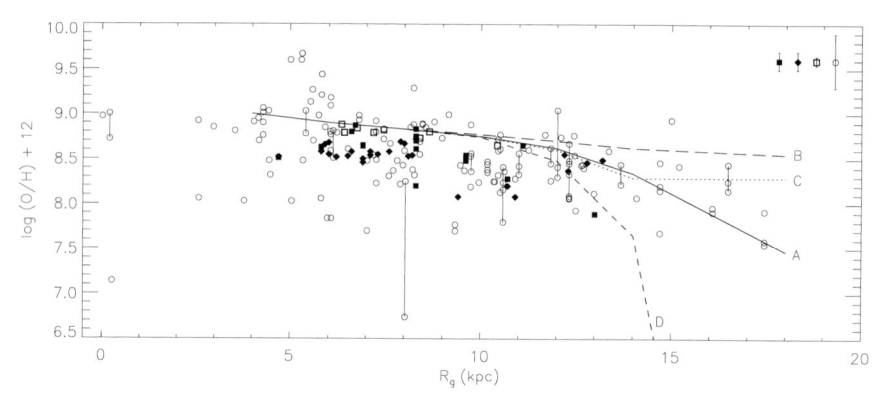

**Figure 1:** Comparison of Galactochemical evolution models (A–D, see Chiappini et al. 2001 for details) with oxygen abundance determinations from H II regions (Esteban et al. 2005: boxes; Rudolph et al. 2006, and references therein: circles) and OB-type stars (Gummersbach et al. 1998: filled boxes; Daflon & Cunha 2004: diamonds), as a function of Galactocentric radius $R_g$. The data indicate the existence of an abundance gradient. Note, however, a systematic offset of the mean observed abundances relative to the model and the enormous spread of the observational data, by $\gtrsim 1$ order of magnitude at almost any $R_g$. Typical error bars are shown in the upper right corner. Symbols are interconnected in cases of double observations. All $R_g$ are scaled to a solar Galactocentric distance of $R_0 = 8.0$ kpc (Reid 1993).

steep abundance gradient and a remarkable range in abundance at any Galactocentric distance $R$. The abundances vary by $\gtrsim 1$ order in magnitude, as in the solar neighbourhood. Individual stars in more distant clusters also show a scatter in abundances by factors $\sim 3$–5 (e.g. Trundle et al. 2007). A systematic offset between observed and predicted mean abundances at every $R$ exists.

**BA-type Supergiants.** OB-type stars develop to supergiants of late B and A-type (BA-type supergiants, BA-SGs) in a later stage of evolution. These are of particular interest for ground-based observations because of their low bolometric corrections. Thus BA-type supergiants are among the visually brightest normal stars in spiral and irregular galaxies. At absolute visual magnitudes up to $M_V \simeq -9.5$ they can rival entire globular clusters and even dwarf spheroidal galaxies in integrated light. Consequently, the present generation of 8–10m-class telescopes and efficient instrumentation allows high-resolution spectroscopy of individual BA-type supergiants in the galaxies of the Local Group to be performed. At medium resolution this can potentially be done even in systems out to distances of the Virgo and Fornax clusters of galaxies (e.g. Kudritzki 1998). This would make all classes of late-type galaxies in the Hubble sequence accessible to detailed studies – in the field, in galaxy groups and in clusters.

The spectra of BA-type supergiants extend the chemical species traced by OB-type stars to the entire iron group and additionally s-process elements, thus covering many of the indicators relevant for understanding cosmic nucleosynthesis. These include, but also largely extend the species traced by gaseous nebulae – in particular H II-regions –, the classical abundance indicators in extragalactic research (see e.g. Garnett 2004). Hence, BA-type supergiants can be used to investigate abundance patterns and gradients in other galaxies to a far greater extent than from the study of gaseous nebulae alone. Thus, in general, crucial observational constraints for theoretical investigations of galactochemical evolution can be derived. In parallel, the metallicity-dependence of stellar winds and stellar evolution can be studied. Finally, BA-type supergiants can act as primary indicators for the cosmic distance scale by application of the wind momentum–luminosity relationship (WLR, Puls et al. 1996; Kudritzki et al. 1999) and by the flux-weighted gravity–luminosity relationship (FGLR, Kudritzki et al. 2003, 2008; Kudritzki & Przybilla 2003), see Sect. 5. In addition to the stellar metallicity, interstellar reddening and extinction can also be accurately determined, so that BA-type supergiants provide significant advantages compared to classical (photometric) distance indicators such as Cepheid and RR Lyrae stars.

Despite this immense potential, quantitative analyses of BA-type supergiants are rather scarce. Only a few of the brightest Galactic objects were studied in an early phase, e.g. Deneb in the pioneering work of Groth (1961) or $\eta$ Leo by Przybylski (1969) and Wolf (1971). The potential of BA-type supergiants for extragalactic research was already recognised in the first analyses of the visually brightest stars in the Magellanic Clouds (e.g. Przybylski 1968, 1972; Wolf 1972, 1973). These studies were outstanding for their time, but also restricted from the present point of view because of simplified model atmospheres, limited accuracy of atomic data, and the lower quality of the observational material (photographic plates). Note, however, that non-LTE model atmospheres were already investigated in this early phase (Kudritzki 1973).

BA-type supergiants have become an active field of research again only recently. Venn (1995a,b), Verdugo et al. (1999) and Takeda & Takada-Hidai (2000, and references therein) conducted systematic studies of larger samples of Galactic BA-type supergiants. Stellar parameters and elemental abundances were derived on the basis of hydrostatic line-blanketed LTE model atmospheres, utilising non-LTE line-formation calculations for key ions in some cases. In general, near-solar abundances were found for the heavier elements, resulting in agreement with present-day abundances derived from unevolved Galactic B-stars. Furthermore, the studies indicated (partial) mixing of the surface layers with CN-cycled gas from the stellar core. However, the uncertainties were often quite large. Discrepancies of up to $\sim$20% in $T_{\text{eff}}$, $\sim$0.5 dex in $\log g$ and $\sim$1 dex in the elemental abundances can be found for individual objects analysed by different authors. A considerable mismatch between observed and model spectra persisted even for the hydrogen line spectrum when hydrodynamic line-blanketed non-LTE model atmospheres were applied, as in the pilot study on Deneb by Aufdenberg et al. (2002).

Apparently, the quantitative spectroscopy of both – unevolved OB-type stars and BA-type supergiants – requires substantial improvements before one can exploit their full potential. This was the motivation for our own work.

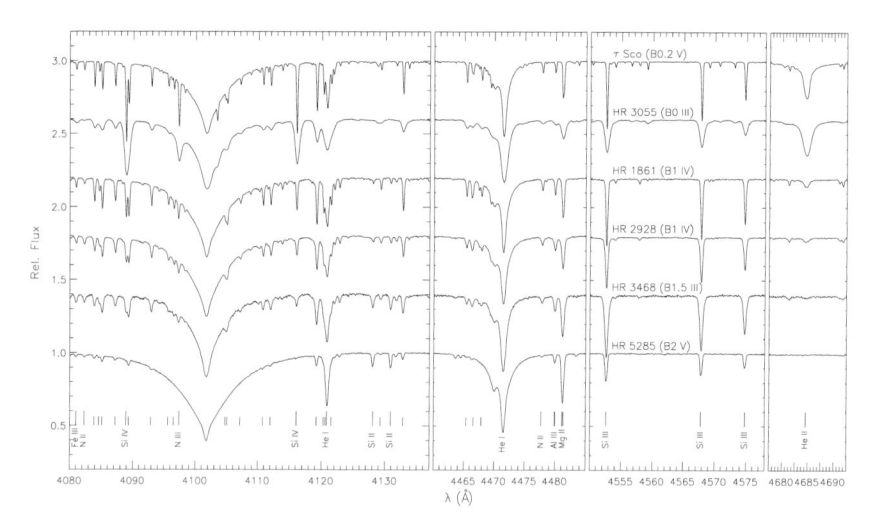

**Figure 2:** Spectral morphology of Galactic B-stars around H$\delta$, He I $\lambda$4471, the Si III triplet $\lambda\lambda$4552–74 and He II $\lambda$4686 Å, arranged in a $T_{\text{eff}}$ sequence. The major spectral features are identified, short vertical marks indicate O II lines. From Nieva (2007).

The general outline of the paper is as follows. Diagnostic challenges for the quantitative spectroscopy of massive stars are discussed in the next section. The focus is on improvements in the modelling and in the analysis methodology. An overview of the present status of our work on the derivation of observational constraints on stellar and Galactochemical evolution is provided in Sects. 3 and 4. Finally, an outlook is given on how these studies can be extended in the framework of the young discipline of extragalactic stellar astronomy to achieve a global understanding of the cosmic cycle of matter, and in particular of galactic evolution.

## 2 Quantitative Spectroscopy

### 2.1 Observations

The quality of the observational data is one factor that determines the accuracy that can be achieved in quantitative analyses. Because of their brightness, high-resolution spectroscopy of Galactic OB-stars and of BA-type supergiants out to the Magellanic Clouds can be accomplished using 2m-class telescopes. Echelle spectrographs like FEROS on the ESO 2.2m or FOCES on the Calar Alto 2.2m telescope are the preferred instruments as they allow all diagnostic lines, which are often scattered throughout the entire optical and near-IR range, to be accessed at once. Moreover, even metal line profiles can be resolved, e.g. for slowly-rotating supergiants. Spectra of bright targets can be obtained at very high signal-to-noise ratios of several hundreds. Such data are required for a *calibration* of the models (see below), that will later be used for the analysis of larger samples of stars, where a $S/N \lesssim 200$ is sufficient. Careful data reduction beyond standard pipeline applications ensures that line profiles – in-

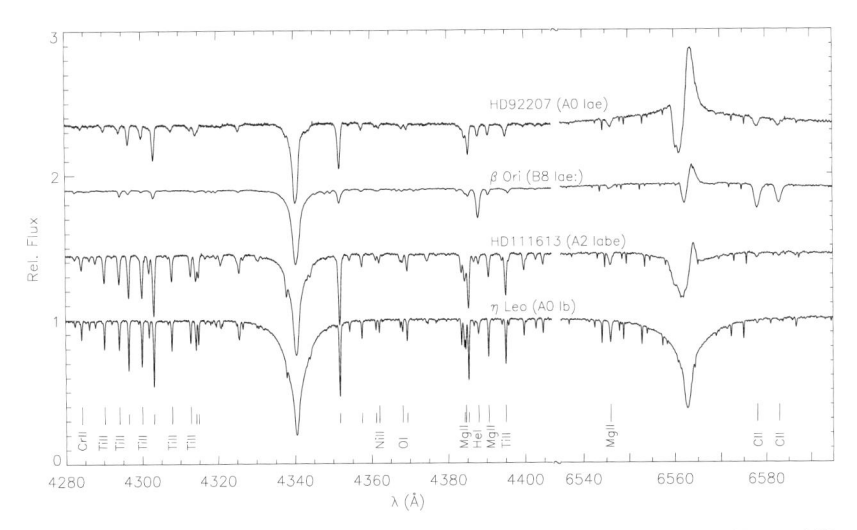

**Figure 3:** Morphology of Galactic BA-type supergiant spectra in regions around H$\gamma$ and H$\alpha$, arranged in luminosity sequence. The major spectral features are identified, short vertical marks indicate Fe II lines. From Przybilla et al. (2006a).

cluding in particular the broad hydrogen features – are recovered accurately, which can be verified by comparison with long slit spectra.

Examples of the quality achieved for the spectra used in the calibration of our models and for establishing the analysis methodology are shown in Figs. 2 and 3 for B-stars and BA-type supergiants, respectively. A good sampling of the parameter space, $T_{\rm eff}$ and $\log g$, is required for a reliable calibration. Only then can the systematic behaviour of the diagnostic indicators be studied, such as the Stark-broadened hydrogen lines, ionization equilibria (among others e.g. He I/II or Si II/III/IV in early B-stars) or various metal features for abundance studies.

Symmetric absorption lines indicate that the photospheres of the stars under study are (close to) hydrostatic. A clear signature of hydrodynamic phenomena is found only in the most luminous supergiants, restricted largely to H$\alpha$, which traces the entire stellar atmosphere because of its strength. The line develops from symmetric absorption into a P-Cygni-profile because of the strengthening stellar wind with increasing luminosity.

Fainter objects, like early-type main sequence stars in the Magellanic Clouds or supergiants in more distant Local Group galaxies, require observations with 8–10m-class telescopes. Echelle spectrographs like UVES on the ESO VLT or HIRES and ESI on Keck are the work horses in this case. However, several hours of observing time per target restrict studies to small samples of stars. Multi-object spectrographs allow these restrictions to be overcome. Instruments like FLAMES/GIRAFFE on the VLT or DEIMOS on Keck provide a multiplex factor of $\sim$100 for a wide field-of-view at a resolving power of $R = \lambda/\Delta\lambda \approx 5000$–$15\,000$, still sufficient for detailed analyses. At larger distances, in galaxies beyond the Local Group, multi-object spectroscopy of even the most luminous supergiants is feasible only at intermediate res-

**Table 1:** Non-LTE model atoms for use with DETAIL/SURFACE

| Ion | Source | Ion | Source |
|---|---|---|---|
| H | Przybilla & Butler (2004a) | Mg I/II | Przybilla et al. (2001a) |
| He I/II | Przybilla (2005) | Si II-IV | Przybilla & Butler (in prep.) |
| C I | Przybilla et al. (2001b) | S II/III | Vrancken et al. (1996), updated |
| C II-IV | Nieva & Przybilla (2006, 2008a) | Ca II | Mashonkina et al. (2007) |
| N I/II | Przybilla & Butler (2001) | Ti II | Becker (1998) |
| O I | Przybilla et al. (2000) | Fe II | Becker (1998) |
| O II | Becker & Butler (1988), updated | | |

olution ($R \approx 1000$–2000) at present. FORS1/2 on the VLT has been successfully employed out to distances of several Mpc.

## 2.2 Models & Analysis Methodology

Quantitative spectroscopy of stars relies on model atmospheres. Ideally, these would account for deviations from local thermodynamic equilibrium (non-LTE effects), line blanketing, spherical extension of the atmosphere and for mass outflow in a comprehensive way, refraining from (almost) any approximation. Such models are in principle available (e.g. via the code CMFGEN, Hillier & Miller 1998), but they are computationally so expensive that a mass-production of models as required for the analysis of larger star samples is still beyond present capabilities. However, in practice such sophisticated models are not required for studies of the photospheres of OB-type dwarfs/giants and BA-type supergiants, as simpler approaches turn out to be equivalent in most aspects (Przybilla et al. 2006a; Nieva & Przybilla 2007), see also below.

We employ a hybrid non-LTE technique, i.e. non-LTE line-formation calculations on the basis of line-blanketed LTE model atmospheres that assume plane-parallel geometry, chemical homogeneity plus hydrostatic and radiative equilibrium. Model atmospheres are computed with the ATLAS9 code (Kurucz 1993). The non-LTE line-formation package DETAIL/SURFACE (Giddings 1981; Butler & Giddings 1985; both updated by K. Butler) is used to address the restricted non-LTE problem. The coupled radiative transfer and statistical equilibrium equations are solved with DETAIL, employing the Accelerated Lambda Iteration scheme of Rybicki & Hummer (1991). This allows even complex ions to be treated in a realistic way, see Table 1 for a summary of model atoms used with DETAIL in this context. Synthetic spectra are calculated on the basis of the resulting level populations with SURFACE, using refined line-broadening theories. Alternatively, the formal solution can also be performed with SURFACE in LTE.

Several conditions need to be met in order to determine reliable level populations (a prerequisite for an accurate analysis) from the solution of the radiative transfer and statistical equilibrium equations (see e.g. Mihalas 1978): I) the local temperatures and particle densities have to be known (i.e. the atmospheric structure) *and* II) the

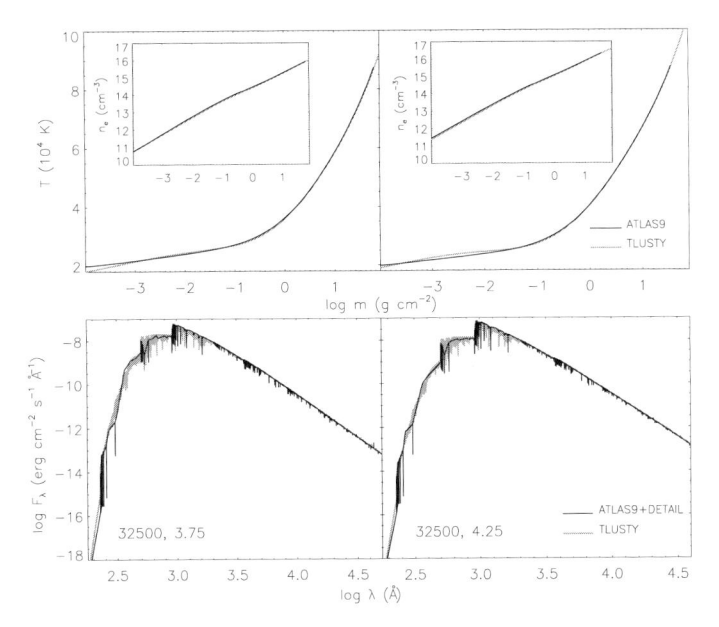

**Figure 4:** Upper panel: Comparison of LTE (ATLAS9) and non-LTE (TLUSTY) model atmosphere structures: temperature and electron density (inset) as a function of column mass for an early B-type dwarf and giant model ($T_{\rm eff}$ and $\log g$ as indicated). Overall, excellent agreement of the model atmospheres is found. The temperature structures deviate by less than 1% in the inner atmosphere, including the regions where the weaker lines and the wings of the stronger features are formed (at column mass $\log m \gtrsim -1$). At the formation depths of the cores of the stronger H and He lines ($-3 \lesssim \log m \lesssim -1.5$) the differences may increase to $\lesssim 2$–3%. Larger deviations may occur only in the outermost parts of the atmosphere, outside the line-formation depths. Lower panel: Comparison of spectral energy distributions, the radiation field computed by DETAIL on the basis of the ATLAS9 atmospheric structure vs. TLUSTY. Excellent agreement is obtained throughout practically the entire wavelength range. Both comparisons together indicate the hybrid non-LTE approach to be equivalent to full non-LTE calculations for this kind of star. From Nieva & Przybilla (2007).

radiation field has to be realistic *and* III) all relevant atomic processes have to be taken into account *and* IV) high-quality atomic data have to be available (i.e. accurate cross-sections for the transitions). In particular, I) and II) require a realistic physical model of the stellar atmosphere and an accurate atmospheric parameter determination. Items III) and IV) are related to the model atoms, i.e. the specifications of atomic input data to be considered in the non-LTE calculations. Shortcomings in any of I)–IV) result in increased uncertainties/errors of the analysis.

Extensive tests have been made to verify the validity of our modelling assumptions in terms of items I) and II) above. One example is shown in Fig. 4, where a comparison of our approach with full non-LTE models (Lanz & Hubeny 2003) in the OB-star regime is made. Excellent agreement is found for the atmospheric structures and the spectral energy distributions, indicating the equivalence of both approaches, see Nieva & Przybilla (2007) for details. Similar tests have been made for BA-type

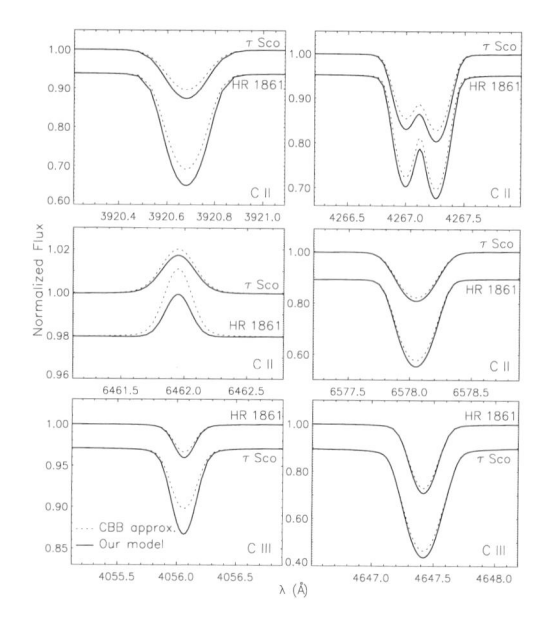

**Figure 5:** Comparison of synthetic C II/III line profiles for two B-type dwarfs ($\tau$ Sco, B0.2 V and HR 1861, B1 IV) using different model atoms. One calculation is based on a well-tested reference model including many collisional data from *ab-initio* calculations while the other adopts standard approximations for evaluating collisional rates for all transitions. Abundance determinations based on the simple model would indicate a significantly larger scatter in the values derived from individual lines. Note that the effects vary from star to star. From Nieva & Przybilla (2008a).

supergiants (Przybilla 2002; Przybilla et al. 2006a). Good agreement with model predictions from the unified (wind+photosphere) non-LTE code FASTWIND (Puls et al. 2005) is found for this kind of star. Notable deviations can occur for the detailed model spectra, in particular for metal lines. However, this is mostly due to different model atoms used in the various codes.

The importance of *realistic* model atoms for quantitative spectroscopy is largely ignored in contemporary astrophysics. This has many reasons. In the past, development was hampered by the lack of the required atomic data. The situation has changed dramatically over the last two decades by the efforts made within the Opacity Project (OP, e.g. Seaton et al. 1994) and the IRON Project (Hummer et al. 1993), and by many developments published in atomic physics literature, which, however, draw little attention within the astronomical community.

In fact, the bulk of stellar studies addresses objects that are in general supposed to show small deviations from detailed equilibrium. There appeared little need to invoke non-LTE techniques for quantitative analyses, a credo being dropped only slowly. For the remainder, the focus was to include huge amounts of atomic data required for the computation of non-LTE model atmospheres in the most efficient way. Large, uniform databases like those provided by e.g. the OP allowed *simple* model

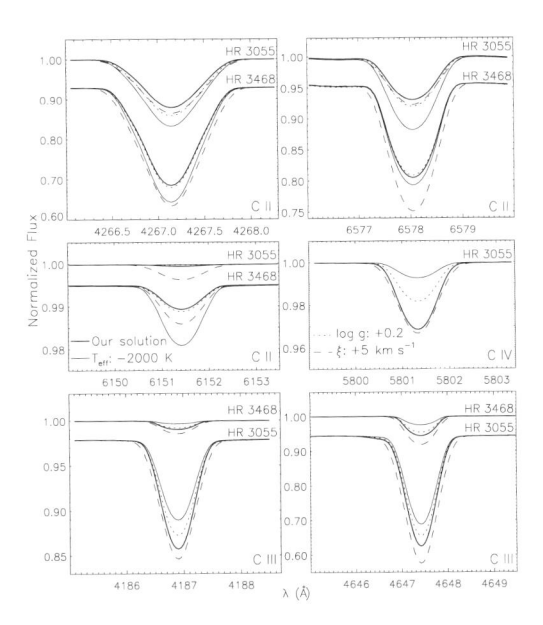

**Figure 6:** Sensitivity of selected C II/III/IV lines to atmospheric parameter variations (as indicated) in two early B-type giants: HR 3055 (B0 III) and HR 3468 (B1.5 III). Our solution corresponds to the final parameters derived for these stars and a constant value of $\varepsilon(C)$ for all lines. The parameter offsets are typical for statistical and systematic uncertainties from published values. All atmospheric parameters need to be constrained well for the ionization equilibria to be established. The star HR 3468 is too cool to show C IV lines. From Nieva & Przybilla (2008a).

atoms to be constructed (nearly) automatically. In reality, model atmosphere computations do not require an exact treatment of the millions of individual lines. A good approximation of the opacities even in a statistical sense is sufficient, masking (most) shortcomings in model atoms. However, for the comparison with the observed line spectra any simplification/inaccuracy in the atomic data *can* matter.

Non-LTE effects prevail whenever radiative processes – which are *non-local* in character (photons may travel wide distances before interacting with the plasma) – dominate the transition rates. Hence, great care is usually exercised to implement accurate oscillator strengths and photoionization cross-sections in model atoms. On the other hand, collision processes thermalise. Collisions do not obey the restrictive selection rules for radiative transitions. In consequence, many processes, such as the coupling between different spin systems in the lighter elements, are determined by collisional rates. However, homogeneous and comprehensive databases on collisional data for ions relevant for stellar atmospheres are still not available. Therefore, simple approximations are usually made to evaluate collision rates, an assumption too crude for accurate quantitative analyses.

In many cases better data for (mostly) electron-impact excitation cross-sections from *ab-initio* computations exist. A comparison of model predictions accounting

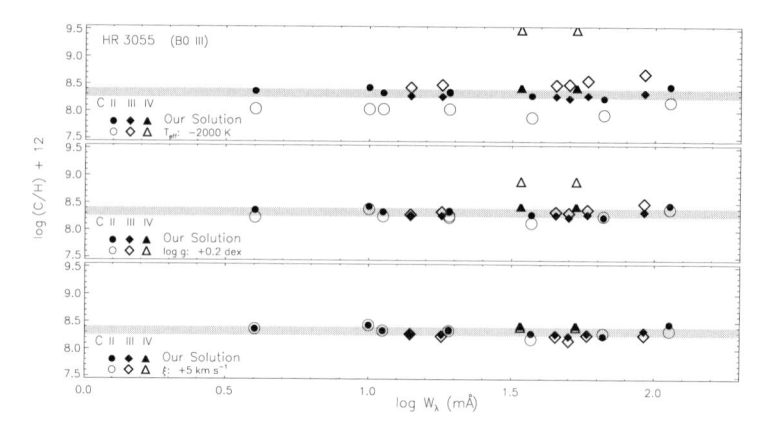

**Figure 7:** Sensitivity of C abundances to stellar parameter variations (for HR 3055, B0 III): $T_{\text{eff}}$ (upper), log $g$ (middle) and microturbulent velocity (lower panel). The offsets of the parameters are displayed in the lower left part of each panel. A large spread in abundance from individual lines of the three ionization stages results, in particular for variations of $T_{\text{eff}}$. Note also the implications of using only few lines of one ion for abundance determinations in the presence of systematic errors in the atmospheric parameters. The grey bands correspond to $1\sigma$-uncertainties of the stellar C abundance in our solution. From Nieva & Przybilla (2008a).

for accurate collisional data and for those using standard approximations in the example of carbon (Fig. 5) shows considerable effects on the spectrum synthesis. Note that the effects may vary with the plasma parameters (temperature, density), as these affect the transition rates. In a similar way the extent of the model atom, e.g. the energy levels or transitions considered, can influence the model predictions. Use of radiative data from different sources may also lead to noticeable effects (see e.g. Nieva & Przybilla 2008a). Therefore, a comparison with observation (for as many spectral lines as possible) is needed to decide which model atom realisation provides the closest match – a procedure called *model atom calibration*. One can take advantage of the rules and regularities of atomic physics for guidance in this process. A wide range of stellar parameters should be considered in the calibration (i.e. stars of different spectral type/luminosity class), as the resulting model atom has to reproduce observation equally well in all objects.

The model atom calibration is complicated by the fact that the stellar parameters of the calibration stars may not be known precisely *a priori*. Examples of the effects of stellar parameter variations on the predicted line profiles are shown in Fig. 6 (for details see Nieva & Przybilla 2008a). Note that the individual spectral lines react in different ways to the variations. Consequently, the determination of stellar parameters (using methods discussed in the next section) and the model atom calibration have to be done hand in hand, requiring iteration to constrain the entire set of parameters. Fulfilment of boundary conditions, like ionization balance (all ionization stages of an element are required to indicate the same abundance), are of crucial importance in this process. The aim is to minimise the abundance scatter from the analysis of the entire line sample, see Fig. 7 for an example. This has to be achieved

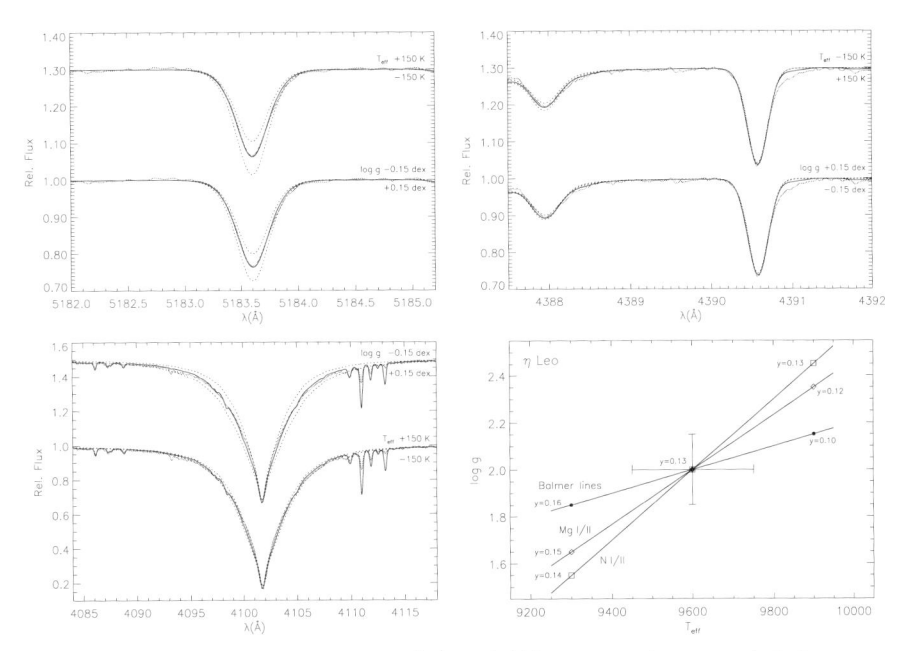

**Figure 8:** Impact of stellar parameter variations (within conservative uncertainties) on non-LTE line profile fits for diagnostic lines in the Galactic A-SG $\eta$ Leo: ionization equilibrium of Mg I (upper left panel) and Mg II (also including He I $\lambda$4388 Å, upper right panel) and the Stark-broadened H$\delta$ line (lower left panel). Spectrum synthesis for the final parameters (thin black line) and varied parameters, as indicated (dotted lines), are compared to observation (grey). Vertical shifts have been applied to the upper profiles. Note the high sensitivity of the minor ionic species Mg I to parameter variations, while Mg II is almost unaffected. Application to the different parameter indicators allows a fit diagram in the $T_{\rm eff} - \log g$-plane to be constructed (lower right panel). The stellar parameters are determined from the intersection of the individual indicators. The curves are parameterised by the helium abundance $y$. From Przybilla et al. (2006a).

for all calibration stars simultaneously, which makes the process highly complex. However, once calibrated, model atoms like those from Table 1 are available for accurate quantitative analyses forthwith.

## 2.3 Stellar Parameter Determination

Our stellar parameter determination relies on a spectroscopic approach. *Multiple non-LTE ionization equilibria and all available Stark-broadened hydrogen lines* (usually from the Balmer and Paschen series) are used rigorously as temperature and surface gravity indicators, analysed via detailed line-profile fits ($\chi^2$-minimisation). An example of the procedure is shown in Fig. 8. Ionization equilibria and the hydrogen lines are both required as the individual indicators are degenerate with respect to simultaneous variations of $T_{\rm eff}/\log g$. However, the intersection of the different loci determines the stellar parameters to high accuracy. The uncertainties in $T_{\rm eff}$ can be

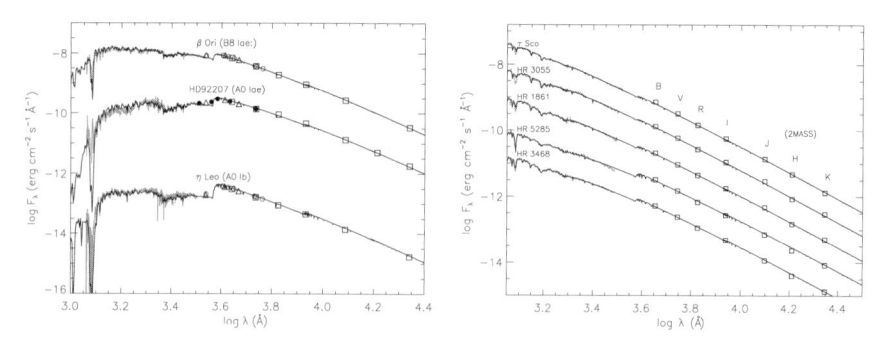

**Figure 9:** Comparison of model fluxes (black lines) and measured SEDs from the UV to the near-IR, for BA-SGs (left panel) and B-type stars (right panel). Displayed are IUE spectrophotometry (grey lines), Johnson (boxes), Strömgren (triangles), Walraven (filled circles) and Geneva photometry (open circles). The observations have been dereddened. From Przybilla et al. (2006a); Nieva & Przybilla (2006).

as low as 1–2% and 0.05–0.10 dex in $\log g$ (Przybilla et al. 2006a; Nieva & Przybilla 2008a). The *redundancy* of matching several indicators simultaneously is one of the main improvements and strengths of our approach.

A further improvement is a self-consistent account of 'secondary' parameters, such as helium abundance (all available helium lines are analysed), metallicity (from the abundance determination, see below) and microturbulence[3], which are usually either approximated or even ignored in the basic parameter determination. These may indeed be of secondary importance for less luminous objects in most cases. However, close to the Eddington limit they can no longer be neglected, as the balance of gravitational forces and radiative acceleration, which strongly impacts the atmospheric structure in supergiants, is delicate (Przybilla et al. 2006a). This complicates the stellar parameter determination enormously, resulting in the need for *iterative* optimisation in a five-dimensional parameter space instead of solving the usual two-dimensional problem. The huge differences in basic parameters of BA-type supergiant studies from the literature (see Sect. 1) can be expected, as none of the investigations accounted for this in full.

An important cross-check is the comparison with the observed spectral energy distribution (SED), i.e. verifying that not only the diagnostic features in the spectra are matched but also the *global* energy output of the star. Examples for the excellent agreement between our model atmosphere computations and observations for BA-type supergiants and OB-stars are displayed in Fig. 9.

Line-profile fitting of high-resolution spectra also allows detailed information on (projected) rotational velocity $v \sin i$ and macroturbulence $\zeta$ to be derived. An example is shown in Fig. 10. Macroturbulence turns out to be widespread among OB-type stars and BA-type supergiants.

---

[3]The microturbulent velocity $\xi$ is determined as usual by demanding elemental abundances to be independent of line equivalent width $W_\lambda$. Note that we do not find a need to assume different values of $\xi$ for individual ions (even of the same element, see e.g. Albayrak 2000), or to assume depth-dependent microturbulent velocities.

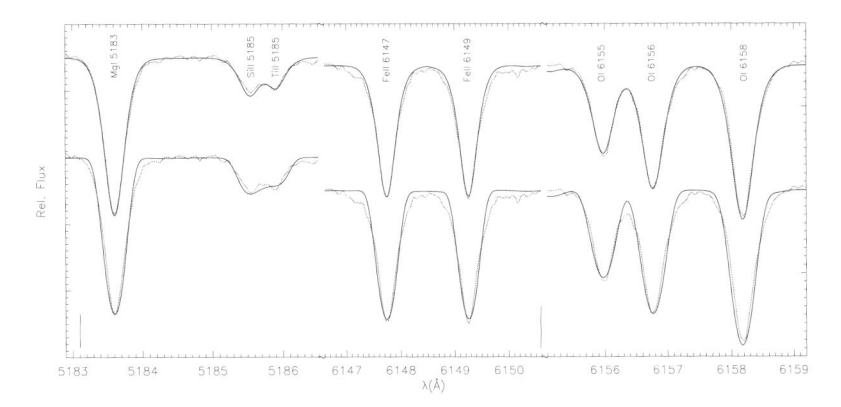

**Figure 10:** Line-profile fits (thin black) to observation of the A-SG $\eta$ Leo (grey lines) accounting for rotation and macroturbulence (upper comparison) and for rotation only (lower comparison). An excellent match is obtained in the first case. The markers are vertically extended by 0.05 units. From Przybilla et al. (2006a).

Finally, when assembling all information (atmospheric parameters, photometry, comparison with stellar evolution models, etc.), a comprehensive view of the stars under study may be developed. This encompasses in particular the fundamental stellar parameters mass $M$, luminosity $L$ and radius $R$.

## 2.4 Elemental Abundances

The chemical composition can be determined with high precision once accurate atmospheric parameters are available. Again, line-profile fitting is preferred over traditional curve-of-growth techniques for the analysis.

An example for the abundance analysis of an A-type supergiant is shown in Fig. 11. Homogeneous abundances are derived from *all* lines in the element spectra, which show little scatter around the mean value. Our rigorous non-LTE analysis thus avoids the systematic trends of abundance with line-strength found in the LTE approximation, e.g. for N I or O I. In other cases, for S II, Ti II and Fe II, the non-LTE abundances are systematically shifted relative to LTE. Also the statistical scatter is reduced in non-LTE, and contrary to common assumption significant non-LTE abundance corrections are found even in the weak-line limit. These can exceed a factor $\sim 2$ for lines with $W_\lambda \lesssim 10$ mÅ, in particular for more luminous objects, where non-LTE effects are amplified. In addition, systematic errors need to be considered because of uncertainties in the stellar parameters, atomic data, and the quality of the spectra. We find that non-LTE abundance uncertainties amount to typically 0.05–0.10 dex (random) and $\sim 0.10$ dex (systematic $1\sigma$-errors) when applying our analysis methodology in combination with realistic model atoms to high-S/N spectra. This applies to OB-stars and BA-type supergiants alike. It is hereby shown that previously reported problems with reproducing observed spectra are only artefacts of inadequate models and/or analysis techniques.

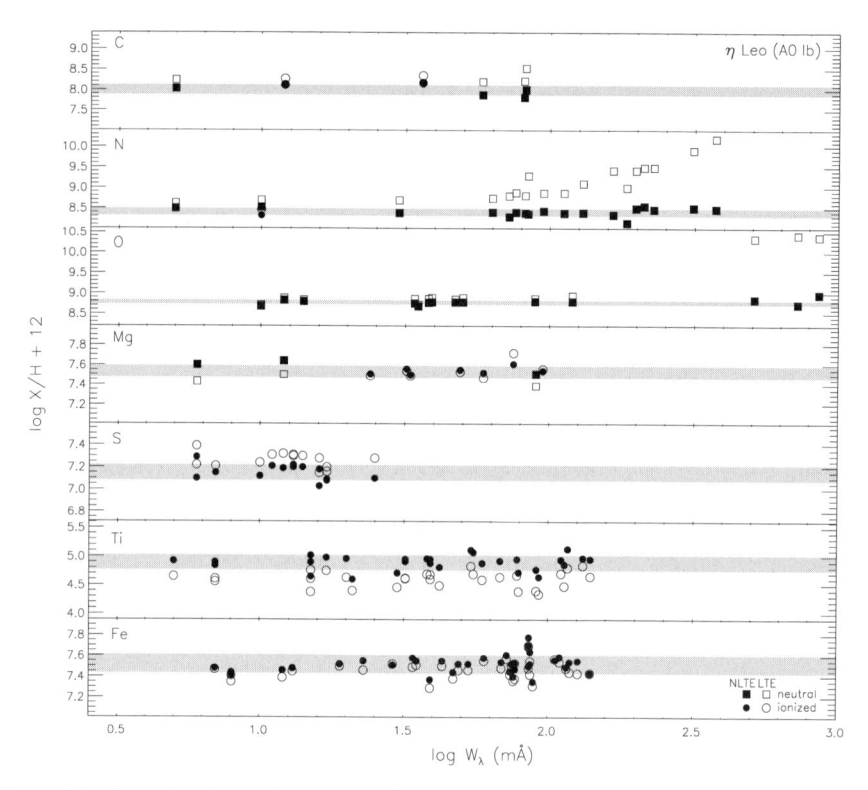

**Figure 11:** Line abundances for several elements in $\eta$ Leo as a function of equivalent width (see legend for symbol identifications). The grey bands cover the $1\sigma$-uncertainty ranges around the non-LTE mean values. Non-LTE calculations reduce the line-to-line scatter and remove systematic trends. Note that even weak lines can show considerable departures from LTE. From Przybilla et al. (2006a).

A comprehensive view of abundances in two Galactic high-luminosity A-type supergiants is given in Fig. 12. The non-LTE analysis gives a scaled solar composition for the heavier elements, and abundance patterns of the lighter elements characteristic for mixing with CN-processed matter. This is consistent with what is expected for evolved massive Population I stars. Inappropriate LTE analyses on the other hand tend to systematically underestimate iron group abundances and overestimate the light and $\alpha$-process element abundances by up to factors of 2–3 on the mean (most noticeable in HD 92207). Taking the LTE results at face value this would imply $\alpha$-enhancement, a characteristic feature of old Population II objects, which, however, is in sharp contrast to the young nature of the supergiants. In other galaxies, where we have less-comprehensive independent information, such LTE analyses may result in misleading conclusions. Note that while many astrophysically important elements are covered by our non-LTE calculations, the investigation of many other species is restricted to LTE, as model atoms are unavailable at present. Considerable efforts will be required to complete the analysis inventory. On the other hand, model atoms for most observed ions in OB-stars are already available.

Massive Stars                                                                 339

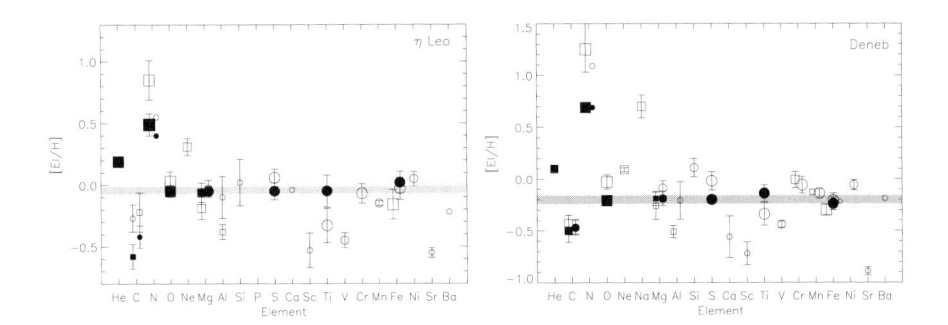

**Figure 12:** Elemental abundance analysis (relative to the solar composition, Grevesse & Sauval 1998) for two bright Galactic A-type supergiants. Symbol identification as in Fig. 11, the symbol size codes the number of spectral lines analysed. The error bars represent $1\sigma$-uncertainties from the line-to-line scatter. The grey shaded area marks the deduced metallicity of the objects within $1\sigma$-errors. Non-LTE abundance analyses reveal a (scaled) solar abundance pattern, except for the light elements which have been affected by mixing with nuclear-processed matter. From Przybilla et al. (2006a); Schiller & Przybilla (2008).

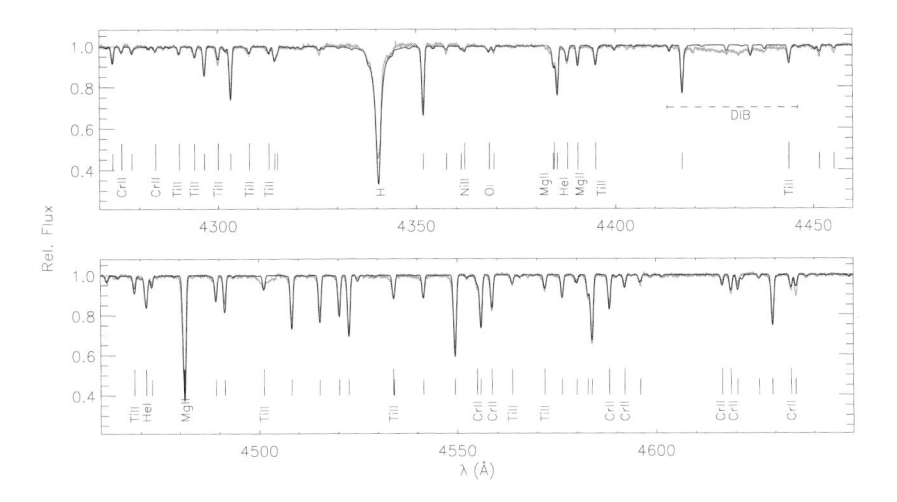

**Figure 13:** Comparison of spectrum synthesis (thin black line) with a high-resolution spectrum of the Galactic A-type supergiant HD 92207 (grey). The major spectral features are identified, short vertical marks designate Fe II lines. The location of the diffuse interstellar band around 4430 Å is indicated. With few exceptions, excellent agreement between theory and observation is found (H$\gamma$ is affected by the stellar wind in this highly luminous star). From Przybilla et al. (2006a).

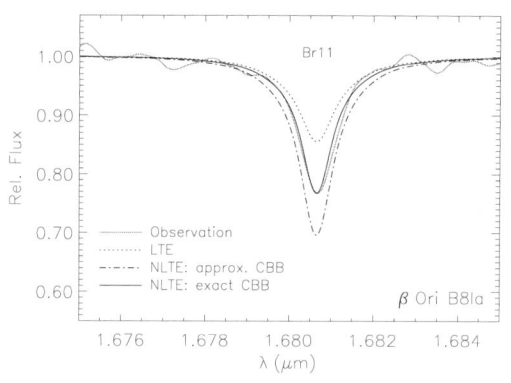

**Figure 14:** Modelling of hydrogen lines in the near-IR, e.g. Br11 in the late B-SGs $\beta$ Ori, using different approaches according to legend. Agreement between observation and theory can be obtained only in non-LTE when electron-impact excitation data from *ab-initio* calculations are adopted for the model atom. Non-LTE computations using approximation formulae for the evaluation of collisional bound-bound rates overpredict equivalent widths by a factor $\sim$2 in this case, while LTE modelling gives much too weak lines. Stellar parameters as determined from the analysis of the optical spectrum have been adopted. All three models reproduce the higher Balmer lines equally well. According to Przybilla & Butler (2004a).

The precision achieved in the individual line analysis results in a highly consistent match of the *entire* synthetic spectrum with observation, see e.g. Fig. 13. Approximately 70% of the total number of features in the optical and near-IR spectra of BA-type supergiants can be considered in non-LTE at present, and the rest in LTE. The coverage of features that can be treated in non-LTE is over 90% in OB-type stars, see e.g. Przybilla et al. (2008b) for an example of the quality that can be achieved in fitting observed spectra.

Such comprehensive spectrum synthesis is a prerequisite for the analysis of medium-resolution spectra, such as provided by multi-object spectrographs. Individual lines can often no longer be resolved in that case. Possible applications include in particular the study of faint objects at larger distances like OB-type main-sequence stars in the Magellanic Clouds or BA-type supergiants in galaxies beyond the Local Group, see Sect. 5 for a discussion.

## 2.5 Near-IR Spectroscopy

Near-IR observations of early-type stars are a relatively new field to astronomy. They are motivated by the desire to study star formation and young stellar populations throughout the Galactic disk and in the Galactic centre in particular – regions where strong extinction prohibits observations in the optical. Moreover, the field will grow in importance once the next generation of extremely large telescopes comes into operation. These will rely on adaptive optics systems in order to facilitate diffraction-limited observations. As a consequence, they will be operated primarily at near-IR wavelengths.

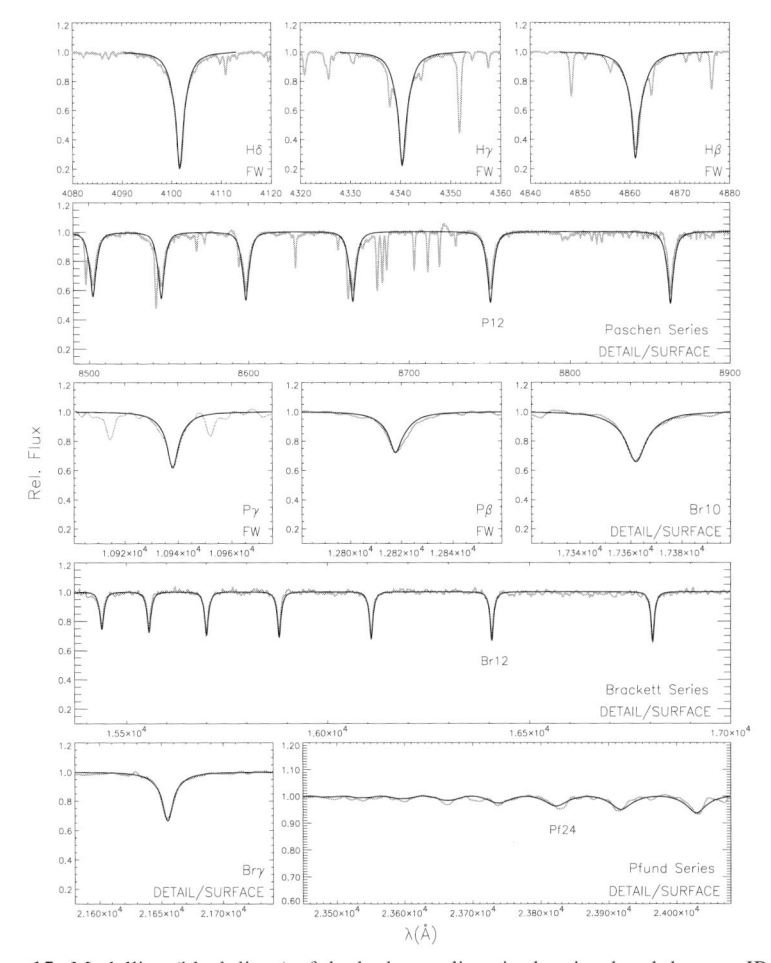

**Figure 15:** Modelling (black lines) of the hydrogen lines in the visual and the near-IR spectrum (grey) of the Galactic A-supergiant Deneb (A2 Ia). Overall, excellent agreement is achieved. The synthetic spectra are calculated with a hybrid non-LTE technique using DETAIL/SURFACE (photospheric lines) or FASTWIND (FW, wind-affected lines), as indicated, for the same stellar parameters. Some lines, as P$\beta$ and H$\beta$, are noticeably affected by the stellar wind. From Schiller & Przybilla (2008).

Little information on quantitative near-IR spectroscopy of normal OB-type dwarfs/giants can be found in modern literature (Lenorzer et al. 2004; Repolust et al. 2005). Even less is known on BA-type supergiants.

Problems with the models were indicated in the pioneering study of the Paschen and in particular the Pfund series in the A-supergiant prototype Deneb (Aufdenberg et al. 2002), which were attributed to deficiencies in our understanding of the physics of stellar atmospheres. It was later shown (Przybilla & Butler 2004a) that this was in fact a result of inaccurate electron-impact excitation cross-sections used in the hydrogen model atom by Aufdenberg et al. (2002). In Fig. 14 the sensitivity of the near-IR lines of hydrogen to the model assumptions is demonstrated. This is because of the amplification of non-LTE effects in the Rayleigh-Jeans tail of the energy distribution of hot stars. The line source function may be written as

$$S_l = \frac{2h\nu^3/c^2}{b_i/b_j \exp(h\nu/kT) - 1},\qquad(1)$$

using standard nomenclature for physical constants. The source function is particularly sensitive to variations in the ratio of the departure coefficients $b_i$

$$|\Delta S_l| = \left|\frac{S_l}{b_i/b_j - \exp(-h\nu/kT)}\Delta(b_i/b_j)\right|$$
$$\underset{h\nu/kT\ll 1}{\approx} \left|\frac{S_l}{(b_i/b_j - 1) + h\nu/kT}\Delta(b_i/b_j)\right|\qquad(2)$$

when $h\nu/kT$ is small, because the denominator can adopt values close to zero, amplifying effects of a varying $\Delta(b_i/b_j)$. This makes near-IR lines in hot stars very susceptible to even small changes in the atomic data and details of the calculation. In fact, the $b_i$ vary by only several percent in the example above, resulting in changes by a factor $\sim 2$ in equivalent width. For the particular case of Deneb, an excellent match of model and observation has been obtained in the meantime (Schiller & Przybilla 2008), see Fig. 15.

Similar sensitivities to details of the model calculations are also found for helium lines in hot stars (Przybilla 2005; Przybilla et al. 2005; Nieva & Przybilla 2007). Examples are shown in Fig. 16. Difficulties were found in earlier non-LTE studies of the He I $\lambda 10\,830$ Å line: the observed trends of $W_\lambda$ as a function of $T_{\text{eff}}$ were not reproduced by the models, which predicted either too strong absorption or strong emission. These problems were related to inaccurate photoionization cross-sections and the neglect of line blocking. Little is known about near-IR metal lines in this kind of star.

The models and analysis methodology introduced in the previous sections are not restricted to the two object classes discussed here. We have already used these with great success for quantitative analyses of main sequence A-stars such as Vega (Przybilla 2002), subluminous B-stars (Przybilla et al. 2006b; Geier et al. 2007), extreme helium stars (Przybilla et al. 2005, 2006c), the Sun and solar-type stars (Przybilla & Butler 2004b; Mashonkina et al. 2007).

## 3 Observational Constraints on the Evolution of Massive Stars

A determination of accurate atmospheric parameters and elemental abundances for a larger sample of objects allows tight observational constraints on evolution models for massive stars to be derived. The most rewarding observational indicators for an empirical verification of the models are surface abundances of the light elements that participate in the main fusion cycle: carbon, nitrogen, helium, and, to a lesser degree,

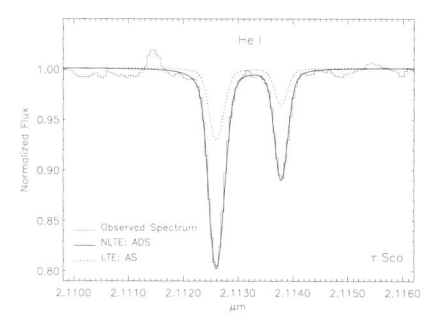

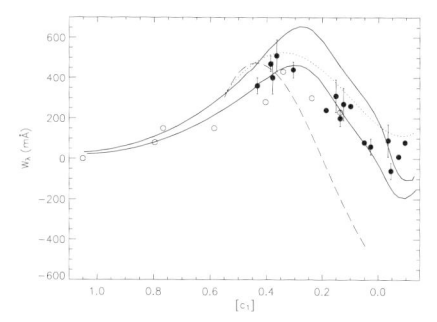

**Figure 16:** Modelling of helium lines in the near-IR. Left panel: non-LTE strengthening of the He I 2.11 μm lines in τ Sco. Stellar parameters as determined from the analysis in the optical have been adopted. From Nieva & Przybilla (2007). Right panel: comparison of observed equivalent widths for He I 10830 Å (Lennon & Dufton 1989: filled circles; Leone et al. 1995: open circles) with non-LTE model predictions (Auer & Mihalas 1973: dotted line; Dufton & McKeith 1980: dashed line; Przybilla 2005: full lines, for $\xi = 0$, lower, and $8\,\mathrm{km\,s^{-1}}$, upper curve). The abscissa is the reddening-free $[c_1]$ index, a temperature indicator. From Przybilla (2005).

oxygen. These allow the otherwise inaccessible (magneto-)hydrodynamic mixing processes in the stellar interior to be traced. These processes redistribute angular momentum, transport burning products from the core to the surface and replenish hydrogen, thus extending lifetimes and increasing the stellar luminosity.

First results from an application of our analysis methodology to B-type dwarfs/giants (Nieva & Przybilla 2008b) and BA-type supergiants (Przybilla et al. 2006a, 2008a; Firnstein & Przybilla 2006) are summarised in Fig. 17 (and Fig. 18, see next section). A comparison with evolution tracks is made. Most of the apparently slow-rotating B-stars show N/C ratios close to the pristine value of ~0.3. A notable exception is τ Sco, which is a truly slow-rotating magnetic star (Donati et al. 2006). The situation is more complex with the BA-SGs. They all exhibit slow rotation because of their expanded envelopes, irrespective of the initial rotational velocity of their progenitors on the main sequence. It appears that the objects at masses below $\sim 15\,\mathrm{M_\odot}$ show larger amounts of nuclear-processed material than those around $\sim 20\,\mathrm{M_\odot}$. Larger N/C ratios are found again for the most massive objects of the sample. Note that the sample objects of $M \gtrsim 30\,\mathrm{M_\odot}$ are located either in the Magellanic Clouds or in M31, i.e. they have a different metallicity than the Galactic stars.

Two important conclusions can already be drawn from this small sample. Let us concentrate first on the objects more massive than $\sim 15\,\mathrm{M_\odot}$. A general trend of increased mixing of nuclear-processed material with increasing stellar mass is found, in accordance with the predictions of evolution models (e.g. Heger & Langer 2000; Maeder & Meynet 2000). Moreover, the strongest mixing signature is found for the most metal-poor object, AzV 475 in the Small Magellanic Cloud, also in agreement with theory (e.g. Maeder & Meynet 2001). However, the mixing efficiency appears to be higher (by a factor ~2) than predicted by current state-of-the-art evolution computations for rotating stars with mass-loss. Stellar evolution models accounting

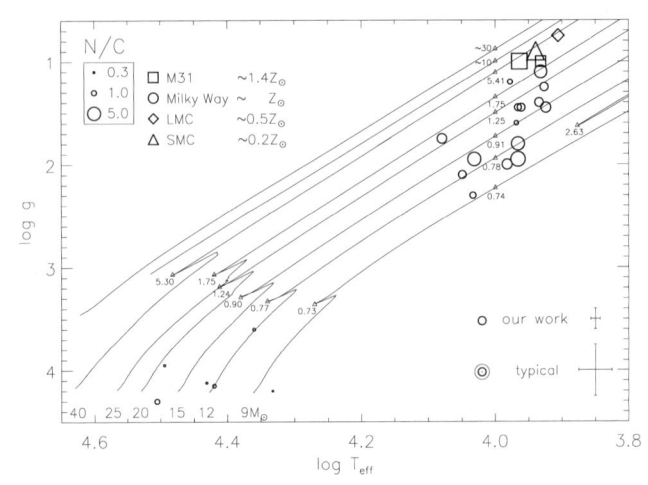

**Figure 17:** Observational constraints on massive star evolution. Displayed are results for the most sensitive indicator for mixing with nuclear-processed matter, N/C, in a homogeneously analysed sample of Galactic BA-type supergiants (circles) and their progenitors on the main sequence, OB dwarfs. The N/C ratios are encoded on a logarithmic scale, with some examples indicated. Error bars ($1\sigma$-statistical & systematic) from our work and those typical for previous work are also indicated. Stellar evolution tracks (Meynet & Maeder 2003) are displayed for rotating stars at solar metallicity $Z_\odot$ (full lines). Starting with an initial N/C$\sim$0.3, theory predicts N/C$\sim$1 for BA-type supergiants evolving to the red, and N/C$\sim$2–3 after the first dredge-up. Since we find N/C values as high as $\gtrsim 6$ the observed mixing efficiency is higher than predicted. Also, blue loops extend to hotter temperatures than predicted. A few results for objects in other Local Group galaxies are also shown.

for the interplay of rotation and magnetic fields may resolve this discrepancy since they find a higher efficiency for chemical mixing (e.g. Maeder & Meynet 2005).

Larger N/C ratios at $M \lesssim 15\,M_\odot$ can be explained if the objects are on a blue loop, i.e. if they have already undergone the first dredge-up during a previous phase as a red supergiant, in addition to rotational mixing[4]. This interpretation is in contrast to earlier findings (e.g. Venn 1995b). Further support for the blue-loop scenario comes from lifetime considerations. Stellar evolution calculations indicate that supergiants spend a much longer time on a blue-loop (with central helium burning) than required for the crossing of the Hertzsprung-Russell diagram (HRD) from the blue to the red (the short phase of core contraction after central hydrogen burning has ceased). E.g., in the case of a rotating $9\,M_\odot$ model of Meynet & Maeder (2003) the difference is about a factor 15. It is well-established that blue loops are required to explain the Cepheid variables, but their extent in the HRD – in particular the upper limits in temperature and stellar mass – are essentially unknown. The blue-loop phase is highly sensitive to the details of the stellar evolution calculations ('... is a sort of magnifying glass, revealing relentlessly the faults of calculations of earlier phases.', Kippenhahn & Weigert 1990).

---

[4] Also in this case an increased mixing efficiency, by about a factor 2, would be required.

Consequently, a systematic study of a larger sample of massive stars could provide the tight observational constraints required for a thorough verification and refinement of the stellar evolution models. Precision analyses of stars covering the relevant part of the HRD are under way.

## 4  Constraints on Galactochemical Evolution

Abundances of the heavier elements, from oxygen on, are unaffected by mixing in the course of stellar evolution. As a consequence, massive stars can be used as tracers of pristine abundances for the variety of elements detectable in their spectra. Gravitational settling or radiative levitation act on too long timescales to affect the surface composition in the stars' short lifetimes.

The results of the abundance analysis for the B-star and BA-type supergiant sample are discussed next, with particular emphasis on the context of Galactochemical evolution. Note that while our B-star sample is located in the solar neighbourhood, the distribution of the BA-SGs is extended, out to distances of $\sim$3 kpc. As a result, the B-stars should be almost unaffected by a Galactic abundance gradient, whereas some effects may be present for the BA-type supergiants. This has to be kept in mind in the following discussion.

The histograms in Fig. 18 compare our results for C, N, O and Mg abundances with previous studies (excluding older LTE work). Let us concentrate on the B-type stars first. The most remarkable characteristic is that our sample indicates *highly homogeneous* present-day abundances in the solar neighbourhood, except for nitrogen (which is affected by mixing with nuclear-processed material). This is in contrast to all previous studies, which show the large range in abundance – and therefore chemical *inhomogeneity* – already described in Sect. 1. Despite the small sample size the findings are significant because of the absence of selection effects: the target stars are randomly distributed in associations and the field and they cover a wide range of stellar parameters. Note that the same six stars show a significant scatter in abundances in one of the most reliable investigations to date, by Kilian (1992, 1994), which we interpret as being due to of residual uncertainties in stellar parameters and in the model atoms. Agreement with other observational indicators like the gas-phase ISM, which also imply homogeneity out to distances of $\sim$1.5 kpc (e.g. Sofia & Meyer 2001), is thus achieved, supporting, in addition, conclusions from theoretical considerations that homogenisation processes should be highly efficient (Edmunds 1975; Roy & Kunth 1995).

Carbon and magnesium abundances are slightly higher on average than derived in most previous analyses. A significantly higher oxygen abundance is found, close to the 'old' solar value (Grevesse & Sauval 1998) – a highly significant result in the ongoing controversy about the impact of the 'new' solar abundances (Asplund et al. 2005) on the solar interior model (e.g. Bahcall et al. 2005). Note that mixing with nuclear-processed material will also affect the carbon abundances. However, as the initial carbon abundance is much larger than the nitrogen abundance, little depletion is needed to produce a noticeable effect in nitrogen. Stellar evolution models (Meynet & Maeder 2003) predict depletion by up to $\sim$0.05 dex for

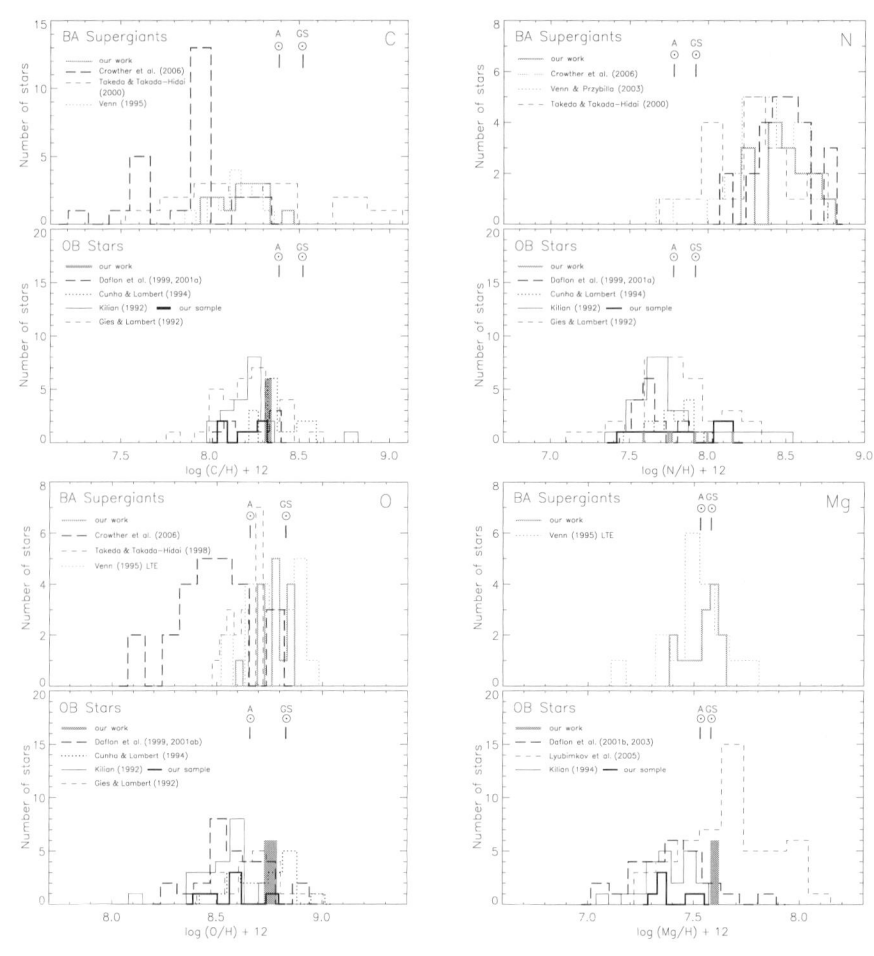

**Figure 18:** Comparison of abundances from B-type dwarfs and giants (in the solar neighbourhood) and their evolved progeny, BA-type supergiants (out to ∼2.5 kpc distance) as obtained in our work and from the literature. In contrast to all previous studies we find highly homogeneous abundances in the unevolved B stars (except for N, which is sensitive to mixing with nuclear-processed matter even on the main sequence). Results from the other studies span a range of ∼1 order of magnitude in abundance (as in Fig. 1). While our B star sample is only small at present, the results are significant as the comparison with one of the most careful analyses so far (Kilian 1992) for the same stars shows. The chemical homogeneity of the ISM in the solar neighbourhood (Sofia & Meyer 2001) is hereby confirmed on the basis of B stars for the first time. Moreover, the B star abundances for elements unaffected by stellar evolution (O, Mg) are confirmed by the study of BA-type supergiants. These show a slightly wider spread because a large volume was sampled such that an influence of the Galactic abundance gradient/intermediate-scale inhomogeneities can become apparent. Solar abundances (⊙) are adopted from Grevesse & Sauval (1998) and Asplund et al. (2005). Bin sizes are related to the 1-$\sigma$ scatter in the individual studies. From Nieva & Przybilla (2008a,b: C & N in OB stars). Preliminary results from Firnstein & Przybilla (in prep.) and Przybilla, Nieva & Heber (in prep.).

stars of average rotation similar to our sample objects. Therefore, a pristine value of $\log C/H + 12 = 8.35 \pm 0.05$ is implied from the sample (Nieva & Przybilla 2008a). The pristine N abundance may be deduced from the stars with the lowest values. Both, C and N abundances, come close to the 'new' solar value of Asplund et al. (2005).

The findings from the B-star sample are supported by the results from the BA-SGs. Despite lacking the high degree of homogeneity (a result of the larger volume sampled), they show a much smaller range in abundance than previous studies. The average abundances for O and Mg are in excellent agreement with the values from the B-star sample. Carbon is depleted and nitrogen enriched as a consequence of the evolved state of the supergiants. The upper limit of C abundances in the BA-SGs is consistent with the pristine value derived from the B-stars, meeting a boundary condition imposed by mixing.

Consistency is achieved from two classes of indicators, which show completely different spectra (BA-SG spectra are dominated by neutral and single-ionized species, B-type stars by higher ionization stages). These first results imply that a re-interpretation of studies of massive, early-type stars in the context of Galactochemical evolution is required. It appears that previous findings of chemical inhomogeneity (Figs. 1 & 18) were a result of the limited accuracy of the models and the analysis methodology used. A much higher degree of homogeneity is indicated from our work, which needs verification for the solar neighbourhood from a larger sample of objects. A wider range of elements should be considered. The work also requires an extension to other regions of the Milky Way to study star clusters and OB associations.

Now we have the tools at hand to determine abundances to the precision necessary for deriving unbiased elemental abundance gradients[5], see Fig. 19 for an example of the dramatic improvements that can be expected. These will provide tight observational constraints on Galactochemical evolution models and may guide future improvements of the modelling.

## 5 Extragalactic Stellar Astronomy

Quantitative spectroscopy of massive, early-type stars is feasible also in nearby galaxies, because of the immense luminosity of OB-type stars and in particular BA-type supergiants. This opens up the possibility to study stellar evolution as a function of metallicity, one of the key quantities of the models.

**Local Group.** Earlier versions of our model atoms were used by Korn et al. (2002, 2005) in an analysis of high-resolution spectra of unevolved B-stars in the Large Magellanic Cloud (LMC). Pristine present-day abundances of the light elements

---

[5]For the most part, relatively steep abundance gradients (see e.g. Fig. 1) are discussed in the literature. Note, however, that the Milky Way has a central bar (e.g. López-Corredoira et al. 2007). Barred spiral galaxies usually show shallow abundance gradients (e.g. Zaritsky et al. 1994), which may be attributed to homogenisation due to large-scale radial gas flows induced by the bar (e.g. Roberts et al. 1979). Apparently, a conundrum needs to be solved.

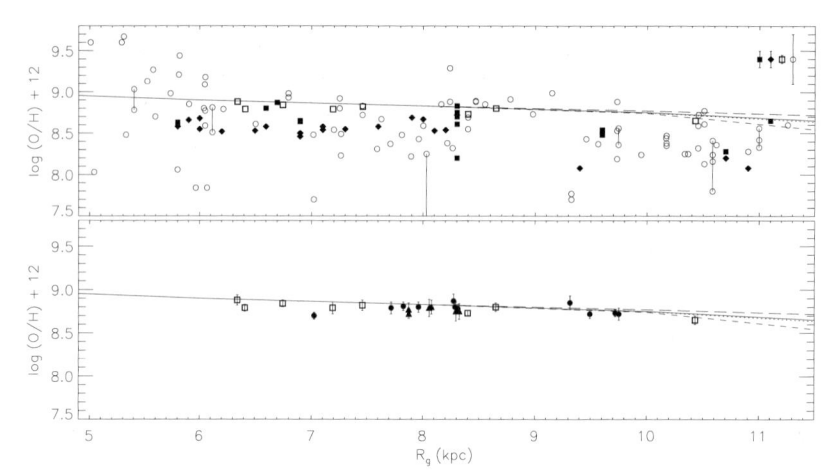

**Figure 19:** The Galactic oxygen abundance gradient. Upper panel: detail of Fig. 1. Lower panel: preliminary results from our work on BA-type supergiants (dots) and B-stars (triangles). Individual error bars are indicated (1-$\sigma$ statistical uncertainties), lines mark model predictions of Chiappini et al. (2001), see Fig. 1. *A clear trend is found and the scatter in abundance is dramatically reduced when typical systematic errors are eliminated.* Excellent agreement with the work of Esteban et al. (2005) on H II regions (boxes) is found after applying a correction of +0.08 dex to account for depletion on dust grains in the nebulae. Data from Przybilla et al. (2006a), Firnstein & Przybilla (in prep.), Przybilla, Nieva & Heber (in prep.).

were obtained from stars for the first time. It was shown that the LMC is indeed nitrogen-poor, as indicated earlier by studies of H II regions. More work on OB-type dwarfs/giants is required to study both slow and fast rotators in order to derive comprehensive observational constraints on the evolution models near the main sequence. Here, the FLAMES survey of massive stars (Evans et al. 2005, 2007) provides a unique database for three galaxies: the Milky Way, the Large, and the Small Magellanic Cloud.

First results from an application of our analysis methodology to high-resolution spectra of selected A-SGs in the Magellanic Clouds and in M 31 are shown in Fig. 17, cf. also Venn & Przybilla (2003). An extension of the work to larger samples of supergiants in these and other galaxies of the Local Group will in particular allow the metallicity-dependent mixing efficiency of the stellar evolution models (e.g. Maeder & Meynet 2001) to be verified.

A-type supergiants as tracers for galactochemical evolution were studied at high spectral resolution in several galaxies of the Local Group. These comprise first steps towards a determination of abundance gradients/patterns in spiral galaxies like M 31 (Venn et al. 2000) or in dwarf irregular (dIrr) galaxies like NGC 6822 (Venn et al. 2001). Gas-rich dIrr galaxies are of particular interest, as they are the closest analogues to the basic building blocks in hierarchical galaxy formation scenarios ('near-field cosmology'). Detailed information on abundances for various elements permits deeper insights in the nucleosynthesis histories of dIrr galaxies beyond the Mag-

**Figure 20:** NGC 3621 at a distance of 6.6 Mpc. The positions of stars (circles) and H II regions (boxes) observed with FORS1 on the ESO VLT are marked on a colour image obtained by combining 5 min $B$, $V$ and $I$-band frames taken with the same instrument in imaging mode. The field of view is approximately $7' \times 7'$. Different coloured markers are used for clarity only. From Bresolin et al. (2001).

ellanic Clouds to be obtained (Venn et al. 2003, and references therein). Progress in this branch of extragalactic stellar astronomy is slow, as high-resolution spectroscopy of supergiants at these distances is costly. Several hours of observing time on 8–10m-class telescopes are required per object.

Quantitative studies at intermediate spectral resolution allow only less-comprehensive information to be derived. However, they also have their advantages: large observational samples can be easily accessed using multi-object spectrographs, and fainter targets become observable, facilitating *quantitative spectroscopy of supergiants in galaxies beyond the Local Group*.

**Beyond the Local Group.** Observations of blue supergiant candidates in NGC 3621 (Fig. 20) at a distance of 6.6 Mpc pushed the capabilities of FORS1 and

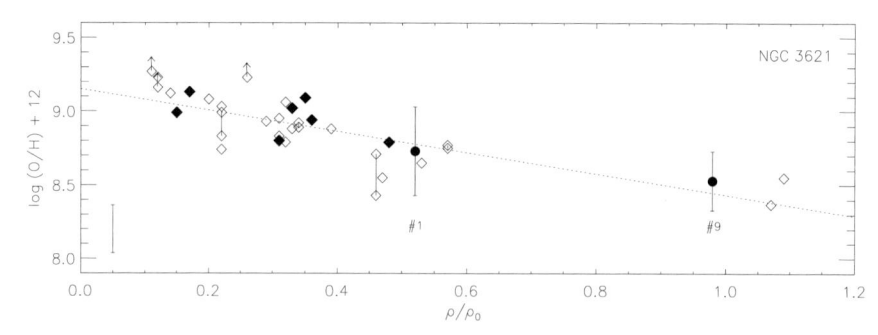

**Figure 21:** The oxygen abundance gradient in NGC 3621 as a function of fractional isophotal radius $\rho/\rho_0$. Data for H II regions are adopted from Ryder (1995, open diamonds) and Zaritsky et al. (1994, filled diamonds). Arrows indicate lower limits, a typical error bar is shown at bottom left. Some H II regions were observed twice, in this case the corresponding symbols are connected. The dotted line denotes the least-squares fit to the data. Stellar abundances are superimposed (#9: Bresolin et al. 2001; #1: Przybilla 2002, dots). Note that oxygen abundances cannot be directly derived from the two available A-type supergiant spectra. Instead, the metallicity estimates are transformed to the given abundance scale, assuming O/H = M/H.

the ESO VLT to their limits. Spectra of 19 objects down to $V \approx 22$ mag were obtained in 10.7 hrs of integration time, confirming many as supergiants (Bresolin et al. 2001). Intermediate-resolution spectra can be quantitatively analysed once a reliable and comprehensive modelling is achieved (Fig. 13, see Przybilla et al. 2006a for thorough tests). However, medium-resolution spectroscopy implies a loss in detail and accuracy of the information that can be extracted from observation. We expect that metallicities can be determined to better than a factor $\sim 2$, i.e. at an accuracy similar to that achieved in published high-resolution studies. Our modelling is therefore *highly competitive* in extragalactic research.

An example is shown in Fig. 21, where data from two stars in NGC 3621 are compared to the oxygen abundance gradient as derived from H II regions. Because individual lines are unresolved in the spectra, the estimated metallicity is transformed into O abundance by setting O/H = M/H – a good approximation, cf. Sect. 2.4. At first glance, excellent agreement is found, but see below.

Such studies can be extended to larger samples of objects covering the entire extent of galaxies. This allows the results from the only indicators used so far, luminous H II nebulae, to be *verified* and *extended*. The most detailed study in this respect is on NGC 300, a spiral galaxy in the nearby Sculptor Group observed within the Araucaria project (Gieren et al. 2005), where in total 30 blue supergiants have so far been analysed (Bresolin et al. 2002; Urbaneja et al. 2005; Kudritzki et al. 2008).

Abundances from nebulae are in most cases derived by strong-line methods, i.e. empirical correlations between line ratios and abundance are used (like the $R_{23}$-index for oxygen), as the spectral features necessary for a direct analysis are often not observed because of their weakness. It turns out that absolute abundances and the abundance gradient from nebulae depend strongly on the calibration. Several of the most-widely used calibrations from the literature *fail* to achieve consistency

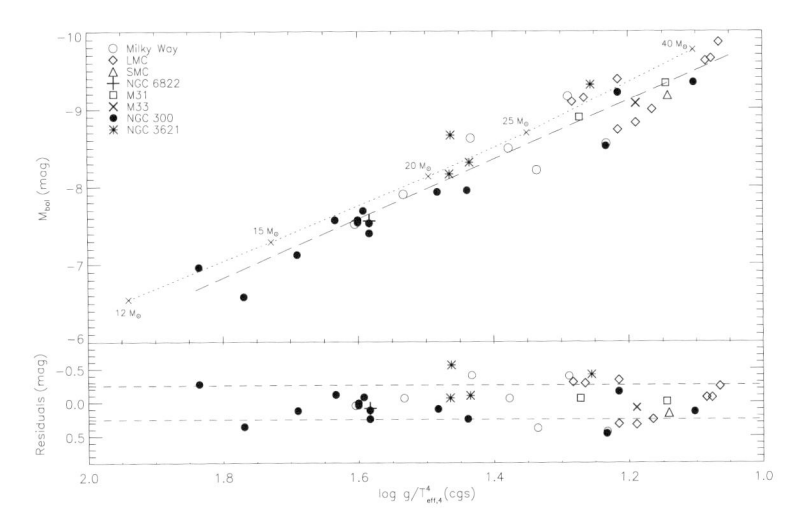

**Figure 22:** Absolute bolometric magnitude vs. logarithm of flux-weighted gravity of B8 to A4 supergiants in eight galaxies (see legend). The linear regression to the observational data is marked by a dashed line. The relationship obtained from stellar evolution models at solar metallicity and accounting for effects of rotation (Meynet & Maeder 2003) is also shown and labelled with the initial zero-age main-sequence masses of the corresponding models. Note that $T_{\rm eff}$ is used in units of $10^4$ K. According to Kudritzki et al. (2003), with updates from Przybilla et al. (2006a, 2008a).

with the stellar indicators: nebular abundances may be too large by up to a factor 2–3, and they may indicate an abundance gradient that is too steep (see Urbaneja et al. 2005 for details). Such discrepancies have also been found in other galaxies, from detailed analyses of H II regions (Kennicutt et al. 2003; Bresolin et al. 2004b). The presence of systematic bias in abundance studies based on photoionization models is also indicated (Stasińska 2005), in particular for metal-rich nebulae.

These examples indicate that the presently available observational constraints on galactochemical evolution may be subject to serious systematic error, implying severe deficits in our preset-day understanding of galaxy evolution. Accurate analyses of extragalactic BA-type supergiants using our models have an enormous potential to improve on this, promising to provide thorough constraints for galactochemical evolution models.

**Blue Supergiants & the Extragalactic Distance Scale.** Blue supergiants allow not only for a determination of stellar metallicities, but also for the derivation of interstellar reddening and extinction in other galaxies. Hence, blue supergiants can *indirectly* contribute to the work on the extragalactic distance scale. Systematic errors due to metallicity and extinction may be reduced in Cepheid studies. However, there is more to gain.

Precise stellar parameters of BA-type supergiants, as obtained from an application of our models, allow the flux-weighted gravity–luminosity relationship (FGLR, Kudritzki et al. 2003, 2008) to be exploited for *direct* distance determinations. The

FGLR is a consequence of the fast evolution of BA-type supergiants from the blue to the red part of the HRD at roughly constant mass and luminosity (e.g. Maeder & Meynet 2000). Then, the stellar gravity and effective temperature are coupled through the relation $g/T_{\text{eff}}^4 = \text{const}$. Assuming that the luminosity scales with stellar mass ($L \propto M^\alpha$, $\alpha \approx 3$), a relationship between absolute bolometric luminosity $M_{\text{bol}}$ and flux-weighted gravity results, of the form

$$-M_{\text{bol}} = a \log(g/T_{\text{eff}}^4) + b\,, \qquad (3)$$

see Kudritzki et al. (2008) for a detailed discussion. After an empirical calibration of the coefficients $a$ and $b$ this promises to become a highly robust distance determination technique[6] (Fig. 22). Many of the systematic uncertainties affecting the Cepheid distances can be avoided via this *spectroscopic* method. Quantitative spectroscopy of BA-type supergiants can therefore be expected to contribute in resolving the ongoing discussion on the value of the Hubble constant (Freedman et al. 2001; Sandage et al. 2006).

**Acknowledgements**

This work is the result of close cooperation with many colleagues. I would like to express my gratitude to all of them, in particular to K. Butler, R. P. Kudritzki, U. Heber, M. F. Nieva, S. R. Becker, F. Bresolin, M. Firnstein, J. Puls, F. Schiller, M. A. Urbaneja and K. A. Venn. The support of U. Heber over the past four years in Bamberg is greatly appreciated. I wish to thank U. Heber, K. Butler and M.F. Nieva for valuable suggestions and careful reading of the manuscript. Part of the research was supported by the Deutsche Forschungsgemeinschaft under grant PR 685/3-1.

## References

Albayrak, B. 2000, A&A, 364, 237

Asplund, M., Grevesse, N., Sauval, A. J. 2005, in: T. G. Barnes III, F. N. Bash (eds.), *Cosmic Abundances as Records of Stellar Evolution and Nucleosynthesis* (ASP, San Francisco), 25

Auer, L. H., Mihalas, D. 1973, ApJS, 25, 433

Aufdenberg, J. P., Hauschildt, P. H., Baron, E., et al. 2002, ApJ, 570, 344

Bahcall, J. N., Basu, S., Pinsonneault, M., Serenelli, A. M. 2005, ApJ, 618, 1049

Becker, S. R. 1998, in: I. D. Howarth (ed.), *Boulder-Munich II: Properties of Hot, Luminous Stars* (ASP, San Francisco), 137

Becker, S. R., Butler, K. 1988, A&A, 201, 232

Bresolin, F., Kudritzki, R. P., Méndez, R. H., Przybilla, N. 2001, ApJ, 548, L159

Bresolin, F., Gieren, W., Kudritzki, R. P., et al. 2002, ApJ, 567, 277

Bresolin, F., Pietrzynski, G., Gieren, W., et al. 2004a, ApJ, 600, 182

---

[6]Distances from the FGLR are not significantly affected by the intrinsic photometric variability of the blue supergiants (Bresolin et al. 2004a).

Bresolin, F., Garnett, D. R., Kennicutt, R. C., Jr. 2004b, ApJ, 615, 228

Burbidge, E. M., Burbidge, G. R., Fowler, W. A., Hoyle, F. 1957, Rev. Modern Phys., 29, 547

Butler, K., Giddings, J. R. 1985, Newsletter on Analysis of Astronomical Spectra, No. 9 (Univ. London)

Cameron, A. G. 1957, PASP, 69, 201

Cescutti, G., Matteucci, F., François, P., Chiappini, C. 2007, A&A, 462, 943

Chiappini, C., Matteucci, F., Romano, D. 2001, ApJ, 554, 1044

Chiappini, C., Romano, D., Matteucci, F. 2003, MNRAS, 339, 63

Crowther, P. A., Lennon, D. J., Walborn, N. R. 2006, A&A, 446, 279

Cunha, K., Lambert, D. L. 1994, ApJ, 426, 170

Daflon, S., Cunha, K. 2004, ApJ, 617, 1115

Daflon, S., Cunha, K., Becker, S. R. 1999, ApJ, 522, 950

Daflon, S., Cunha, K., Becker, S. R., Smith, V. V. 2001a, ApJ, 552, 309

Daflon, S., Cunha, K., Butler, K., Smith, V. V. 2001b, ApJ, 563, 325

Daflon, S., Cunha, K., Smith, V. V., Butler, K. 2003, A&A, 399, 525

Donati, J.-F., Howarth, I. D., Jardine, M. M., et al. 2006, MNRAS, 370, 629

Dufton, P. L., McKeith, C. D. 1980, A&A, 81, 8

Edmunds, M. G. 1975, Ap&SS, 32, 483

Edvardsson, B., Andersen, J., Gustafsson, B., et al. 1993, A&A, 275, 101

Esteban, C., García-Rojas, J., Peimbert, M., et al. 2005, ApJ, 618, L95

Evans, C. J., Smartt, S. J., Lee J. K., et al. 2005, A&A, 437, 467

Evans, C. J., Lennon, D. J., Smartt, S. J., Trundle, C. 2007, A&A, 464, 289

Firnstein, M., Przybilla, N. 2006, Proceedings of Science, PoS(NIC-IX)095

Freedman, W. L., Madore, B. F., Gibson, B. K. 2001, ApJ, 553, 47

Fuhrmann, K. 2004, AN, 325, 3

Garnett, D. R. 2004, in: Esteban, C., García López, R. J., Herrero, A., Sánchez, F. (eds.), *Cosmochemistry* (Cambridge: Cambridge University Press), 171

Geier, S., Nesslinger, S., Heber, U., et al. 2007, A&A, 464, 299

Giddings, J. R. 1981, Ph. D. Thesis, Univ. London

Gieren, W., Pietrzynski, G. Bresolin, F., et al. 2005, ESO Messenger, 121, 23

Gies, D. R., Lambert, D. L. 1992, ApJ, 387, 673

Grevesse, N., Sauval, A. J. 1998, Space Sci. Rev., 85, 161

Groth, H. G. 1961, ZAp, 51, 231

Gummersbach, C. A., Kaufer, A., Schaefer, D. R., et al. 1998, A&A, 338, 881

Heger, A., Langer, N. 2000, ApJ, 544, 1016

Heger, A., Woosley, S. E., Spruit, H. C. 2005, ApJ, 626, 350

Hillier, D. J., Miller, D. L. 1998, ApJ, 496, 407

Hou, J. J., Prantzos, N., Boissier, S. 2000, A&A, 362, 921

Hummer, D. G., Berrington, K. A., Eissner, W., et al. 1993, A&A, 279, 298

Kennicutt, R. C., Jr., Bresolin, F., Garnett, D. R. 2003, ApJ, 591, 801

Kilian, J. 1992, A&A, 262, 171

Kippenhahn, R., Weigert, A. 1990, *Stellar Structure and Evolution* (Springer, Berlin)

Korn, A. J., Keller, S. C., Kaufer, A., et al. 2002, A&A, 385, 143

Korn, A. J., Nieva, M. F., Daflon, S., Cunha, K. 2005, ApJ, 633, 899

Kudritzki, R.P. 1973, A&A, 28, 103

Kudritzki, R. P. 1998, in: Aparicio, A., Herrero, A., Sánchez, F. (eds.), *Stellar Astrophysics for the Local Group* (Cambridge University Press, Cambridge), 149

Kudritzki, R. P., Przybilla, N. 2003, in: Alloin, D., Gieren, W. (eds.), *Stellar Candles for the Extragalactic Distance Scale* (Springer Verlag, Berlin), 123

Kudritzki, R. P., Puls, J., Lennon, D. J., et al. 1999, A&A, 350, 970

Kudritzki, R. P., Bresolin, F., Przybilla, N. 2003, ApJ, 582, L83

Kudritzki, R. P., Urbaneja, M. A., Bresolin, F., et al. 2008, ApJ, 681, 269

Kurucz, R. L. 1993, Kurucz CD-ROM No. 13 (SAO, Cambridge, Mass.)

Lanz, T., Hubeny, I. 2003, ApJS, 146, 417

Lennon, D. J., Dufton, P. L. 1989, A&A, 225, 439

Lenorzer, A., Mokiem, M. R., de Koter, A., Puls, J. 2004, A&A, 422, 275

Leone, F., Lanzafame, A. C., & Pasquini, L. 1995, A&A, 293, 457

López-Corredoira, M., Cabrera-Lavers, A., Mahoney, T. J., et al. 2007, AJ, 133, 154

Lyubimkov, L. S., Rostopchin, S. I., Rachkovskaya, T., et al. 2005, MNRAS, 358, 193

Maeder, A., Meynet, G. 2000, ARA&A, 38, 143

Maeder, A., Meynet, G. 2001, A&A, 373, 555

Maeder, A., Meynet, G. 2005, A&A, 440, 1041

Mashonkina, L., Korn, A. J., Przybilla, N. 2007, A&A, 461, 261

Matteucci, F. 2001, *The Chemical Evolution of the Galaxy* (Springer, Berlin)

Meynet, G., Maeder, A. 2003, A&A, 404, 975

Mihalas, D. 1978, *Stellar Atmospheres*, 2nd edition (Freeman, San Francisco)

Nieva, M. F. 2007, Ph. D. Thesis, Univ. Erlangen-Nuremberg & ON, Rio de Janeiro

Nieva, M. F., Przybilla, N. 2006, ApJ, 639, L39

Nieva, M. F., Przybilla, N. 2007, A&A, 467, 295

Nieva, M. F., Przybilla, N. 2008a, A&A, 481, 199

Nieva, M. F., Przybilla, N. 2008b, Rev. Mex. AA Conf. Ser., in press (arXiv:0712.0511)

Oey, M. S. 2003, in: K. van der Hucht, A. Herrero, C. Esteban (eds.), *A Massive Star Odyssey: From Main Sequence to Supernova* (ASP, San Francisco), 620

Pagel, B. E. J. 1997, *Nucleosynthesis and Chemical Evolution of Galaxies* (Cambridge University Press, Cambridge)

Przybilla, N. 2002, Ph. D. Thesis, Univ. Munich

Przybilla, N. 2005, A&A, 443, 293

Przybilla, N., Butler, K. 2001, A&A, 379, 955

Przybilla, N., Butler, K. 2004a, ApJ, 609, 1181

Przybilla, N., Butler, K. 2004b, ApJ, 610, L61

Przybilla, N., Butler, K., Becker, S. R., et al. 2000, A&A, 359, 1085

Przybilla, N., Butler, K., Becker, S. R., Kudritzki, R. P. 2001a, A&A, 369, 1009

Przybilla, N., Butler, K., Kudritzki, R. P. 2001b, A&A, 379, 936

Przybilla, N., Butler, K., Heber, U., Jeffery, C. S. 2005, A&A, 443, L25

Przybilla, N., Butler, K., Becker, S. R., Kudritzki, R. P. 2006a, A&A, 445, 1099

Przybilla, N., Nieva, M. F., Edelmann, H. 2006b, Baltic Astronomy, 15, 107

Przybilla, N., Nieva, M. F., Heber, U., Jeffery, C. S. 2006c, Baltic Astronomy, 15, 163

Przybilla, N., Butler, K., Kudritzki, R. P. 2008a, in: G. Israelian, G. Meynet (eds.), *The Metal Rich Universe* (Cambridge University Press, Cambridge), 335

Przybilla, N., Nieva, M. F., Heber, U., et al. 2008b, A&A, 480, L37

Przybylski, A. 1968, MNRAS, 139, 313

Przybylski, A. 1969, MNRAS, 146, 71

Przybylski, A. 1972, MNRAS, 159, 155

Puls, J., Kudritzki, R. P., Herrero, A., et al. 1996, A&A, 305, 171

Puls, J., Urbaneja, M. A., Venero, R., et al. 2005, A&A, 435, 669

Reid, M. J. 1993, ARA&A, 31, 345

Repolust, T., Puls, J., Hanson, M. M., et al. 2005, A&A, 440, 261

Roberts, W. W., Jr., Huntley, J. M., van Albada, G. D. 1979, ApJ, 233, 67

Roy, J.-R., Kunth, D. 1995, A&A, 294, 432

Rudolph, A. L., Fich, M., Bell, G. R., et al. 2006, ApJS, 162, 346

Rybicki, G. B., Hummer, D. G. 1991, A&A, 245, 171

Ryder, S. D. 1995, ApJ, 444, 610

Sandage, A., Tammann, G. A., Saha, A., et al. 2006, ApJ, 653, 843

Schiller, F., Przybilla, N. 2008, A&A, 479, 849

Seaton, M. J., Yan, Y., Mihalas, D., Pradhan, A. K. 1994, MNRAS, 266, 805

Sofia, U. J., Meyer, D. M. 2001, ApJ, 554, L221

Stasińska, G. 2005, A&A, 434, 507

Takeda, Y., Takada-Hidai, M. 1998, PASJ, 50, 629

Takeda, Y., Takada-Hidai, M. 2000, PASJ, 52, 113

Trundle, C., Dufton, P. L., Hunter, I., et al. 2007, A&A, 471, 625

Urbaneja, M. A., Herrero, A., Bresolin, F., et al. 2005, ApJ, 622, 862

Venn, K. A. 1995a, ApJS, 99, 659

Venn, K. A. 1995b, ApJ, 449, 839

Venn, K. A., Przybilla, N. 2003, in: C. Charbonnel, D. Schaerer, G. Meynet (eds.), *CNO in the Universe* (ASP, San Francisco), 20

Venn, K. A., McCarthy, J. K., Lennon, D. J., et al. 2000, ApJ, 541, 610

Venn, K. A., Lennon, D. J., Kaufer, A., et al. 2001, ApJ, 547, 765

Venn, K. A., Kaufer, A., Tolstoy, E., et al. 2003, in: K. van der Hucht, A. Herrero, C. Esteban (eds.), *A Massive Star Odyssey: From Main Sequence to Supernova* (ASP, San Francisco), 30

Verdugo, E., Talavera, A., Gómez de Castro, A. I. 1999, A&A, 346, 819

Vrancken, M., Butler, K., Becker, S. R. 1996, A&A, 311, 661

Wolf, B. 1971, A&A, 10, 383

Wolf, B. 1972, A&A, 20, 275

Wolf, B. 1973, A&A, 28, 335

Zaritsky, D., Kennicutt, R. C., Jr., Huchra, J. P. 1994, ApJ, 420, 87

# Formation and Evolution of Brown Dwarfs

Alexander Scholz

SUPA, School of Physics and Astronomy
University of St. Andrews
North Haugh, St. Andrews, Fife KY16 9SS, United Kingdom
as110@st-andrews.ac.uk

## Abstract

Brown dwarfs – intermediate in mass between dwarf stars and giant planets – exhibit a hybrid nature, partly stellar, partly planetary, in their observational properties. Hence, they provide us with an unique testbed to probe our understanding of the formation and evolution of stars and planets. In this contribution, I will discuss the hybrid nature of brown dwarfs based on highlight results of our ongoing observational programs: a) All brown dwarfs older than 10 Myr are observed to be fast rotators, indicating that they do not possess a mechanism for long-term angular momentum disposal by magnetic winds – which is one of the key properties of solar-type stars. b) Like stars, young brown dwarfs are surrounded by disks of gas and dust, implying that they form in a way similar to stars, and may also host 'miniature' planetary systems.

## 1 Introduction

Brown dwarfs are substellar objects straddling the mass domains of stars and planets. Their defining property in comparison with stars is their inability to maintain stable H-burning in their core. Lacking a long-term energy source, they will cool down rapidly as they age and will never reach the main-sequence. This definition establishes the *Hydrogen Burning Mass Limit* (HBML, $\sim 0.07$–$0.09\ M_\odot$) as the borderline between stars and brown dwarfs (see the review by Oppenheimer et al. 2000).

The potential existence of objects with masses below the HBML has been pointed out in the 1960s (Kumar 1963, Hayashi & Nakano 1963). After more than 30 years of unsuccessful searches, the first bona-fide brown dwarfs have been discovered in 1995: the Pleiades member Teide 1 (Rebolo et al. 1995) and Gliese 229B, a companion to a red dwarf star (Nakajima et al. 1995). The emergence of sensitive CCD detectors and 8-m class telescopes for follow-up spectroscopy in combination with improved model predictions for brown dwarf spectra led to a rapid increase in the number of confirmed substellar objects. Today, hundreds of brown dwarfs have been identified in star forming regions (e.g., Comeron et al. 2000, Natta et al. 2002), open clusters (e.g., Bejar et al. 1999, Pinfield et al. 2000), and the field (e.g., Kirkpatrick et al. 1999, Phan-Bao et al. 2001). The availability of large samples of brown dwarfs has now shifted the focus from the survey work to a more in depth investigation of their physical properties.

Since the mass of the objects fundamentally determines their physics, brown dwarfs are hybrids in their observational properties – partly stellar, partly planetary in nature. This makes brown dwarfs ideal test cases for our understanding of formation and evolution of stars and planets. Simply speaking, the substellar mass range – from $\sim 0.01$ to $\sim 0.08\,M_\odot$ – can serve as a testbed: it allows us to take our theories about stars or planets to the extreme. By probing the physics of brown dwarfs, we are effectively interpolating between planetary and stellar astrophysics.

For this conference contribution, I have aimed to highlight the hybrid nature of brown dwarfs based on recent observational results in two of the most intensely debated subjects in current brown dwarf research. In the first part, I will focus on the rotational evolution of very low mass objects (Sect. 2). On long timescales, the angular momentum history of brown dwarfs is very similar to that of giant planets, and very different from solar-mass stars. In the second part, I will focus on the properties of disks around very young brown dwarfs (Sect. 3). It will become clear that during their earliest evolutionary stages, brown dwarfs share many properties with young stars.

## 2 Rotation of very low mass objects

Rotation is a key parameter of stellar evolution. The main phases in the rotational evolution of solar-type stars are as follows (see Herbst et al. 2007, Bouvier et al. 1997, Mathieu 2003): a) In the T Tauri phase, the rotation is braked by magnetic coupling between star and circumstellar disk. b) While conserving angular momentum, stars spin up in the pre-main sequence phase as a consequence of the hydrostatical contraction. c) On long timescales, solar-type stars lose angular momentum due to magnetically driven stellar winds. Thus, rotation is intrinsically connected to basic stellar physics, in particular to the properties of stellar magnetic fields.

While this scenario is well-established for solar-mass stars, we still know little about the processes regulating the rotational history of very low mass (VLM) stars and brown dwarfs. There is good reason to believe that the mechanisms for angular momentum loss change in this mass regime: It is widely believed that the large-scale magnetic fields of solar-type stars are generated by an $\alpha\omega$-type dynamo operating in the shear layer between radiative core and convective envelope. VLM objects, however, are fully convective throughout their evolution, thus their dynamo action, their magnetic properties, and in turn their rotational evolution may be different from what we know from more massive stars.

This is motivation for us to carry out a long-term program to monitor VLM objects in young open clusters with ages ranging from 3 to 750 Myr. The main goal of the project is to measure photometric rotation periods, providing a observational database to constrain models of rotational evolution in this mass range. The project, initiated in 1999 and still ongoing, relies mostly on photometric time series obtained with the 1.23 m telescope at the German-Spanish Astronmical Centre on

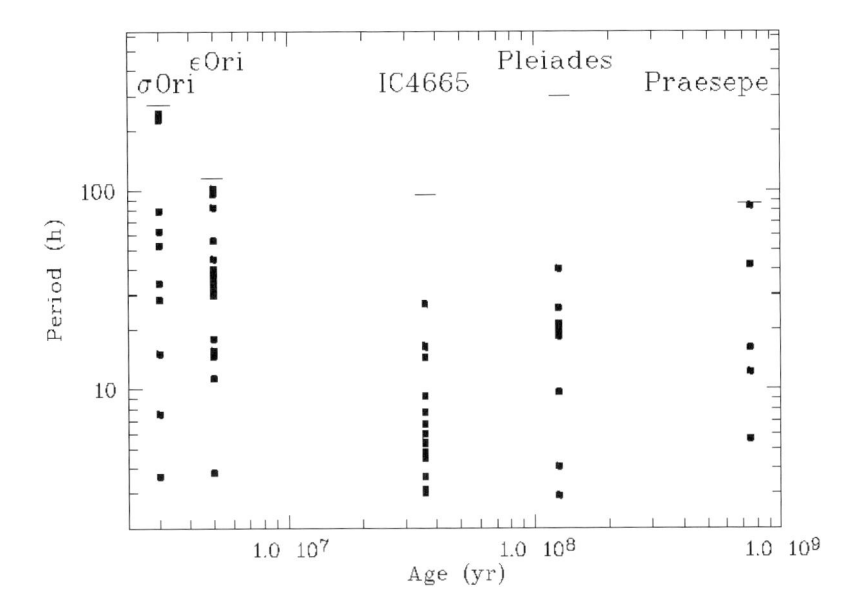

**Figure 1:** Rotation periods for very low mass objects as a function of age, compiled from Scholz & Eislöffel (2004a, 2004b, 2005, 2007) and Scholz (2004). Horizontal bars show the detection limits in the period search for all given clusters.

Calar Alto/Spain, the ESO/MPG WFI at the ESO 2.2 m telescope in La Silla, and the 2 m Schmidt telescope at the Thüringer Landessternwarte Tautenburg/Germany.

The main outcome of the project is the measurement of 80 rotation periods for VLM objects with masses $\leq 0.4\,M_\odot$, including 20 periods for probable brown dwarfs with masses $\leq 0.08\,M_\odot$. Together with the recent publications from the Monitor collaboration (e.g., Irwin et al. 2007), we have now a substantial database of VLM periods. In Fig. 1, we show all our VLM periods as a function of cluster age. The plot features datapoints for the young open clusters around $\sigma$ Ori (age about 3 Myr) and $\epsilon$ Ori (5 Myr), for IC4665 (40 Myr), the Pleiades (125 Myr), and Praesepe (750 Myr). The full results of the program are published in Scholz & Eislöffel (2004a, 2004b, 2005, 2007) and Scholz (2004); in the following we will focus on three highlights from this work.

## 2.1 Evidence for disk-locking in the VLM regime

For solar-mass stars, there is now ample observational evidence that star-disk interactions regulate the angular momentum in the T Tauri phase. For the two clusters with the largest database of rotation periods, the ONC and NGC2264, a connection between the presence of a disk and slow rotation has been established based on near/mid-infrared surveys (see review by Herbst et al. 2007, also Rebull et al. 2005,

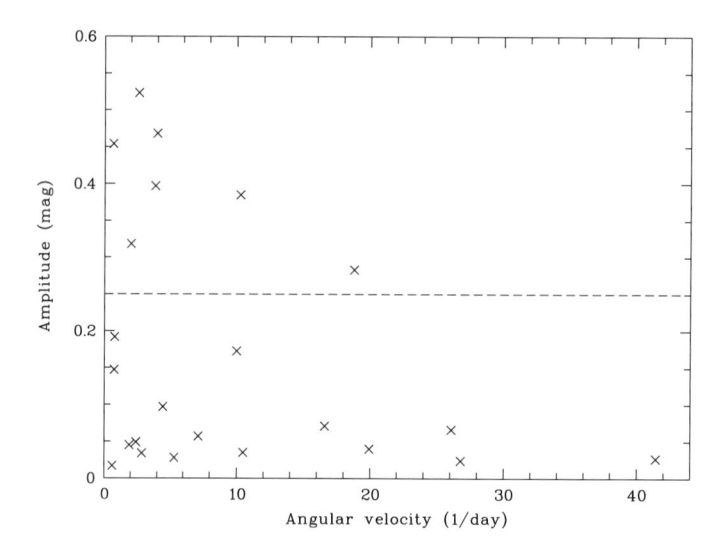

**Figure 2:** Angular velocity versus photometric amplitude for VLM objects in the very young $\sigma$ Ori cluster (from Scholz & Eislöffel 2004). The dashed line delineates the separation between likely accretors (upper part of the plot) and likely non-accretors (lower part).

Cieza & Baliber 2007). Theoretically, it is not clear yet what kind of mechanism is responsible for the rotational braking, the leading candidates being 'disk-locking' (e.g., Koenigl 1991) and accretion-powered winds (e.g., Matt & Pudritz 2006).

Our period database allows us to probe if a signature of 'disk braking' can also be seen at very low masses. In Fig. 2, we plot angular velocity vs. photometric amplitude for VLM objects in the 3 Myr old $\sigma$ Ori cluster. As argued in Scholz & Eislöffel (2004a), photometric variability with amplitudes exceeding $\sim 0.25$ mag is unlikely to be explained with cool spots only; instead we are likely seeing the effects of hot spots co-rotating with the objects, usually attributed to an accretion shockfront. Complementary observations for those objects (near-infrared photometry, low-resolution spectroscopy) confirm this interpretation. Thus, for objects in the upper part of the diagram (above the dashed line) we have evidence of accretion, and thus interaction between star and disk. For objects in the lower part of the plot, both the variability characteristic and near-infrared photometry are fully consistent with a diskless object.

Albeit hampered by small number statistics, Fig. 2 shows that accretors are exclusively slow rotators with angular velocities $< 20 \, d^{-1}$, while non-accretors show a broad range of rotation rates. This is precisely the disk braking signature seen for solar-mass stars in the ONC and in NGC2264 (see Rebull et al. 2005, Cieza & Baliber 2007). Thus, we have found the first tentative evidence for a rotational braking by the disk (or more accurately: by object-disk interaction) in the VLM regime. A very similar result has been obtained by Mohanty et al. (2005) based on spectroscopically determined accretion signatures and rotational velocities. Taken together, these findings can be interpreted as indication for the presence of large-scale, stable magnetic fields in accreting VLM objects, which contribute to the rotational braking – similarly as in solar-mass stars.

A cautious note should be added: The rotational database in the VLM regime is still sparse compared with more massive stars; thus the accretion/disk vs. rotation connection cannot be established with high confidence. Future studies based on deep photometric monitoring and Spitzer data (see Sect. 3) may provide more definite answers.

## 2.2 Evolution of VLM periods: the first 100 Myr

In the first 100 Myr of the stellar evolution, stars and brown dwarfs undergo a spin-up due to pre-main sequence contraction. This effect can clearly be seen in Fig. 1, where the upper limit of the periods drops significantly between 5 and 125 Myr. In addition, there may be rotational braking due to stellar winds. In Scholz & Eislöffel (2004b) we probe the effect of stellar winds in the VLM regime by comparing the periods in $\sigma$ Ori ($\sim 3$ Myr) and the Pleiades (age 125 Myr) with models of rotational evolution. We describe the period evolution as follows:

$$P_f = \alpha \times (R_f/R_i)^2 \times P_i \quad (1)$$

Here $i$-indices refer to parameters at 3 Myr and $f$-indices to parameters at 125 Myr. The radii are taken from the Chabrier & Baraffe (1997) evolutionary tracks.

The effect of wind braking is parametrized in the term $\alpha$. We consider three different scenarios:

- $\alpha = 1$: No rotational braking (model A), i.e. angular momentum is fully conserved.

- $\alpha = (t_f/t_i)^{1/2}$: Angular momentum loss through stellar winds following the Skumanich law (model B). This is the classical braking law which has been empirically established for solar-mass main sequence stars (e.g., Skumanich 1972, Barnes 2001).

- $\alpha = \exp((t_f - t_i)/\tau_C)$: Exponential rotational braking through stellar winds (model C). This type of braking is applicable to fast rotating solar-mass stars in the 'saturated' regime of the rotation activity relationship (e.g., Terndrup et al. 1999). Compared with the Skumanich law, it provides a weak rotational braking on the considered timescale.

We use these three scenarios to produce evolutionary tracks for the rotation periods, starting with the periods in the $\sigma$ Ori cluster, which have been found to be representative for the VLM period distribution at this particular age (see Scholz 2004). Thus, we are calculating the periods *forward in time*, to predict the period distribution at the age of the Pleiades. In Fig. 3, we show the periods in $\sigma$ Ori and the Pleiades as well as the evolutionary tracks. Model A is shown with dotted, model B with dash-dotted, and model C with dashed lines.

As can be seen in this plot, model A without any rotational braking does not fit the Pleiades periods very well. The predicted upper period limit at 125 Myr is less

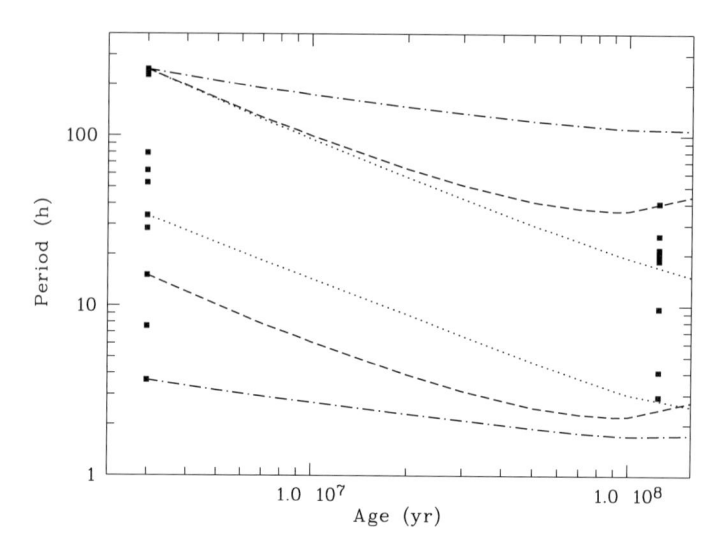

**Figure 3:** Rotational evolution of VLM objects: periods in the $\sigma$ Ori cluster (age 3 Myr) and the Pleiades (125 Myr) compared with rotational evolutionary tracks (from Scholz & Eislöffel 2004b). Model A (dotted lines) – no braking; model B (dash-dotted lines) – Skumanich braking; model C (dashed lines) – exponential braking.

than 20 h, while five out of nine periods measured in the Pleiades exceed this value. Thus, some rotational braking due to winds has to be included for VLM objects. On the other hand, the tracks for Skumanich braking (model B) do not fit either: They predict an upper period limit $> 100$ h at the age of the Pleiades, while we do not see any rotation periods $> 40$ h. Note that the observed upper period limit in the Pleiades is fully consistent with the value inferred from spectroscopic $v \sin i$ studies (Terndrup et al. 1999) and it thus unlikely to be strongly affected by a bias in our period sample. As discussed in Scholz & Eislöffel (2004b), the currently available period and $v \sin i$ data firmly establishes the upper period limit in the Pleiades at $\sim 50$ h – a factor of two lower than the prediction for Skumanich braking.

Therefore, a more modest form of rotational braking has to be invoked for VLM objects, as it is done in model C with exponential braking. As can be seen in Fig. 3, model C provides the best fit to the data, when the spin-down timescale is adjusted to $\sim 150$ Myr, consistent with published values for this parameter (Terndrup et al. 1999). (Given the lack of understanding for the underlying physics for the exponential rotational braking, model C should be treated as an ad hoc solution to provide moderate braking rather than an accurate physical model.)

In summary, the comparison of rotation periods at 3 and 125 Myr provides clear evidence for some rotational braking due to stellar winds in the VLM regime. However, the well-established Skumanich law is not applicable in this mass regime, instead a weaker type of braking, for example a exponential braking law, is needed for VLM objects.

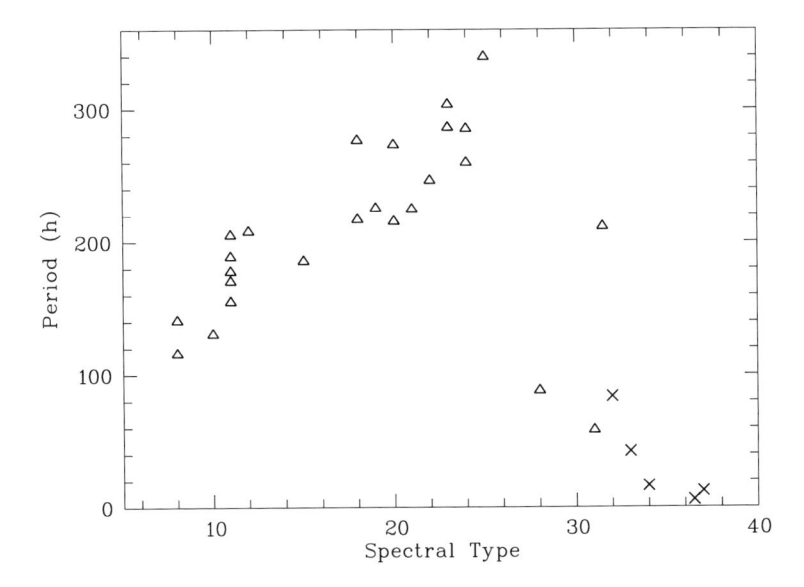

**Figure 4:** Rotation period vs. spectral type (as indicator for mass), from Scholz & Eislöffel (2007): Crosses are the five VLM periods in Praesepe from this work, triangles are periods in the coeval Hyades from Radick et al. (1987) and Prosser et al. (1995). Spectral type is parameterised as follows: G0 – 10, K0 – 20, M0 – 30.

## 2.3 The long-term rotational evolution for VLM objects

The aforementioned Skumanich braking is responsible for a rapid spin-down of solar-type stars on the main sequence. To investigate the long-term effect of rotational braking in the VLM range, we have to probe objects which are significantly older than 100 Myr. The open cluster Praesepe is the ideal laboratory to test wind braking: At an age of 750 Myr, all stars have finished their initial contraction, allowing us to isolate the effects of wind braking on the rotation. Moreover, Praesepe is sufficiently nearby and dense for wide-field monitoring. In the recent paper, we have published the first rotation periods for Praesepe members, all for stars with masses $< 0.5\,M_\odot$ (Scholz & Eislöffel 2007). Our five periods range from a few hours up to about four days and likely provide a realistic first glimpse of the VLM period distribution in Praesepe.

In Fig. 4 we compare our VLM periods (crosses) with rotational data for more massive stars in the Hyades (triangles), a coeval cluster for which Radick et al. (1987) and Prosser (1995) have published a substantial sample of periods. The periods are plotted vs. spectral type (10 – G0, 20 – K0, 30 – M0); as all objects are coeval, this serves as a proxy for object mass. The Hyades periods clearly show that for F-K main-sequence stars the periods increase towards later spectral types. According to Radick et al. (1987), these periods can be explained in terms of the correlation between magnetic activity and the inverse Rossby number $Ro$, the ratio between rotation period and convective turnover time-scale (e.g. Noyes et al. 1984).

This relation basically implies that the magnetic field amplification mainly depends on convection properties and rotation – supporting the idea of an $\alpha\omega$ dynamo operating in F–K stars.

As it is apparent from Fig. 4, this sequence breaks down roughly at spectral types K8-M2, corresponding to a mass of about 0.4–0.6 $M_\odot$. Without clear transition regime, the periods drop by almost one order of magnitude between K8 and M2. Even if the VLM period distribution is not complete, the simple fact that the fast rotators do exist among M dwarfs and not in the G-K spectral range indicates a fundamental change in stellar rotation at the given mass limit.

This solidifies the results reported in Sect. 2.2 – VLM objects are subject to a different wind braking than solar-type stars. While already apparent in the analysis of the Pleiades data, the Praesepe periods now show the full effect of this dramatic change in the rotational braking. The underlying physical reason for the breakdown of the Skumanich braking is not clear. It may be related to a change in wind properties, surface magnetic field or dynamo action. It is tempting to search for the origin of this behaviour in the change in interior structure: Objects with 0.5 $M_\odot$ are still thought to harbour a substantial radiative core (Chabrier & Baraffe 1997). Going to lower masses, however, the bottom of the convective envelope drops quickly until the objects are fully convective. Since this also eliminates the shear layer, where the solar-type dynamo is thought to operate, this may cause a significant change in the efficiency of the rotational braking.

However, structural models of low-mass stars essentially agree on the fact that this transition is supposed to occur at masses between 0.3 and 0.4 $M_\odot$ (e.g. Chabrier & Baraffe 1997, Montalbán, D'Antona & Mazzitelli 2000) or even below (Mullan & MacDonald 2001), that is, at lower masses than the observed drop in rotation periods apparent in Fig. 4. Thus, it is questionable if the break in the mass-period relation is entirely to explain with the change to fully convective objects. A more complete period sample for main-sequence clusters as well as complementary data (e.g., Zeeman Dopper Imaging, magnetic field measurements, magnetic activity data) are highly desirable to clarify the situation.

Figure 4 thus presents new compelling evidence for a fundamental change in the angular momentum regulation in the VLM regime. In contrast to solar-mass stars, VLM objects and brown dwarfs do not spin down quickly as they get older as a consequence of efficient braking due to magnetic winds. Instead, they continue to be fast rotators for timescales in the order of magnitude of 1 Gyr. In this respect, brown dwarfs resemble giant planets – lacking an efficient means to carry away angular momentum, they stay locked at typical rotation periods shorter than one day.

## 3  Substellar disks

The study of brown dwarf disks has been one of the hot topics in cool stars research over the past years. A primary motivation for detailed observational studies of substellar disks is to obtain clues about brown dwarf origins. The underlying problem: Stars and planets can be distinguished based on their formation history, implying that the formation process is a strong function of object mass. In this context, brown

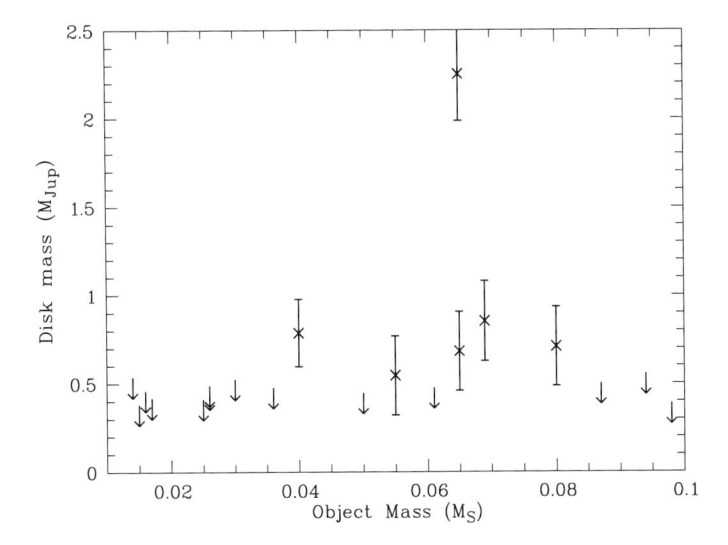

**Figure 5:** Disk masses vs. object masses for 20 brown dwarfs in Taurus (from Scholz et al. 2006). The errorbars are computed from the $1\sigma$ uncertainties for the fluxes and do not take into account the mostly systematical effects of uncertainties in dust opacity, dust temperature, and distance. $2\sigma$ upper limits are shown for objects without significant mm emission. Objects with very similar masses have been separated on the x-axis for clarity.

dwarfs constitute an interesting intermediate case – are they rather ultra-low mass stars or ejected giant planets or something in between? This question and its various implications have been intensely debated since the discovery of brown dwarfs; recent reviews of the observational and theoretical status of the research have been published by Luhman et al. (2007) and Whitworth et al. (2007).

The majority of the suggested formation scenarios can be summarized in two main classes: a) *In-situ formation* is the formation by collapse of an isolated ultra-low mass cloud core – a star-like origin of brown dwarfs (e.g., Padoan & Nordlund 2005). According to this scenario, young brown dwarfs are supposed to show T Tauri-like features, for example long-lived accretion disks. b) *Ejection models* describe brown dwarfs as object which undergo an ejection due to a gravitational encounter, either from a multiple stellar system or from a massive circumstellar disk (e.g., Bate et al. 2002, 2003). If ejection is an indispensable part of the brown dwarf formation process, we would expect to see its imprint in observational properties of young substellar objects – for example as 'truncated' disks. This is one of the reasons why brown dwarf disks are intriguing.

## 3.1 Disk masses and radii

The existence of dusty disks around brown dwarfs has been demonstrated convincingly by a number of groups in recent years. Starting with near-infrared surveys carried out by Muench et al. (2001), Jayawardhana et al. (2003), Liu et al. (2003)

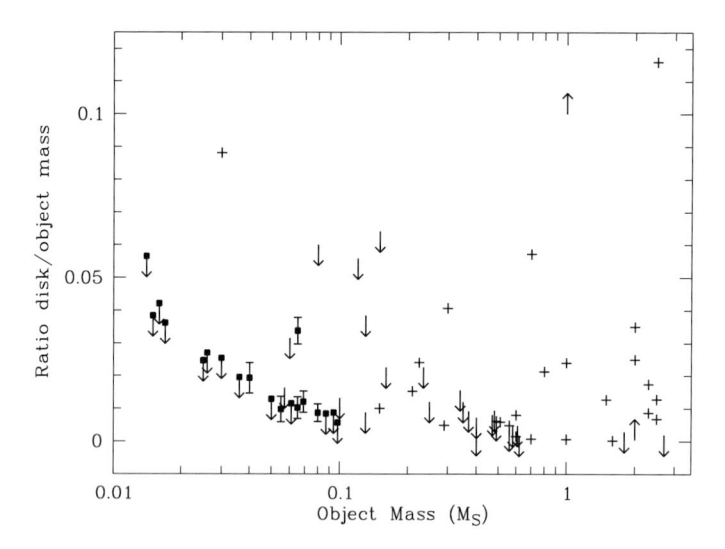

**Figure 6:** Ratio disk mass to object mass vs. object mass (from Scholz et al. 2006). The plots includes our measurements for brown dwarfs (marked with filled small squares) as well as data from various literature sources (+). Arrows show upper or lower limits. For the Taurus brown dwarfs, upper limits are based on $2\sigma$ flux upper limits. Objects with very similar masses have been separated on the x-axis for clarity.

and others, it has been established that disk fractions and lifetimes in the brown dwarf regime are not vastly different from stars. However, all these studies probed only the inner part of the disk; to constrain scenarios for brown dwarf formation, we need to determine global disk properties, i.e. disk masses and radii.

The established way to put limits on the total dust mass in the disk is imaging in the submm/mm regime, where the disks are optically thin. Building upon the results of Klein et al. (2003), we have thus carried out a comprehensive mm survey of young brown dwarfs in Taurus, covering in total 20 sources (Scholz, Jayawardhana, & Wood 2006). Using the MAMBO bolometer at the 30-m IRAM telescope, we achieved a $1\sigma$ sensitivity of $\sim 0.7$ mJy at 1.3 mm for most of our sources. This is about the optimum sensitivity that can be achieved with current instrumentation in this wavelength regime. From our 20 sources, 6 are detected at 1.3 mm on a 2.5–$10\sigma$ level, while for the remaining 14 we obtain useful upper limits. Using plausible assumptions for dust opacity and temperature, the mm fluxes can directly be related to the total dust mass in the disk. In Fig. 5 we plot the inferred disk masses (based on a gas-to-dust ratio of 100) for the 20 brown dwarfs in Taurus.

As can be seen in the plot, the disk masses for the six detections range from fractions of a Jupiter mass up to 2.3 $M_{Jup}$. Although the uncertainties are substantial, particularly in the dust properties, the plot demonstrates that a significant fraction of brown dwarfs harbours disks with total masses of the order of one Jupiter mass.

It may be more instructive to look at the brown dwarf disk masses in comparison with published disk masses for solar-type stars. In Fig. 6 we combine our new measurements with results from various literature sources (e.g., Osterloh & Beckwith 1995, Nuernberger et al. 1998, 1999, Klein et al. 2003, see Scholz et al. 2006 for a full list of references). For this figure, we choose to plot the *relative* disk mass, i.e. the ratio between disk and object mass. The diagram illustrates nicely that the relative disk mass does not show any trend with object mass – at any given object mass, the disk masses ranges between < 1% and 5% (with very few outliers). Thus, brown dwarf disks can be considered to be scaled-down T Tauri disks.

The mm detections of brown dwarf disks become more valuable when combined with Spitzer mid-infrared photometry. Spitzer with its unprecedented sensitivity at wavelengths 3–24 μm had major impact on this field – it allows us for the first time to assess the mid-infrared properties of significant samples of brown dwarf disks. Full coverage of the spectral energy distribution from the mid-infrared to the mm regime is indispensable for a full understanding of the SED, to control the numerous degeneracies inherent to radiative transfer SED models.

For five brown dwarfs with detection at 1.3 mm and coverage by Spitzer IRAC/MIPS surveys, we have analysed the SED using Monte Carlo radiative transfer code provided by Kenneth Wood (see Scholz et al. 2006 for details on the modeling and full results). One main goal here was to put limits on the outer disk radius. Specifically, we aimed to constrain the minimum outer disk radius which gives a SED consistent with the observed one. We find that for all five sources at least a 10 AU disk is required to match the observed SED and to reproduce the measured mm fluxes.

In the ejection scenario for brown dwarf formation, disks are expected to be 'truncated', i.e. they should have low masses and small radii compared with stellar disks. The simulations by Bate et al. (2002, 2003) do not exclude the existence of large and massive disks, but they are expected to be rare. Bate et al. (2002) predict that in an ejection scenario only less or about $\sim 5\%$ brown dwarfs harbour disks with radii larger of $\sim 10$ AU or larger.

From our mm observations, we conclude that at least 5 out of 20 objects (25%) have disks with radii of 10 AU or larger.[1] This results corresponds to ages of 2 Myr and thus cannot be directly compared to the outcomes of the simulations, which do not follow the evolution up to this point. Truncated disks are expected to evolve viscously to larger radii after the ejection process (Bate et al. 2003), and thus may end up having radii larger than 10 AU. In this case, however, they will have very low masses. The combined results of our mm study – brown dwarfs disks with radii > 10 AU and relative masses comparable to those of stars – do not agree with the currently available predictions for ejection models.

In our survey brown dwarfs appear to harbor scaled-down T Tauri disks. The disk properties found in our paper and in the literature (e.g., Pascucci et al. 2003, Klein et al. 2003, Mohanty et al. 2004) are completely consistent with a scenario in which brown dwarfs form in situ, i.e., from isolated molecular cloud cores with very low masses. By applying Occam's razor, we conclude that there is no need to invoke an ejection process. On the other hand, if ejection plays a role in brown dwarf formation, it is unlikely to be the dominant formation mode.

---

[1] The remaining 15 sources lack a mm detection, thus we cannot put limits on the disk radius

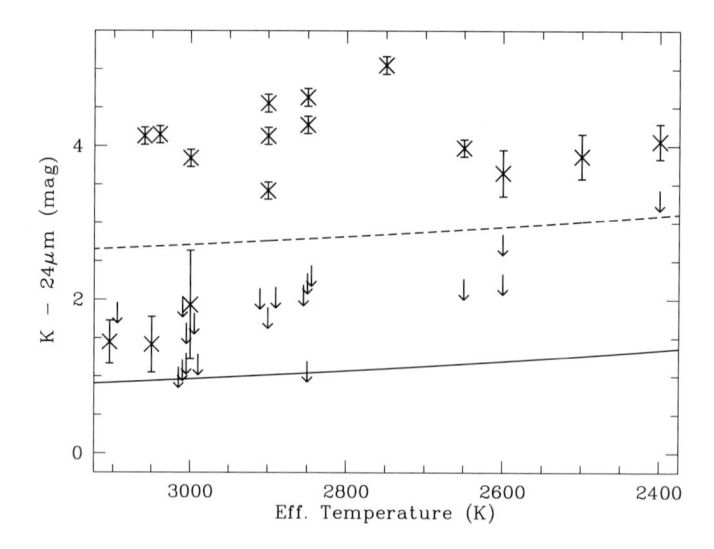

**Figure 7:** K-band minus 24 µm colours for UpSco brown dwarfs calculated from MIPS fluxes vs. effective temperatures (from Scholz et al. 2007). Upper limits correspond to $2\sigma$ upper limits at 24 µm. The solid line show the photospheric colours assuming blackbody radiation; the dashed lines show colours assuming five times the photospheric emission at 24 µm and thus significant excess.

## 3.2 Lifetimes of brown dwarf disks

Probing the longevity of substellar disks is motivated by two main problems: On the one side, it allows us again to test brown dwarf formation scenarios. If ejection processes play a key role in the early history of brown dwarfs, we might expect to see only very few long-lived disks. On the other side, disk lifetimes of a few Myrs are a prerequisite for planet formation processes. Thus, searching for long-lived disks can put limits on the chances that brown dwarfs might harbour their own planetary systems.

Therefore we have conducted a large-scale Spitzer survey covering 35 confirmed substellar objects in the Upper Scorpius (UpSco) star forming region with ages of $\sim 5$ Myr. Using IRS spectroscopy and MIPS photometry, we register the SED from 8 to 12 and at 24 µm and compare the observed SEDs with models to probe for dust settling and grain growth. The full results of this survey are published in Scholz et al. (2007); here we focus on one particular finding.

In Fig. 7 we plot the K-band minus 24 µm colour for our large sample of UpSco brown dwarfs. Since K-band photometry is dominated by photospheric flux, while the 24 µm flux is mostly produced in the inner disk regions (radii $< 10$ AU), this colour gives an excellent indicator for the presence of a disk. The solid line shows the expected photospheric colours. As can be seen in this plot, our targets clearly fall in two groups: More than one third of the objects ($37 \pm 9\%$) show enhanced emission at 24 µm, indicative of the existence of a dusty disk. The remaining objects have colours consistent with pure photospheric flux levels. These objects either have no dusty disks or disks with large inner holes. Both groups are separated by a gap of more than 1 magnitude, indicating a rapid clearing of the inner disk regions and a short transition phase ($\sim 10^5$ yr).

The disk frequency in the brown dwarf regime (37%) is somewhat higher than what has been derived previously for K0-M5 stars in the same region (Carpenter et al. 2006), on a $1.8\sigma$ confidence level, suggesting increasing disk lifetimes towards lower object masses. Thus, our results clearly demonstrate that a large fraction of brown dwarf disks survive for at least 5 Myr. Disk lifetimes among brown dwarfs are similar or larger, but definitely not shorter than for more massive stars. Again, this can be interpreted as indication that ejection processes are unlikely to play a dominant role in the early evolution of brown dwarfs.

The implications for planet formation theory from this study are interesting as well: In UpSco, we clearly see disks in an advanced stage of evolution. Processes like dust settling to the disk mid-plane and grain growth are ubiquitous. A number of disks in our sample is 'passive' in the sense that no significant gas accretion is observed, yet a clear signature of an inner dusty disks (see Scholz et al. 2007). Together with the longevity of the disks, these are arguments for planet forming processes in brown dwarf disks.

## 3.3 Dusty disks at the bottom of the IMF

Ultradeep surveys in young open clusters have revealed the existence of a population of isolated objects with masses below the Deuterium burning limit ($\sim 15\,M_{\rm Jup}$), inconsistently called 'planemos', 'free-floating planets', 'sub-brown dwarfs', or 'isolated planetary mass objects' in the literature. (In the following, we choose the latter term, abbreviated IPMO.) The largest samples of IPMOs have been identified in the 3 Myr old $\sigma$ Ori cluster (Zapatero Osorio et al. 2000) and in the 1 Myr old ONC (Lucas & Roche 2000). In these two clusters, the mass function of isolated objects seems to extend down to 5–10 $M_{\rm Jup}$ or even further, naturally extending the brown dwarf regime to lower masses.

Most of the problems discussed for the formation of brown dwarfs apply even more acutely for IPMOs, see Whitworth et al. (2007) for a review. For example, it is under debate whether IPMOs can form directly from the collapse of ultra-low mass cores; instead, they may be giant planets ejected from circumstellar disks or stellar embryos ejected from mini-clusters or decaying multiple systems. If formed similar to stars, IPMOs pose a severe challenge to star formation theory, because their masses are close to or even beyond the theoretical predictions for the opacity limit for fragmentation (see review by Bonnell et al. 2007). Thus, probing the nature and origin of IPMOs is the natural next step towards a more complete understanding of the IMF at very low masses.

We have carried out a comprehensive disk survey targeting the full sample of spectroscopically confirmed IPMOs in the $\sigma$ Ori cluster (Scholz & Jayawardhana 2008). The project is based on ultradeep dedicated Spitzer/IRAC imaging at wavelengths 3.6–8.0 µm. All our 18 targets are detected at 3.6 and 4.5 µm, 16 at 5.8 µm, and 13 at 8.0 µm. Figure 8 illustrates our results: Plotted are the IRAC colours [3.6]–[8.0] µm vs. the $I - J$ colour, which can be taken as a proxy for effective temperature

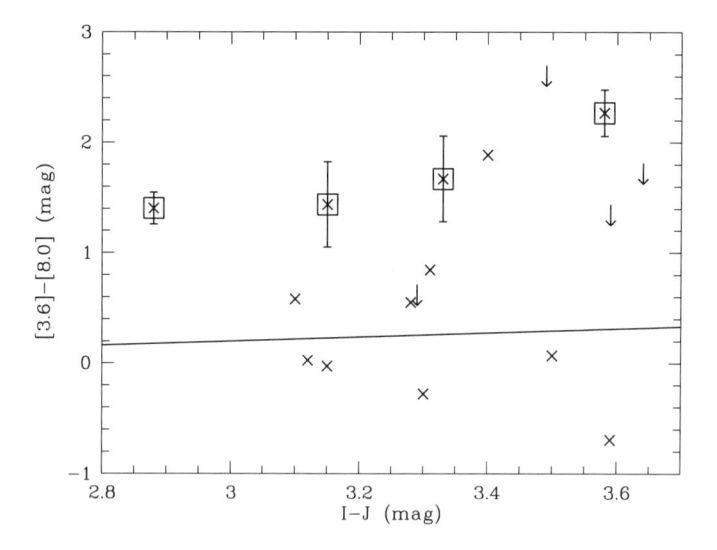

**Figure 8:** IRAC colours for IMPOs in $\sigma$ Ori (crosses) in comparison with field M/L dwarfs (solid line, Patten et al. 2006), from Scholz & Jayawardhana (2008). Objects with $> 3\sigma$ excess with respect to the field dwarfs are marked with squares.

and thus mass. The plot contains objects covering masses of 8 to 20 $M_{\rm Jup}$. The solid line in the plot is a linear fit to the colours of field L dwarfs (Patten et al. 2006), which delineates the expected photospheric flux level. All object which show significant colour excess in comparison with the photospheric level are marked with squares.

We derive excess frequencies of $24^{+13}_{-11}\%$ (4 out of 17 objects) at 5.8 μm and $29^{+16}_{-13}\%$ (4 out of 14 objects) at 8.0 μm. We put more emphasis on the latter result, which is illustrated in Fig. 8, since disk excesses may appear only beyond 5.8 μm. We interpret the colour excess – more than 1 mag at 8.0 μm – as thermal emission from dusty disks. The disk frequencies in the IPMO regime are fully consistent with what has been found for stars and brown dwarfs in the same cluster ($\sim 30\%$, Hernandez et al. 2006).

Our survey establishs for the first time that a substantial fraction of IPMOs harbours disks with lifetimes of at least 3 Myr (the likely age of the cluster). This fits into previous claims for a T Tauri-like phase in the planetary mass regime (Natta et al. 2002, Luhman et al. 2005, Jayawardhana & Ivanov 2006, Allers et al. 2006). Disk fractions and thus lifetimes are similar for objects spanning more than two orders of magnitude in mass (0.008 to 2 $M_\odot$), possibly indicating that stars, brown dwarfs, and IPMOs share a common origin.

We are still in the very early stages of the characterisation of IPMOs in young open clusters; more definite results can be expected from the next generation of telescopes. Based on the currently available results, however, IPMOs are unlikely to be ejected giant planets; instead, they appear to be the natural extension of the stellar/brown dwarf mass function into the planetary mass regime.

# 4 Conclusions

In this contribution, I present highlight results from recent observational studies of brown dwarfs and very low mass stars, focused on a) the rotational evolution and b) the disk properties in the substellar regime. The main goal of this paper is to illustrate that brown dwarfs provide an adequate testbed for stellar and planetary physics – they exhibit a hybrid nature in their observational properties. In the following, I will summarize some of our results based on this underlying principle:

- The early evolution of brown dwarfs follows the blueprint established for solar-mass stars. They undergo a T Tauri phase, harbouring dusty disks with lifetimes that are compatible with those of more massive objects. The global properties of brown dwarf disks – masses and radii – indicate that they are scaled-down versions of circumstellar T Tauri-like disks. The analogies of young stars and brown dwarfs include evidence for rotational disk braking and the potential for planet forming processes. As of today, no clear observational signature has been found for a significant difference in the formation modes of solar-mass stars and brown dwarfs. These conclusions seem to hold even for objects with masses of 8–20 $M_{\rm Jup}$ – objects at the bottom of the IMF. The lifetimes of the inner disks are in fact very similar over more than two orders of magnitude in object mass, indicating that stars, brown dwarfs, and even isolated planetary mass objects may have a common origin. Thus, *young* brown dwarfs appear to be miniature stars in their observational properties.

- As brown dwarfs age, they rapidly evolve to ultracool objects, which are more comparable to giant planets than to stars. One example is the lack of rotational braking due to stellar winds. Solar-mass stars quickly spin down on the main sequence due to angular momentum losses by magnetically induced winds. This mechanism of efficient rotational braking on long timescales breaks down in the very low mass regime, as it has been shown by recent observations. The origin of this behaviour is not understood yet; candidate explanations are changes in magnetic field configuration or wind properties. As a result, brown dwarfs and very low mass stars continue to be fast rotators over timescales in the order of Gyrs. In their long-term rotational evolution, substellar objects are thus comparable to giant planets in our solar system.

**Acknowledgements**

I would like to thank my main collaborators Jochen Eislöffel (TLS Tautenburg) and Ray Jayawardhana (University of Toronto) for their continuous support over the past years. This work has been partly supported by the *Deutsche Forschungsgemeinschaft* (DFG) through grants Ei 409/11-1 and -2 to J. Eislöffel, by University of Toronto and by NSERC through grants to R. Jayawardhana.

# References

Allers, K. N., Kessler-Silacci, J. E., Cieza, L. A., Jaffe, D. T., 2006, ApJ, 644, 364

Barnes, S. A., 2001, ApJ, 561, 1095

Bate, M. R., Bonnell, I. A., Bromm, V., 2002, MNRAS, 332, 65

Bate, M. R., Bonnell, I. A., Bromm, V., 2003, MNRAS, 339, 577

Béjar, V. J. S., Zapatero Osorio, M. R., Rebolo, R. 1999, ApJ, 521, 671

Bonnell, I. A., Larson, R. B., Zinnecker, H., 2007, in Reipurth B., Jewitt D., Keil K., eds, Protostars and Planets V. University of Arizona Press, Tucson, p. 149

Bouvier, J., Forestini, M., Allain, S. 1997, A&A, 326, 1023

Carpenter, J. M., Mamajek, E. E., Hillenbrand, L. A., Meyer, M. R., 2006, ApJ, 651, 49

Chabrier, G., Baraffe, I., 1997, A&A, 327, 1039

Cieza, L., Baliber, N., 2007, ApJ, in press

Comerón, F., Neuhäuser, R., Kaas, A. A., 2000, A&A, 359, 269

Hayashi, C., Nakano, T., 1963, PThPh, 30, 460

Herbst, W., Eislöffel, J., Mundt, R., Scholz, A., 2007, in Reipurth B., Jewitt D., Keil K., eds, Protostars and Planets V. University of Arizona Press, Tucson, p. 297

Hernández, J., Hartmann, L., Megeath, T., Gutermuth, R., Muzerolle, J., Calvet, N., Vivas, A. K., Briceño, C., Allen, L., Stauffer, J., Young, E., Fazio, G., 2007, ApJ, 662, 1067

Irwin, J., Hodgkin, S., Aigrain, S., Hebb, L., Bouvier, J., Clarke, C., Moraux, E., Bramich, D. M., 2007 MNRAS, 377, 741

Jayawardhana, R., Ardila, D. R., Stelzer, B., Haisch, K. E., Jr., 2003, AJ, 126, 1515

Jayawardhana, R., Ivanov, V. D., 2006, ApJ, 647, 167

Kirkpatrick, J. D., Reid, I. N., Liebert, J., Cutri, R. M., Nelson, B., Beichman, Ch. A., Dahn, C. C., Monet, D. G., Gizis, J. E., Skrutskie, M. F., 1999, ApJ, 519, 802

Klein, R., Apai, D., Pascucci, I., Henning, Th., Waters, L. B. F. M., 2003, ApJ, 593, 57

Königl, A. 1991, ApJ, 370, 39

Kumar, Sh. S., 1963, ApJ, 137, 1121

Liu, M. C., Najita, J., Tokunaga, A. T., 2003, ApJ, 585, 372

Luhman, K. L., Joergens, V., Lada, C., Muzerolle, J., Pascucci, I., White, R., 2007, in Reipurth B., Jewitt D., Keil K., eds, Protostars and Planets V. University of Arizona Press, Tucson, p. 443

Luhman, K. L., Adame, L., D'Alessio, P., Calvet, N., Hartmann, L., Megeath, S. T., Fazio, G. G., 2005, ApJ, 635, 93

Mathieu, R. D. 2003, IAUS, 215

Matt, S., Pudritz, R. E., 2005, ApJ, 632, 135

Mohanty, S., Jayawardhana, R., Natta, A., Fujiyoshi, T., Tamura, M., Barrado y Navascués, D., 2004, ApJ, 609, 33

Mohanty, S., Jayawardhana, R., Basri, G., 2005, MmSAI, 76, 303

Montalbán, J., D'Antona, F., Mazzitelli, I., 2000, A&A, 360, 935

Muench, A. A., Alves, J., Lada, Ch. J., Lada, E. A., 2001, ApJ, 558, 51

Mullan, D. J., MacDonald J., 2001, ApJ, 559, 353

Nakajima, T., Oppenheimer, B. R., Kulkarni, S. R., Golimowski, D. A., Matthews, K., Durrance, S. T., 1995, Nature, 378, 463

Natta, A., Testi, L., Comerón, F., Oliva, E., D'Antona, F., Baffa, C., Comoretto, G., Gennari, S., 2002, A&A, 393, 597

Noyes, R. W., Hartmann, L. W., Baliunas, S. L., Duncan, D. K., Vaughan, A. H., 1984, ApJ, 279, 763

Nuernberger, D., Brandner, W., Yorke, H. W., Zinnecker, H. 1998, A&A, 330, 549

Nuernberger, D., Chini, R., Zinnecker, H. 1997, A&A, 324, 1036

Oppenheimer, B. R., Kulkarni, S. R., Stauffer, J. R., in Mannings, V., Boss, A.P., Russell, S. S., eds., Protostars and Planets IV, University of Arizona Press, Tucson, p. 1313

Osterloh, M., Beckwith, S. V. W., 1995, ApJ, 439, 288

Padoan, P., Nordlund, A., 2004, ApJ, 617, 559

Pascucci, I., Apai, D., Henning, Th., Dullemond, C. P., 2003, ApJ, 590, 111

Patten, B. M., Stauffer, J. R., Burrows, A., Marengo, M., Hora, J. L., Luhman, K. L., Sonnett, S. M., Henry, T. J., Raghavan, D., Megeath, S. T., Liebert, J., Fazio, G. G., 2006, ApJ, 651, 502

Phan-Bao, N., Guibert, J., Crifo, F., Delfosse, X., Forveille, T., Borsenberger, J., Epchtein, N., Fouqué, P., Simon, G., 2001, A&A, 380, 590

Pinfield, D. J., Hodgkin, S. T., Jameson, R. F., et al. 2000, MNRAS, 313, 347

Prosser, C. F. et al., 1995, PASP, 107, 211

Radick, R. R., Thompson, D. T., Lockwood, G. W., Duncan, D. K., Baggett, W. E., 1987, ApJ, 321, 459

Rebolo, R., Zapatero Osorio, M. R., Martín, E. L, 1995, Nature, 377, 129

Rebull, L. M., Stauffer, J. R., Megeath, S. T., Hora, J. L., Hartmann, L., 2006, ApJ, 646, 297

Reipurth, B., Clarke, C., 2001, AJ, 122, 432

Scholz, A., 2004, PhD Thesis, Faculty for Physics and Astronomy of the Friedrich-Schiller-University, Jena/Germany

Scholz, A., Eislöffel, J., 2004a, A&A, 419, 249

Scholz, A., Eislöffel, J., 2004b, A&A, 421, 259

Scholz, A., Eislöffel, J., 2005, A&A, 429, 1007

Scholz, A., Jayawardhana, R., Wood, K., 2006, ApJ, 645, 1498

Scholz, A., Jayawardhana, R., Wood, K., Meeus, G., Stelzer, B., Walker, Ch., O'Sullivan, M., 2007, ApJ, 660, 1517

Scholz, A., Eislöffel, J., 2007, MNRAS, 381, 1638

Scholz, A., Jayawardhana, R., 2008, ApJL, 672, 49

Skumanich, A., 1972, ApJ, 171, 565

Terndrup, D. M., Krishnamurthi, A., Pinsonneault, M. H., Stauffer, J. R., 1999, AJ, 118, 1814

Whitworth, A., Bate, M. R., Nordlund, Å., Reipurth, B., Zinnecker, H., 2007, in Reipurth B., Jewitt D., Keil K., eds, Protostars and Planets V. University of Arizona Press, Tucson, p. 459

# Astroparticle Physics in Europe: Status and Perspectives

Christian Spiering

DESY

15738 Zeuthen, Platanenallee 6, Germany

christian.spiering@desy.de

### Abstract

*Astroparticle physics has evolved as an interdisciplinary field at the intersection of particle physics, astronomy and cosmology. Over the last two decades, it has moved from infancy to technological maturity and is now envisaging projects on the 100 Million Euro scale. This price tag requires international coordination, cooperation and convergence to a few flagship projects. The Roadmap Committee of ApPEC (Astroparticle Physics European Coordination) has recently released a roadmap covering the next ten years. This talk describes status and perspectives of astroparticle physics in Europe and reports the recommendations of the Roadmap Committee.*

## 1 Introduction

Although the notation "astroparticle physics" was coined only 25 years ago, and although it has been widely used only since the nineties, its roots go back to the early years of the last century, when Victor Hess discovered cosmic rays. The origin of these particles – mostly protons, light and heavy nuclei – was unknown, and it is going to be solved only now, 100 years later. However, there was a long period when cosmic rays served as the source of most newly discovered elementary particles: 1932, the first anti-particle, the positron, was recorded in cosmic rays, followed by the muon in 1936 and in 1947 by the pion, the first of the vast family of mesons. Until the beginning of the fifties, cosmic rays remained the main source of new particles and laid the ground for the "particle zoo", which a decade later was explained by the quark model. It was only in the mid of the fifties that particle accelerators started their triumphal development and cosmic rays lost their role in particle physics. With a single exception, cosmic particle physics disappeared from the screen of most particle physicists.

The exception was the long-lasting attempt to detect solar neutrinos. It was pioneered by Ray Davis in the sixties who measured the $^8$B neutrino flux with the radio-chemical ClAr method – detecting however only a third of the predicted flux.

The deficit was confirmed in the eighties by Kamiokande, a water Cherenkov detector in Japan. The following measurements of pp neutrinos with GaGe detectors in the Russian Baksan laboratory (SAGE experiment) and in the Italian Gran Sasso Laboratory (GALLEX experiment) corroborated the suggestion that the solution to the neutrino deficit was not given by a different solar model but by neutrino oscillations. This solution was eventually confirmed by the SNO experiment in Canada (Bahcall 2005, McDonald et al. 2004).

In 2002, Ray Davis and Masatoshi Koshiba were awarded the Nobel Prize in Physics for opening the neutrino window to the Universe, specifically for the detection of neutrinos from the Sun and the Supernova SN1987A in the Large Magellanic Cloud. Their work was a unique synthesis of particle physics and astrophysics since solar neutrinos also provided the first clear evidence that neutrinos have mass. It represents a classical illustration of the interdisciplinary field at the intersection of particle physics, astronomy and cosmology which now is known as astroparticle physics. One may note that Koshiba's Kamiokande detector was originally built to detect proton decay – another bracketing of astrophysics and particle physics by a single technique.

The detection of solar and Supernova neutrinos is not the only new window to the Universe opened by astroparticle physics. Another one is that of high energetic gamma rays recorded by ground based Cherenkov telescopes. From the first source detected in 1989, three sources known in 1996, to nearly 70 sources identified by the end of 2007, the high energy sky has revealed a stunning richness of new phenomena and puzzling details (see Fig. 11 below). Other branches of astroparticle physics did not yet provide such gold-plated discoveries but moved into unprecedented sensitivity regions with rapidly increasing discovery potential, like the search for dark matter particles, the search for decaying protons or the attempt to determine the absolute values of neutrino masses.

## 2 Basic questions

The Roadmap Committee of ApPEC (Astroparticle Physics European Coordination) has recently released a roadmap covering the next ten years. Recommendations of the committee (http://www.aspera-eu.org) have been formulated by addressing a set of basic questions:

a) What are the constituents of the Universe? In particular: What is dark matter?

b) Do protons have a finite life time?

c) What are the properties of neutrinos? What is their role in cosmic evolution?

d) What do neutrinos tell us about the interior of the Sun and the Earth, and about Supernova explosions?

e) What is the origin of cosmic rays ? What is the view of the sky at extreme energies ?

f) What will gravitational waves they tell us about violent cosmic processes and about the nature of gravity?

An answer to any of these questions would mark a major break-through in understanding the Universe and would open an entirely new field of research on its own.

## 3 Dark Matter and Dark Energy

Over the last decade, the content of the Universe has been measured with unprecedented precision. Whereas normal baryonic matter contributes only about 4%, the dominant constituents are unknown forms of matter and energy: Dark Matter (22%) and Dark Energy (74%).

Whereas the concept of Dark Energy was introduced only recently – in response to a negative pressure driving cosmic expansion –, Dark Matter has been discussed for decades. The prevalent view is that Dark Matter consists of stable relic particles from the Big Bang, and that nearly all of it is in the form of Cold Dark Matter (CDM). In the early Universe, CDM particles would have already cooled to non-relativistic velocities when decoupling from the expanding and cooling Universe. Hot dark matter (HDM) has been relativistic at the time of decoupling. Neutrinos are typical HDM particles; their contribution to the total matter budget, however, is small.

### 3.1 The Search for Dark Matter

The favoured candidate for dark matter is a Weakly Interacting Massive Particle (WIMP) related to new physics at the TeV scale (Steigman & Turner 1985, Jungmann et al. 1996). Among the various WIMP candidates, the lightest supersymmetric (SUSY) particle in the Minimal SuperSymmetric Model (MSSM) is favoured – likely the *neutralino*. Another theoretically well-founded dark matter candidate is the axion (Peccei & Quinn 1997, Raffelt 2006). Even though axions would be much lighter than WIMPs, they still could constitute CDM, since they are have not been produced in thermal equilibrium and would be non-relativistic.

*Direct WIMP searches*

"Direct" WIMP searches focus on the detection of nuclear recoils from WIMPs interacting in underground detectors (Gaitskell 2004, Baudis 2005, Sadoulet 2007). No WIMP candidate has been found so far. Assuming that all Dark Matter is made of WIMPs, present experiments with a several-kilogram target mass can therefore exclude WIMPS with an interaction cross section larger than $10^{-43}$ cm$^2$ (i.e. $10^{-7}$ picobarn). MSSM predictions for neutralino cross sections range from $10^{-5}$ to $10^{-12}$ pb (see Fig. 1).

Experimental sensitivities will be boosted to $10^{-8}$ pb in a couple of years and may reach, with ton-scale detectors, $10^{-10}$ pb in 7–10 years. Therefore, there is a fair chance to detect dark matter particles in the next decade – provided the progress in background rejection can be realized and provided CDM is made of super-symmetric particles.

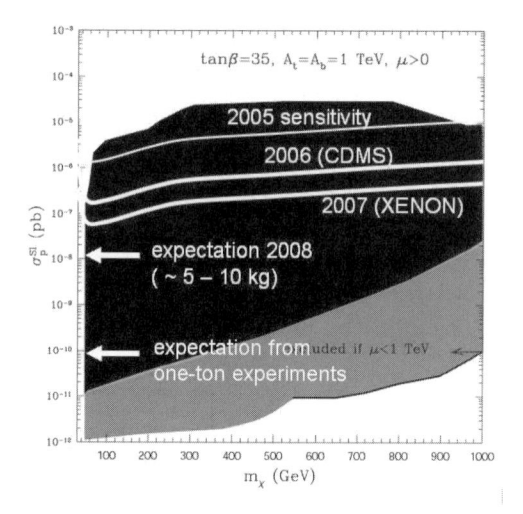

**Figure 1:** Spin-independent WIMP cross section vs. WIMP mass for an MSSM prediction (dark area), with parameters fixed to the values shown at top right (Kim et al. 2002). The upper curve represents the limits obtained by 2005. CDMS, a bolometric detector operated in the US, achieved the 2006 record limit and was surpassed in 2007 by XENON, a liquid xenon detector operated in the Italian Gran Sasso Underground Laboratory (Angle 2007). The arrows indicate the sensitivities expected in about a year from now, and within a decade from now with one-ton experiments. All results assume the WIMP to be a neutralino in the standard MSSM formulation.

Presently, there are two favoured detection techniques:

- *Bolometric* detectors are operated at a temperature of 10–20 mK and detect the feeble heat, ionization and scintillation signals from WIMP interactions in crystals made, e.g., from germanium, silicon or $CaWO_4$. Present flagship experiments are CDMS in the USA, and CRESST (Gran Sasso Laboratory, Italy) and EDELWEISS (Fréjus Laboratory, France) in Europe.

- *Noble liquid* detectors record ionization and scintillation from nuclear recoils in liquid xenon, argon or neon. XENON (Gran Sasso) and ZEPLIN (Boulby mine, UK) use liquid xenon targets of about 10 kg mass, while WARP (Gran Sasso) and ArDM (Canfranc, Spain) operate, or prepare, liquid argon detectors. Actually the most recent significant step in the race for better sensitivities has been made by XENON (see Angle 2007 and Fig. 1).

A variety of presently more than 20 dark matter experiments worldwide (see for a review Baudis 2007) must, within several years, converge to two or three few ton-scale experiments with negligible background. In Europe, there are two large initiatives towards experiments on the ton scale: EURECA, joining most players of the bolometric approach, and ELIXIR, joining most of the liquid xenon experts. R&D on alternative methods will be continued, but most of the resources will naturally be focused to ton-scale flagship projects and the corresponding underground infrastructures. Figures 2 and 3 sketch a scenario towards ton-scale experiments with negligible background (different to ton-scale experiments attempting to identify an annual signal variation on top of a large background, like the DAMA experiment, see below). Figure 2 shows the possible development of limits and sensitivities as a function of time, assuming a standard MSSM WIMP with spin-independent coupling. A $10^{-8}$ pb sensitivity can be reached within the next couple of years. Improvements by further two orders of magnitude require more massive detectors. The coloured area for >2009 indicates the range of projections given by different experiments, most of them envisaging an intermediate step at the 100 kg scale. Note that this scenario is made from a 2007 perspective and that initial LHC results may substantially influence the design of the very few "ultimate" detectors. Needless to say that all these plans stand or fall with the capability to reduce the background, even for ton scale masses, to less than a very few events per year.

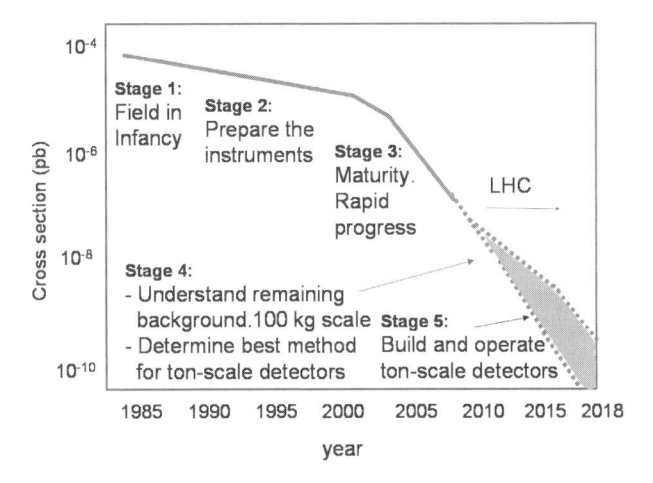

**Figure 2:** Possible development of limits and sensitivities as a function of time (see text for explanations).

Figure 3 shows a first tentative projection of investment expenses for ELIXIR and EURECA, including the cost for construction of a suitable low-background, deep underground infrastructure[1]. Estimates will be made more precise within design phases covering the next three years. Note that a possible prioritisation within given funding envelopes may change this picture substantially.

Following identification of a proper signal, clearly distinguished against background, one would like to get a final confirmation for the nature of the signal, by observing a "smoking gun" signature which ensures that the signal is due to WIMPs and not due to something else, such as backgrounds. There are three such signatures: *a)* annual modulation, *b)* directionality, *c)* target dependence.

The annual modulation signature reflects the periodic change of the WIMP velocity in the detector frame due to the motion of the Earth around the Sun. The variation

---
[1]This figure may serve as an illustration for the kind of information which has been collected for all astroparticle experiments in Europe.

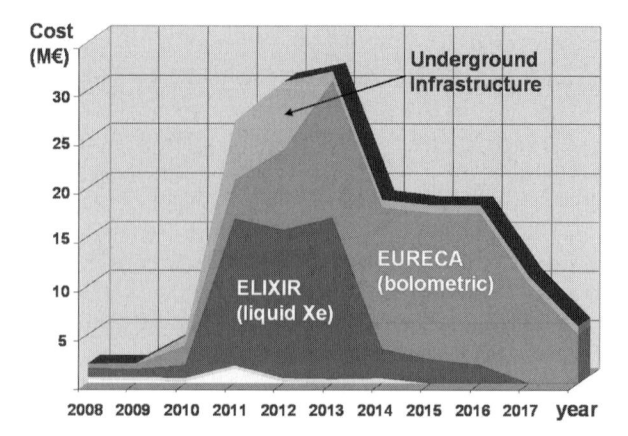

**Figure 3:** Present projection of the investment expenses for the ton scale versions of ELIXIR and EURECA, including the cost for an low-background underground infrastructure housing the detectors. Since detectors will be built in a modular way, this scenario includes a stepwise approach to the ton scale, via 100-kg stages.

is only of a few percent of the total WIMP signal, therefore large target masses are needed to be sensitive to the effect. Indeed, the DAMA experiment (Bernabei 2004), recording the scintillation light in NaI crystals, has reported an observation of this signature in its data from 100 kg NaI (see Fig. 4), but the interpretation remains controversial. The collaboration is presently running a 250 kg version of the experiment, with first results expected in 2008, and is asking for ton-scale ressources in a next step.

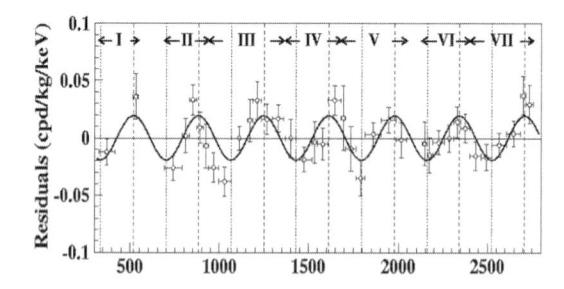

**Figure 4:** Seasonal variations of the counting rate residuals observed by the DAMA experiment over a period of seven years (Bernabei 2004).

The target dependence signature follows from the different interactions of WIMPs with different nuclei – both in rate and in spectral shape. The directionality signature would clearly distinguish the WIMP signal from a terrestrial background and search for a large forward/backward asymmetry. It requires detectors capable of measuring the nuclear recoil direction, a condition potentially only met by gaseous detectors. At present, these detectors are in an R&D phase.

*Indirect Dark Matter Search*

The mentioned *direct* searches are flanked by *indirect searches*. These would identify charged particles, gamma rays or neutrinos from WIMP annihilation in the cores of galaxies, the Sun or the Earth. WIMPs can be gravitationally trapped in celestial bodies and eventually accumulate until their density becomes large enough that the self-annihilation of WIMPs would be in equilibrium with WIMP capture. Earth-bound or satellite detectors would then detect the decay products of these annihilations: an excess of neutrinos from the centre of the Earth or Sun, a gamma signal from the centre of the Galaxy, or an excess in positrons and anti-protons from galactic plane or halo. Anti-particles such as positrons and anti-protons produced in WIMP annihilation would be trapped in the galactic magnetic fields and be detected as an excess over the background generated by other well understood processes.

Present and planned detectors capable to contribute to indirect Dark Matter detection are described in section 7: satellite detectors (Pamela and AMS for charged cosmic rays, Agile and GLAST for gamma rays) and earth bound telescopes (Magic, H.E.S.S., Veritas, CTA for gamma rays, Baikal, IceCube, Antares, KM3NeT for neutrinos).

Direct and indirect methods give complementary information. For instance, gravitational trapping works best for slow WIMPs, making indirect searches most sensitive. In the case of direct detection, however, the higher energy of recoil nuclei makes fast WIMPs easier to detect. For indirect searches, the annihilation rates would depend on all the cosmic history of WIMP accumulation and not only on the present density, providing another aspect of complementarity to direct searches.

*Synthesis of direct, indirect and accelerator signatures*

While astroparticle physicists are searching dark matter, particle physicists are preparing searches for super-symmetric particles at the Large Hadron Collider in Geneva, which is expected to start operation end of 2008 and may provide first physics results on SUSY searches in 2010 or 2011. A detection of SUSY particles at the LHC would certainly considerably boost dark matter searches. Eventually, only the synthesis of all three observations – direct and indirect detection of cosmic candidates for dark matter and identification of the neutralino at the LHC – would give sufficient confidence about the character of the observed particles. (see the recent review of Bertone 2007 for a description of a multidisciplinary approach to Dark Matter search).

## 3.2 Dark Energy

Evidence of Dark Matter and Dark Energy has emerged from astronomical observations. Astronomical studies of galactic dynamics, gravitational lensing, large scale structures and CMBR anisotropies provide arguments for Dark Matter. Combining these observations with the observation that the universe is accelerating (SNIa methods), that the Universe is flat (from CMBR measurements) and that Dark Matter alone cannot provide the critical density (from large scale structures) establishes the need for something like "Dark Energy".

However, whereas Dark Matter may consist of distinct particles and can be searched by the methods of astroparticle physics, Dark Energy may be a continuous phenomenon. "Particle search strategies" equivalent to the Dark Matter case do not exist. Dark Energy can primarily explored through its influence on cosmic evolution. Observations in this area traditionally use astronomical techniques (see e.g. Perlmutter & Schmidt 2003, Spergel et al. 2007). Particle physicists have joined this new field and are playing a major role – for instance by contributing with their experience in processing large amounts of data.

The next generation of experiments relevant for Dark Energy search includes the European Planck mission on a satellite, and the ground based Dark Energy Survey, DES, the Low Frequency Array, LOFAR, and the ALMA-Pathfinder APEX. Projects proposed to be started after 2013 include various survey telescopes, most notably the Large Synoptic Survey Telescope, LSST, and the Panoramic Survey Telescope & Rapid Response System, PanSTARRS. Space based missions include the the wide field space imager DUNE and the spectroscopic all-sky cosmic explorer, SPACE, the SuperNova/Acceleration Probe, SNAP, and the James Webb Space Telescope, JWST. Needless to say that survey telescopes serve a variety of standard astronomical tasks and are not bounded to Dark Energy search. This is also true for the European Extremely Large Telescope, E-ELT, and the Square Kilometer Array, SKA. The inclusion of SKA in the ESFRI list demonstrates a high European priority.

Given the deep implications for fundamental physics, Dark Energy missions find the strongest support from the astroparticle physics community. Recommendations are formulated in the European ASTRONET Roadmap.

## 4  Proton decay and low energy neutrino astronomy

Grand Unified Theories (GUTs) of particle physics predict that the proton has a finite lifetime. The related physics may be closely linked to the physics of the Big Bang and the cosmic matter-antimatter asymmetry. Data from the Super-Kamiokande detector in Japan constrain the proton lifetime to be larger than $10^{34}$ years, tantalizingly close to predictions of various SUSY-GUT predictions. A sensitivity improvement of an order of magnitude requires detectors on the $10^5$–$10^6$ ton scale. The discovery of proton decay would be one of the most fundamental discoveries for physics and cosmology and certainly merits a worldwide coherent effort.

Proton decay detectors do also detect cosmic neutrinos. Figure 5 shows a "grand unified neutrino spectrum". Solar neutrinos, burst neutrinos from SN1987A, reactor neutrinos, terrestrial neutrinos and atmospheric neutrinos have been already detected. They would be also in the focus of a next-stage proton decay detector. Another guaranteed, although not yet detected, flux is that of neutrinos generated in collisions of ultra-energetic protons with the 3K cosmic microwave background (CMB), the so-called GZK (Greisen-Zatsepin-Kuzmin) neutrinos. Whereas GZK neutrinos as well as neutrinos from Active Galactic Nuclei (AGN) are likely to be detected by neutrino telescopes in the next decade (see below), no realistic idea exists how to detect 1.9 K cosmological neutrinos, the analogue to the 2.7 K microwave radiation.

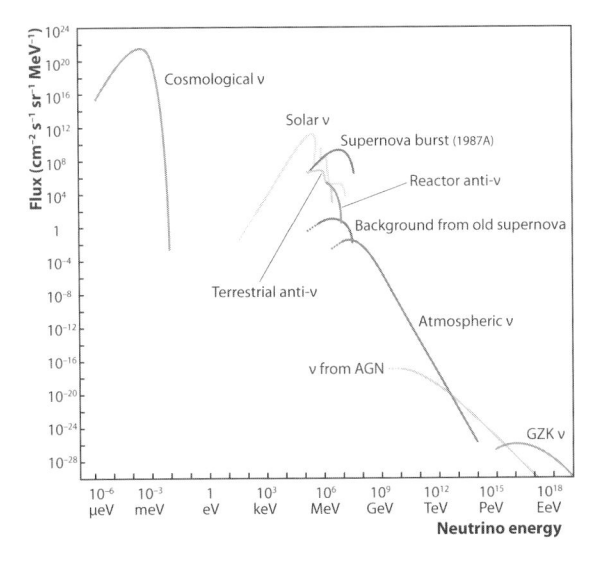

**Figure 5:** The "grand unified" neutrino spectrum

Solar neutrinos, detection of neutrino oscillations by solar and atmospheric neutrinos, neutrinos from the supernova SN1987A, geo-neutrinos – large underground detectors have produced an extremely rich harvest of discoveries (McDonald et al. 2004). We note the most recent event in the series of solar neutrino measurements: the first real-time detection of solar $^7$Be neutrinos by the BOREXINO experiment (Arpesella et al. 2008) – an impressive success, eventually achieved after a long troublesome process of internal, and in particular externally imposed, delays.

The triumphal legacy of underground neutrino physics is intended to be continued by worldwide one or two multi-purpose detectors on the mass scale of 100–1000 kilotons. The physics potential of such a large multi-purpose facility would cover a large variety of questions:

a) The proton decay sensitivity would be improved by one order of magnitude.

b) A galactic Supernova would result in $10^4$–$10^5$ neutrino events, compared to only 20 events for SN1987A. This would provide incredibly detailed information on the early phase of the Supernova explosion.

c) The diffuse flux from past supernovae would probe the cosmological star formation rate.

d) The details of the processes in the solar interior can be studied with high statistics and the details of the Standard Solar Model determined with percent accuracy.

e) The high-statistics study of atmospheric neutrinos could improve our knowledge on the neutrino mass matrix and provide unique information on the neutrino mass hierarchy.

f) Our understanding of the Earth interior would be improved by the study of geo-neutrinos.

g) The study of neutrinos of medium energy from the Sun and the centre of the Earth could reveal signs for dark matter.

h) Last but not least, a large underground detector could detect artificially produced neutrinos from nuclear reactors or particle accelerators, over a long baseline between neutrino source and detector.

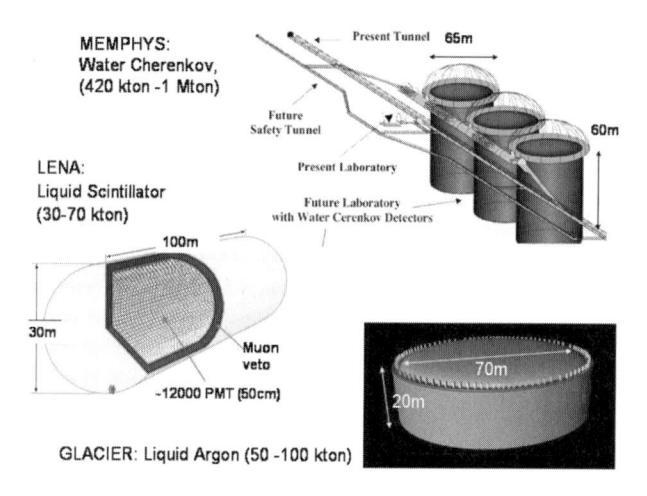

**Figure 6:** Representative projects for the three proposed methods: MEMPHYS (420–1000 kton water), LENA (30–70 kton scintillator) and GLACIER (50–100 kton liquid argon).

Three detection techniques are currently studied: Water-Cherenkov detectors (like Super-Kamiokande), liquid scintillator detectors (like BOREXINO) and liquid argon detectors (a technique pioneered by the Italian ICARUS collaboration). The present prominent European projects under study are MEMPHYS, a Megaton-scale water detector (de Bellefon et al. 2006), LENA (Wurm et al. 2007), a 30–70 kiloton liquid scintillator detector, and GLACIER (Ereditato & Rubbia 2005), a 50–100 kiloton liquid argon detector (see Fig. 6).

The European LAGUNA consortium (Autiero et al. 2007) has been awarded a FP7 Design Study grant in order *a)* to explore and compare the capabilities of the three methods and *b)* to evaluate the possibilities of excavation of deep large cavities and the accompanying infrastructures. The design study should converge, on a time scale of 2010, to a common proposal. The total cost depends on the method and the actual size and is estimated to be between 400 and 800 Million Euro. Civil engineering may start in 2013. The cost would be shared internationally for such a Mega project.

## 5 Neutrino properties

In the context of astroparticle physics, neutrinos, rather than being the subject of research, mainly play the role of messengers: from the Sun, from a Supernova, from Active Galaxies and other celestial objects. Still, some of their intrinsic properties remain undetermined. From the oscillatory behaviour of neutrinos we can deduce that the weak eigenstates of neutrinos (flavour eigenstates) are not identical with their mass eigenstates, that the neutrino masses are non-zero and that they differ from each other. That is why the flavour eigenstates, electron neutrino $\nu_e$, muon neutrino $\nu_\mu$ and tau neutrino $\nu_\tau$, oscillate between each other. From oscillations we can determine how strong the states mix and the mass differences. But what are the absolute values of the masses? Further: are neutrinos their own antiparticles ("Majorana particles")?

Many of the projects devoted to these questions are not really *astro*-particle experiments, but often share certain aspects with "typical" astroparticle experiments: the infrastructure (like low-background, deep caverns), the methods (like high purity liquid scintillator techniques), or sometimes just the scientist's community. That is why they are addressed by the ApPEC roadmap.

The main sources of information on neutrino parameters are the following:

a) oscillation experiments using neutrinos from accelerators or nuclear reactors as well as atmospheric or solar neutrinos. They provide information on mixing parameters, on a possible CP violation and on mass differences, but not absolute masses.

Absolute masses can be derived from three types of data or experiments:

b) cosmological data (see for a review Turner 2007),

c) end-point measurement of electron energy spectra in $\beta$-decays,

d) neutrino-less double beta decay experiments. They are the only experiments which can also prove the Majorana nature of neutrinos.

Neutrino oscillations impose a lower limit on the heaviest neutrino mass of about 0.05 eV (since this occurs to be the mass difference between the heaviest and the second heaviest mass state). This implies that neutrinos contribute at least 0.1% of cosmic matter. Neutrinos with a small finite mass contribute to hot dark matter, which suppresses the power spectrum of density fluctuations in the early Universe at "small" scales, of the order of one to ten Mega-parsec. The recent high precision measurements of density fluctuations in the Cosmic Microwave Background (WMAP) and the observations of the Large Scale Structure distribution of galaxies (2dFGRS and SDSS), combined with other cosmological data, yield an upper limit of about 1.5% on the amount of hot dark matter in the Universe, corresponding to an upper limit of about 0.6–0.7 eV on the sum of all three neutrino masses (see e.g. Hannestad & Raffelt 2006). The future sensitivity of cosmological measurements with Large Scale Surveys and with the CMB mission Planck, combined with the weak gravitational lensing of radiation from background galaxies and of the CMB is expected to reach a value of $\approx 0.1$ eV.

*Direct mass measurement*

The only laboratory technique for the direct measurement of a small neutrino mass (without additional assumptions on the character of the neutrino) is the precise measurement of the electron spectrum in $\beta$-decays. Here, the neutrino mass (or an upper limit to it) is inferred from the shape of the energy spectrum near its kinematical end point. The present upper limit is at 2.3 eV (Kraus et al., 2005, Lobashov et al. 2003). The KATRIN experiment in Karlsruhe will improve the sensitivity of past experiments down to 0.2 eV (Robertson 2007). Operation of KATRIN is expected to start in 2009/2010.

Given the cosmological sensitivities, one may ask for the competitiveness of the KATRIN experiment. Precision cosmology yields an upper limit of 0.7 eV for the sum of all three neutrino masses (or 0.7 eV/3 $\approx$ 0.23 eV with respect to the lightest mass state). This does not seem to leave much room for a device with 0.2-eV sensitivity. However, one must keep in mind that the cosmological limit, despite the impressive success of precision cosmology, has to be derived within a system of assumptions and interpretations, and is not obtained directly. Considering the importance of the neutrino mass question, and the difficulty in associating the cosmological limit to a precise systematic confidence level, it is therefore important to pursue direct measurements up to their eventual technological – and financial – limits. There is only one way to move beyond KATRIN sensitivity: using calorimetric instead of spectrometric methods. The potential of these methods is presently explored.

*Neutrino-less double beta decay*

The observation of neutrino-less double beta decay may allow going to even lower masses than end-point measurements of the KATRIN type. However, it requires the neutrino to be a *Majorana* particle, i.e. representing the only fermion which is its own anti-particle. Implications of massive neutrinos for models beyond the Standard Model differ for Majorana and Dirac neutrinos. Therefore the answer to the question whether nature took the "Majorana option" is essential.

In a neutrino-less double beta decay, a nucleus $(A, Z)$ would turn into another $(A, Z + 2)$ by transforming two neutrons into protons and emitting two electrons: $(A, Z) \rightarrow (A, Z + 2) + 2e^-$. This differs from "normal" double-beta decay (second order process of the weak interaction), which is rare but has been detected and studied: $(A, Z) \rightarrow (A, Z + 2) + 2e^- + 2\nu$. Neutrino-less double beta decay is possible only for massive Majorana neutrinos. The observed lifetime would be inversely proportional to the neutrino mass squared. Corresponding experiments are performed in low radioactivity environments deep underground, in order to suppress fake events (see for overviews Elliot & Vogel 2002 and Avignone, Elliot & Engel 2007).

Searches for double beta have been performed since the 1950s, but it was the discovery of neutrino oscillations which eventually led to a renaissance of the early enthusiasm and enormously boosted the existing efforts. Present best limits are at 0.3–0.8 eV (Avignone, Elliot & Engel 2007), with the uncertainty in the mass limit reflecting the limited knowledge on the nuclear matrix elements. Figure 7 shows the allowed effective neutrino masses (i.e. the linear combination of masses of the three mass states which is measured in double beta experiments) vs. the mass of the lightest neutrino. Constrains come from the mentioned double-beta limit and from cosmological observations.

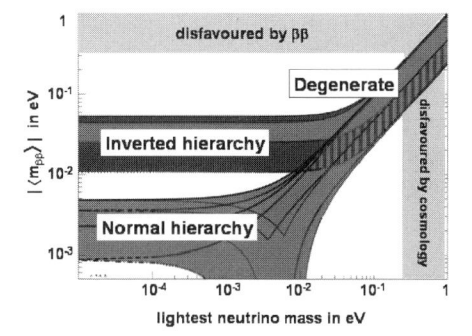

**Figure 7:** Allowed effective neutrino mass (as measured in double beta decay experiments) vs. mass of the lightest mass state (adopted from Petcov 2005). Different lines and colours for theoretical predictions correspond to various assumptions on CP violating phases. Degenerate scenarios correspond to nearly equal neutrino masses, for different masses one distinguishes "normal" hierarchies and "inverted" hierarchies.

One single experiment (Klapdor-Kleingrothaus et al. 2004) claims a positive observation and derives a mass of 0.2–0.6 eV. This claim is highly controversial but can be tested with next generation experiments which will reach a sensitivity of better than 0.1 eV between 2009 and 2013. At present there are three European flagship projects: CUORICINO, a bolometric detector in the Gran Sasso Lab, uses 41 kg $^{130}$Te and plans to operate a 740 kg detector (CUORE) in the next decade. NEMO is a detector with both tracking and calometry capabilities using 8 kg $^{130}$Mo and $^{62}$Se in the Fréjus underground laboratory. A 100 kg $^{62}$Se detector (Super-NEMO) for the next decade is in the design phase. GERDA is a $^{76}$Ge detector. It will be operated in the Gran Sasso Lab from 2009 on and stepwise upgraded from 15 kg to 35 kg. In another approach, called COBRA, the use of CdTe is tested. US projects include MAJORANA, a germanium project like GERDA, and EXO, a xenon detector. These experiments could possibly "scrape" the mass range of the inverted hierarchy scenarios in Fig. 7).

Europe is currently clearly leading the field of double beta decay searches and is in the strategic position to play a major role in next generation experiments. To cover a mass range of 20–50 meV, i.e. most of the range suggested by the inverted mass scale scenarios, one needs detectors with an active mass of order one ton, good resolution and very low background. Construction of such detectors might start in 2014–2017. Different nuclear isotopes and different experimental techniques are needed to establish the effect and extract a neutrino mass value. The price tag for one of these experiments is at the 100–200 Million Euro scale, with a large contribution from the production cost for isotopes. The priority and urgency with which these experiments will be tackled will depend on the background rejection achieved in the currently prepared stages, on the available funding, and on the future bounds on the neutrino mass from cosmological observations.

## 6 Underground Laboratories

Proton decay, neutrino-less double beta decay or interaction of dark matter particles are extremely rare processes and the effects are extremely feeble. The signals from solar or geo-neutrinos are similarly weak. The study of these processes requires a low-background environment, shielded against processes which may fake a true signal. This environment is provided by special underground laboratories. The various tasks require different characteristics of the site. Double beta experiments and dark matter searches need housing for ton-scale detectors and very low radioactive background. Detectors for solar neutrino neutrinos need a moderate depth between 1000 and 2000 meter water equivalent and much larger caverns. For proton decay experiments, neither large depth nor extremely low radioactivity is needed, but the cavern for a Megaton detector naturally has to be huge.

There are five European underground laboratories which have been used in the past and are used presently for astroparticle physics deep underground experiments, with depths ranging between one and nearly five kilometer water equivalent: The Laboratori Nazionali del Gran Sasso (LNGS) along a motorway tunnel in the Apennines (Italy), the Laboratoire Souterrain de Modane, LSM, located along the Fréjus Road tunnel connecting Italy and France, the Laboratorio subterráneo de Canfranc, LSC, arranged along a tunnel connecting Spain and France, the Boulby Underground Laboratory in an operational potash and rock-salt mine on the North-East coast of England. Russia operates the Baksan Neutrino Observatory, BNO, in a dedicated tunnel in the Caucasus. A sixth very deep site in Finland is under discussion and some additional shallow locations are considered for special applications or as test sites (see Coccia 2006 for a review of European underground laboratories and Bettini 2007 for a recent compendium of underground laboratories worldwide).

The following years will lead to clearer picture of a task distribution between European sites. ApPEC will help finding solutions in case of conflicting national preferences and prioritizing possible extensions of the underground labs in accordance with the actual needs in Europe and worldwide.

## 7 The high energy universe

Much of classical astronomy and astrophysics deals with thermal radiation, emitted by hot or warm objects such as stars or dust. The hottest of these objects, such as hot spots on the surfaces of neutron stars, emit radiation in the range of some $10^3$ to $10^4$ eV, about thousand times more energetic than visible light. We know, however, that non-thermal phenomena, involving much higher energies, play an important role in the cosmos. First evidence for such phenomena came with the discovery of cosmic rays by Victor Hess in 1912. Hess measured radiation levels during balloon flights and found a significant increase with height, which he correctly attributed to a hitherto unknown penetrating radiation from space. In 1938, Pierre Auger proved the existence of extensive air showers – cascades of elementary particles – initiated by primary particles with energies above $10^{15}$ eV by simultaneously observing the arrival of secondary particles in Geiger counters many meters a part. Modern cosmic-ray detectors reveal a cosmic-ray energy spectrum extending to $10^{20}$ eV and beyond (see Fig. 8). That are breath-taking energies, a hundred million times above that of terrestrial accelerators (Watson 2006, Olinto 2007).

How can cosmic accelerators boost particles to these energies? What is the nature of the particles? Do the particles at the very highest energies originate from the decay of super-heavy particle rather than from acceleration processes (top-down versus bottom-up scenarios)?

The mystery of cosmic rays is going to be solved by an interplay of detectors for high energy gamma rays, charged cosmic rays and neutrinos.

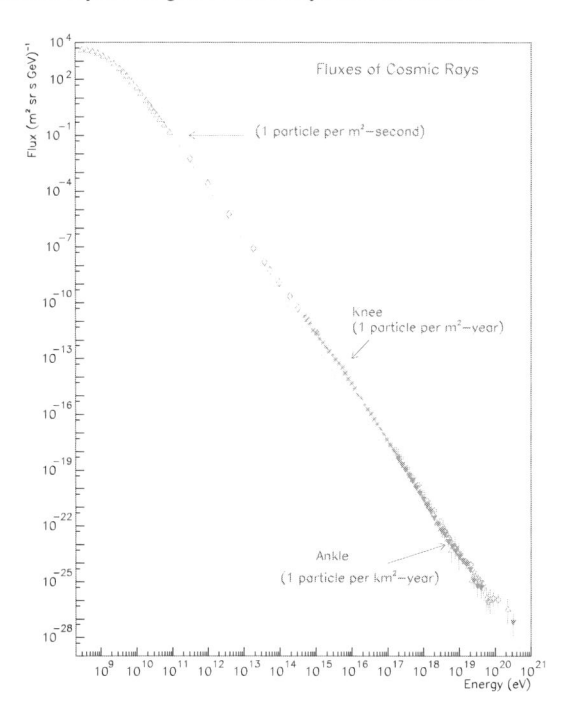

**Figure 8:** The spectrum of cosmic rays (courtesy S.P. Swordy). The region below $10^{14}$ eV is the domain of balloon and satellite experiments, at higher energies ground based techniques take over. Galactic supernova remnants can accelerate particles to energies of $10^{16}$–$10^{17}$ eV, well above the "knee". Galactic sources are believed to run out of power at $10^{17}$–$10^{18}$ eV. Highest observed energies dwarf the Large Hadron Collider at CERN which will accelerate protons to $10^{13}$ eV.

## 7.1 Charged cosmic rays

*The highest energies*

The present flagship in the search for sources of ultra-high energy cosmic rays is the Southern Pierre Auger Observatory in Argentina (Abraham et al. 2004), a 3000-km$^2$

array of water tanks, flanked by air fluorescence telescopes, which measure direction and energy of giant air showers (see Fig. 9).

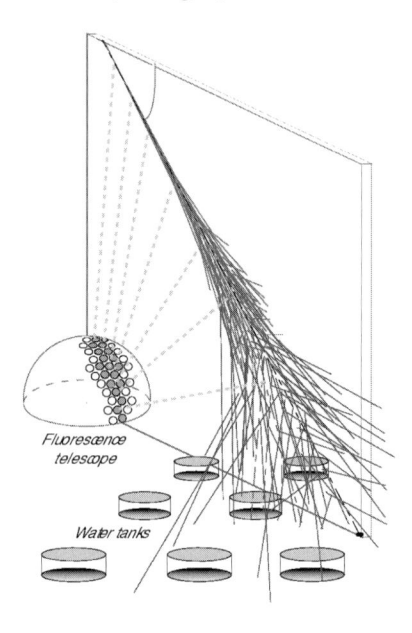

**Figure 9:** The Pierre-Auger detection principles: Fluorescence light from air showers is recorded by telescopes, particles at ground level are recorded by Cherenkov water tanks.

Even at energies above $10^{19}$ eV, where the cosmic flux is only about one particle per year and square kilometer, the Auger Observatory can collect a reasonable number of events. Starting with these energies, the deflection of charged particles in cosmic magnetic fields is going to be negligible and source tracing becomes possible. Very recently, the Auger collaboration has published a first sky map of events with energies above $10^{19.6}$ eV (Abraham et al. 2007, see Fig. 10). There is a clear correlation of events with the super-galactic plane. Also, the authors report a correlation with positions of Active Galactic Nuclei (chance probability of 0.17%). Such a correlation would be in agreement with theoretical expectations which classify only two objects to be able to accelerate particles up to $10^{20}$ eV or higher: the jets of AGN and Gamma Ray Bursts. Whether the interpretation of AGN as sources of the observed cosmic rays will withstand further tests and higher statistics has to be seen. If confirmed, it would mark the first step into astronomy with charged cosmic rays.

Full-sky coverage would be obtained by a Northern observation site, which has been determined to be in Colorado/ USA. This array would significantly exceed the Southern detector in size, possibly at the expense of low energy sensitivity, i.e. using a larger spacing of detector elements which would result in a higher energy threshold. European groups will play a significant role to establish the scientific case, and after its full demonstration make a significant contribution to the design and construction of Auger-North. The cost is estimated at 90 Million Euro, with a 45% European contribution, start of construction is conceived for 2010.

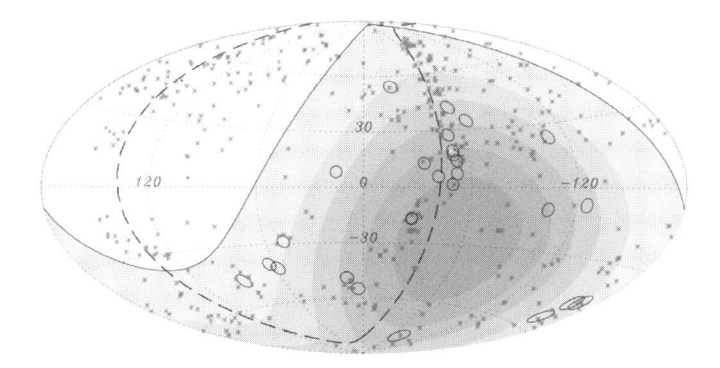

**Figure 10:** Sky map in galactic coordinates of 27 cosmic rays with highest energies detected by the Pierre Auger Observatory (black circles), compared to positions of 472 quasars and active nuclei with redshift $z \leq 0.018$. The dashed line marks the super-galactic plane.

*Between knee and ankle*

The subjects of the Pierre Auger Observatory are extragalactic sources. There are hardly sources in our own Galaxy which could accelerate particles above $10^{19}$ eV. Still, it remains interesting at which energy galactic sources actually run out of power. The energy range between $10^{16}$ and $10^{18}$ eV has been covered by very few experiments. Energy spectra determined by different experiments differ significantly, mostly due to the problems in proper energy calibration. On the other hand the region above $10^{16}$ eV is of crucial importance for our understanding of the origin and propagation of cosmic rays in the Galaxy. What is the mass composition? Is this region dominated by sources other than supernova remnants? Is there an early onset of an extragalactic component? What is the relation between cut-off effects due to leakage out of the Galaxy and cut-off effects due to maximum energies in sources? Three experiments (Kascade-Grande in Karlsruhe/Germany, Tunka-133 in Siberia and IceCube/IceTop at the South Pole), each with about 1 km$^2$ area, are exploring this energy range. They will yield a precise measurement of the energy spectrum as well as improved knowledge about the mass composition (see e.g. Kampert 2006).

*Below the knee*

At energies below the knee, one notes the recent successful launch of the Pamela satellite experiment. It will hopefully be followed in a few years by the launch of the much larger AMS spectrometer and its operation on the International Space Station ISS. The plans for AMS are strongly affected by the unclear situation for Space Shuttle missions. Pamela and AMS, as well as future balloon missions, will search for anti-nuclei with much increased sensitivity and also measure the energy spectrum of different nuclei below the knee, up to energies $10^{12}$–$10^{15}$ eV/nucleon (see for a science summary Picozza & Morselli 2006). A large satellite mission (Nucleon) extending the direct particle measurements to close to the knee is planned in Russia, with some Italian participation.

## 7.2 TeV gamma rays

In contrast to charged cosmic rays, gamma rays propagate straight; compared to neutrinos, they are easy to detect. This has made them a powerful tracer of cosmic processes.

Among all the different techniques developed so far for gamma detection, primarily two have succeeded in providing catalogues with reliable source detections and spectral measurements: *satellite detectors* and ground based *Imaging Atmospheric Cherenkov Telescopes* (IACTs).

The first steps of cosmic particle acceleration are studied with satellite detectors for MeV energies like INTEGRAL, and for MeV-GeV energies, where the EGRET satellite has revealed more than 300 sources of radiation. In 2008, the GLAST detector will be launched and is expected to provide an even richer view of the universe at energies up to several $10^{13}$ eV.

Due to the small area of detectors on satellites, at energies above a few tens of GeV they run out of statistics. The higher energies are the domain of ground-based Cherenkov telescopes, covering the range above hundred GeV with extremely large sensitivities. They record the Cherenkov light from air showers originating from gamma ray interactions in the atmosphere. Large dishes focus the light to arrays of photomultipliers ("cameras"). From the shower image, direction, energy and character of the primary particle (hadron versus gamma ray) can be derived.

The IACT technique was pioneered in the USA with the development of the Whipple Telescope. Actually, it took the Whipple group nearly 20 years to eventually detect in 1989 a first source, the Crab Nebula (Weekes et al. 1989). During the last decade, European groups have been leading the development of IACTs and the field of ground-based high-energy gamma ray astronomy. Figure 11 shows a comparison of the TeV gamma sky map in 1996 and 2006. It illustrates the tremendous progress achieved within ten years. Most of the new sources in Fig. 11 have been established by H.E.S.S., an array of four Cherenkov telescopes in Namibia, and MAGIC, a large telescope at La Palma (Voelk 2006, Aharonian2007). Both telescopes are being upgraded.

IACTs have by now discovered more than seventy emitters of gamma rays at the $10^{11}$ to $10^{13}$ eV scale, many of them lining the Milky Way and revealing a complex morphology (see e.g. Fig. 12, taken from Aharonian et al. 2005). Most of the TeV sources correspond to known objects like binary stellar systems or supernova remnants. Others are still entirely unknown at any other wavelength and obviously emit most of their energy in the TeV range ("dark accelerators"). Going outside our own Galaxy, a large number of Active Galactic Nuclei have been observed and their fast variability demonstrated.

While the results achieved with current instruments are already very impressive, they just give a taste of the TeV cosmos. The detailed understanding of the underlying processes and the chance to cover more than the "tip of the iceberg" can be improved dramatically by a much larger array of telescopes based on now well established techniques and observation strategies. A proto-collaboration has been formed to work jointly towards the design and realisation of such an instrument, which was christened CTA (Cherenkov Telescope Array, see Hermann et al. 2007). It involves

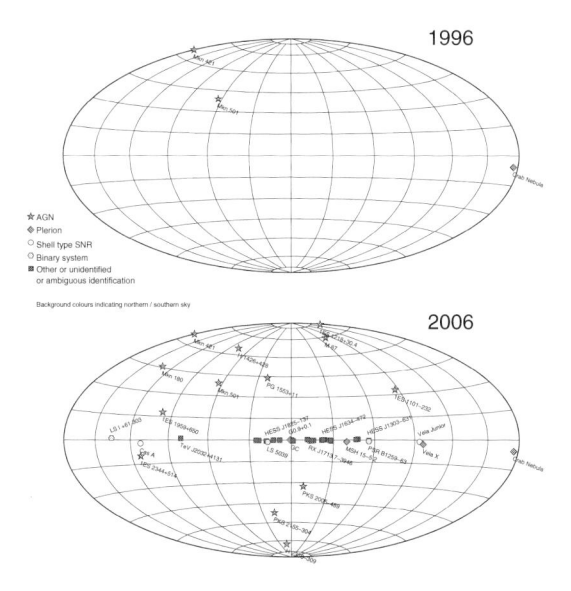

**Figure 11:** The TeV gamma sky in 1996 (top) and 2006 (bottom); courtesy K. Bernlöhr.

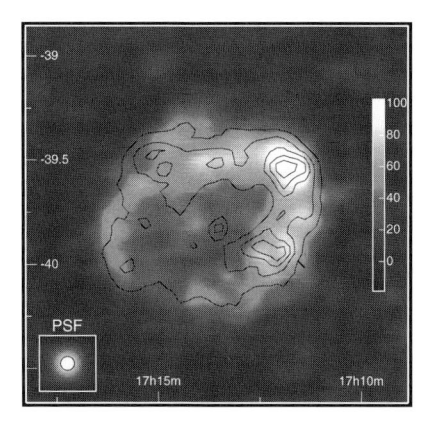

**Figure 12:** Supernova remnant RXJ1713.7-3946 at radio wavelength (contours) and TeV energies (coloured regions).

all European groups currently participating in IACTs, as well as a large number of additional new partners from particle physics and astrophysics. Actually, CTA is on the list of emerging projects compiled by the *European Strategy Forum for Research Infrastructures (ESFRI)* and has been proposed by the ApPEC steering committee to be promoted to the status of a full ESFRI entry.

The goal of CTA is simultaneously increasing the energy bandwidth towards lower and higher energies, improving the sensitivity at currently accessible energies, and providing large statistics of highly constrained and very well reconstructed events (see Fig. 13, taken from Hermann et al. 2007). CTA will likely consist of a few very large central dishes providing superb efficiency below 50 GeV, embedded in an array of medium dishes giving high performance around a TeV, the latter being surrounded by a few-km$^2$ array of small dishes to catch the bright but rare showers at 100 TeV: altogether 40–70 telescopes. A similar concept (AGIS) is being discussed in the USA, and the need for cooperation and coordination is obvious.

CTA is conceived to cover both hemispheres, with one site in each. The field of view of the Southern site includes most of the Galaxy, the Northern telescope would instead focus to extragalactic objects. At energies above a few tens of TeV and over Mega-parsec distances, gamma rays are absorbed by the cosmic infrared light fields, and above a few hundreds of TeV by the 3K background. Therefore high energy sensitivity of the Northern ("extra-galactic") site is less important than for the Southern ("center of the galaxy") site. For the Southern site, emphasis would be put to high-energy sensitivity and excellent angular resolution in order to study the morphology of galactic objects.

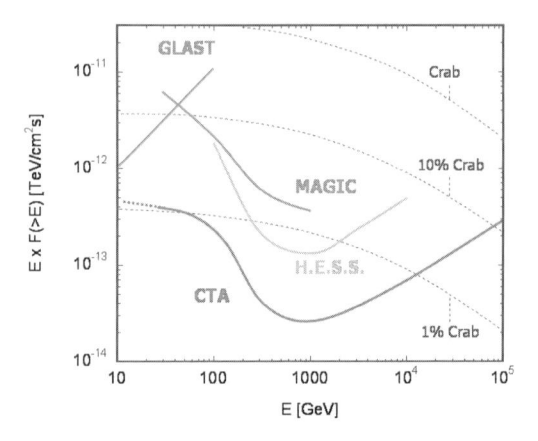

**Figure 13:** Sensitivity of present-generation Cherenkov instruments (H.E.S.S. and MAGIC) in comparison to that of the next-generation satellite experiment GLAST, and to the envisaged sensitivity of a next-generation Cherenkov instrument. The final values for CTA will depend on the actual layout. For reference, the gamma ray flux from the Crab Nebula is shown.

CTA, with its many telescopes of small field of view, will likely be complemented by wide-angle devices like HAWC, the successor of the MILAGRO experiment in the USA (Sinnis, Smith & McEnery 2004). This is a large water pool with photomultipliers detecting the light from air shower particles entering the water. Compared to IACTs, HAWC would have a higher threshold and worse flux sensitivity but – due to its large field of view – better survey capabilities and better sensitivity to extended sources. From the low energy side, the US-initiated GLAST instrument would overlap with CTA. Since parallel observations are desirable and GLAST will be launched very soon, CTA construction should start as early as possible.

## 7.3 High energy neutrinos

The physics case for high energy neutrino astronomy is obvious: neutrinos can provide an uncontroversial proof of the hadronic character of the source; moreover they can reach us from cosmic regions which are opaque to other types of radiation (Waxman 2007). However, whereas neutrino astronomy in the energy domain of MeV has been established with the impressive observation of solar neutrinos and neutrinos from supernova SN-1987A, neutrinos with energies of Giga electron volts and above, which must accompany the production of high energy cosmic rays, still await discovery. Detectors underground have turned out to be too small to detect the corresponding feeble fluxes. The high energy frontier of TeV and PeV (1 PeV= $10^{15}$ eV) is currently being tackled by much larger, expandable arrays constructed in deep, open water or ice. They consist of photomultipliers detecting the Cherenkov light from charged particles produced by neutrino interactions (see Fig. 14). Flux estimations from astrophysical sources suggest that detectors on the cubic kilometre size scale are required for clear discoveries.

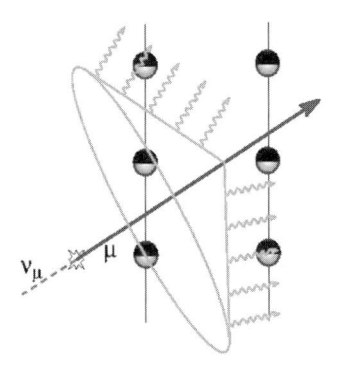

**Figure 14:** Neutrino telescopes consist of large arrays of photomultiplier tubes underwater or under ice. They detect the Cherenkov light emitted by charged particles which have been produced in neutrino interactions – here from an up-going muon which stems from a neutrino having crossed the Earth.

European physicists have played a key role in construction and operation of the two pioneering large neutrino telescopes, NT200 in Lake Baikal (Belolaptikov et al. 1997) and AMANDA at the South Pole (Andres et al. 2001), and are also strongly involved in AMANDA's successor, IceCube (Ahrens et al. 2003, Spiering 2005).

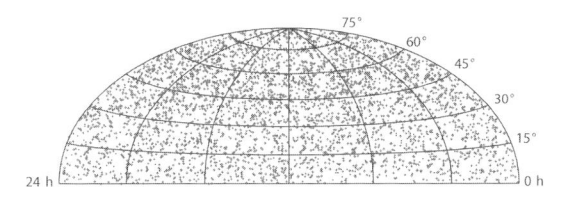

**Figure 15:** Sky map of 4282 events recorded by AMANDA in 2000–2004.

Figure 15 shows a sky plot of the 4282 events recorded by AMANDA over five years (Achterberg et al. 2007). Even with this highest statistics of high energy neutrino events ever collected, no point source signal could yet be identified, motivating the construction of detectors more than one order of magnitude beyond AMANDA size. Such a cubic kilometre detector, IceCube, is presently being deployed at the South Pole (Ahrens et al. 2003). Completion is foreseen in January 2011; it then will consist of 4800 photomultipliers arranged in 80 strings (see Fig. 16) half of which have been installed by February 2008.

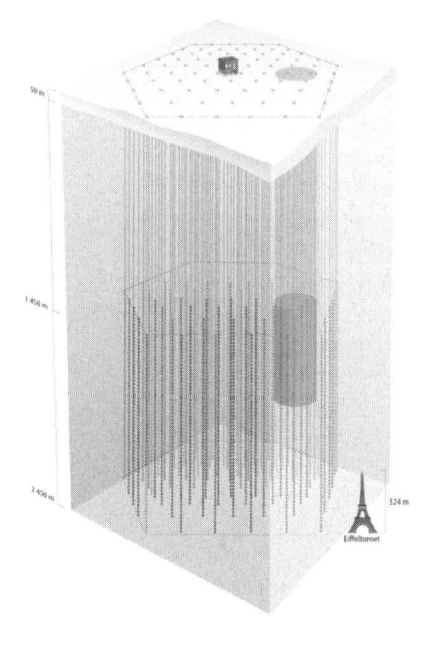

**Figure 16:** IceCube schematic view. IceCube consists of 80 strings each equipped with 60 photomultipliers between 1400 and 2400 meters depth. AMANDA (small cylinder) is integrated into IceCube. IceCube is complemented by a surface air shower array, IceTop, which records air showers and greatly enhances the physics capabilities of the deep ice detector.

Using the Earth as a filter, IceCube observes the Northern sky. Complete sky coverage, in particular of the central parts of the Galaxy with many promising source candidates, requires a cubic kilometre detector in the Northern hemisphere (Halzen 2007). A prototype installation of AMANDA size, ANTARES, is presently being installed close to Toulon/France (Kouchner 2007), with 10 of a total of 12 strings already operating. R&D work towards a cubic kilometer detector is also pursued at two other Mediterranean sites, the one (NEMO) close to Sicily, the other (NESTOR) close to the Peloponnesus (Spiering 2003, Katz 2006, Amore 2007).

Resources for a cubic kilometre Mediterranean detector will be pooled in a single, optimized large research infrastructure. An EU-funded 3-year study (KM3NeT) is in progress to work out the technical design of a neutrino observatory in the Mediterranean, with construction envisaged to start in 2011 (Katz 2006). ESFRI has included KM3NeT in the *European Roadmap for Research Infrastructures*, thus assigning high priority to this project. Start of the construction of KM3NeT is going to be preceded by the successful operation of small scale or prototype detector(s) in the Mediterranean. Its design should also incorporate the improved knowledge on galactic sources as provided by gamma ray observations, as well as initial results from IceCube – including e.g. the possibility to construct, for similar cost, a 3 or 5 times larger array with higher energy threshold. Still, the time lag between IceCube and KM3NeT should be kept as small as possible. Figure 17 shows an example for a possible KM3NeT configuration, based on a "hollow cylinder" structure.

**Figure 17:** Artists view by M. Kraan (NIHKEF) of a possible design of KM3NeT.

*Techniques for extremely high energies:*

Emission of Cherenkov light in water or ice provides a relatively strong signal and hence a relatively low energy threshold for neutrino detection. However, the limited light transmission in water and ice requires a large number of light sensors to cover the required detection volume. Towards higher energies, novel detectors focus on other signatures of neutrino-induced charged particle cascades, which can be detected from a larger distance (see Böser & Nahnhauer 2006 for a review). Methods include recording the Cherenkov radio emission or acoustic signals from neutrino-induced showers, as well as the use of air shower detectors responding to showers with a "neutrino signature". The very highest energies will be covered by balloon-borne detectors recording radio emission in terrestrial ice masses, by ground-based radio antennas sensitive to radio emission in the moon crust, or by satellite detectors searching for fluorescence light from neutrino-induced air showers. Taken all together, these detectors cover an energy range of more than twelve decades, starting at $10^{13}$–$10^{14}$ eV (10–100 GeV) and extending beyond $10^{22}$ eV. Limits come i.e. from air shower detection with optical methods, from radio searches for particle showers in the Moon, and from radio searches for particle showers in ice. All of them have been derived within the last decade. Exploitation of the full potential of these methods needs large-scale R&D work.

*Summary on high energy neutrino detection*

Within the last five years, experimental sensitivities over the whole energy range have improved by more than an order of magnitude, much faster than during the previous decades. Over the next 7–10 years, flux sensitivities are expected to move further down by a factor of 30–50, over the entire range from tens of TeV to hundreds of EeV. This opens up regions with high discovery potential.

## 8 Gravitational Waves

Gravitational waves would provide us with information on strong field gravity through the study of immediate environments of black holes. Typical examples are coalescences of binary systems of compact objects like neutron stars (NS) or black holes (BH). Even more spectacular events could be observed from galaxy collisions and the subsequent mergers of super-massive black holes residing in the centres of the galaxies. Further expected sources are compact objects spiralling into super-massive black holes, asymmetric supernovae, and rotating asymmetric neutron stars such as pulsars. Processes in the early Universe, on the time and length scales of inflation, must also produce gravitational waves.

Since the expected wavelengths are of the order of the source size, frequencies range from below a milli-Hertz to above a kilo-Hertz. Study of the full diversity of the gravitational wave sky therefore requires complementary approaches: Earth-based detectors are typically sensitive to high-frequency waves, while space-borne detectors sample the low-frequency regime.

Pioneers of direct observations have been using resonant bar detectors, and some (significantly improved) bar detectors are still in operation. However, the most advanced tools for gravitational wave detection are interferometers with kilometre-long arms. The passage of a gravitational wave differently contracts space along the two directions of the arms and influences the light travel time (Hong, Rowan & Sathyaprkash 2005).

At present, the world's most sensitive interferometer is LIGO (USA), the others being Virgo in Italy, GEO600 in Germany and the smaller TAMA in Japan. Given our current understanding of the expected event rates, gravitational wave detection is not very likely with these initial interferometers. Thus a mature plan exists for upgrades to the existing detectors systems to create *enhanced* and after that *advanced* detector systems, such that the observation of gravitational waves within the first weeks or months of operating the advanced detectors at their design sensitivity is expected (Fig. 18)

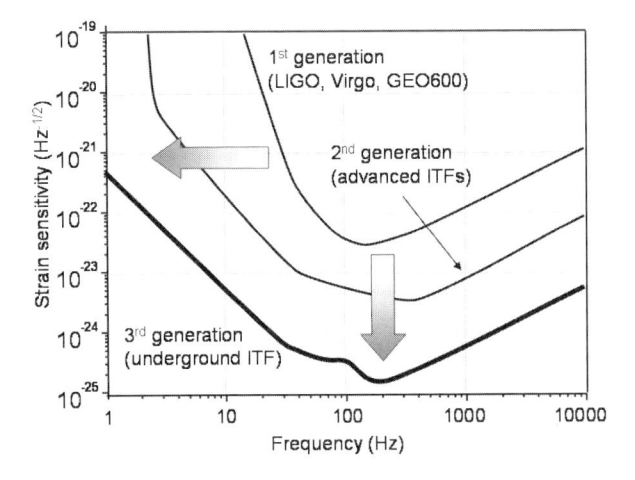

**Figure 18:** Current and expected sensitivities for ground-based gravitational waves detectors (Courtesy H. Lück). Not shown are the curves for GEO-HF which will is the GEO600 interferometer tuned to high frequencies and for DUAL, a future medium bandwidth detector called DUAL. The third generation interferometer curve is a very preliminary estimate.

The European ground interferometers (GEO and Virgo) are turning to observation mode with a fraction of their time dedicated to their improvement (GEO-HF, Virgo+ and Advanced Virgo) – see Fig. 19. Predicted event rates, e.g. for mergers of neutron star/black hole systems (BH-BH, NS-NS, NS-BH) are highly uncertain and range between 3 and 1000 for the "advanced" detectors planned to start data taking in about 5 years.

**Figure 19:** Timeline of current detector operation and planned detector upgrades. The solid lines for the existing detectors indicate data taking times. In the regions of dotted lines the mode of operation is not yet defined. In the scenario shown, LISA would be launched in 2018. The *3rd* generation plans start with a 3 year design study in 2008, followed by a 4 year preparatory construction phase. Construction and commissioning will last for 6 years and allow data taking from 2021 onwards.

Even the advanced versions of the present interferometers will start reaching some fundamental limits, e.g. due to the seismic environment. Therefore the European Gravitational Wave Community envisages a *3rd* generation interferometer as seismically quiet underground facility. The sensitivity target is an order of magnitude better than that of Advanced LIGO and Virgo (three orders of magnitude in event rate) with the seismic cut off going down to less than 1 Hz (see Fig. 18). This new facility (the "Einstein Telescope", *E.T.*) would be a dramatic step and allow Europe to play a key role in what will then be the field of gravitational-wave observational astronomy. E.T. would have a guaranteed rate of many thousand events per year and would move gravitational wave detectors in the category of astronomical observatories. A network of third and second generation detectors would measure to a few percent the masses, sky positions and distances of binary black holes with stellar- and intermediate (i.e. a few hundred times solar) mass, out to a redshift of $z = 2$ and $z = 0.5$, respectively.

E.T has been approved as a FP7 design study in 2007. The outcome of this work will be a conceptual design of the facility (including a selection of possible sites), followed by a more detailed preparatory construction phase to be in a position to start construction around 2015 (see Fig. 19). The design study will include conceptual aspects of the observatory to show that the envisaged sensitivity can be reached with the techniques, the funding and on the timescales foreseen. Cost for E.T. would be on the 500 Million Euro scale.

*Gravitational wave astronomy from space*

The frequency domain much below one Hz can be only explored from space. There is currently an ESA-NASA mission, LISA, which is scheduled for a launch in 2018. LISA would be ideally suited for the study of super-massive black holes mergers, galactic compact binaries (see Fig. 20) and potentially for the signatures of new physics beyond the standard model (Hughes 2007).

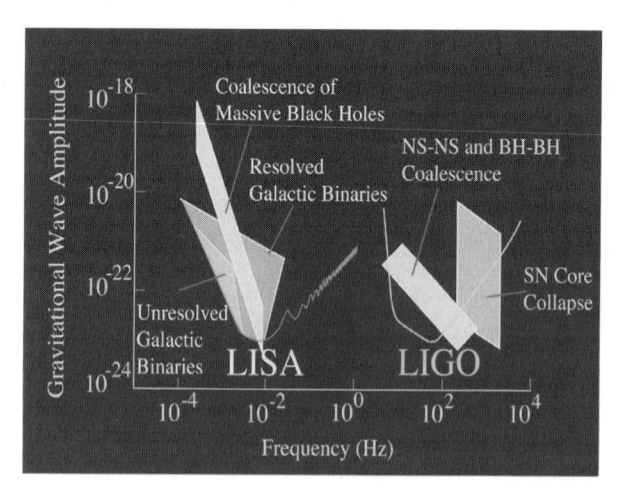

**Figure 20:** Comparison of the sensitivity regions defining the science potential and the science targets of earth-bound and space interferometers, respectively (Courtesy H. Lück).

After transit to the final orbit, LISA will be ready for data taking in 2020, roughly coinciding with the Einstein Telescope, ET. LISA involves three spacecraft flying approximately 5 million kilometres apart in an equilateral triangle formation. These very long arms allow to cover a frequency range of $3 \cdot 10^{-5}$ to 1 Hz, complementary to the frequency window covered by ground-based instruments. Prior to LISA, the *LISA Pathfinder* mission, to be launched in 2010 by ESA, must test some of the critical new technology required for the instrument and proof the feasibility of the concept.

The advanced ground based detectors will assure the detection of gravitational waves within a few weeks or months. This is a necessary condition to move ahead and to aquire the substantial funding for E.T. and/or LISA. These two detectors then would move gravitational wave instruments eventually into the league of true astronomical observatories.

## 9 The Big Picture

In its strategy paper, the ApPEC roadmap committee argues that astroparticle physics is likely at the dawn of a golden age, as traditional astrophysics was two to three decades ago. The enormous discovery potential of the field stems from the fact that attainable sensitivities are improving with a speed exceeding that of the previous two decades. Improvement of sensitivities alone is arguably not enough to raise expectations. But on top of this, we are entering territories with a high discovery potential, as predicted by theoretical models.

The increasing speed of progress is illustrated by Fig. 2, with a strongly higher gradient of sensitivity improvement for Dark Matter search than any time before. Curves of nearly identical shape can be drawn e.g. for detection of high energy neutrinos or charged cosmic rays. For the first time experimental and theoretical techniques allow – or are going to allow – forefront questions to be tackled with the necessary sensitivity. A long pioneering period during which methods and technologies have been prepared is expected to pay off over the next 5–15 years.

The price tag of frontline astroparticle projects requires international collaboration, as does the realization of the infrastructure. Cubic-kilometre neutrino telescopes, large gamma ray observatories, Megaton detectors for proton decay, or ultimate low-temperature devices to search for dark matter particles or neutrino-less double beta decay are in the range of 50–800 MEuro. Cooperation is the only way *a)* to achieve the critical scale for projects which require budgets and manpower not available to a single nation and *b)* to avoid duplication of resources and structures. Astroparticle physics is therefore facing a similar concentration process as particle physics since several decades.

The following list of experiments on the >50 MEuro scale cost for investment is the result of a first strategic approach of European Astroparticle Physics (Phase-I of the ASPERA Roadmap).

- **High energy gamma astronomy:**
  A large Cherenkov Telescope Array (CTA). Desirably two sites, one South, one North. Overall cost 150–170 MEuro. Prototypes 2011, start of construction in 2012.

- **High energy neutrino astronomy:**
  A kilometer scale neutrino telescope KM3NeT in the Mediterranean, complementing IceCube on the opposite hemisphere. Cost scale 200 MEuro. Start construction after a preparatory phase, in 2011 or 2012.

- **High energy cosmic ray astronomy:**
  Auger-North, complementing the Pierre-Auger Site in Argentina. Cost about 90 MEuro, 45% of that from Europe. Start construction in 2010 or 2011.

- **Direct Dark Matter searches:**
  Two "zero-background" dark matter experiments on the ton scale, with a cost estimate of 150–180 MEuro for both experiments and the related infrastructure together. Two different nuclei and techniques (e.g. bolometric and noble liquid). Decision in 2010/2011.

- **Masses and possible Majorana nature of neutrinos by double beta decay experiments:**
  Next generation experiments are GERDA, CUORE, Super-NEMO. Decision about an "ultimate" ton-scale experiment in the next decade, start of construction not before 2013. Cost on the 150 MEuro scale. Share with non-European countries.

- **Proton decay and low energy neutrino astrophysics:**
  A detector on the Megaton scale. Worldwide collaboration, close coordination with USA and Japan. Cost between 400 and 600 MEuro. Decision on technology after the end of the design phase, 2010–2012.

- **Gravitational waves:**
  The third generation Gravitational Wave interferometer, located underground. Cost at the 500 MEuro scale. Need detection of Gravitational Waves with "advanced" interferometers before construction would be approved. Coordination with space plans (LISA) is important.

Naturally, there must be room for initiatives below the 50 Million Euro level. The Roadmap committee suggests that about 20% of astroparticle funding should be reserved for smaller initiatives, for participation in overseas experiments with non-European dominance, and for R&D. Technological innovation has been a prerequisite of the enormous progress made over the last two decades and enabled maturity in most fields of astroparticle physics. It is also a prerequisite for future progress towards greater sensitivity and lower cost and must be supported with significant funds.

With ApPEC and the related ERA-Net ASPERA, the process of coherent approaches within Europe has already successfully started. ApPEC represents nearly two thousand European scientists involved in the field. ApPEC helped to launch IL-IAS, an Integrated Infrastructure Initiative with leading European infrastructures in Astroparticle physics. ILIAS covers experiments on double beta decay, dark matter searches and gravitational wave detection as well as theoretical astroparticle physics. ApPEC has also actively promoted the approval of KM3NeT as ESFRI project and

FP6 design study, of CTA as emerging ESFRI project, of the Megaton neutrino detector study LAGUNA and the *3rd* generation gravitational interferometer E.T. as FP7 design studies and also the FP7 support for the preparatory phase of KM3NeT. ApPEC will also play an important role in forming a coherent landscape of the necessary infrastructures, in particular of the underground laboratories.

Phase-I of the roadmap (http://www.aspera-eu.org) describes physics case and status of astroparticle physics and formulates recommendations for each of the subfields. In a second phase (Phase-II), detailed information on time schedule and cost have been collected from all experiments. There is also a census on the present funding level collected from the national funding agencies. Phase-III has just started. During this phase, a precise calendar for milestones and decisions will be prepared. Also, priorities will be formulated, based on different funding scenarios. A Phase-III roadmap paper will be released in autumn 2008. This work will provide the necessary input for the decisions on the large projects of the list above. Clearly, the required resources exceed the present funding level. The roadmap committee of ApPEC is convinced that the prospects in this field merit a substantially increased support.

**Acknowledgements**

I thank my co-authors of the ApPEC Roadmap committee: F. Avignone, J. Bernabeu, L. Bezrukov, P. Binetruy, H. Blümer, K. Danzmann, F. v. Feilitzsch, E. Fernandez, W. Hofmann, J. Iliopoulos, U.Katz, P.Lipari, M. Martinez, A. Masiero, B. Mours, F. Ronga, A. Rubbia, S. Sarkar, G. Sigl, G. Smadja, N. Smith and A.Watson. I also acknowledge discussions with and support of members of the ApPEC Steering committee, in particular T. Berghöfer, M. Bourquin and S. Katsanevas.

# References

Abraham, J. et al.: 2004, Nucl. Instr. Meth. A523, 50.

Abraham, J. et al. (The Pierre Auger Collaboration): 2007, Science 318, 939.

Achterberg, A. et al.: 2007, Phys.Rev.D75, 102001.

Aharonian, F.: 2006, Astronom. & Astrophys. 449, 223.

Aharonian, F.: 2007, Science 315, 70.

Ahrens, J. et al.: 2003, Astropart. Phys. 20, 507.

Amore I et al.: 2007, IJMPA 22, 3509.

Andres, E. et al.: 2001, Nature 410, 441.

Angle, J. et al.: 2007, arXiv:astro-ph/0706.0039.

Arpesella C. et al.: 2008, Phys. Rev. Lett. B658, 101, and arXiv:0708.2251.

Autiero, D. et al.: 2007, arXiv:0705.0116.

Avignone, F., Elliot, S., Engel, J.: 2007, arXiv:0708.1033, to appear in Rev. Mod. Phys.

Bahcall, J.N.: 2005, Phys. Scripta T121, 46.

Baudis, L.: 2005, arXiv:astro-ph/0511.805.

Baudis, L.: 2007, arXiv:astro-ph/0511.3788.

Belolaptikov, I. et al.: 1997, Astropart. Phys. 7, 263.

Bernabei, R.: 2004, Int. Journ. Mod. Phys. D13, 2127.

Bertone, G.: 2007, arXiv:0710.5603.

de Bellefon, A. et al.: 2006, arXiv: hep-ex/0607026.

Bettini, A.: 2007, arXiv:0712/1051.

Böser, S. & Nahnhauer, R. (eds.): 2006, Proc. ARENA workshop, Zeuthen, Germany.

Coccia, E.: 2006, Journal of Physics, Conf. Series 39, 497.

Elliot, S. & Vogl, P., 2002, Ann. Rev. Nucl.Part. Sci. 52, 115.

Ereditato, A. & Rubbia, A.: 2005, arXiv:hep-ph/0509022.

Gaitskell, R.J.: 2004, Ann. Rev. Nucl. Part. Sci. 54, 315.

Halzen, F.: 2007, Science 315, 66.

Hannestad, S. & Raffelt, G.: 2006, JCAP 0611, 016.

Hermann G. et al.: 2007, Contribution to the 30th Int. Conf. on Cosmic Rays, Merida, Mexico, arXiV:0709.2048.

Hong, J., Rowan, S. & Sathyaprkash, B.: 2005, arXiv: gr-qc/0501007.

Hughes, S.: 2007, arXiv:0711/0188.

J. Jungmann, M.Kamionkowski & K. Gried: 1996, Phys. Rep. 267, 195.

Kampert, K.H.: 2006, arXiV:astrop-ph/0611884.

Katz, U.: 2006, Nucl. Instr. Meth. A567, 457.

Klapdor-Kleingrothaus, H.V. et al.: 2004, Nucl. Inst. Meth. A522, 371.

Kouchner, A. et al: 2007, arXiv:0710.0272.

Kraus, Ch. et al.: 2005, Eur. Phys. J. C40, 447.

Lobashev, V.M. et al.: 2003, Nucl.Phys.A 719, 153.

McDonald, A. et al.: 2004, Rev. Sci. Instr. 75, 293.

Olinto, A.: 2007, Science 135, 68.

Peccei, R.D. & Quinn, H.R: 1977, Phys. Rev. D16, 1791.

Perlmutter, S. & Schmidt, B.P.: 2003, in Lecture Notes in Physics, Berlin Springer Verlag, Vol.589, ed. K. Weiler, 195.

Picozza P. & Morselli, A.: 2006, arXiv:astrop-ph/0608697

Raffelt G.: 2006, arXiv:hep-ph/0611118.

Robertson, R. et al.: 2007, arXiv:0712/3893.

Sadoulet, B.: 2007, Science 315, 61.

Sinnis,G., Smith, J. & McEnery E., 2004: arXiv:astro-ph/0403096.

Spergel et al.: 2007, ApJS, 170, 377.

Spiering, C.: 2003: Journ. Phys. G29, 843.

Spiering, C.: 2005: Phys. Scripta T121, 112.

G. Steigman & M.S. Turner: 1985, Nucl. Phys. B253, 375. Turner, M.: 2007, Science 315, 59.

Watson, A.: 2006, Journ. Phys. Conf. Series 39, 365 and arXiv: astro-ph/0511800.

Voelk, H.: 2006, arXiv: astro-ph/0603501.

Weekes, T. et al.: 1989, Astrophys. Journ. 342, 379.

Waxman, E.: 2007, Science 315, 63.

Wurm, M. et al.: 2007, Phys.Rev. D75, 023007.

# Index of Contributors

| | | | | |
|---|---|---|---|---|
| Bartelmann, Matthias | 92 | | Kramer, Michael | 255 |
| Beuther, Henrik | 15 | | | |
| Bœhm, Céline | 107 | | Mastropietro, Chiara | 228 |
| Bromm, Volker | 307 | | Meisenheimer, Klaus | 279 |
| Burkert, Andreas | 228 | | | |
| | | | Przybilla, Norbert | 323 |
| Ceverino, Daniel | 64 | | | |
| | | | Reiners, Ansgar | 40 |
| Hörandel, Jörg R. | 198 | | | |
| Horns, Dieter | 167 | | Scholz, Alexander | 357 |
| | | | Spiering, Christian | 375 |
| Kippenhahn, Rudolf | 1 | | | |
| Klypin, Anatoly | 64 | | Tinker, Jeremy | 64 |
| Kokkotas, Konstantinos D. | 140 | | | |

# General Table of Contents

## Volume 1 (1988): Cosmic Chemistry

Geiss, J.: Composition in Halley's Comet:
  Clues to Origin and History of Cometary Matter .......................... 1/1
Palme, H.: Chemical Abundances in Meteorites ................................. 1/28
Gehren, T.: Chemical Abundances in Stars ........................................ 1/52
Omont, A.: Chemistry of Circumstellar Shells ................................... 1/102
Herbst, E.: Interstellar Molecular Formation Processes ........................ 1/114
Edmunds, M.G.: Chemical Abundances in Galaxies .............................. 1/139
Arnould, M.: An Overview of the Theory of Nucleosynthesis .................... 1/155
Schwenn, R.: Chemical Composition and Ionisation States of the
  Solar Wind – Plasma as Characteristics of Solar Phenomena ............... 1/179
Kratz, K.-L.: Nucear Physics Constraints to Bring the Astrophysical
  R-Process to the "Waiting Point" ........................................ 1/184
Henkel, R., Sedlmayr, E., Gail, H.-P.: Nonequilibrium Chemistry
  in Circumstellar Shells ................................................. 1/231
Ungerechts, H.: Molecular Clouds in the Milky Way: the Columbia-Chile
  CO Survey and Detailed Studies with the KOSMA 3 m Telescope ........... 1/210
Stutzki, J.: Molecular Millimeter and Submillimeter Observations ............ 1/221

## Volume 2 (1989)

Rees, M.J.: Is There a Massive Black Hole in Every Galaxy?
  (19th Karl Schwarzschild Lecture 1989) .................................. 2/1
Patermann, C.: European and Other International Cooperation
  in Large-Scale Astronomical Projects .................................... 2/13
Lamers, H.J.G.L.M.: A Decade of Stellar Research with IUE .................... 2/24
Schoenfelder, V.: Astrophysics with GRO ........................................ 2/47
Lemke, D., Kessler, M.: The Infrared Space Observatory ISO .................... 2/53
Jahreiß, H.: HIPPARCOS after Launch!?
  The Preparation of the Input Catalogue .................................. 2/72
Ip, W.H.: The Cassini/Huygens Mission ......................................... 2/86
Beckers, J.M.: Plan for High Resolution Imaging with the VLT .................. 2/90
Rimmele, Th., von der Luehe, O.: A Correlation Tracker
  for Solar Fine Scale Studies ............................................ 2/105
Schuecker, P., Horstmann, H., Seitter, W.C., Ott, H.-A., Duemmler, R.,
  Tucholke, H.-J., Teuber, D., Meijer, J., Cunow, B.:
  The Muenster Redshift Project (MRSP) .................................... 2/109
Kraan-Korteweg, R.C.: Galaxies in the Galactic Plane .......................... 2/119
Meisenheimer, K.: Synchrotron Light from Extragalactic Radio Jets
  and Hot Spots ........................................................... 2/129
Staubert, R.: Very High Energy X-Rays from Supernova 1987A ................... 2/141

Hanuschik, R.W.: Optical Spectrophotometry
of the Supernova 1987A in the LMC .................................... 2/148
Weinberger, R.: Planetary Nebulae in Late Evolutionary Stages ..................... 2/167
Pauliny-Toth, I.I.K., Alberdi, A., Zensus, J A., Cohen, M.H.:
Structural Variations in the Quasar 2134+004 ............................ 2/177
Chini, R.: Submillimeter Observations
of Galactic and Extragalactic Objects .................................... 2/180
Kroll, R.: Atmospheric Variations in Chemically Peculiar Stars ..................... 2/194
Maitzen, H.M.: Chemically Peculiar Stars of the Upper Main Sequence ............ 2/205
Beisser, K.: Dynamics and Structures of Cometary Dust Tails ..................... 2/221
Teuber, D.: Automated Data Analysis .......................................... 2/229
Grosbol, P.: MIDAS ......................................................... 2/242
Stix, M.: The Sun's Differential Rotation ....................................... 2/248
Buchert, T.: Lighting up Pancakes –
Towards a Theory of Galaxy-formation ................................. 2/267
Yorke, H.W.: The Simulation of Hydrodynamic Processes
with Large Computers .................................................. 2/283
Langer, N.: Evolution of Massive Stars
(First Ludwig Biermann Award Lecture 1989) ........................... 2/306
Baade, R.: Multi-dimensional Radiation Transfer
in the Expanding Envelopes of Binary Systems .......................... 2/324
Duschl, W.J.: Accretion Disks in Close Binarys .................................. 2/333

## Volume 3 (1990): Accretion and Winds

Meyer, F.: Some New Elements in Accretion Disk Theory .......................... 3/1
King, A.R.: Mass Transfer and Evolution in Close Binaries ........................ 3/14
Kley, W.: Radiation Hydrodynamics of the Boundary Layer
of Accretion Disks in Cataclysmic Variables ............................. 3/21
Hessman, F.V.: Curious Observations of Cataclysmic Variables ..................... 3/32
Schwope, A.D.: Accretion in AM Herculis Stars .................................. 3/44
Hasinger, G.: X-ray Diagnostics of Accretion Disks ............................... 3/60
Rebetzky, A., Herold, H., Kraus, U., Nollert, H.-P., Ruder, H.:
Accretion Phenomena at Neutron Stars .................................. 3/74
Schmitt, D.: A Torus-Dynamo for Magnetic Fields
in Galaxies and Accretion Disks ........................................ 3/86
Owocki, S.P.: Winds from Hot Stars ............................................ 3/98
Pauldrach, A.W.A., Puls, J.: Radiation Driven Winds
of Hot Luminous Stars. Applications of Stationary Wind Models ........... 3/124
Puls, J., Pauldrach, A.W.A.: Theory of Radiatively Driven Winds
of Hot Stars: II. Some Aspects of Radiative Transfer ...................... 3/140
Gail, H.-P.: Winds of Late Type Stars ........................................... 3/156

Hamann, W.-R., Wessolowski, U., Schmutz, W., Schwarz, E.,
    Duennebeil, G., Koesterke, L., Baum, E., Leuenhagen, U.:
    Analyses of Wolf-Rayet Stars ............................................. 3/174

Schroeder, K.-P.: The Transition of Supergiant CS Matter from
    Cool Winds to Coronae – New Insights with X AUR Binary Systems ....... 3/187

Dominik, C.: Dust Driven Mass Lost in the HRD ................................ 3/199

Montmerle, T.: The Close Circumstellar Environment
    of Young Stellar Objects ................................................. 3/209

Camenzind, M.: Magnetized Disk-Winds
    and the Origin of Bipolar Outflows ....................................... 3/234

Staude, H.J., Neckel, Th.: Bipolar Nebulae Driven by the Winds
    of Young Stars ........................................................... 3/266

Stahl, O.: Winds of Luminous Blue Variables .................................. 3/286

Jenkner, H.: The Hubble Space Telescope Before Launch:
    A Personal Perspective ................................................... 3/297

Christensen-Dalsgaard, J.: Helioseismic Measurements
    of the Solar Internal Rotation ........................................... 3/313

Deiss, B.M.: Fluctuations of the Interstellar Medium ......................... 3/350

Dorfi, E.A.: Acceleration of Cosmic Rays in Supernova Remnants ............... 3/361

## Volume 4 (1991)

Parker, E.N.: Convection, Spontaneous Discontinuities,
    and Stellar Winds and X-Ray Emission
    (20th Karl Schwarzschild Lecture 1990) ................................... 4/1

Schrijver, C.J.: The Sun as a Prototype
    in the Study of Stellar Magnetic Activity ................................ 4/18

Steffen, M., Freytag, B.: Hydrodynamics of the Solar Photosphere:
    Model Calculations and Spectroscopic Observations ........................ 4/43

Wittmann, A.D.: Solar Spectroscopy with a $100 \times 100$ Diode Array ....... 4/61

Staude, J.: Solar Research at Potsdam:
    Papers on the Structure and Dynamics of Sunspots ......................... 4/69

Fleck, B.: Time-Resolved Stokes V Polarimetry
    of Small Scale Magnetic Structures on the Sun ............................ 4/90

Glatzel, W.: Instabilities in Astrophysical Shear Flows ...................... 4/104

Schmidt, W.: Simultaneous Observations with a Tunable Filter
    and the Echelle Spectrograph of the Vacuum Tower Telescope
    at Teneriffe ............................................................. 4/117

Fahr, H.J.: Aspects of the Present Heliospheric Research ..................... 4/126

Marsch, E.: Turbulence in the Solar Wind ..................................... 4/145

Gruen, E.: Dust Rings Around Planets ......................................... 4/157

Hoffmann, M.: Asteroid-Asteroid Interactions – Dynamically Irrelevant? ....... 4/165

Aschenbach, B.: First Results from the X-Ray Astronomy Mission ROSAT ........ 4/173

Wicenec, A.: TYCHO/HIPPARCOS A Successful Mission! .......................... 4/188

Spruit, H.C.: Shock Waves in Accretion Disks .................................... 4/197
Solanki, S.K.: Magnetic Field Measurements on Cool Stars ...................... 4/208
Hanuschik, R.W.: The Expanding Envelope of Supernova 1987A
  in the Large Magellanic Cloud
  (2nd Ludwig Biermann Award Lecture 1990) ............................. 4/233
Krause, F., Wielebinski, R.: Dynamos in Galaxies ............................... 4/260

## Volume 5 (1992): Variabilities in Stars and Galaxies

Wolf, B.: Luminous Blue Variables; Quiescent and Eruptive States ................... 5/1
Gautschy, A.: On Pulsations of Luminous Stars .................................... 5/16
Richter, G.A.: Cataclysmic Variables – Selected Problems ......................... 5/26
Luthardt, R.: Symbiotic Stars ..................................................... 5/38
Andreae, J.: Abundances of Classical Novae ....................................... 5/58
Starrfield, S.: Recent Advances in Studies of the Nova Outburst ................... 5/73
Pringle, J.E.: Accretion Disc Phenomena .......................................... 5/97
Landstreet, J.D.: The Variability of Magnetic Stars .............................. 5/105
Baade, D.: Observational Aspects of Stellar Seismology ........................... 5/125
Dziembowski, W.: Testing Stellar Evolution Theory
  with Oscillation Frequency Data ............................................ 5/143
Spurzem, R.: Evolution of Stars and Gas in Galactic Nuclei ....................... 5/161
Gerhard, O.E.: Gas Motions in the Inner Galaxy
  and the Dynamics of the Galactic Bulge Region .............................. 5/174
Schmitt, J.H.M.M.: Stellar X-Ray Variability
  as Observed with the ROSAT XRT ............................................. 5/188
Notni, P.: M82 – The Bipolar Galaxy .............................................. 5/200
Quirrenbach, A.: Variability and VLBI Observations
  of Extragalactic Radio Surces .............................................. 5/214
Kollatschny, W.: Emission Line Variability in AGN's .............................. 5/229
Ulrich, M.-H.: The Continuum of Quasars and Active Galactic Nuclei,
  and Its Time Variability ................................................... 5/247
Bartelmann, M.: Gravitational Lensing by Large-Scale Structures .................. 5/259

## Volume 6 (1993): Stellar Evolution and Interstellar Matter

Hoyle, F.: The Synthesis of the Light Elements
  (21st Karl Schwarzschild Lecture 1992) ..................................... 6/1
Heiles, C.: A Personal Perspective of the Diffuse Interstellar Gas
  and Particularly the Wim ................................................... 6/19
Dettmar, R.-J.: Diffuse Ionized Gas and the Disk-Halo Connection
  in Spiral Galaxies ......................................................... 6/33
Williams, D.A.: The Chemical Composition of the Interstellar Gas ................. 6/49
Mauersberger, R., Henkel, C.: Dense Gas in Galactic Nuclei ....................... 6/69
Krabbe, A.: Near Infrared Imaging Spectroscopy of Galactic Nuclei ................ 6/103

Dorschner, J.: Subject and Agent of Galactic Evolution .......................... 6/117

Markiewicz, W.J.: Coagulation of Interstellar Grains in a
    Turbulent Pre-Solar Nebula: Models and Laboratory Experiments .......... 6/149

Goeres, A.: The Formation of PAHs in C-Type Star Environments ................ 6/165

Koeppen, J.: The Chemical History of the Interstellar Medium .................... 6/179

Zinnecker, H., McCaughrean, M.J., Rayner, J.T., Wilking, B.A.,
    Moneti, A.: Near Infrared Images of Star-Forming Regions ................ 6/191

Stutzki, R.: The Small Scale Structure of Molecular Clouds ...................... 6/209

Bodenheimer, P.: Theory of Protostars ......................................... 6/233

Kunze, R.: On the Impact of Massive Stars on their Environment –
    the Photoevaporation by H II Regions .................................... 6/257

Puls, J., Pauldrach, A.W.A., Kudritzki, R.-P., Owocki, S.P., Najarro, F.:
    Radiation Driven Winds of Hot Stars – some Remarks on Stationary
    Models and Spectrum Synthesis in Time-Dependent Simulations
    (3rd Ludwig Biermann Award Lecture 1992) ............................ 6/271

## Volume 7 (1994)

Wilson, R.N.: Karl Schwarzschild and Telscope Optics
    (22nd Karl Schwarzschild Lecture 1993) ................................. 7/1

Lucy, L.B.: Astronomical Inverse Problems .................................... 7/31

Moffat, A.F.J.: Turbulence in Outflows from Hot Stars ........................... 7/51

Leitherer, C.: Massive Stars in Starburst Galaxies
    and the Origin of Galactic Superwinds ................................... 7/73

Mueller, E., Janka, H.-T.:
    Multi-Dimensional Simulations of Neutrino-Driven Supernovae ............ 7/103

Hasinger, G.: Supersoft X-Ray Sources ........................................ 7/129

Herbstmeier, U., Kerp, J., Moritz, P.:
    X-Ray Diagnostics of Interstellar Clouds ................................. 7/151

Luks, T.: Structure and Kinematics of the Magellanic Clouds ..................... 7/171

Burkert, A.: On the Formation of Elliptical Galaxies
    (4th Ludwig Biermann Award Lecture 1993) ............................ 7/191

Spiekermann, G., Seitter, W.C., Boschan, P., Cunow, B., Duemmler, R.,
    Naumann, M., Ott, H.-A., Schuecker, P., Ungruhe, R.:
    Cosmology with a Million Low Resolution Redshifts:
    The Muenster Redshift Project MRSP .................................. 7/207

Wegner, G.: Motions and Spatial Distributions of Galaxies ....................... 7/235

White, S.D.M.: Large-Scale Structure .......................................... 7/255

## Volume 8 (1995): Cosmic Magnetic Fields

Trümper, J.E.: X-Rays from Neutron Stars
    (23rd Karl Schwarzschild Lecture 1994) ................................. 8/1

Schuessler, M.: Solar Magnetic Fields ......................................... 8/11

Keller, Ch.U.: Properties of Solar Magnetic Fields from Speckle Polarimetry
  (5th Ludwig Biermann Award Lecture 1994) ............................. 8/27
Schmitt, D., Degenhardt, U.:
  Equilibrium and Stability of Quiescent Prominences ...................... 8/61
Steiner, O., Grossmann-Doerth, U., Knoelker, M., Schuessler, M.:
  Simulation oif the Interaction of Convective Flow
  with Magnetic Elements in the Solar Atmosphere ......................... 8/81
Fischer, O.: Polarization by Interstellar Dust –
  Modelling and Interpretation of Polarization Maps ...................... 8/103
Schwope, A.D.: Accretion and Magnetism – AM Herculis Stars .................. 8/125
Schmidt, G.D.: White Dwarfs as Magnetic Stars ............................... 8/147
Richtler, T.: Globular Cluster Systems of Elliptical Galaxies ..................... 8/163
Wielebinski, R.: Galactic and Extragalactic Magnetic Fields ..................... 8/185
Camenzind, M.: Magnetic Fields and the Physics of Active Galactic Nuclei ......... 8/201
Dietrich, M.:
  Broad Emission-Line Variability Studies of Active Galactic Nuclei ......... 8/235
Böhringer, H.: Hot, X-Ray Emitting Plasma, Radio Halos,
  and Magnetic Fields in Clusters of Galaxies ............................. 8/259
Hopp, U., Kuhn, B.:
  How Empty are the Voids? Results of an Optical Survey ................... 8/277
Raedler, K.-H.: Cosmic Dynamos ............................................. 8/295
Hesse, M.: Three-Dimensional Magnetic Reconnection
  in Space- and Astrophysical Plasmas and its Consequences
  for Particle Acceleration ............................................... 8/323
Kiessling, M.K.-H.: Condensation in Gravitating Systems as Pase Transition ........ 8/349

## Volume 9 (1996): Positions, Motions, and Cosmic Evolution

van de Hulst, H.:
  Scaling Laws in Multiple Light Scattering under very Small Angles
  (24th Karl Schwarzschild Lecture 1995) .................................... 9/1
Mannheim, K.: Gamma Rays from Compact Objects
  (6th Ludwig Biermann Award Lecture 1995) ............................. 9/17
Schoenfelder, V.:
  Highlight Results from the Compton Gamma-Ray Observatory .............. 9/49
Turon, C.: HIPPARCOS, a new Start
  for many Astronomical and Astrophysical Topics .......................... 9/69
Bastian, U., Schilbach, E.:
  GAIA, the successor of HIPPARCOS in the 21st century .................... 9/87
Baade, D.: The Operations Model for the Very Large Telescope .................... 9/95
Baars, J.W.M., Martin, R.N.: The Heinrich Hertz Telescope –
  A New Instrument for Submillimeter-wavelength Astronomy ................ 9/111
Gouguenheim, L., Bottinelli, L., Theureau, G., Paturel, G.,
  Teerikorpi, P.: The Extragalactive Distance Scale
  and the Hubble Constant: Controversies and Misconceptions ............... 9/127

Tammann, G.A.: Why is there still Controversy on the Hubble Constant? ........... 9/139

Mann, I.: Dust in Interplanetary Space:
    a Component of Small Bodies in the Solar System ....................... 9/173

Fichtner, H.: Production of Energetic Particles at the Heliospheric Shock –
    Implications for the Global Structure of the Heliosphere ................. 9/191

Schroeder, K.-P., Eggleton, P.P.: Calibrating Late Stellar Evolution
    by means of zeta AUR Systems – Blue Loop Luminosity
    as a Critical Test for Core-Overshooting ................................. 9/221

Zensus, J.A., Krichbaum, T.P., Lobanov, P.A.:
    Jets in High-Luminosity Compact Radio Sources ......................... 9/221

Gilmore, G.: Positions, Motions, and Evolution
    of the Oldest Stellar Populations ......................................... 9/263

Samland, M., Hensler, G.: Modelling the Evolution of Galaxies ................... 9/277

Kallrath, J.: Fields of Activity for Astronomers and Astrophysicists
    in Industry – Survey and Experience in Chemical Industry – ............... 9/307

## Volume 10 (1997): Gravitation

Thorne, K.S.: Gravitational Radiation – a New Window Onto the Universe
    (25th Karl Schwarzschild Lecture 1996) ................................... 10/1

Grebel, E.K.: Star Formation Histories of Local Group Dwarf Galaxies
    (7th Ludwig Biermann Award Lecture 1996 (i)) ........................... 10/29

Bartelmann, M.L.: On Arcs in X-Ray Clusters
    (7th Ludwig Biermann Award Lecture 1996 (ii)) .......................... 10/61

Ehlers, J.: 80 Years of General Relativity ........................................ 10/91

Lamb, D.Q.: The Distance Scale To Gamma-Ray Bursts ......................... 10/101

Meszaros, P.: Gamma-Ray Burst Models ........................................ 10/127

Schulte-Ladbeck, R.: Massive Stars – Near and Far ............................. 10/135

Geller, M.J.: The Great Wall and Beyond –
    Surveys of the Universe to $z < 0.1$ ..................................... 10/159

Rees, M.J.: Black Holes in Galactic Nuclei ...................................... 10/179

Mueller, J., Soffel, M.: Experimental Gravity and Lunar Laser Ranging .......... 10/191

Ruffert, M., Janka, H.-Th.: Merging Neutron Stars ............................. 10/201

Werner, K., Dreizler, S., Heber, U., Kappelmann, N., Kruk, J., Rauch, T.,
    Wolff, B.: Ultraviolet Spectroscopy of Hot Compact Stars ................ 10/219

Roeser, H.-J., Meisenheimer, K., Neumann, M., Conway, R.G., Davis, R.J.,
    Perley, R.A.: The Jet of the Quasar 3C 273/ at High Resolution ........... 10/253

Lemke, D.: ISO: The First 10 Months of the Mission ........................... 10/263

Fleck, B.: First Results from SOHO ............................................. 10/273

Thommes, E., Meisenheimer, K., Fockenbrock, R., Hippelein, H.,
    Roeser, H.-J.: Search for Primeval Galaxies
    with the Calar Alto Deep Imaging Survey (CADIS) ...................... 10/297

Neuhaeuser, R.: The New Pre-main Sequence Population
    South of the Taurus Molecular Clouds ................................. 10/323

## Volume 11 (1998): Stars and Galaxies

Taylor, J.H. jr.: Binary Pulsars and General Relativity
    (26th Karl Schwarzschild Lecture 1997 – *not published*) .................... 11/1

Napiwotzki, R.: From Central Stars of Planetary Nebulae to White Dwarfs
    (9th Ludwig Biermann Award Lecture 1997) ............................ 11/3

Dvorak, R.: On the Dynamics of Bodies in Our Planetary System ................. 11/29

Langer, N., Heger, A., García-Segura, G.: Massive Stars:
    the Pre-Supernova Evolution of Internal and Circumstellar Structure ......... 11/57

Ferguson, H.C.: The Hubble Deep Field ......................................... 11/83

Staveley-Smith, L., Sungeun Kim, Putman, M., Stanimirović, S.:
    Neutral Hydrogen in the Magellanic System ............................ 11/117

Arnaboldi, M., Capaccioli, M.: Extragalactic Planetary Nebulae
    as Mass Tracers in the Outer Halos of Early-type Galaxies ................ 11/129

Dorfi, E.A., Häfner, S.: AGB Stars and Mass Loss ............................. 11/147

Kerber, F.: Planetary Nebulae:
    the Normal, the Strange, and Sakurai's Object .......................... 11/161

Kaufer, A.: Variable Circumstellar Structure of Luminous Hot Stars:
    the Impact of Spectroscopic Long-term Campaigns ..................... 11/177

Strassmeier, K.G.: Stellar Variability as a Tool in Astrophysics.
    A Joint Research Initiative in Austria ................................... 11/197

Mauersberger, R., Bronfman, L.: Molecular Gas in the Inner Milky Way .......... 11/209

Zeilinger, W.W.: Elliptical Galaxies ............................................ 11/229

Falcke, H.: Jets in Active Galaxies: New Results from HST and VLA ............. 11/245

Schuecker, P., Seitter, W.C.: The Deceleration of Cosmic Expansion ............. 11/267

Vrielmann, S.: Eclipse Mapping of Accretion Disks ............................ 11/285

Schmid, H.M.: Raman Scattering
    and the Geometric Structure of Symbiotic Stars ........................ 11/297

Schmidtobreick, L., Schlosser, W., Koczet, P., Wiemann, S., Jütte, M.:
    The Milky Way in the UV ............................................ 11/317

Albrecht, R.: From the Hubble Space Telescope
    to the Next Generation Space Telescope ............................... 11/331

Heck, A.: Electronic Publishing in its Context
    and in a Professional Perspective ...................................... 11/337

## Volume 12 (1999):
## Astronomical Instruments and Methods at the Turn of the 21st Century

Strittmatter, P.A.: Steps to the Large Binocular Telescope – and Beyond
    (27th Karl Schwarzschild Lecture 1998) ................................. 12/1

Neuhäuser, R.: The Spatial Distribution and Origin
    of the Widely Dispersed ROSAT T Tauri Stars
    (10th Ludwig Biermann Award Lecture 1998) .......................... 12/27

Huber, C.E.: Space Research at the Threshold of the 21st Century –
    Aims and Technologies ................................................. 12/47
Downes, D.: High-Resolution Millimeter and Submillimeter Astronomy:
    Recent Results and Future Directions .................................... 12/69
Röser, S.: DIVA – Beyond HIPPARCOS and Towards GAIA .................... 12/97
Krabbe, A., Röser, H.P.:
    SOFIA – Astronomy and Technology in the 21st Century ................ 12/107
Fort, B.P.: Lensing by Large-Scale Structures .................................. 12/131
Wambsganss, J.: Gravitational Lensing as a Universal Astrophysical Tool ......... 12/149
Mannheim, K.: Frontiers in High-Energy Astroparticle Physics ................... 12/167
Basri, G.B.: Brown Dwarfs: The First Three Years .............................. 12/187
Heithausen, A., Stutzki, J., Bensch, F., Falgarone, E., Panis, J.-F.:
    Results from the IRAM Key Project:
    "Small Scale Structure of Pre-Star-forming Regions" ...................... 12/201
Duschl, W.J.: The Galactic Center ............................................ 12/221
Wisotzki, L.: The Evolution of the QSO Luminosity Function
    between $z = 0$ and $z = 3$ ............................................. 12/231
Dreizler, S.: Spectroscopy of Hot Hydrogen Deficient White Dwarfs .............. 12/255
Moehler, S.: Hot Stars in Globular Clusters .................................... 12/281
Theis, Ch.: Modeling Encounters of Galaxies: The Case of NGC 4449 ............ 12/309

Volume 13 (2000): New Astrophysical Horizons

Ostriker, J.P.: Historical Reflections
    on the Role of Numerical Modeling in Astrophysics
    (28th Karl Schwarzschild Lecture 1999) .................................. 13/1
Kissler-Patig, M.: Extragalactic Globular Cluster Systems:
    A new Perspective on Galaxy Formation and Evolution
    (11th Ludwig Biermann Award Lecture 1999) ............................ 13/13
Sigwarth, M.: Dynamics of Solar Magnetic Fields –
    A Spectroscopic Investigation ........................................... 13/45
Tilgner, A.: Models of Experimental Fluid Dynamos ............................ 13/71
Eislöffel, J.: Morphology and Kinematics of Jets from Young Stars ............... 13/81
Englmaier, P.: Gas Streams and Spiral Structure in the Milky Way ................ 13/97
Schmitt, J.H.M.M.:
    Stellar X-Ray Astronomy: Perspectives for the New Millenium ........... 13/115
Klose, S.: Gamma Ray Bursts in the 1990's –
    a Multi-wavelength Scientific Adventure ................................. 13/129
Gänsicke, B.T.: Evolution of White Dwarfs in Cataclysmic Variables ............. 13/151
Koo, D.: Exploring Distant Galaxy Evolution: Highlights with Keck .............. 13/173
Fritze-von Alvensleben, U.:
    The Evolution of Galaxies on Cosmological Timescales ................... 13/189
Ziegler, B.L.: Evolution of Early-type Galaxies in Clusters ...................... 13/211

Menten, K., Bertoldi, F.:
  Extragalactic (Sub)millimeter Astronomy – Today and Tomorrow ......... 13/229
Davies, J.I.: In Search of the Low Surface Brightness Universe ................... 13/245
Chini, R.: The Hexapod Telescope – A Never-ending Story ..................... 13/257

## Volume 14 (2001): Dynamic Stability and Instabilities in the Universe

Penrose, R.: The Schwarzschild Singularity:
  One Clue to Resolving the Quantum Measurement Paradox
  (29th Karl Schwarzschild Lecture 2000) .................................... 14/1
Falcke, H.: The Silent Majority –
  Jets and Radio Cores from Low-Luminosity Black Holes
  (12th Ludwig Biermann Award Lecture 2000) ........................... 14/15
Richter, P. H.: Chaos in Cosmos ................................................. 14/53
Duncan, M.J., Levison, H., Dones, L., Thommes, E.:
  Chaos, Comets, and the Kuiper Belt ..................................... 14/93
Kokubo, E.: Planetary Accretion: From Planitesimals to Protoplanets ............. 14/117
Priest, E. R.: Surprises from Our Sun ........................................... 14/133
Liebscher, D.-E.: Large-scale Structure – Witness of Evolution ................... 14/161
Woitke, P.: Dust Induced Structure Formation ................................... 14/185
Heidt, J., Appenzeller, I., Bender, R., Böhm, A., Drory, N., Fricke, K. J.,
  Gabasch, A., Hopp, U., Jäger, K., Kümmel, M., Mehlert, D.,
  Möllenhoff, C., Moorwood, A., Nicklas, H., Noll, S., Saglia, R.,
  Seifert, W., Seitz, S., Stahl, O., Sutorius, E., Szeifert, Th.,
  Wagner, S. J., and Ziegler, B.: The FORS Deep Field ..................... 14/209
Grebel, E. K.: A Map of the Northern Sky:
  The Sloan Digital Sky Survey in Its First Year .......................... 14/223
Glatzel, W.:
  Mechanism and Result of Dynamical Instabilities in Hot Stars ............ 14/245
Weis, K.: LBV Nebulae: The Mass Lost from the Most Massive Stars ............. 14/261
Baumgardt, H.: Dynamical Evolution of Star Clusters .......................... 14/283
Bomans, D. J.: Warm and Hot Diffuse Gas in Dwarf Galaxies .................... 14/297

## Volume 15 (2002): JENAM 2001 – Five Days of Creation: Astronomy with Large Telescopes from Ground and Space

Kodaira, K.: Macro- and Microscopic Views of Nearby Galaxies
  (30th Karl Schwarzschild Lecture 2001) ................................... 15/1
Komossa, S.: X-ray Evidence for Supermassive Black Holes
  at the Centers of Nearby, Non-Active Galaxies
  (13th Ludwig Biermann Award Lecture 2001) ........................... 15/27
Richstone, D. O.: Supermassive Black Holes ................................... 15/57
Hasinger, G.: The Distant Universe Seen with Chandra and XMM-Newton ......... 15/71
Danzmann, K. and Rüdiger, A.:
  Seeing the Universe in the Light of Gravitational Waves ................... 15/93
Gandorfer, A.: Observations of Weak Polarisation Signals from the Sun .......... 15/113

Mazeh, T. and Zucker, S.: A Statistical Analysis of the Extrasolar Planets
and the Low-Mass Secondaries ........................................ 15/133

Hegmann, M.: Radiative Transfer in Turbulent Molecular Clouds ................ 15/151

Alves, J. F.: Seeing the Light through the Dark:
the Initial Conditions to Star Formation ............................... 15/165

Maiolino, R.: Obscured Active Galactic Nuclei ................................ 15/179

Britzen, S.: Cosmological Evolution of AGN – A Radioastronomer's View ........ 15/199

Thomas, D., Maraston, C., and Bender, R.: The Epoch(s)
of Early-Type Galaxy Formation in Clusters and in the Field .............. 15/219

Popescu, C. C. and Tuffs, R. J.: Modelling the Spectral Energy Distribution
of Galaxies from the Ultraviolet to Submillimeter ...................... 15/239

Elbaz, D.: Nature of the Cosmic Infrared Background
and Cosmic Star Formation History: Are Galaxies Shy? ................. 15/259

## Volume 16 (2003): The Cosmic Circuit of Matter

Townes, C. H.: The Behavior of Stars Observed by Infrared Interferometry
(31th Karl Schwarzschild Lecture 2002) ................................ 16/1

Klessen, R. S.: Star Formation in Turbulent Interstellar Gas
(14th Ludwig Biermann Award Lecture 2002) .......................... 16/23

Hanslmeier, A.: Dynamics of Small Scale Motions in the Solar Photosphere ........ 16/55

Franco, J., Kurtz, S., García-Segura, G.:
The Interstellar Medium and Star Formation: The Impact of Massive Stars .. 16/85

Helling, Ch.: Circuit of Dust in Substellar Objects ............................. 16/115

Pauldrach, A. W. A.: Hot Stars: Old-Fashioned or Trendy? ..................... 16/133

Kerschbaum, F., Olofsson, H., Posch, Th., González Delgado, D., Bergman, P.,
Mutschke, H., Jäger, C., Dorschner, J., Schöier, F.:
Gas and Dust Mass Loss of O-rich AGB-stars .......................... 16/171

Christlieb, N.: Finding the Most Metal-poor Stars of the Galactic Halo
with the Hamburg/ESO Objective-prism Survey ........................ 16/191

Hüttemeister, S.: A Tale of Bars and Starbursts:
Dense Gas in the Central Regions of Galaxies .......................... 16/207

Schröder, K.-P.: Tip-AGB Mass-Loss on the Galactic Scale ..................... 16/227

Klaas, U.: The Dusty Sight of Galaxies:
ISOPHOT Surveys of Normal Galaxies, ULIRGS, and Quasars ............. 16/243

Truran, J. W.: Abundance Evolution with Cosmic Time ....................... 16/261

Böhringer, H.: Matter and Energy in Clusters of Galaxies as Probes
for Galaxy and Large-Scale Structure Formation in the Universe ........... 16/275

## Volume 17 (2004): The Sun and Planetary Systems – Paradigms for the Universe

Boehm-Vitense, E.: What Hyades F Stars tell us about Heating Mechanisms
in the outer Stellar Atmospheres
(32th Karl Schwarzschild Lecture 2003) ................................ 17/1

Bellot Rubio, L. R.: Sunspots as seen in Polarized Light
(15th Ludwig Biermann Award Lecture 2003) .......................... 17/21

Stix, M.: Helioseismology .................................................... 17/51
Vögler, A. Simulating Radiative Magneto-convection in the Solar Photosphere ...... 17/69
Peter, H.: Structure and Dynamics of the Low Corona of the Sun ................. 17/87
Krüger, H.: Jupiter's Dust Disk – An Astrophysical Laboratory ................... 17/111
Wuchterl, G.: Planet Formation – Is the Solar System misleading? ................ 17/129
Poppe, T.: Experimental Studies on the Dusty History of the Solar System ......... 17/169
Ness, J.-U.: High-resolution X-ray Plasma Diagnostics of Stellar Coronae
    in the XMM-Newton and Chandra Era ................................. 17/189
Fellhauer, M.: $\omega$ Cen – an Ultra Compact Dwarf Galaxy? ....................... 17/209
Leibundgut, B.: Cosmology with Supernovae ..................................... 17/221
Beckers, J. M.: Interferometric Imaging in Astronomy: A Personal Retrospective ... 17/239
Stenflo, J. O.: The New World of Scattering Physics
    Seen by High-precision Imaging Polarimetry ........................... 17/269

## Volume 18 (2005): From Cosmological Structures to the Milky Way

Giacconi, R.: The Dawn of X-Ray Astronomy
    (33rd Karl Schwarzschild Lecture 2004) ............................... 18/1
Herwig, F.: The Second Stars
    (16th Ludwig Biermann Award Lecture 2004) ........................... 18/21
Kraan-Korteweg, R.: Cosmological Structures behind the Milky Way .............. 18/49
Schuecker, P.: New Cosmology with Clusters of Galaxies ........................ 18/77
Böhm, A., Ziegler, B. L.:
    The Evolution of Field Spiral Galaxies over the Past 8 Gyrs ............... 18/109
Palouš, J.: Galaxy Collisions, Gas Striping and Star Formation in the Evolution
    of Galaxies ....................................................... 18/129
Ferrari, C.: Star Formation in Merging Galaxy Clusters ......................... 18/153
Recchi, S., Hensler, G.:
    Continuous Star Formation in Blue Compact Dwarf Galaxies ............. 18/171
Brunthaler, A.: The Proper Motion and Geometric Distance of M33 ............... 18/187
Schödel, R., Eckart, A., Straubmeier, C., Pott, J.-U.:
    NIR Observations of the Galactic Center ............................... 18/203
Ehlerová, S.: Structures in the Interstellar Medium ............................. 18/213
Joergens, V.: Origins of Brown Dwarfs ......................................... 18/225

## Volume 19 (2006): The Many Facets of the Universe – Revelations by New Instruments.

Tammann, G. A.: The Ups and Downs of the Hubble Constant
    (34th Karl Schwarzschild Lecture 2005) ............................... 19/1
Richter, P.: High-Velocity Clouds and the Local Intergalactic Medium
    (17th Ludwig Biermann Award Lecture 2005) ........................... 19/31
Baschek, B.: Physics of stellar atmospheres – new aspects of old problems
    (Talk in honor of Albrecht Unsöld's 100th anniversary) ................... 19/61

Olofsson, H.: The circumstellar environment of asymptotic giant branch stars ....... 19/75
Hirschi, R. et al.: Stellar evolution of massive stars at very low metallicities ........ 19/101
Röpke, F. K.: Multi-dimensional numerical simulations of
    type Ia supernova explosions ............................................. 19/127
Heitsch, F.: The Formation of Turbulent Molecular Clouds: A Modeler's View ..... 19/157
Herbst, E.: Astrochemistry and Star Formation: Successes and Challenges ......... 19/167
Kley, W.: Protoplanetary Disks and embedded Planets ........................... 19/195
Horneck, G.: Search for life in the Universe –
    What can we learn from our own Biosphere? ............................ 19/215
Guenther, E. W.: GQ Lup and its companion ..................................... 19/237
Posch, T., et al.: Progress and Perspectives in Solid State Astrophysics –
    From ISO to Herschel .................................................. 19/251
Brüggen, M., Beck, R. & Falcke, H.:
    German LOFAR - A New Era in Radio Astronomy ...................... 19/277
Stutzki, J.: SOFIA: The Stratospheric Observatory for Infrared Astronomy ........ 19/293
Sargent, A., Bock, D.: Astronomy with CARMA – Raising Our Sites ............. 19/315

Volume 20 (2008): Cosmic Matter.

Kippenhahn, R.: Als die Computer die Astronomie eroberten
    (35th Karl Schwarzschild Lecture 2007) ................................... 20/1
Beuther, H.: Massive Star Formation: The Power of Interferometry
    (18th Ludwig Biermann Award Lecture 2007 (i)) ........................ 20/15
Reiners, A.: At the Bottom of the Main Sequence
    Activity and Magnetic Fields Beyond the Threshold to Complete Convection
    (18th Ludwig Biermann Award Lecture 2007 (ii)) ........................ 20/40
Klypin, A., Ceverino, D., and Tinker, J.: Structure Formation in the
    Expanding Universe: Dark and Bright Sides ............................ 20/64
Bartelmann, M.: From COBE to Planck ......................................... 20/92
Boehm, C.: Thirty Years of Research in Cosmology, Particle Physics
    and Astrophysics and How Many More to Discover Dark Matter? ......... 20/107
Kokkotas, K. D.: Gravitational Wave Astronomy ................................ 20/140
Horns, D.: High-(Energy)-Lights – The Very High Energy Gamma-Ray Sky ....... 20/167
Hörandel, J. R.: Astronomy with Ultra High-Energy Particles .................... 20/198
Mastropietro, C. and Burkert, A.: Hydrodynamical Simulations of the Bullet Cluster 20/228
Kramer, M.: Pulsar Timing – From Astrophysics to Fundamental Physics ......... 20/255
Meisenheimer, K.: The Assembly of Present-Day Galaxies
    as Witnessed by Deep Surveys ......................................... 20/279
Bromm, V.: The First Stars ................................................... 20/307
Przybilla, N.: Massive Stars as Tracers for Stellar and Galactochemical Evolution .. 20/323
Scholz, A.: Formation and Evolution of Brown Dwarfs ........................ 20/357
Spiering, C.: Status and Perspectives of Astroparticle Physics in Europe .......... 20/375